IUTAM Symposium on Dynamics of
Advanced Materials and Smart Structures

SOLID MECHANICS AND ITS APPLICATIONS
Volume 106

Series Editor: G.M.L. GLADWELL
Department of Civil Engineering
University of Waterloo
Waterloo, Ontario, Canada N2L 3GI

Aims and Scope of the Series

The fundamental questions arising in mechanics are: *Why?, How?,* and *How much?* The aim of this series is to provide lucid accounts written by authoritative researchers giving vision and insight in answering these questions on the subject of mechanics as it relates to solids.

The scope of the series covers the entire spectrum of solid mechanics. Thus it includes the foundation of mechanics; variational formulations; computational mechanics; statics, kinematics and dynamics of rigid and elastic bodies: vibrations of solids and structures; dynamical systems and chaos; the theories of elasticity, plasticity and viscoelasticity; composite materials; rods, beams, shells and membranes; structural control and stability; soils, rocks and geomechanics; fracture; tribology; experimental mechanics; biomechanics and machine design.

The median level of presentation is the first year graduate student. Some texts are monographs defining the current state of the field; others are accessible to final year undergraduates; but essentially the emphasis is on readability and clarity.

For a list of related mechanics titles, see final pages.

IUTAM Symposium on Dynamics of Advanced Materials and Smart Structures

Proceedings of the IUTAM Symposium
held in Yonezawa, Japan, 20–24 May 2002

Edited by

K. WATANABE

*Yamagata University,
Yonezawa, Japan*

and

F. ZIEGLER

*Technical University of Vienna,
Vienna, Austria*

KLUWER ACADEMIC PUBLISHERS
DORDRECHT / BOSTON / LONDON

A C.I.P. Catalogue record for this book is available from the Library of Congress.

ISBN 1-4020-1061-3

Published by Kluwer Academic Publishers,
P.O. Box 17, 3300 AA Dordrecht, The Netherlands.

Sold and distributed in North, Central and South America
by Kluwer Academic Publishers,
101 Philip Drive, Norwell, MA 02061, U.S.A.

In all other countries, sold and distributed
by Kluwer Academic Publishers,
P.O. Box 322, 3300 AH Dordrecht, The Netherlands.

Cover illustration: Designed by Yoshihisa Watanabe.

Printed on acid-free paper

All Rights Reserved
© 2003 Kluwer Academic Publishers
No part of this work may be reproduced, stored in a retrieval system, or transmitted
in any form or by any means, electronic, mechanical, photocopying, microfilming, recording
or otherwise, without written permission from the Publisher, with the exception
of any material supplied specifically for the purpose of being entered
and executed on a computer system, for exclusive use by the purchaser of the work.

Printed in the Netherlands.

CONTENTS

Preface   xiii

Closing Address   xv

Committees and Sponsors   xix

List of Participants   xxi

Symposium Program   xxvii

Exact Solution for a Thermoelastic Problem of Wave Propagations in a Piezoelectric Plate
F. Ashida and T. R. Tauchert   1

Simulation of Impact-Induced Martensitic Phase-Transition Front Propagation in Thermoelastic Solids
A. Berezovski and G. A. Maugin   9

Wave Scattering and Attenuation in Polymer-Based Composites: Analysis and Measurements
S. Biwa, Y. Watanabe, S. Idekoba and N. Ohno   19

Dynamics of Structural Systems with Devices Driven by Fuzzy Controllers
F. Casciati and R. Rossi   29

Model Reduction for Complex Adaptive Structures
W. Chang and V. V. Varadan 41

Transient Analysis of Smart Structures Using a Coupled Piezoelectric-Mechanical Theory
A. Chattopadhyay, R. P. Thornburgh and A. Ghoshal 53

Dynamic Behavior of Shape Memory Alloy Structural Devices: Numerical and Experimental Investigation
L. Faravelli and S. Casciati 63

Overall Design and Simulation of Smart Structures
U. Gabbert, H. Koppe, F. Seeger, and T. N. Trajkov 73

Bio-Mimetic Smart Microstructures: Attachment Devices in Insects as a Possible Source for Technical Design
S. N. Gorb 85

Stress-Focusing Effects in a Spherical Inclusion Embedded in an Infinite Medium Caused by Instantaneous Phase Transformation
T. Hata 95

Free Large Vibrations of Buckled Laminated Plates
R. Heuer 105

High-Performance Impact Absorbing Materials- The Concept, Design Tools and Applications
J. Holnicki-Szulc and P. Pawlowski 115

Maysel's Formula for Small Vibrations Superimposed upon Large
Static Deformations of Piezoelastic Bodies
H. Irschik and U. Pichler   125

The Analysis of Transient Thermal Stresses in Piezothermoelastic Semi-
infinite Body with an Edge Crack
M. Ishihara, O. P. Niraula and N. Noda   137

Transient Dynamic Stresses around a Rectangular Crack in a Nonhomogeneous
Layer Between Two Dissimilar Elastic Half-Spaces
S. Itou   147

Remote Smart Damage Detection via Internet with Unsupervised
Statistical Diagnosis
A. Iwasaki, A. Todoroki and T. Sugiya   157

Smart Actuation from Coupling between Active Polymer Gels and
Fibrous Structures
G. Jeronimidis   167

Thermally Induced Vibration of an Inhomogeneous Beam due to a Cyclic
Heating
R. Kawamura, Y. Tanigawa, and R. B. Hetnarski   177

Application of Optical Fiber Sensors to Smart Structures
S.-H. Kim, D.-C. Seo and J.-J. Lee   187

Application of Stress and Strain Control to Living Tissues
V. Kiryukhin and Y. Nyashin   197

Unfolding of Morning Glory Flower as a Deployable Structure
H. Kobayashi, M. Daimaruya and H. Fujita — 207

Mechanics of Plasma Membrane Vesicles in Cells
T. Kosawada — 217

Control of Structures by Means of High-Frequency Vibration
A. Kovaleva — 227

Numerical Modeling of Smart Devices
R. Lerch, H. Landes and M. Kaltenbacher — 237

A Review of Simulation Methods for Smart Structures with Piezoelectric Materials
G. R. Liu, C. Cai, K. Y. Lam and V. K. Varadan — 251

Application of Transfer Matrix Method in Analyzing the Inhomogeneous Initial Stress Problem in Prestressed Layered Piezoelectric Media
H. Liu, Z. B. Kuang and Z. M. Cai — 263

Infinitesimal Mechanism Modes of Tensegrity Modules
H. Murakami and Y. Nishimura — 273

Shape and Stress Control in Elastic and Inelastic Structures
Y. Nyashin and V. Kiryukhin — 285

Transient Piezothermoelasticity for a Cylindrical Composite Panel
Y. Ootao and Y. Tanigawa — 297

Active Damping of Torsional Vibration in a Piezoelectric Fiber Composite Shaft
P. M. Przybylowicz   307

High-Performance PZT and PNN-PZT Actuators
J. Qiu, J. Tani and H. Takahashi   317

Energy Release Rate Criteria for Piezoelectric Solids
N. Rajapakse and S. X. Xu   327

Integral Approach for Velocity Feedback Control in a Thin Plate with Piezoelectric Patches
S. Sadek, J.C. Bruch, Jr., J.M. Sloss, and S. Adali   337

Non-Parametric Representations of MR Linear Damper Behaviour
B. Sapinski   347

Active Control of Smart Structures using Port Controlled Hamiltonian Systems
K. Schlacher and K. Zehetleitner   357

Numerical Simulation for Control of Progressive Buckling with Defects on Axisymmetric Shell Structure
Y. Shibuya and S. Watanabe   367

Wave Propagation in Piezoelectric Circular plate under Thermo-Electro-Mechanical Loading
N. Sumi   377

Modeling of Piezoelectric/Magnetostrictive Materials for Smart Structures
M. Sunar 387

Control of Thermally-Induced Structural Vibration via Piezoelectric Pulses
T. R. Tauchert and F. Ashida 397

Active Damping of Parametric Vibrations of Mechanical Disturbed Systems
A. Tylikowski 409

Finite Element Models for Linear Electroelastic Dynamics
F. Ubertini 419

Exact Thermoelasticity Solution for Cylindrical Bending Deformations of Functionally Graded Plates
S. S. Vel and R. C. Batra 429

Shape Memory: Heterogeneity and Thermodynamics
D. Vokoun and V. Kafka 439

Complex Variable Solution of Plane Problem for Functionally Graded Materials
X. Wang and N. Hasebe 449

Green's Function for Two-Dimensional Waves in a Radially Inhomogeneous Elastic Solid
K. Watanabe and T. Takeuchi 459

Index of authors 469

May 22, 2002 IUTAM Symposium on Dynamics of Advanced Materials and Smart Structures

Participants in front of Yonezawa Conference Hall (Den-Koku no Mori).

Preface

Two key words for mechanical engineering in the future are *Micro* and *Intelligence*. It is well known that the leadership in the intelligence technology is a matter of vital importance for the future status of industrial society, and thus national research projects for intelligent materials, structures and machines have started not only in advanced countries, but also in developing countries. Materials and structures which have self-sensing, diagnosis and actuating systems, are called *intelligent* or *smart*, and are of growing research interest in the world. In this situation, the IUTAM symposium on *Dynamics of Advanced Materials and Smart Structures* was a timely one.

Smart materials and structures are those equipped with sensors and actuators to achieve their designed performance in a changing environment. They have complex structural properties and mechanical responses. Many engineering problems, such as interface and edge phenomena, mechanical and electro-magnetic interaction/coupling and sensing, actuating and control techniques, arise in the development of intelligent structures. Due to the multi-disciplinary nature of these problems, all of the classical sciences and technologies, such as applied mathematics, material science, solid and fluid mechanics, control techniques and others must be assembled and used to solve them.

IUTAM well understands the importance of this emerging technology. An IUTAM symposium on *Smart Structures and Structronic Systems* (Chaired by U. Gabbert and H.-S. Tzou) was held in Magdeburg, in 2000. Since this symposium, much progress has been made in the field of intelligence. The symposium on *Dynamics of Advanced Materials and Smart Structures* is the second in a row of IUTAM Symposia on the technology, and aims at a fusion of advanced materials and smart structures. The symposium not only reflects the progress made in the last two years, but also includes material science, and extends its scope to new fields, which will give many fresh suggestions for future research. In addition to the regular sessions on smart materials and structures, a session on bio-mimetic structures and active natural flora was organized.

The symposium was held at Yonezawa Conference Hall from May 20-24. 78 participants came to Yonezawa from 17 countries, and stayed for a week. The 50 papers were presented in five consecutive sessions: (1) Advanced materials, (2) Sensors and actuators, (3) Smart structure concept, (4) Controllability for shape and vibration and (5) Bio-mimetic structures. Nine keynote lectures, 33 contributed lectures and 8 posters were presented. All the papers maintained a high academic standard, and active discussion followed each presentation. The organizers believe that every participant enjoyed not only the fruitful discussion, but also the beauty surrounding the Japanese symposium site. As an outcome from the symposium we realized that the ultimate smart structure is the living bio-structure. Much more work should be performed on bio-structures, and the information drawn from biological systems should be channelled into technical development.

Finally, the editors would like to thank all speakers and participants in the symposium for their invaluable contributions to the field of advanced materials and smart structures. They also wish to express their heartfelt gratitude to all the members of the Scientific Committee, the Advisory and the Local Organizing Committees, for their cooperation, crucial advice and cordial encouragement, and to Professor U. Gabbert who assisted us as the chair of the related IUTAM symposium. The kind advice given by the editorial staff of Kluwer Academic Publishers and by Professor G. M. L. Gladwell, for preparing these proceedings is highly appreciated.

October 20, 2002
Kazumi Watanabe
Franz Ziegler

IUTAM SYMPOSIUM
DYNAMICS OF ADVANCED MATERIALS AND SMART STRUCTURES

YONEZAWA, JAPAN
May 20 – 24, 2002

CONCLUSIONS

ULRICH GABBERT

Otto-von-Guericke-Universität Magdeburg,
Universitätsplatz 2, 39106 Magdeburg, Germany

IUTAM regarded the joint proposal of Professors Kazumi and Ziegler of the symposium as excellent and well founded; it was readily accepted and adopted by the General Assembly of IUTAM. Undoubtedly, IUTAM considers the new interdisciplinary area of knowledge-based advanced materials and smart structures an important field of mechanics.

The five-days symposium was aimed at responding to the rapid developments in this field, and at providing a forum for discussing recent research progress, future directions, and trends. The symposium focused on fundamental mechanics and electromechanics of advanced materials and smart structures in dynamic applications; these applications raise new topics involving a number of disciplines of mechanics.

Researchers from Asia, Europe, North America and Africa came to present their latest discoveries. The symposium gave us the opportunity to share the results of many different research groups, and enter into a dialog to widen both our technical and social understanding.

The main topics of the symposium were covered by nine outstanding keynote lectures. The symposium comprised a total number of 33 lectures providing the participants with new theoretical findings as well as the latest developments in advanced materials and smart structures. As the time was limited, 8 papers were presented in a poster session on Tuesday afternoon, where the authors had the opportunity to explain and comment on their work.

1. TECHNICAL OBSERVATIONS

The symposium covered many related fields in this wide interdisciplinary subject: advanced materials, actuators, sensors, structures, data processing, and control electronics (see Fig. 1). Multi-field aspects such as coupled elastic, electric, magnetic, temperature light phenomena, as well as control effectiveness and other related topics were also discussed.

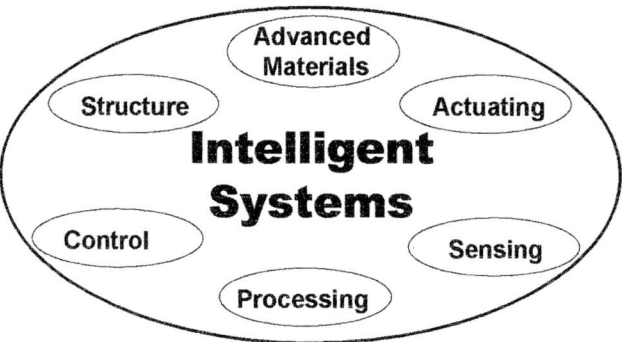

Fig. 1: Inherent parts of intelligent systems

1.1. Advanced materials

The development of new advanced materials and material systems with enhanced properties is considered to be a key issue of smart structure technology. The symposium focused on (i) piezoelectric materials, (ii) shape memory alloys, (iii) electro-active polymers, (iv) active fluids, (v) functionally graded materials, and (vi) biomaterials. Papers and discussions provided better understanding of the physical properties of active materials, presented advanced mathematical models, and elaborated new technologies to increase the performance of the materials.

A highly challenging discussion was held about the behavior of biomaterials, such as bones and cell materials, and their technological and medical effects.

1.2. Advanced smart structural concepts

The principal aspects which were discussed intensively during the symposium can be summarized as follows: (i) better mathematical understanding of the sensing, actuation and systems behaviour, including control as the basis for designing and analysing smart structures, (ii) optimisation, such as topology, shape, actuator-sensor and control design, as well as the optimisation of the close-loop behaviour of systems, (iii) development of overall virtual models, as well as new analytical and numerical methods for design purposes, (iv) solution of fully coupled multi-physics models, including non-linear effects, damping, high frequencies, and failure mechanisms, such as fatigue, damage, and cracks, (v) application of experimental methods for health monitoring, identification of material properties, systems and parameter identification, and model updating, (vi) solution of multi-physics fields by coupling different software tools (CAD, FEM, BEM, Matlab/Simulink), (vii) new industrial applications of interest.

1.3. *Biomimetic smart structures*

Although only one session dealt with bio-mimetic investigations, it is my firm conviction that this field deserves attention in future - also under the umbrella of IUTAM. We should try to learn more from nature, and increase our efforts in (i) understanding biological systems, and (ii) transforming established biological concepts into technical solutions. The very long evolutionary process of nature has yielded highly integrated, extremely intelligent and very efficient systems, in particular from an energetic point of view. Papers dealing with biological systems such as friction and contact techniques applied by insects, or the investigation of the unfolding of a *morning glory flower*, gave interesting insight in this challenging field.

2. ACKNOWLEDGEMENTS

The open and friendly atmosphere at the Symposium provided an excellent platform for holding intensive discussions, and exchanging ideas among the participants. All presentations exuded enthusiasm, and culminated in interesting, intense and exciting discussions, most of which continued during the breaks. It is almost impossible to summarize the research progress here, but the *Symposium Proceedings* present the symposium highlights, serve as a milestone of this new emerging field, and promote the technology in both scientific research and practical applications.

Finally, I must say that sponsoring a scientific meeting is one thing, but organizing it is another. The Chairman, the Co-Chairman and their associates in charge of the scientific program and local arrangements have done a great job. Everybody who has ever been faced with such a challenge knows the efforts needed to organize a successful meeting such as this one. For this reason, we extend our thanks to the International Scientific Committee, the Chairman, Professor Kazumi Watanabe, and his associates who assisted him in carrying this heavy load and responsibility, as well as the Co-Chairman Professor Franz Ziegler. Besides the excellent scientific program, the cultural program organized by our hosts should be mentioned. The participants enjoyed the lecture about Japanese history, Yonezawa folk art, and the banquet in the wonderful garden of the Uesugi Kinen-Kan – the former residence of Count Uesugi.

I would also like to remind you of the excellent speech of the President of Japan NCTAM, Professor T. Kambe, who gave us also a brief introduction into the Haiku poetry. He cited the famous Haiku, which Matsuo Basho wrote in 1689 when he was visiting Yamadera near Yonezawa:

> Shizukasa ya
> iwa ni shimi-iru
> semi no koe
> *M. Basho*

On behalf of IUTAM I express my thanks again to our hosts - the Yamagata University in Yonezawa and Professor Kazumi Watanabe and his staff - for this significant scientific event. But I would also like to thank all the participants who made it a success. In my opinion, we had an outstanding symposium with a lot of excellent lectures, questions and answers - one of the best symposia IUTAM has ever held.

I wish all of you a good trip home, and hope to see you again next time.

Sayonara!

Committees and Sponsors

Scientific Committee

    U. Gabbert (Universitaet Magdeburg, Germany)

    A. S. Kovaleva (Russian Academy of Sciences, Russia)

    N. Noda (Shizuoka University, Japan)

    W. Schiehlen, (University of Stuttgart, Germany) (IUTAM Bureau)

    V. V. Varadan (The Penn - State University, U. S. A.)

    K. Watanabe (Yamagata University, Japan)---chair

    F. Ziegler (Technical University of Vienna, Austria---co-chair)

Advisory Board

T. Inoue (Kyoto University)   T. Kambe (President, NCTAM, Japan)

I. Narisawa (Yamagata University)   J. Tani (Tohoku University)

G. Yagawa (Tokyo University)

Local organizing committee

K. Adachi (Yamagata University)   F. Ashida (Shimane University)---co-chair

S. Biwa (Nagoya University)   Y. Furuya (Hirosaki University)

T. Hata (Shiuzuoka University)   K. Hayashi (Tohoku University)

H. Iizuka (Yamagata University)   S. Ito (Kanagawa University)

A. Kamitani (Yamagata University)   T. Kosawada (Yamagata University)

M. Kurashige (Iwate University)   M. Kuroda (Yamagata University)

T. Ohyoshi (Akita University)   Y. Shibuya (Akita University)

Y. Sugano (Iwate University)   N. Sumi (Shizuoka University)

Y. Tanigawa (Osaka Pref. University)   S. Ueda (Osaka Institute of Technology)

K. Watanabe (Yamagata University)---chair

Supported by

- Science Council of Japan
- Japan Society of Mechanical Engineers (JSME)

Sponsored by

- International Union of Theoretical and Applied Mechanics (IUTAM)
- Yamagata University
- Japan Ministry of Education, Culture, Sports and Technology
- Yamagata Prefecture
- Yonezawa City
- Yonezawa Chamber of Commerce and Industry
- The Asahi Glass Foundation
- Commemorative Association for the Japan World Exposition (1970)
- Intelligent Cosmos Academic Foundation
- The Iwatani Naoji Foundation
- The Mikiya Science and Technology Foundation
- Nippon Sheet Glass Foundation for Materials Science and Engineering
- Suzuki Foundation
- Yonezawa Kogyo-Kai
- Yoshida Foundation for Science and Technology

List of Participants

Abe, S., New Products Div., NOK Co., Fujisawa, Kanagawa, 251-0042, Japan
  abeshin@nok.co.jp
Adachi, K., Cooperative Research Centre, Yamagata University,
  Yonezawa, Yamagata 992-8510, Japan
  kadachi@yz.yamagata-u.ac.jp
Akasaka, T., Aoki 3-9-2-810, Kawaguchi, Saitama, 332-0032, Japan
Ashida, F., Department of Electrical and Control Systems Engineering,
  Shimane University, Matsue, Shimane, 690-8504, Japan
  ashida@ecs.shimane-u.ac.jp
Batra, R. C., Department of Engineering Science and Mechanics, M/C 0219,
  Virginia Polytechnic Institute and State University,
  Blacksburg, VA 24061, U. S. A.
  rbatra@vt.edu
Berezovski, A., Institute of Cybernetics at Tallinn Technical University,
  Department of Mechanics and Applied Mathematics,
  Akadeemia tee 21, 12618, Tallinn, Estonia
  Arkadi.Berezovski@cs.ioc.ee
Biwa, S., Department of Micro System Engineering, Nagoya University,
  Chikusa, Nagoya 464-8603, Japan
  biwa@everest.mech.nagoya-u.ac.jp
Casciati, F., Department of Structural Mechanics, University of Pavia,
  Via Ferrata 1, 27100, Pavia, Italy
  Fabio@dipmec.unipv.it
Chattopadhyay, A., Department of Mechanical & Aerospace Engineering,
  Arizona State University,
  P.O. Box 876106, Tempe, AZ 85287-6106, U. S. A.
  aditi@asu.edu
Chen, D. H., Department of Mechanical Engineering, Tokyo University of Science,
  Kagurazaka 1-3, Shinjyuku, Tokyo 162-8601, Japan
  chend@rs.kagu.sut.ac.jp
Faravelli, L., Department of Structural Mechanics, University of Pavia,
  Via Ferrata 1/ I27100, Pavia, Italy
  lucia@dipmec.unipv.it
Feng, J., Department of Mathematics and Statistics, University of Massachusetts-
  Amherst, Amherst, MA 01002, U. S. A.
  feng@math.umass.edu
Fujimoto, T., Group #1, CAE Div., Toyota Communication Systems,
  Susono, Shizuoka 410-1193 Japan
  f-moto@sannet.ne.jp
Gabbert, U., Institut für Mechanik, Otto-von-Guericke-Universität Magdeburg,
  Universitätsplatz 2, 39106 Magdeburg, Germany
  ulrich.gabbert@mb.uni-magdeburg.de

Gorb, S. N., Evolutionary Biomaterials Group, Max-Planck-Institut fuer Metallforschung,
Heisenbergstr. 3, D-70569 Stuttgart, Germany
s.gorb@mf.mpg.de

Govindjee, S., Structural Engineeing, Mechanics and Materials, Civil and Enviromental Engineering, University of California at Berkeley,
709 Davis Hall, Berkeley, CA 94720-1710, U. S. A.
sanjay@ce.berkeley.edu

Hasebe, N., Department of Civil Engineering, Nagoya Institute of Technology,
Showa, Nagoya, 466-8555, Japan
hasebe@kozo4.ace.nitech.ac.jp

Hasegawa, H., Department of Mechanical Engineering, Meiji University,
Tama-ku, Kawasaki, Kanagawa 214-8571, Japan
ae00008@isc.meiji.ac.jp

Hata, T., Faculty of Education, Shizuoka University, Shizuoka, 422-8529, Japan
eithata@ipc.shizuoka.ac.jp

Heuer, R., Civil Engineering Department, Technical University of Vienna,
Wiedner Hauptstr. 8-10/E201, A-1040, Vienna, Austria
rh@hp720.allmech.tuwien.ac.at

Holnicki-Szulc, J., Institute of Fundamental Technological Research,
Swietokrzyska 21, 00-049 Warsaw, Poland
holnicki@ippt.gov.pl

Iizuka, H., Department of Mechanical Engineering, Yamagata University,
Yonezawa, Yamagata 992-8510, Japan
h-iizuka@yz.yamagata-u.ac.jp

Imai, K., Department of Mechanical Engineering, Iwate University,
Morioka, Iwate 020-8551, Japan
imai@iwate-u.ac.jp

Irschik, H., Division of Technical Mechanics, Johannes Kepler University of Linz,
A-4040 Linz-Auhof, Austria
irschik@mechatronik.uni-linz.ac.at

Ishihara, M., Department of Mechanical Engineering, Shizuoka University,
Johoku 3-5-1, Hamamatsu, Shizuoka, 432-8561, Japan
tmmishi@ipc.shizuoka.ac.jp

Itou, S., Department of Mechanical Engineering, Kanagawa University,
Rokkaku-Bashi, Kanagawa, Yokohama 221-8686, Japan
itous001@kanagawa-u.ac.jp

Iwasaki, A., Department of Mechanical Sciences and Engineering, Tokyo Institute of Technology, Oh-Okayama, Meguro-ku, Tokyo, 152-8552, Japan
aiwasaki@ginza.mes.titech.ac.jp

Jeronimidis, G., Centre for Biomimetics, Department of Engineering, Reading University, Whitenights, Reading, RG6 2AY, U. K.
G.Jeronimidis@reading.ac.uk

Kabe, K., Computational Mechanics Lab., Tire Tech. Div., Yokohama Rubber Co.,
  Hiratsuka, Kanagawa, 254-8601, Japan
  kabe@hpt.yrc.co.jp
Kambe, T., Higashiyama 2-11-3, Meguro, Tokyo 153-0043, Japan
  kambe@gate01.com
Kamitani, A., Department of Infomatics, Yamagata University,
  Yonezawa, Yamagata 992-8510 Japan
  kamitani@emperor.yz.yamagata-u.ac.jp
Kaunda, M. A. E., School of Engineering, University of Durban-Westville,
  Durban, 4000, South Africa
  mkaunda@pixie.udw.ac.za
Kawamura, R., Department of Mechanical Systems Engineering, Osaka Prefecture
  University, Sakai, Osaka, 599-8531, Japan
  kawamura@mecha.osakafu-u.ac.jp
Kikuchi, H., Technical CAE, Engineering Polymer, DuPont K. K.,
  Kiyohara, Utsunomya, Tochigi 321-3231 Japan
  Hiroyuki.kikuchi@jpn.dupont.com
Kiryukhin, V., Theoretical Mechanics Department, Perm State Technical
  University, Komsomolsky str., 29a, Perm, 614600, Russia
  kvy@theormech.,pstu.ac.ru
Kobayashi, H., Dept. Mechanical Engineering, Muroran Institute of Technology,
  27-1, Mizumoto, Muroran, Hokkaido, 050-8585, Japan
  kobayasi@mmm.muroran-it.ac.jp
Kobayashi, S., System Dept. #1, Corporate IT Div., Toyota Motor Co.,
  Toyota, Aichi, 471-8571 Japan
  kobayashi@mail.toyota.co.jp
Khono, Y., Bridgestone Co., Ogawahigashi-cyo, Kodaira, Tokyo, 187-8531 Japan
  kouno-y@bridgestone.co.jp
Kosawada, T., Department of Mechanical Engineering, Yamagata University,
  Yonezawa, Yamagata 992-8510, Japan
  kosawada@yz.yamagata-u.ac.jp
Kovaleva, A., Mechanical Engineering Research Institute, Russian Academy of
  Science, Kavkazsky blv. 44-3-17, Moscow 115516, Russia
  a.kovaleva@ru.net
Kuang, Z. B., Department of Engineering Mechanics, Shanghai Jiaotong University,
  Shanghai, 200240, China
  ZBKuang@mail.SJTU.edu.cn
Kurashige, M., Department of Mechanical Engineering, Iwate University,
  Ueda 4-3-5, Morioka, Iwate 020-8551, Japan
  kurashige@iwate-u.ac.jp
Kuroda, M., Department of Mechanical Engineering, Yamagata University,
  Yonezawa, Yamagata, 992-8510 Japan
  kuroda@yz.yamagata-u.ac.jp
Lee, J. J., Korean Advanced Institute of Science and Technology,
  373-1 Gusong-dong, Yusong-ku, Taejon, 305-701, Korea
  jjlee@mail.kaist.ac.kr

Lerch, R., Universität Erlangen-Nürnberg, Lehrstuhl für Sensorik,
Paul-Gordan-Str. 3/5, D-91052 Erlangen, Germany
reinhard.lerch@lse.e-technik.uni-erlangen.de

Liu, G. R., Dept. of Mechanical Engineering, National University of Singapore,
Engineering Drive 1, S117576, Singapore
mpeliugr@nus.edu.sg

Matsuo, T., Department of Mechanical Engineering, Fukushima Technical College,
Iwaki, Fukushima, 970-8034, Japan
matsuo@fukushima-nct.ac.jp

Murakami, H., Department of Mechanical and Aerospace Engineering,
University of California at San Diego
9500 Gilman Drive, La Jolla, CA 2093-0411, U. S. A.
murakami@mae.ucsd.edu

Narisawa, I., Kita-Yaroku 414-22, Kuroiso, Tochigi 329-3132, Japan
narisawa@gamma.ocn.ne.jp

Noda, N., Department of Mechanical Engineering, Shizuoka University,
Hamamatsu, Shizuoka, 432-8561, Japan
tmnnoda@ipc.shizuoka.ac.jp

Nozaki, H., Faculty of Education, Ibaragi University,
Bunnkyo 2-1-1, Mito, Ibaragi 310-8512, Japan
nozaki@ipc.ibaraki.ac.jp

Nyashin, Y., Theoretical Mechanics Department, Perm State Technical University,
Komsomolsky str., 29a, Perm, 614600, Russia
nyashin@thermech.pstu.ac.ru

Obata, Y., Inst. Struct. and Eng. Mat., NIAIST, Kita-ku, Nagoya, 462-8510 Japan
y-obata@aist.go.jp

Ootao, Y., Department of Mechanical Systems Engineering, Osaka Prefecture
University, Sakai, Osaka, 599-8531, Japan
ootao@mecha.osakafu-u.ac.jp

Przybyłowicz, P. M., Warsaw University of Technology, Institute of Machine
Design Fundamentals, Narbutta 84, 02-524 Warsaw, Poland
pmp@chello.pl

Qui, J., Institute of Fluid Science, Tohoku University,
Aoba, Sendai, 980-8577, Japan
qiu@ifs.tohoku.ac.jp

Rajapakse, N., Dept. of Mechanical Engineering, University of British Columbia,
Vancouver, BC V6T 1Z4, Canada
rajapakse@mech.ubc.ca

Sadek, I. S., Department of Computer Science, Mathematics and Statistics,
American University of Sharjah, P. O. Box 26666, Sharjah, U. A. E.
sadek@aus.ac.ae

Saito, M., Denki-Kogyo Co., Kanuma, Tochigi, 322-0014, Japan
mi-saito@denkikogyo.co.jp

Sapinski, B., Department of Process Control, University of Mining and Metallurgy,
   al. Mickiewicza 30- 059 Cracow, Poland
   deep@uci.agh.edu.pl
Schlacher, K., Johannes Kepler University of Linz,
   Altenbergerstrasse 69, A-4040 Linz, Austria
   kurt.schlacher@jku.at
Seki, A., Denki-Kogyo Co., Kanuma, Tochigi, 322-0014, Japan
   a-seki@denkikogyo.co.jp
Shibuya, Y., Department of Mechanical Engineering, Akita University,
   Akita 010-8502, Japan
   shibuya@ipc.akita-u.ac.jp
Sumi, N., Faculty of Education, Shizuoka University, Shizuoka, 422-8529, Japan
   einsumi@ipc.shizuoka.ac.jp
Sunar, M., Mechanical Engineering Department, King Fahd University of
   Petroleum and Minerals, Dhahran 31261, Saudi Arabia
   mehmets@kfupm.edu.sa
Takahashi, H., R&D Dept., Fuji Ceramics Corp., 2320-11 Yamamiya, Fujinomiya,
   Shizuoka, 418-0111, Japan
   LEN06236@nifty.ne.jp
Takahashi, K., R. & D. Dept., Taketoyo-Plant, NOF Corp.,
   Taketoyo, Chita, Aichi, 470-2398, Japan
   katsuhiko_takahashi@nof.co.jp
Tanigawa, Y., Department of Mechanical Systems Engineering, Osaka Prefecture
   University, Sakai, Osaka, 599-8531, Japan
   tanigawa@mecha.osakafu-u.ac.jp
Tauchert, T. R., Department of Engineering Mechanics, University of Kentucky,
   Lexington, KY, 40506-0046, U. S. A.
   tauchert@engr.uky.edu
Tylikowski, A., Institute of Machine Design Fundamentals, Warsaw University of
   Technology, Narbutta 84, 02-524 Warsaw, Poland
   aty@simr.pw.edu.pl
Ubertini, F., DISTART, Universita degli Studi di Bologna,
   Viale Risorgimento 2, 40136 Bologna, Italy
   francesco.ubertini@mail.ing.unibo.it
Ueda, S., Department of Mechanical Engineering, Osaka Inst. Tech.,
   Oomiya, Asahi, Osaka, 535-8585 Japan
   ueda@med.oit.ac.jp
Varadan, V. V., Department of Engineering Science & Mechanics,
   The Pennsylvania State University, University Park, PA 16802, U. S. A.
   vvvesm@engr.psu.edu
Varadan, V. K., Center for the Engineering of Electronic and Acoustic Materials,
   The Pennsylvania State University, State College, PA 16801, U. S. A.
Vokoun, D., National Tsing Hua University, Department of Materials Science and
   Engineering, 101, Section 2, Kuang-Fu Road, Hsinchu 300, Taiwan
   davidvokoun@yahoo.com.tw

Wang, X., Department of Civil Engineering, Nagoya Institute of Technology, Showa, Nagoya, 466-8555, Japan
xfwang@kozo4.ace.nitech.ac.jp

Watanabe, K., Department of Mechanical Engineering, School of Engineering, Yamagata University, Yonezawa, Yamagata 992-8510, Japan
kazy@yz.yamagata-u.ac.jp

Ziegler, F., Civil Engineering Department, Technical University of Vienna, Wiedner Hauptstr. 8-10/E201, A-1040, Vienna, Austria
franz.ziegler@tuwien.ac.at

Symposium program

*speaker

May 20 (Monday)

8:30-9:00	**Registration**
9:00-9:15	**Opening**
9:20-10:00	**Keynote lecture (1)** [Chairperson: V. V. Varadan (USA)] "Shape and Stress Control in Elastic and Inelastic Structures" Y. Nyashin*(Russia) and V. Kiryukhin
10:00-10:10	< Break >
10:10-12:20	**< Sensor and Actuator >** Chairpersons: U. Gabbert (Germany) and M. Sunar (Saudi Arabia)
10:10	"Smart Actuation from Coupling between Active Polymer Gels and Fibrous Structures," G. Jeronimidis (UK)
10:40	"Application of Newly Developed Transmission-type EFPI Optical Fiber Sensors to Smart Structures," J.-J. Lee*(Korea), S.-H. Kim and D.-C. Seo
11:10-11:20	< Break >
11:20	"Remote Smart Damage Detection via Internet with Unsupervised Statistical Diagnosis," A. Iwasaki*(Japan), A. Todoroki and T. Sugiya
11:50	"High-Performance PZT and PNN-PZT Actuators," J. Qiu*(Japan), J. Tani and H. Takahashi
12:20-14:00	< Lunch >
14:00-14:40	**Keynote lecture (2)** [Chairperson: A. Kovaleva (Russia)] "Dynamics of Structural Systems with Devices Driven by Fuzzy Controllers," F. Casciati (Italy)
14:40-14:50	< Break >
14:50-15:50	**<Active Control of Smart Structures>** Chairpersons: T. R. Tauchert (USA) and I.S. Sadek(UAE)
14:50	"Active Control of Smart Structures using Port Controlled Hamiltonian Systems," K. Schlacher*(Austria) and A. Kugi
15:20	"Control of Structures by Means of High-Frequency Vibration" A. Kovaleva (Russia)
15:50-16:00	< Break >
16:00-16:40	**Keynote lecture (3)** [Chairperson: Z. B. Kuang (China)] "Model Reduction and Robust Controllers for Complex Adaptive Structures," W. Chang and V. V. Varadan* (USA)
16:40-16:50	< Break >
16:50-18:20	**< Control and modeling >** Chairpersons: N. Hasebe (Japan) and A. Tylikowski (Poland)
16:50	"Integral Approach for Velocity Feedback Control in Thin Plate with Piezoelectric Patches," I.S. Sadek*(UAE), J.C. Bruch, Jr., J.M. Sloss, and S. Adali
17:20	"Large Deviations, Hamilton-Jacobi Equations and Stochastic Modeling of Surface Processes," J. Feng (USA)

17:50 "Modeling of Piezoelectric/Magnetostrictive Materials for Smart Structures," M. Sunar (Saudi Arabia)
18:20-18:30 < move to welcome party >
18:30-21:00 Welcome party (Jyosi-Enn)

May 21 (Tuesday)
9:00-9:40 **Keynote lecture (4)** [Chairperson: J. Holnicki-Szulc (Poland)]
"Numerical Modeling of Sensing and Actuating Electromechanical Transducer," R. Lerch (Germany)
9:40-9:50 < Break >
9:50-13:00 < Dynamics of Advanced Materials >
Chairpersons: R. Batra (USA) and A. Berezovski (Estonia)
9:50 "Shape Memory: Heterogeneity: Thermodynamics"
V. Kafka and D. Vokoun*(Czech)
10:20 "A Model for the Constitutive Law of Shape Memory Alloy Structural Components under Dynamic Loading"
L. Faravelli (Italy)
10:50 "Energy Release Rate Criteria for Piezoelectric Solids"
N. Rajapakse*(Canada) and S. X. Xu
11:20-11:30 < Break >
11:30 "Wave Scattering and Attenuation in Polymer-Based Composites: Analysis and Measurements," S. Biwa*(Japan), Y. Watanabe, S. Idekoba and N. Ohno
12:00 "Free Large Vibrations of Buckled Laminated Plates,"
R. Heuer (Austria)
12:30 "Application of Transfer Matrix Method in Analyzing the Inhomogeneous Initial Stress Problem in Prestressed Layered Piezoelectric Media," H. Liu, Z. B. Kuang*(China) and Z. M. Cai,
13:00-14:30 < Lunch >
14:30-17:00 < Poster presentation >
14:30-15:30 "5 minutes" poster appeal
Chairperson: S. N. Gorb (Germany)
(P1) "Wave Propagation in Piezoelectric Circular plate under Thermo-Electro-Mechanical Loading," N. Sumi (Japan)
(P2) "Transient Piezothermoelasticity for Cylindrical Composite Panel," Y. Ootao*(Japan) and Y. Tanigawa
(P3) "The Analysis of Transient Thermal Stresses in Piezothermoelastic Semi-infinite Body with an Edge Crack," M. Ishihara*(Japan), O. P. Niraula and N. Noda
(P4) "Transient Dynamic Stress Intensity Factors around a Rectangular Crack in a Nonhomogeneous Interfacial Layer Between Two Dissimilar Elastic Half-Spaces,"
S. Itou (Japan)
(P5) "A Green Function for a Radially Inhomogeneous Elastic Solid," K. Watanabe (Japan) and T. Takeuchi

	(P6) "Exact Solution for a Thermoelastic Problem of Wave Propagations in a Piezoelectric Plate," F. Ashida (Japan) and T. R. Tauchert
	(P7) "Stress-Focusing Effect in a Spherical Inclusion Embedded in an Infinite Medium Caused by Instantaneous Phase Transform," T. Hata (Japan)
	(P8) "Anisotropy in Packing Structure and Elasticity of Sintered Spherical Particles," M. Kurashige*(Japan), H. Kato, C. Matsunaga and K. Imai
15:30-17:00	**Poster discussions**
17:00-18:00	< **Museum guide** >
18:30-20:00	< Light supper >

May 22 (Wednesday)

9:00-9:40	**Keynote (5)**[Chairperson: U. Gabbert (Germany)] "Infinitesimal Mechanism Modes of Tensegrity Modules" H. Murakami*(USA) and Y. Nishimura
9:40-9:50	< Break >
9:50-12:00	< **Biomimetic Smart Structures** > Chairpersons: G. Jeronimidis (UK) and M. Kurashige (Japan)
9:50	"Mechanics of Plasma Membrane Vesicles in Cells," T. Kosawada (Japan)
10:20	"Unfolding of Morning Glory Flower as a Deployable Structure" H. Kobayashi*(Japan), M. Daimaruya and H. Fujita
10:50-11:00	< Break >
11:00	"Bio-Mimetic Smart Microstructures: Attachment Devices in Insects as Possible Source for Technical Design," S. N. Gorb (Germany)
11:30	"Application of Stress and Strain Control Theory to Living Tissues," V. Kiryukhin*(Russia) and Y. Nyashin
12:00-14:00	< Lunch >
14:00-18:00	**Excursion** (short trip around Yonezawa city)
18:30-21:00	**Banquet** (Count Uesugi's house garden)

May 23 (Thursday)

9:00-9:40	**Keynote lecture (6)** [Chairperson: Chattopadhyay (USA)] "Control of Thermally-Induced Structural Vibration via Piezoelectric Pulses," T. R. Tauchert*(USA) and F. Ashida
9:40-9:50	< Break >
9:50-12:00	<**Control of Thermal Vibration** > Chairpersons: S. Govindjee (USA) and R. Heuer (Austria)
9:50	"Thermally Induced Vibration of an Inhomogeneous Beam due to a Cyclic Heating," Y. Tanigawa*(Japan), R. Kawamura and R. B. Hetnarski,
10:20	"Exact Solution for Cylindrical Thermoelastic Deformations of Functionally Graded Thick Plates," S. S. Vel and R. C.

	Batra*(USA)
10:50-11:00	< Break >
11:00	"Complex Variable Solution of Plane Problem for Functionally Graded Materials," X. Wang*(Japan) and N. Hasebe
11:30	"Non-Parametric Representation of MR Linear Damper Dynamic Behavoir," B. Sapinski (Poland)
12:00-14:00	< Lunch >
14:00-14:40	**Keynote lecture (7)** [Chairperson: F. Ziegler (Austria)] "Overall Design and Simulation of Smart Structures " U. Gabbert*(Germany), H. Koppe, F. Seeger, and T. N. Traijkov
14:40-14:50	< Break >
14:50-16:20	< **Simulation of Smart Structures** > Chairpersons: J. J. Lee (Korea) and L. Faravelli (Italy)
14:50	"Numerical Simulation for Control of Progressive Plastic Buckling with Defects on Axisymmetric Shell Structure" Y. Shibuya*(Japan) and S. Watanabe
15:20	"Transient Analysis of Smart Structures Using a Coupled Piezoelectric- Mechanical Theory," A. Chattopadhyay*(USA), R. P. Thornburgh and A. Ghoshal
15:50	"Simulation of Impact-Induced Martensitic Phase-Transition Front Propagation in Thermoelastic Solids," A. Berezovski*(Estonia) and G. A. Maugin
16:20-16:30	< Break >
16:30-17:10	**Keynote lecture (8)**[Chairperson: F. Ubertini (Italy)] "A Review of Simulation Methods for Smart Structures with Piezoelectric Material," G. R. Liu*(Singapore), C. Cai, K. Y. Lam and V. K. Varadan
17:10-17:20	< Break >
17:20-18:20	< **Modeling of Advanced Materials** > Chairperson: M. A. E. Kaunda (S. Africa) and F. Ashida (Japan)
17:20	"Finite Element Models for Linear Electroelastic Dynamics" F. Ubertini (Italy)
17:50	"Application of Quasi-Convexity in Evolutionary Modeling and Simulation of Shape Memory Alloys," S. Govindjee (USA)
18:30-20:00	< Light supper >

May 24 (Friday)

9:00-9:40	**Keynote lecture (9)**[Chairperson: K. Watanabe] "Maysel's Formula for Small Vibrations Superimposed upon Large Static Deformations of Piezoelastic Structures" H. Irschik*(Austria) and U. Pichler
9:40-9:50	< Break >
9:50-12:00	< **Active Damping**> Chairpersons: K. Schlacher (Austria) and N. Rajapakse (Canada)
9:50	"Active Damping of Parametric Vibrations of Mechanical

10:20	Disturbed Systems," A. Tylikowski (Poland) "High-Performance Impact Absorbing Materials- The Concept, Design Tools and Applications," J. Holnicki-Szulc (Poland)
10:50-11:00	< Break >
11:00	"A Self-Sensing Active Constrained Layer Damping Treatment for Composite Structures and Determination of Lame's Constants," M. A. E. Kaunda (S. Africa)
11:30	"Active Damping of Torsional Vibration in a Piezoelectric Fiber Composite Shaft," P. M. Przybylowicz (Poland)
12:00-12:15	**Closing**
12:15-	< Lunch >
18:00-21:00	**Farewell Party** (Tokyo Dai-Ichi Hotel Yonezawa)

EXACT SOLUTION FOR A THERMOELASTIC PROBLEM OF WAVE PROPAGATIONS IN A PIEZOELECTRIC PLATE

FUMIHIRO ASHIDA
Department of Electronic and Control Systems Engineering
Shimane University, Matsue, Shimane 690-8504 Japan
E-mail: ashida@ecs.shimane-u.ac.jp

THEODORE R. TAUCHERT
Department of Mechanical Engineering, University of Kentucky
Lexington, Kentucky 40506-0046 USA
E-mail: tauchert@engr.uky.edu

1. INTRODUCTION

Many papers have treated dynamic problems of various electromechanical structures. However, most of these papers dealt with vibration problems, such as the detection and/or control of harmonic vibrations in various host structures by utilizing attached piezoelectric sensors and/or actuators [1]. On the other hand, there are a few papers that have discussed wave propagation in piezoelectric materials. For example, Wang [2] analyzed an isothermal problem of wave propagation in an electro-elastic structure, when the host structure was vibrated harmonically by means of attached piezoelectric actuators.

It is expected that piezoceramic thin films of nano-order thickness will be put to practical use in the near future. When thermal loads act on such thin films, effects of both relaxation time and the inertia term in the thermoelastic field may be significant and should not be neglected.

Therefore, the present paper deals with the dynamic thermoelastic problem of a thin piezoelectric plate of crystal class 6mm, when not only a relaxation time in the temperature field but also an inertial effect in the elastic field is taken into account. One boundary surface of the thin plate is exposed to a uniform ambient temperature, whereas the other boundary surface is kept at zero temperature. An exact solution to this problem is obtained by employing the Laplace transform technique. Numerical calculations have been carried out for a thin PZT-5A plate, and the numerical results are illustrated graphically.

2. PROBLEM STATEMENT

Let us consider a thin circular piezoelectric plate of crystal class 6mm. The radius and thickness of the plate are denoted by a and b.

2.1 Temperature Field

It is assumed that the thin piezoelectric plate, initially at zero temperature, is suddenly subjected to a uniform ambient temperature T_c on the top surface; the bottom surface is kept at zero temperature and the cylindrical edge is thermally insulated. In this case, the initial and boundary conditions are given by

$$T = T_{,t} = 0 \quad \text{at} \quad t = 0 \tag{1}$$

$$T_{,z} + h T = h T_c \quad \text{on} \quad z = b \tag{2}$$

$$T = 0 \quad \text{on} \quad z = 0 \tag{3}$$

$$T_{,r} = 0 \quad \text{on} \quad r = a \tag{4}$$

where T is temperature, t is time and h is the relative surface heat transfer coefficient. The heat conduction equation with a relaxation time t_0 is expressed by

$$\lambda_z T_{,zz} = \rho C \left(T_{,t} + t_0 T_{,tt} \right) \tag{5}$$

where λ_z is the coefficient of thermal conductivity, ρ is density, and C is specific heat.

2.2 Elastic and Electric Fields

For the elastic and electric fields of the thin piezoelectric plate exposed to the thermal conditions (1) ~ (4), it is assumed that the cylindrical edge of the plate is smoothly constrained against radial deformation and free of electric charge. Then,

$$u_r = 0, \quad \sigma_{rz} = 0 \tag{6}$$

$$E_r = 0, \quad D_r = 0 \tag{7}$$

while the other stresses and electric displacement are expressed by

$$\sigma_{rr} = \sigma_{\theta\theta} = c_{13}\varepsilon_{zz} - e_1 E_z - \beta_1 T, \quad \sigma_{zz} = c_{33}\varepsilon_{zz} - e_3 E_z - \beta_3 T \tag{8}$$

$$D_z = e_3 \varepsilon_{zz} + \eta_3 E_z + p_3 T \tag{9}$$

where u_i are elastic displacements, σ_{ik} are stresses, E_i are electric field intensities, D_i are electric displacements, c_{ik} are elastic moduli, e_i are piezoelectric coefficients, β_i are stress-temperature coefficients, η_3 is dielectric permittivity, and p_3 is pyroelectric constant.

In this case, the elastic and electric fields are governed by

$$\sigma_{zz,z} = \rho u_{z,tt} \tag{10}$$

$$D_{z,zt} = 0. \tag{11}$$

The initial conditions are assumed to be

$$u_z = u_{z,t} = 0 \quad \text{at} \quad t = 0. \tag{12}$$

When the top and bottom surfaces of the thin plate are considered to be free of both traction and electric charge, the boundary conditions are taken to be

$$\sigma_{zz} = 0 \quad \text{on} \quad z = 0, b \tag{13}$$

$$D_z = 0 \quad \text{on} \quad z = 0, b. \tag{14}$$

3. ANALYSIS

In order to solve the governing equations (10) and (11), we introduce a displacement potential Ω and an electric potential Φ as follows:

$$u_z = \Omega_{,z} \tag{15}$$

$$E_z = -\Phi_{,z}. \tag{16}$$

The governing equations (10) and (11) are satisfied providing the displacement and electric potentials satisfy the equations:

$$\Omega_{,zz} - \frac{1}{v_e^2}\Omega_{,tt} = \xi_1 T \tag{17}$$

$$\Phi_{,z} = \delta \Omega_{,zz} + \xi_2 T \tag{18}$$

where

$$v_e^2 = \frac{1}{\rho}\left(c_{33} + \frac{e_3^2}{\eta_3}\right), \quad \xi_1 = \frac{\eta_3 \beta_3 - e_3 p_3}{\eta_3 c_{33} + e_3^2}, \quad \delta = \frac{e_3}{\eta_3}, \quad \xi_2 = \frac{p_3}{\eta_3}. \tag{19}$$

Applying the Laplace transform with respect to the time variable, the exact solution to Eq. (5) which satisfies the initial and boundary conditions (1) ~ (4) is obtained. The temperature is expressed by

$$\begin{aligned}
T = A_0 \frac{z}{b} + \sum_{m=1}^{\infty} &\sin\left(\frac{\gamma_m z}{\lambda}\right) \exp\left(-\frac{t}{2t_0}\right) \\
&\times \left[H(\alpha_m^2) A_m \left\{ \alpha_m \cos\left(\frac{\alpha_m t}{2t_0}\right) + \sin\left(\frac{\alpha_m t}{2t_0}\right) \right\} \right. \\
&\left. + H(\varepsilon_m^2) B_m \left\{ \varepsilon_m \cosh\left(\frac{\varepsilon_m t}{2t_0}\right) + \sinh\left(\frac{\varepsilon_m t}{2t_0}\right) \right\} \right]
\end{aligned} \tag{20}$$

where $H(x)$ is the Heaviside unit step function and

$$\left.\begin{aligned}
A_0 = \frac{hbT_c}{1+hb}, \quad A_m = \frac{C_m}{\alpha_m}, \quad B_m = \frac{C_m}{\varepsilon_m}, \\
C_m = \frac{2\lambda h T_c}{\gamma_m \left\{ (1+hb)\cos\left(\frac{\gamma_m b}{\lambda}\right) - \frac{\gamma_m b}{\lambda}\sin\left(\frac{\gamma_m b}{\lambda}\right) \right\}}
\end{aligned}\right\} \tag{21}$$

$$\alpha_m = \sqrt{4\kappa t_0 \gamma_m^2 - 1}, \quad \varepsilon_m = \sqrt{1 - 4\kappa t_0 \gamma_m^2}. \tag{22}$$

Also, γ_m are roots of the equation:

$$\gamma_m \cos\left(\frac{\gamma_m b}{\lambda}\right) + \lambda h \sin\left(\frac{\gamma_m b}{\lambda}\right) = 0 \tag{23}$$

where $\lambda^2 = \lambda_z / \lambda_r$, and κ is thermal diffusivity.

Applying the same approach as in the case of the temperature field, the exact solution to Eqs. (17) and (18) which satisfies the initial and boundary conditions (12) ~ (14) can be obtained. The displacement potential is expressed by

$$\begin{aligned}
\Omega = F_0 \frac{z^3 - b^2 z}{b^3} &+ \exp\left(-\frac{t}{2t_0}\right) \\
\times \sum_{m=1}^{\infty} &\left[H(\alpha_m^2) \left\langle F_{1m} \left\{ F_{2m} \cos\left(\frac{\alpha_m t}{2t_0}\right) - F_{3m} \sin\left(\frac{\alpha_m t}{2t_0}\right) \right\} \sin\left(\frac{\gamma_m z}{\lambda}\right) \right.\right. \\
&+ \left\{ F_{4m} \cos\left(\frac{\alpha_m t}{2t_0}\right) + F_{5m} \sin\left(\frac{\alpha_m t}{2t_0}\right) \right\} \sinh\left(\frac{z}{2v_e t_0}\right) \cos\left(\frac{\alpha_m z}{2v_e t_0}\right) \\
&- \left.\left.\left\{ F_{5m} \cos\left(\frac{\alpha_m t}{2t_0}\right) - F_{4m} \sin\left(\frac{\alpha_m t}{2t_0}\right) \right\} \cosh\left(\frac{z}{2v_e t_0}\right) \sin\left(\frac{\alpha_m z}{2v_e t_0}\right) \right\rangle \right.
\end{aligned}$$

$$+ H(\varepsilon_m^2) \left\langle G_{1m} \left\{ G_{2m} \cosh\left(\frac{\varepsilon_m t}{2 t_0}\right) - G_{3m} \sinh\left(\frac{\varepsilon_m t}{2 t_0}\right) \right\} \sin\left(\frac{\gamma_m z}{\lambda}\right) \right.$$

$$+ \left\{ G_{4m} \cosh\left(\frac{\varepsilon_m t}{2 t_0}\right) + G_{5m} \sinh\left(\frac{\varepsilon_m t}{2 t_0}\right) \right\} \sinh\left(\frac{z}{2 v_e t_0}\right) \cosh\left(\frac{\varepsilon_m z}{2 v_e t_0}\right)$$

$$- \left\{ G_{5m} \cosh\left(\frac{\varepsilon_m t}{2 t_0}\right) + G_{4m} \sinh\left(\frac{\varepsilon_m t}{2 t_0}\right) \right\} \cosh\left(\frac{z}{2 v_e t_0}\right) \sinh\left(\frac{\varepsilon_m z}{2 v_e t_0}\right) \right\rangle \right]$$

$$- \sum_{j=1}^{\infty} K_{1j} \left\{ K_{2j} \cos(v_e \omega_j t) + K_{3j} \sin(v_e \omega_j t) \right\} \sin(\omega_j z) \tag{24}$$

where $\omega_j = j\pi/b$, and the coefficients F_0, F_{im}, G_{im} and K_{ij} are known, but the expressions for these coefficients are omitted here for brevity. Application of Eq. (18) leads to the electric potential. The stresses σ_{rr}, $\sigma_{\theta\theta}$ and σ_{zz} are derived from Eqs. (15), (16) and (8).

4. NUMERICAL RESULTS

Numerical calculations have been carried out for a PZT-5A plate. The material constants are taken to be

$$\rho = 7750 \text{ kg m}^{-3}, \ \lambda_r = \lambda_z = 1.5 \text{ W m}^{-1} \text{K}^{-1}, \ \alpha_r = 5.1 \times 10^{-6} \text{ K}^{-1},$$

$$Y_r = 61.0 \times 10^9 \text{ N m}^{-2}, \ c_{13} = 75.4 \times 10^9 \text{ N m}^{-2}, \ c_{33} = 111.0 \times 10^9 \text{ N m}^{-2},$$

$$\beta_1 = 1.52 \times 10^6 \text{ N K}^{-1} \text{m}^{-2}, \ \beta_3 = 1.53 \times 10^6 \text{ N K}^{-1} \text{m}^{-2},$$

$$\eta_1 = 8.11 \times 10^{-9} \text{ C}^2 \text{N}^{-1} \text{m}^{-2}, \ \eta_3 = 7.35 \times 10^{-9} \text{ C}^2 \text{N}^{-1} \text{m}^{-2},$$

$$d_1 = -171 \times 10^{-12} \text{ C N}^{-1}, \ e_1 = -5.4 \text{ C m}^{-2}, \ e_3 = 15.8 \text{ C m}^{-2},$$

$$p_3 = -452 \times 10^{-6} \text{ C K}^{-1} \text{m}^{-2}$$

where α_r is the coefficient of linear thermal expansion, Y_r is Young's modulus, d_1 is piezoelectric coefficient, and the values of α_r, β_1, β_3 and p_3 are assumed. In order to investigate the effect of the piezoelectricity on the stresses, numerical calculations also have been performed for a material with thermoelastic properties identical to those of PZT-5A, but without the piezoelectric effect ($e_1 = e_3 = p_3 = 0$).

For convenience in presentation of numerical results, we introduce the following dimensionless quantities:

$$\bar{z} = \frac{z}{b}, \ \bar{t} = \frac{v_{e0} t}{b}, \ \bar{t}_0 = \frac{v_{e0} t_0}{b}, \ B_i = h b, \ \bar{c}_e = \frac{\kappa}{v_{e0} b}, \ \bar{T} = \frac{T}{T_c}, \ \bar{\sigma}_{ik} = \frac{\sigma_{ik}}{\alpha_r Y_r T_c}$$

where v_{e0} is the velocity of propagation of the stress wave in the non-piezoelectric material.

Biot's number and inertia parameter are taken to be

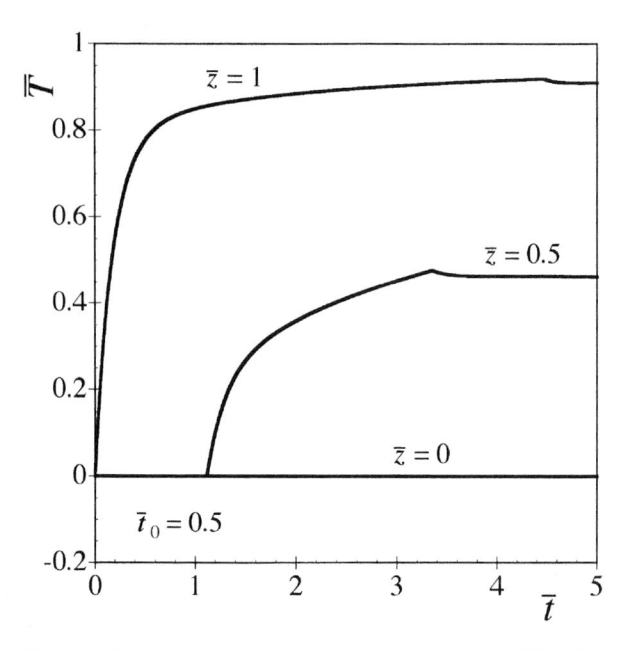

Figure 1. Time histories of temperature in the case of $\bar{t}_0 = 0.5$.

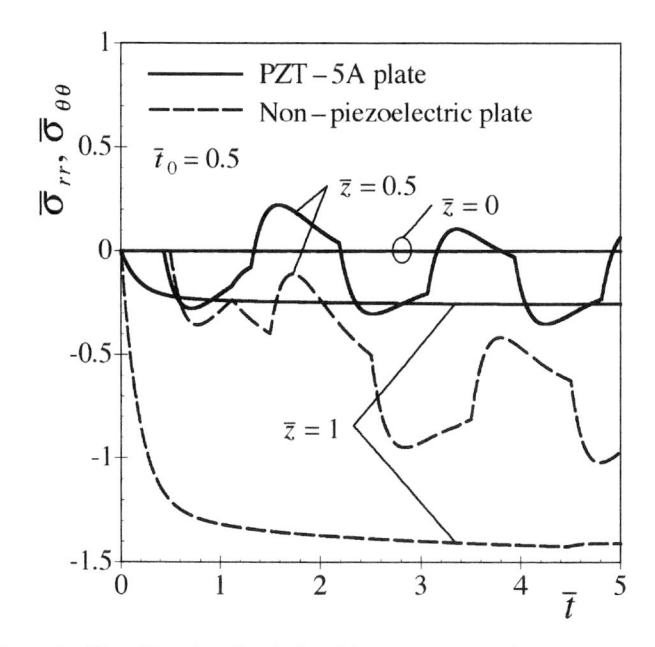

Figure 2. Time histories of radial and hoop stresses in the case of $\bar{t}_0 = 0.5$.

$$B_i = 10, \quad \bar{c}_e = 0.1$$

and two values of the dimensionless relaxation time are selected, namely

$$\bar{t}_0 = 0.05, 0.5 .$$

The thermal wave propagates more quickly than the stress wave when $\bar{t}_0 = 0.05$; conversely, the stress wave travels faster than the thermal wave when $\bar{t}_0 = 0.5$.

Figures 1 and 2 illustrate the time histories of temperature and radial and hoop stresses on the bottom, middle and top surfaces of the thin plate in the case of $\bar{t}_0 = 0.5$, whereas Figs. 3 and 4 illustrate corresponding results in the case of $\bar{t}_0 = 0.05$. In Figs. 2 and 4, the solid lines represent the results obtained for the PZT-5A plate, whereas the broken lines denote the corresponding results derived for the non-piezoelectric plate.

These figures show that the relaxation time exerts a remarkable influence on the radial and hoop stresses, but it has little effect on the temperature. Comparing Fig. 2 with Fig. 4, peaks of the stresses in the case of $\bar{t}_0 = 0.05$ are much sharper and values at the peaks are far larger than those in the case of $\bar{t}_0 = 0.5$.

Figures 2 and 4 show that the maximum values of the absolute radial and hoop stresses in the PZT-5A plate are smaller than those in the non-piezoelectric plate. However, it is seen from Fig. 2 that a tensile stress occurs in the case of PZT-5A

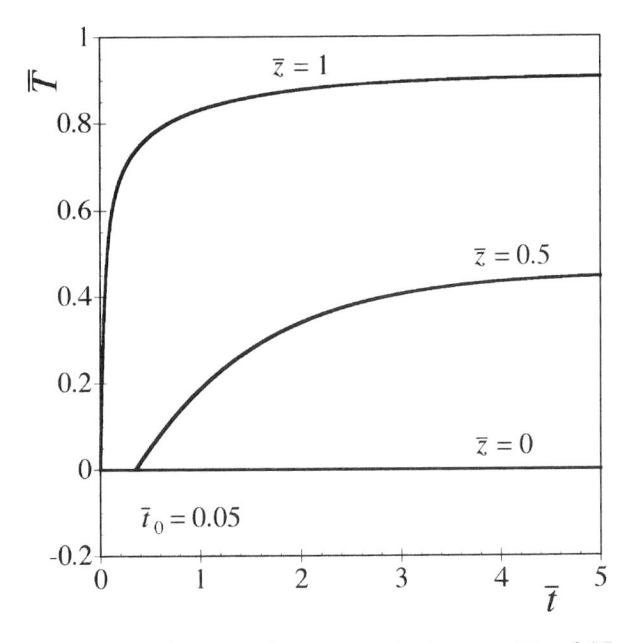

Figure 3. Time histories of temperature in the case of $\bar{t}_0 = 0.05$.

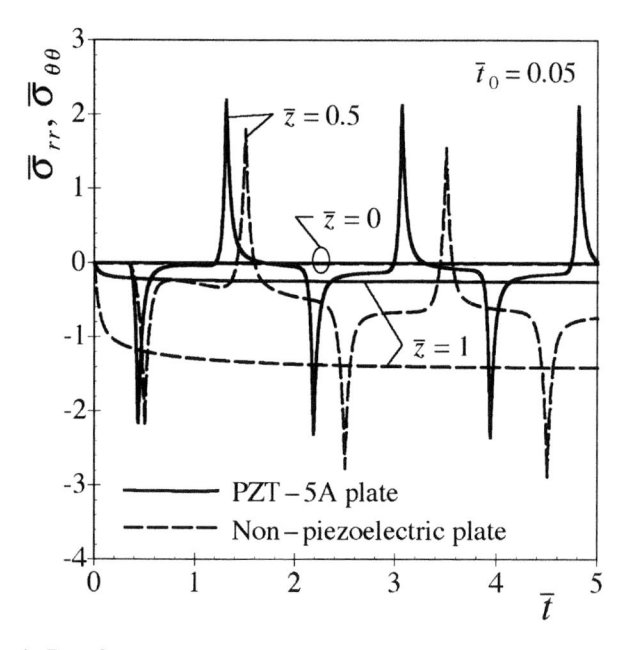

Figure 4. Time histories of radial and hoop stresses in the case of $\bar{t}_0 = 0.05$.

plate. Also, Fig. 4 indicates that the maximum values of tensile stresses in the PZT-5A plate are larger than those in the non-piezoelectric plate.

5. CONCLUDING REMARKS

The present paper has discussed the dynamic thermoelastic problem of a thin circular piezoceramic plate of crystal class 6mm, when not only the relaxation time in the temperature field but also the inertial effect in the elastic field is taken into account. The exact solution to this problem was obtained by employing the Laplace transform technique. Numerical calculations were carried out for the PZT-5A and non-piezoelectric plates. The numerical results have shown that the relaxation time has a remarkable influence on the stresses, although it had little effect on the temperature. It was shown that there were significant differences in the radial and hoop stresses between the PZT-5A and non-piezoelectric plates.

6. REFERENCES

1. Sunar, M. and Rao, S. S. (1999), Recent Advances in Sensing and Control of Flexible Structures via Piezoelectric Materials Technology, *Applied Mechanics Review*, **52**, 1-16.
2. Wang, X. D. and Huang, G. L. (2001), Wave Propagation in Electromechanical Structures: Induced by Surface-Bonded Piezoelectric Actuators, *Journal of Intelligent Material Systems and Structures*, **12**, 105-115.

SIMULATION OF IMPACT-INDUCED MARTENSITIC PHASE-TRANSITION FRONT PROPAGATION IN THERMOELASTIC SOLIDS

ARKADI BEREZOVSKI

Centre for Nonlinear Studies, Institute of Cybernetics at Tallinn Technical University, Akadeemia tee 21, 12618 Tallinn, Estonia
E-mail: Arkadi.Berezovski@cs.ioc.ee

GERARD A. MAUGIN

Laboratoire de Modélisation en Mécanique, Université Pierre et Marie Curie, UMR 7607, Tour 66 4 Place Jussieu, Case 162, 75252, Paris Cédex 05, France
E-mail: gam@ccr.jussieu.fr

1. INTRODUCTION

Most experiments in martensitic phase transformations are performed under quasi-static loading of a specimen. The results of the quasi-static experiments usually characterize the bulk properties of the material in the specimen, but not the local behavior of phase transition fronts. The only well-documented experimental investigation concerning the impact-induced austenite-martensite phase transformations is given by Escobar and Clifton [1, 2]. As Escobar and Clifton noted, measured velocity profiles provide a difference between the particle velocity and the transverse component of the projectile velocity. This velocity difference, in the absence of any evidence of plastic deformation, is indicative of a stress induced phase transformation that propagates into the crystals from the impact face. But the determination of this velocity difference is most difficult from the theoretical point of view. In fact, the above mentioned velocity difference depends on the velocity of a moving phase boundary. However, the extensive study of the problem of moving phase boundaries shows that the velocity of a moving phase boundary cannot be determined in the framework of classical continuum mechanics without any additional hypothesis [3]-[9]. What continuum mechanics is able to determine is the so-called driving force acting on the phase boundary. The propagation of a phase boundary is thus expected to be described by a kinetic relation between the driving traction and the rate at which the transformation proceeds.

We develop a thermomechanical approach to the modeling of phase transition front propagation based on the balance laws of continuum mechanics in the reference configuration [10] and the thermodynamics of discrete systems [11]. Phase boundaries are treated as discontinuity surfaces of zero thickness. Jump conditions following from continuum mechanics are fulfilled at the interface between two phases. We introduce the so-called contact quantities for the description of non-equilibrium states of the discrete elements representing a continuous body. The values of contact quantities for adjacent elements are connected by means of so-called thermodynamic consistency conditions. The thermodynamic consistency manifests itself only at the discrete level of description (e.g., in numerical approximation). It simply means that the thermodynamic state of any discrete element (grid cell) of the computational domain should be consistent with the corresponding state of its sub-elements (sub-cells). These thermodynamic consistency conditions are different for processes with and without entropy production. The latter consideration dictates us the rule of application of the consistency conditions: one is used in the bulk and another at the phase boundary (where the entropy is produced). A thermodynamic criterion for the initiation of the phase transition process follows from the simultaneous satisfaction of both homogeneous and heterogeneous thermodynamic consistency conditions at the phase boundary. A critical value of the driving force is determined that corresponds to the initiation of the phase transition process. It is shown that the developed model captures the experimentally observed particle velocity difference.

2. UNIAXIAL MOTION OF A SLAB

In order to explain some of the key ideas with a minimal mathematical complexity, it is convenient to work in an essentially one-dimensional setting. Consider a slab, which in an unstressed reference configuration occupies the region $0 < x_1 < L$, $-\infty < x_2, x_3 < \infty$, and consider uniaxial motion of the form

$$u_i = u_i(x,t), \quad x = x_1, \tag{1}$$

where t is time, x_i are spatial coordinates, u_i are components of the displacement vector. In this case, we have only three non-vanishing components of the strain tensor

$$\varepsilon_{11} = \frac{\partial u_1}{\partial x}, \quad \varepsilon_{12} = \varepsilon_{21} = \frac{1}{2}\frac{\partial u_2}{\partial x}, \quad \varepsilon_{13} = \varepsilon_{31} = \frac{1}{2}\frac{\partial u_3}{\partial x}. \tag{2}$$

Without loss of generality, we can set $\varepsilon_{13} = 0, v_3 = 0$. Then we obtain uncoupled systems of equation for longitudinal and shear components which express the balance of linear momentum and the time derivative of the Duhamel-Neumann thermoelastic constitutive equation, respectively. We focus our attention on the system of equations for shear components because martensitic phase transformation is expected to be induced by shear.

$$\frac{\partial(\rho_0(x)v_2)}{\partial t} - \frac{\partial \sigma_{12}}{\partial x} = 0, \qquad \frac{\partial}{\partial t}\left(\frac{\sigma_{12}}{\mu(x)}\right) - \frac{\partial v_2}{\partial x} = 0. \tag{3}$$

Here σ_{12} is the shear component of the Cauchy stress tensor, v_2 is transversal particle velocity, ρ_0 is the density, μ is the Lamé coefficient. The indicated explicit dependence on the point x means that the body is materially inhomogeneous in general.

2.1. Jump relations

To consider the possible irreversible transformation of a phase into another one, the separation between the two phases is idealized as a sharp, discontinuity surface $\mathcal{S}$ across which most of the fields suffer finite discontinuity jumps. Let $[A]$ and $<A>$ denote the jump and mean value of a discontinuous field A across $\mathcal{S}$, the unit normal to $\mathcal{S}$ being oriented from the "minus" to the "plus" side:

$$[A] := A^+ - A^-, \qquad <A> := \frac{1}{2}(A^+ + A^-). \tag{4}$$

Let $\tilde{V}$ be the material velocity of the geometrical points of $\mathcal{S}$. The material velocity V is defined by means of the inverse mapping $X = \chi^{-1}(x,t)$, where X denotes the material points [10]. The phase transition fronts considered are *homothermal* (no jump in temperature; the two phases coexist at the same temperature) and *coherent* (they present no defects such as dislocations). Consequently, we have the following continuity conditions [8]-[9]:

$$[V] = 0, \qquad [\theta] = 0 \quad at \quad \mathcal{S}, \tag{5}$$

where θ is temperature. Jump relations associated with the conservation laws in the bulk are formulated according to the theory of *weak solutions* of hyperbolic systems. Thus the jump relations associated with the balance of linear momentum and balance of entropy read [8]-[9]

$$V_N[\rho_0 v_2] + [\sigma_{12}] = 0, \qquad V_N[S] + \left[\frac{k}{\theta}\frac{\partial \theta}{\partial x}\right] = \sigma_\mathcal{S} \geq 0, \tag{6}$$

where $V_N = \tilde{V}$ is the normal speed of the points of $\mathcal{S}$, and $\sigma_\mathcal{S}$ is the entropy production at the interface. As it was shown in [8]-[9], the entropy production can be expressed in terms of the so-called "material" driving force $f_\mathcal{S}$

$$f_\mathcal{S} V_N = \theta_\mathcal{S} \sigma_\mathcal{S} \geq 0, \tag{7}$$

where θ_S is the temperature at S. In addition, the *balance of "material" forces* at the interface between phases can be specified to the form [7]- [9]

$$f_S = -[W] + <\sigma_{ij}>[\varepsilon_{ij}], \qquad (8)$$

where W is the free energy per unit volume.

The surface "balance" equation (8) follows from the balance law for pseudomomentum [10] and generalizes the equilibrium conditions at the phase-transition front to the dynamical case (c.f. [3]- [9]).

2.2. Dynamic loading

In a dynamic problem we shall seek piecewise smooth velocity and stress fields $v_2(x,t), \sigma_{12}(x,t)$ for inhomogeneous thermoelastic materials, which conform the following initial and boundary conditions:

$$\sigma_{12}(x,0) = v_2(x,0) = 0, \quad \text{for} \quad x > 0, \qquad (9)$$

$$v_2(0,t) = v_0(t), \quad \text{or} \quad \sigma_{12}(0,t) = \sigma_0(t) \quad \text{for} \quad t > 0, \qquad (10)$$

and satisfy the field equations (3) and jump conditions (5),(6), and (8). The system of equations (3) is a system of conservation laws which is suitable for a numerical solution, for example, by the wave-propagation algorithm [12].

3. WAVE-PROPAGATION ALGORITHM

The system of equations for one-dimensional elastic wave (3) can be represented in the form of conservation law

$$\frac{\partial q}{\partial t} + \frac{\partial f(q,x)}{\partial x} = 0, \qquad (11)$$

where

$$q(x,t) = \begin{pmatrix} \rho(x)v(x,t) \\ \sigma(x,t)/\mu(x) \end{pmatrix}, \quad f(q,x) = \begin{pmatrix} -\sigma(\varepsilon,x) \\ -v(x) \end{pmatrix}.$$

In the standard wave-propagation algorithm [12], the cell average

$$Q_i^n \approx \frac{1}{\Delta x} \int_{x_{i-1/2}}^{x_{i+1/2}} q(x,t_n)\,dx \qquad (12)$$

is updated in each time step as follows

$$Q_i^{n+1} = Q_i^n - \frac{\Delta t}{\Delta x}(F_{i+1}^n - F_i^n), \qquad (13)$$

where F_i approximates the time average of the exact flux taken at the interface between the cells, i.e.

$$F_i \approx \frac{1}{\Delta t}\int_{t_n}^{t_{n+1}} f(q(x_{i-1/2},t))dt. \qquad (14)$$

3.1. Thermodynamic consistency conditions

The finite-volume algorithm (13) can also be represented in terms of contact quantities [13]-[14]:

$$Q_i^{n+1} = Q_i^n - \frac{\Delta t}{\Delta x}\left(C_i^+(Q_i^n) - C_i^-(Q_i^n)\right), \qquad (15)$$

where $C^\pm$ denote corresponding contact quantities,

$$C^\pm(Q_i) = \begin{pmatrix} \Sigma^\pm(Q_i) \\ \mathcal{V}^\pm(Q_i) \end{pmatrix}. \qquad (16)$$

The contact quantities $\Sigma^\pm$ and $\mathcal{V}^\pm$ are determined by means of thermodynamic consistency conditions [13], [14]. In the *thermoelastic case*, the parameters of the adjacent non-equilibrium elements of a thermoelastic continuum should satisfy the thermodynamic consistency conditions, which can be called "homogeneous" (valid for all processes with no entropy production)

$$\left[-\theta\left(\frac{\partial \sigma_{ij}}{\partial \theta}\right)_{\varepsilon_{ij}} + \sigma_{ij} - \theta\left(\frac{\partial \Sigma_{ij}}{\partial \theta}\right)_{\varepsilon_{ij}} + \Sigma_{ij}\right] = 0, \qquad (17)$$

and "heterogeneous" (corresponds to any inhomogeneity accompanied by entropy production)

$$\left[\theta\left(\frac{\partial S}{\partial \varepsilon_{ij}}\right)_{\sigma_{ij}} + \sigma_{ij} + \theta\left(\frac{\partial S^{int}}{\partial \varepsilon_{ij}}\right)_{\sigma_{ij}} + \Sigma_{ij}\right] = 0. \qquad (18)$$

First we apply the homogeneous consistency condition (17) to determine the values of the contact quantities in homogeneous medium.

3.2. Contact quantities in the bulk

The dynamic part of the homogeneous consistency condition

$$[\sigma_{ij} + \Sigma_{ij}] = 0, \quad \Rightarrow \quad (\Sigma_{12}^+)_{i-1} - (\Sigma_{12}^-)_i = (\sigma_{12})_i - (\sigma_{12})_{i-1}, \qquad (19)$$

should be complemented by the kinematic condition [10] which can be rewritten in the small-strain approximation as follows

$$[v + \mathcal{V}] = 0, \quad \Rightarrow \quad (\mathcal{V}_2^+)_{i-1} - (\mathcal{V}_2^-)_i = (v_2)_i - (v_2)_{i-1}. \qquad (20)$$

The two relations (19) and (20) can be expressed in the vectorial form as follows:

$$C_{i-1}^+(Q_{i-1}^n) - C_i^-(Q_i^n) = f_i(Q_i) - f_{i-1}(Q_{i-1}). \tag{21}$$

It is easy to see, that the last expression is nothing more than the characteristic property for the conservative wave-propagation algorithm [15]. Thus, the thermodynamic consistency conditions and kinematic conditions at the cell edge automatically lead to the conservative wave-propagation algorithm. From another point of view, this means that the wave-propagation algorithm is thermodynamically consistent. However, phase transitions are always accompanied by the production of entropy. This is why we need to apply the heterogeneous consistency condition at the phase boundary.

3.3. *Contact quantities at the phase boundary*

We propose to apply the heterogeneous consistency conditions (18) for the calculation of the contact stresses at the phase boundary. Further, we suppose that the jump of the entropy of interaction is equal to the jump of entropy at the phase boundary

$$[S] = [S^{int}]. \tag{22}$$

For the computation of the entropy jump at the phase boundary, we will exploit the jump relation corresponding to the balance of the entropy $(6)_2$ and the expression for the entropy production in terms of the driving force (7). It follows from (22) and $(6)_2$ that the entropy of interaction has both thermal and dynamic contributions

$$S^{int} = S_{dyn}^{int} + S_{therm}^{int}. \tag{23}$$

As previously, we divide the heterogeneous consistency condition (18) into dynamic and thermal parts. For the dynamic part we obtain

$$[\sigma_{ij}] + [\Sigma_{ij}] = -\left[\theta \left(\frac{\partial S_{dyn}^{int}}{\partial \varepsilon_{ij}}\right)_\sigma\right]. \tag{24}$$

We can compute the derivatives of the entropy of interaction with respect to thermodynamic variables ε_{ij} by extending of definition of the entropy of interaction on every point of the body by similarity to $(6)_2$:

$$S_{dyn}^{int} = \frac{f}{\theta}, \quad f = -W + <\sigma_{ij}> \varepsilon_{ij} + f_0, \quad [f_0] = 0. \tag{25}$$

In the uniaxial case we have then for the shear contact stresses

$$(\Sigma_{12}^+)_{p-1} - (\Sigma_{12}^-)_p = 0. \tag{26}$$

This relation should be complemented by coherency condition (5), which can be expressed in terms of contact velocities as follows

$$(\mathcal{V}_2^+)_{p-1} - (\mathcal{V}_2^-)_p = 0. \tag{27}$$

The contact velocities are still connected with contact stresses by the relations along characteristic lines. However, all the considerations are valid only after the initiation of the phase transformation process.

3.4. *A thermodynamic initiation criterion*

We propose to expect the initiation of the stress-induced phase transition if both heterogeneous and homogeneous consistency conditions (17) and (18) are fulfilled at the phase boundary simultaneously.

Eliminating the jumps of stresses from (17), (18), we again consider the dynamic part of the combined consistency condition

$$\left[\theta \left(\frac{\partial}{\partial \theta} \left(\theta \frac{\partial S_{dyn}^{int}}{\partial \varepsilon_{ij}} \right) \right)_\varepsilon - \theta \left(\frac{\partial S_{dyn}^{int}}{\partial \varepsilon_{ij}} \right)_\sigma \right] = 0. \tag{28}$$

For the shear component, the dynamic part of the combined consistency condition (28) leads to the continuity of the shear stress at the phase boundary

$$[\sigma_{12}] = 0. \tag{29}$$

This inconvenient for the criterion of the initiation of phase transformation process and we do not use it. Therefore we should check also the combined consistency condition (28) for the normal components. Thus, for the pure shear wave, we obtain for the driving force at the interface

$$f_S = [f] = -\frac{\theta_0^2}{2} [\alpha(3\lambda + 2\mu)] \left\langle \frac{\alpha(3\lambda + 2\mu)}{\lambda + 2\mu} \right\rangle. \tag{30}$$

The right hand side of the latter relation can be interpreted as a critical value for the driving force. Therefore, the proposed criterion for the initiation of the stress-induced phase-transition in the case of uniaxial shear waves is the following one:

$$|f_S| \geq |f_{critical}| \equiv \left| \frac{\theta_0^2}{2} [\alpha(3\lambda + 2\mu)] \left\langle \frac{\alpha(3\lambda + 2\mu)}{\lambda + 2\mu} \right\rangle \right|. \tag{31}$$

The material velocity at the interface is determined by means of jump relation for linear momentum (6)

$$V_N^2 = \frac{[\sigma_{12}]}{2 <\rho_0> [\varepsilon_{12}]}, \tag{32}$$

The direction of the front propagation is determined by the positivity of the entropy production (7). Now we can compute all contact quantities and determine the driving force and the material velocity at the phase boundary. The obtained relations at the phase boundary are used in the described numerical scheme for the simulation of phase-transition front propagation.

4. NUMERICAL RESULTS

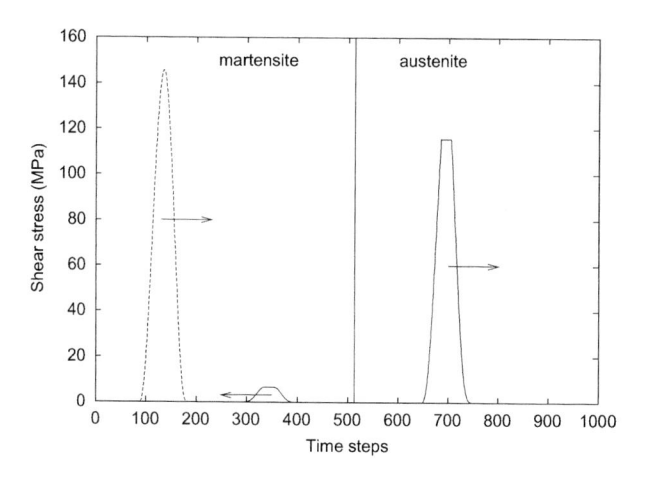

Figure 1. Shear wave after interaction with phase boundary.

First we characterize the interaction of a shear stress wave with the phase boundary. To compare further the results of numerical simulation with experimental data by Escobar and Clifton [1], [2], we extract the properties of austenite phase of the Cu-14.44Al-4.19Ni shape-memory alloy from their paper: the density $\rho = 7100\,kg/m^3$, the elastic modulus $E = 120\,GPa$, the shear wave velocity $c_s = 1187\,m/s$, the dilatation coefficient $\alpha = 6.75 \cdot 10^{-6}\,1/K$. As it was recently reported [16], elastic properties of martensitic phase of Cu-Al-Ni shape-memory alloy after impact loading are very sensitive to the amplitude of loading. Therefore, for the martensitic phase we choose, respectively, $E = 60\,GPa$, $c_s = 1055\,m/s$, with the same density and dilatation coefficient as above.

We simulate the wave propagation induced by an impulsive loading of a slab by shear stress of the shape shown in the left part of Fig. 1. After interaction with the phase boundary, we obtain transmitted and reflected waves, amplitudes of which are cut due to the martensitic phase transformation (Fig. 1). In addition, we observe a displacement of the phase boundary into the previous austenitic region.

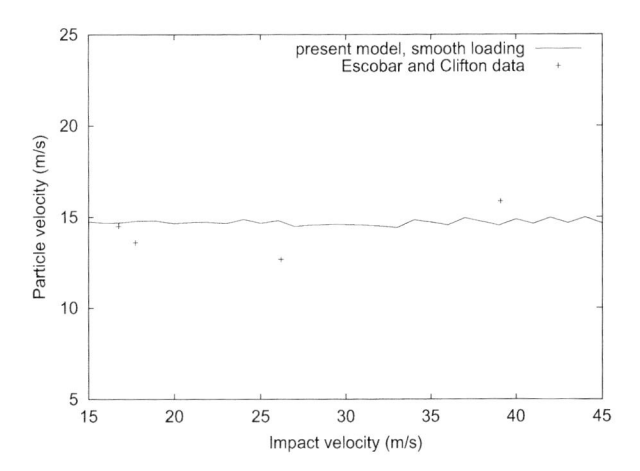

Figure 2. Particle velocity versus impact velocity. Smooth loading.

Now we will compare our simulations with experimental data for dynamic loading described by Escobar and Clifton [1], [2]. In their experiments, Escobar and Clifton used thin plate-like specimens of Cu-14.44Al-4.19Ni shape-memory alloy single crystal. One face of this austenitic specimen was subjected to an oblique impact loading, generating both shear and compression. The conditions of the experiment were carefully designed so as to lead to plane wave propagation in the direction of the specimen surface normal. The orientation of the specimen relative to the lattice was chosen to activate only a single variant of martensite. The temperature changes during Escobar and Clifton's experiments are thought to be relatively unimportant.

As Escobar and Clifton noted, measured velocity profiles provide a difference between the measured particle velocity and the transverse component of the projectile velocity. This velocity difference, in the absence of any evidence of plastic deformation, is indicative of a stress induced phase transformation. To compare the results of modeling with experimental data by Escobar and Clifton, the calculations of the particle velocity were performed for different impact velocities. The results of the comparison are given in Fig. 2, where we can see that the computed particle velocity is practically independent of the impact velocity. Thus, our simulations capture the experimentally observed difference between tangential impact velocity and transversal particle velocity, which is indicative for the existence of phase transformation.

References

1. Escobar, J.C. and Clifton, R.J. (1993) On pressure-shear plate impact for studying the kinetics of stress-induced phase-transformations, *Mat. Sci. & Engng.* **A170** 125-142.
2. Escobar, J.C. and Clifton, R.J. (1995) Pressure-shear impact-induced phase transformations in Cu-14.44Al-4.19Ni single crystals, in: *Active Materials and Smart Structures*, SPIE Proceedings, **2427** 186-197.
3. Truskinovsky, L. (1987) Dynamics of nonequilibrium phase boundaries in a heat conducting nonlinear elastic medium, *J. Appl. Math. Mech.* (PMM) **51** 777-784.
4. Abeyaratne, R. and Knowles, J.K. (1990) On the driving traction acting on a surface of strain discontinuity in a continuum, *J. Mech. Phys. Solids* **38** 345-360.
5. Abeyaratne, R. and Knowles, J.K.(1994) Dynamics of propagating phase boundaries: adiabatic theory for thermoelastic solids, *Physica D* **79** 269-288.
6. Abeyaratne, R. and Knowles, J.K. (1997) On the kinetics of an austenite-martensite phase transformation induced by impact in a Cu-Al-Ni shape-memory alloy, *Acta Mater.* **45** 1671-1683.
7. Maugin, G.A. and Trimarco, C. (1997) Driving force on phase transition fronts in thermoelectroelastic crystals, *Math. Mech. Solids* **2** 199-214.
8. Maugin, G.A. (1997) Thermomechanics of inhomogeneous - heterogeneous systems: application to the irreversible progress of two- and three-dimensional defects, *ARI* **50** 41-56.
9. Maugin, G.A. (1998) On shock waves and phase-transition fronts in continua, *ARI* **50** 141-150.
10. Maugin, G.A. (1993) *Material Inhomogeneities in Elasticity*, Chapman and Hall, London.
11. Muschik, W. (1993) Fundamentals of non-equilibrium thermodynamics, in: *Non-Equilibrium Thermodynamics with Application to Solids*, edited by Muschik W., Springer, Wien 1-63.
12. LeVeque, R.J. (1997) Wave propagation algorithms for multidimensional hyperbolic systems, *J. Comp. Physics* **131** 327-353.
13. Berezovski, A. and Maugin, G.A. (2001) Simulation of thermoelastic wave propagation by means of a composite wave-propagation algorithm, *J. Comp. Physics* **168** 249-264.
14. Berezovski, A., Engelbrecht, J. and Maugin, G.A. (2000) Thermoelastic wave propagation in inhomogeneous media, *Arch. Appl. Mech.* **70** 694-706.
15. Bale, D.S., LeVeque, R.J., Mitran S. and Rossmanith, J.A. (2002) A wave propagation method for conservation laws and balance laws with spatially varying flux functions, (to be published in SIAM J. Sci. Computing)
16. Emel'yanov, Y. et al, (2000) Detection of shock-wave-induced internal stresses in Cu-Al-Ni shape memory alloy by means of acoustic technique, *Scripta mater.* **43** 1051-1057.

WAVE SCATTERING AND ATTENUATION IN POLYMER-BASED COMPOSITES: ANALYSIS AND MEASUREMENTS

S. BIWA, Y. WATANABE, S. IDEKOBA and N. OHNO
Department of Micro System Engineering, Nagoya University,
Nagoya 464-8603 Japan
E-mail: biwa@mech.nagoya-u.ac.jp

1. INTRODUCTION

Polymer-based composite materials are widely used in application, including various particulate and fibrous composites. In this regard, evaluation of their dynamical properties as well as monitoring of their manufacturing processes become highly important. Ultrasonic waves offer useful means to this purpose [1]. Especially, the velocity measurements have long been practiced to assess elastic properties of composites. In addition, there has been growing interest in complementing the evaluation with the attenuation measurements of ultrasound. There is certain advantage in utilizing the attenuation for material characterization in that the changes in material properties appear more sensitively in the attenuation than in the velocity. Attenuation properties have been explored for polymer-based composites to monitor their processing, and to evaluate their deterioration due to mechanical/thermal loading as well as moisture absorption [2-4]. In order to enhance nondestructive characterization and smart monitoring of polymer-based composites by ultrasonic attenuation, however, it is important to understand the propagation of ultrasound in these composites and to interpret the measured attenuation qualitatively as well as quantitatively.

Many investigators have studied wave propagation in various polymer-matrix composites experimentally and theoretically, c.f. [5-8]. For wave velocity and velocity dispersion, the theory and experiments are well correlated in these studies. As far as the attenuation behavior of polymer-matrix composites is concerned, however, few of these foregoing investigators carried out comparison between the theoretical predictions and the experimental data. Wave attenuation characteristics of polymer-matrix composites are greatly influenced by the viscoelastic properties of the constituent phases in addition to the scattering loss. Theoretically, Beltzer and Brauner [9] put forward a formulation of wave scattering in two-phase composites with explicit incorporation of the lossy nature of the matrix. In contrast to other elastodynamic multiple scattering theories that rely on the configurational averaging of the wave field over statistically distributed inclusions, c.f. [7], their theory rests on the evaluation of the single-inclusion scattering characteristics combined with the differential (incremental) scheme of micromechanics.

Recently, the present authors re-examined the above theory with some modification, and extended it to analyze more general case of longitudinal and transverse waves propagating in unidirectional polymer-matrix composites [10]. The theory turned out to give predictions in fair conformity with the actual measurements of attenuation. The aim of this paper is to discuss the further developments of the theoretical modeling in comparison to the measurements. After summarizing the theoretical background, the model is first applied to analyze wave attenuation in particle-reinforced polymer composites, where dilute particle concentration is assumed to keep simplicity of the analysis. Second, the analysis of wave attenuation in unidirectional fiber composites is elaborated by using a differential scheme in order to incorporate high concentration of fibers. The computed results for attenuation spectra are compared to the experimental results for the three representative wave modes, which are longitudinal wave as well as transverse waves with polarization parallel and perpendicular to the fiber direction.

2. THEORETICAL BACKGROUND

2.1 Independent Scattering Model

As schematically shown in Fig.1(a), a composite is assumed to consist of isotropic elastic inclusions (spherical particles or circular-cylindrical fibers) of radius a in an isotropic viscoelastic matrix subjected to time-harmonic wave motion. The number of inclusions in a unit volume of the composite is denoted by n_s, and the volume fraction by ϕ. The spatial decay of the time-averaged energy flux density $\langle e \rangle$ of the plane wave propagating into the x_1-direction is written in the form [9,10]

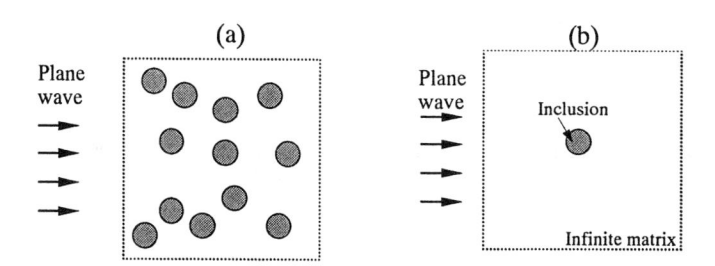

Figure 1. (a) Matrix/inclusion composite and (b) a single inclusion in infinite matrix subjected to plane-wave incidence.

$$\frac{d\langle e \rangle}{dx_1} = -\langle I^{sca} \rangle - \langle I^{mat} \rangle, \quad (1)$$

where $\langle I^{sca} \rangle$ and $\langle I^{mat} \rangle$ are the scattering loss rate and the absorption loss rate in the matrix, respectively, which are time-averaged for unit volume of the composite.

In order to evaluate the scattering loss, a simplified problem is considered for plane wave scattering by a single inclusion in an infinitely extended matrix as shown in Fig.1(b). When the distribution of inclusions is dilute, mutual interaction among the inclusions can be neglected by invoking the notion of the single and independent scattering [11], and the scattering loss by n_s inclusions in a unit volume of the composite can be estimated additively based on the single-inclusion scattering,

$$\langle I^{sca} \rangle = n_s \gamma^{sca} \langle e \rangle_0, \quad (2)$$

where γ^{sca} is the scattering cross-section of the inclusion. It is noted that since $\langle e \rangle$ decays spatially, the reference value of the plane-wave energy flux density $\langle e \rangle$ is chosen as its value evaluated at the center of the particle, and denoted by $\langle e \rangle_0$.

The absorption loss in the matrix is approximately formulated as the energy dissipation rate associated to the incident wave motion in the net region occupied by the matrix phase. It has been found [9,10] that this can be reduced to the following form to a first-order approximation,

$$\langle I^{mat} \rangle = 2\alpha_1 \langle e \rangle_0 - 2\phi\alpha_1 \langle e \rangle_0, \quad (3)$$

where α_1 is the attenuation coefficient for the incident wave in the matrix.

Substitution of eqs.(2) and (3) into eq.(1) and identification of $\langle e \rangle$ with $\langle e \rangle_0$ yield a first-order ordinary differential equation for $\langle e \rangle(x_1)$. From its solution, the attenuation coefficient of the composite α is obtained as

$$\alpha = (1-\phi)\alpha_1 + \frac{1}{2}n_s\gamma^{sca}. \quad (4)$$

2.2 Differential Scheme

For non-dilute composites, Beltzer and Brauner [9] and later Biwa et al. [12] applied the differential (incremental) scheme, which employs the independent scattering formalism at each incremental step with updated properties of the surrounding medium around a single inclusion. Namely, the composite material with the inclusion volume fraction ϕ is replaced by a homogeneous viscoelastic material with equivalent macroscopic properties. Then, inclusions amounting to infinitesimal volume fraction increment $d\tilde{\phi}$ are added to this equivalent material, and the changes of its macroscopic properties are described in incremental forms. First by a law of mixture, the composite density ρ changes by

$$d\rho = (\rho_2 - \rho)d\tilde{\phi}, \tag{5}$$

where ρ_2 is the density of the inclusion. The above-defined $d\tilde{\phi}$ is the relative increment of the inclusion volume fraction and related to the absolute increment $d\phi$,

$$d\phi = (1-\phi)d\tilde{\phi}. \tag{6}$$

The changes in the macroscopic acoustic properties also need to be formulated. In the case of isotropic composites, these are the phase velocities, c_L and c_T, and the attenuation coefficients, α_L and α_T, of longitudinal and transverse waves, respectively. The changes in these quantities are formulated differently for different reinforcement configurations, so they will not be given explicitly here. For velocity changes, appropriate micromechanical stiffness models can be used together with eq.(5), since the frequency dependence of the velocities in the practical frequency range is often insignificant and may be neglected as a first approximation [12]. For the changes in attenuation coefficients, eq.(4) can be used in differential forms, i.e.

$$d\alpha_L = \left(-\alpha_L + \frac{\gamma_L^{sca}}{2\pi a^2}\right)d\tilde{\phi}, \quad d\alpha_T = \left(-\alpha_T + \frac{\gamma_T^{sca}}{2\pi a^2}\right)d\tilde{\phi}, \tag{7}$$

where γ_L^{sca} and γ_T^{sca} are the scattering cross-sections of the inclusion embedded in the equivalent matrix subjected to the longitudinal and transverse wave incidence, respectively. These differential relations are solved with the conditions that $\phi = 0$ and the macroscopic parameters are equal to those of the matrix when $\tilde{\phi} = 0$. At each increment of $d\tilde{\phi}$, the surrounding material has updated equivalent properties of the composite, so the effect of the neighboring inclusions on the scattering response is taken into account in an indirect and approximate sense.

3. ANALYSIS

3.1. Properties of Matrix and Inclusion

The theoretical model described above is applied to two representative composite material systems, i.e., glass-particle-reinforced epoxy composite and unidirectional carbon-fiber-reinforced epoxy composite. In these composite systems, the matrix materials are described as viscoelastic, and the inclusions (particles or fibers) are assumed elastic. In the following, the material parameters associated to the matrix are expressed with the subscript 1, and those for the inclusion with the subscript 2.

For isotropic viscoelastic materials, the complex Lamé parameters λ and μ can be obtained from the wave velocity and attenuation as function of the frequency f.

Table 1. Material parameters used for numerical analysis.

	Complex or elastic moduli			Density
Epoxy matrix	$\lambda_1+2\mu_1$ (GPa) 7.61-0.283i	μ_1 (GPa) 1.58-0.128i		ρ_1 (kg/m³) 1180
Glass particle	$\lambda_2+2\mu_2$ (GPa) 77.8	μ_2 (GPa) 26.0		ρ_2 (kg/m³) 2470
Carbon fiber	κ_2 (GPa) 14.99	μ_{T2} (GPa) 5.02	μ_{A2} (GPa) 24.0	ρ_2 (kg/m³) 1670

The ultrasonic phase velocities c_{L1}, c_{T1} and attenuation coefficients α_{L1}, α_{T1} of epoxy have been taken from the measurements reported by Kinra et al. [13] and Biwa et al. [10]. The attenuation coefficients of epoxy show almost linear frequency dependence as commonly observed for solid polymers. The corresponding phase velocities are revealed to exhibit only weak dispersion in the present frequency range. It has proved that for a limited frequency range it suffices for the numerical modeling to approximate the real and imaginary parts of the complex Lamé parameters as frequency-independent constants. Indeed, when c_{L1} and c_{T1} are frequency-independent (dispersion-free) and α_{L1} and α_{T1} are proportional to the frequency f, the complex moduli become frequency-independent constants. Thus λ_1 and μ_1 for epoxy have been fitted by complex constants shown in Table 1.

The elastic properties of glass particle have been taken from Kinra et al. [13]. The transversely isotropic stiffness of carbon fiber is from the existing literature [14]. Among the five independent moduli of the transversely isotropic fiber, three are needed in the present analysis: they are the transverse plane-strain bulk modulus κ_2, the shear modulus perpendicular to the fiber axis μ_{T2}, and the shear modulus parallel to the fiber axis μ_{A2}. These two shear moduli are distinguished according to the particular polarization mode in issue. Furthermore, the elastic Lamé parameters λ_2 are replaced by $\kappa_2 + \mu_{T2}$, which enables the use of the framework for isotropic solids to transversely isotropic ones.

3.2. Scattering Response of Single Inclusion

The wave scattering by a spherical particle or a circular-cylindrical fiber can be analyzed based on the eigenfunction expansion technique in polar coordinates [15]. The scattering cross-section of the inclusion is computed as

$$\gamma^{sca} = \frac{1}{\langle e \rangle_0} \frac{1}{T} \int_0^T \int_\Gamma (-\sigma_{ij}^{sca} \dot{u}_j^{sca}) n_i dSdt, \qquad (8)$$

In the above expressions, the product of the stress and the velocity of the scattered wave in the form of $-\sigma^{sca} \cdot u^{sca}$ is the acoustic Poynting vector representing the scattered energy flux [16]. The unit vector n_i is the outward normal to the matrix-inclusion boundary Γ, and $T=1/f$ is the period.

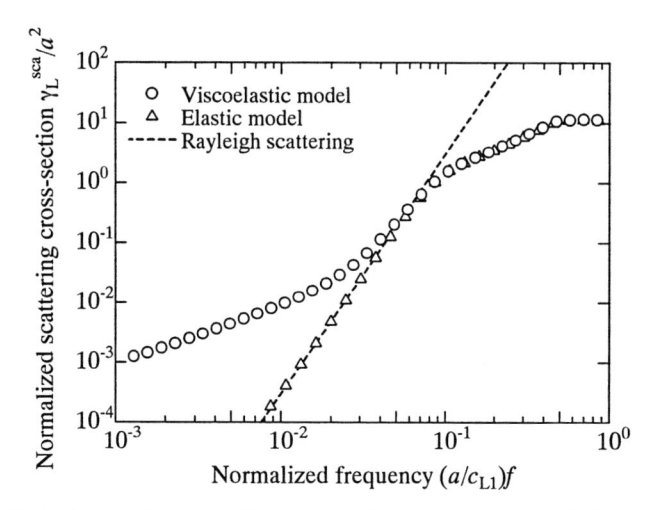

Figure 2. Variation of the normalized scattering cross-section of glass particle in epoxy matrix with the normalized frequency.

For demonstration, Fig.2 shows the computed normalized scattering cross-section γ_L^{sca}/a^2 of the glass particle in epoxy for longitudinal wave incidence (circles) as a function of the normalized frequency $(a/c_{L1})f$ in logarithmic plot. For comparison, the same cross-section when the matrix is instead modeled as an elastic solid characterized by the real parts of complex moduli in Table 1 is also shown in Fig.2 (triangles), and compared to the result when the matrix is properly modeled as viscoelastic (circles). In classical Rayleigh scattering theory, it is well known that the scattering cross-section of a particle scales to the fourth-power of the frequency in the low-frequency (long-wavelength) limit, which explicitly reads

$$\frac{\gamma_L^{sca}}{a^2} = \frac{4\pi}{9} g_C \left(\frac{2\pi f a}{c_{L1}}\right)^4, \qquad (9)$$

where the coefficient g_C is given in terms of material properties of matrix and particle. The expression of eq.(9) is also plotted in Fig.2 as a dashed line, which conforms to the result of elastic modeling (triangles) in the low-frequency region of $(a/c_{L1})f > 0.05$, say. When the matrix is treated as viscoelastic, however, the cross-section appears to depend linearly on $(a/c_{L1})f$ in this frequency range.

4. COMPARSION OF ANALYSIS AND MEASUREMENTS

4.1 Glass-Particle-Reinforced Epoxy Composite

Figure 3 reproduces the longitudinal attenuation coefficient of glass-particle

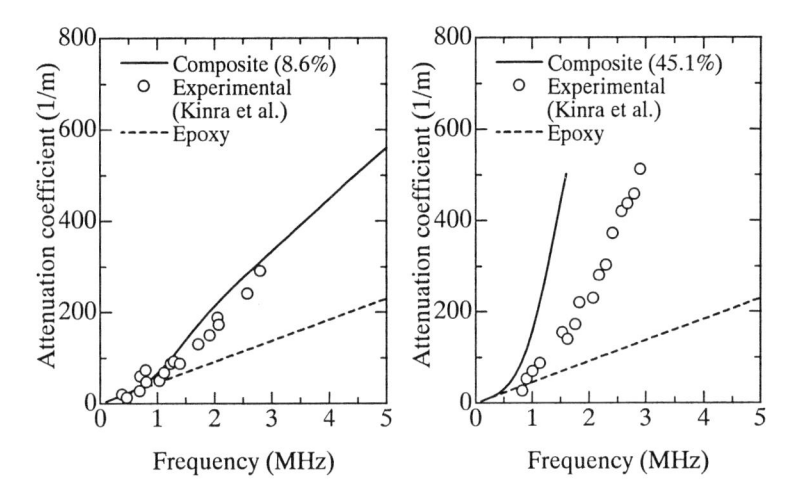

Figure 3. Attenuation spectra of glass/epoxy composites, for the particle volume fraction 8.6% and 45.1% (circles: measurements by Kinra et al.[13]).

reinforced epoxy composites as measured by Kinra et al. [13] using ultrasonic burst waves. The particle radius is 150 microns, and the experimental results are for the case when the particle volume fraction is 8.6% and 45.1%. The theoretical predictions in the corresponding cases are also shown in Fig.3. For this composite, the frequency up to 5 MHz roughly corresponds to the normalized frequency $(a/c_{L1})f$ of 0.3. Therefore, the scattering loss is significant and as a consequence, the composite shows greater attenuation than the epoxy matrix.

For the case of $\phi = 0.086$, apparently the theoretical result simulates the measured attenuation spectrum very well. One possible reason for good agreement between the theoretical prediction and the measurement is the relatively low volume fraction of the particles. For the case of $\phi = 0.451$, the numerical results considerably exceed the measurements. This discrepancy is considered to be due to interaction between particles neglected in the analysis.

4.2 Unidirectional Carbon-Fiber-Reinforced Epoxy Composite

The ultrasonic attenuation characteristics were measured for unidirectional carbon-fiber-reinforced epoxy composites (CFRP) as well as epoxy resin by the aid of normal-incidence piezoelectric transducers based on the buffer-rod method [17]. The attenuation spectra were obtained via the spectral analysis of the ultrasonic echoes. For the three modes of (a) the longitudinal wave, (b) the transversely polarized transverse wave, and (c) the axially polarized transverse wave, the attenuation coefficients of a unidirectional CFRP sample of nominal fiber volume fraction 60% are depicted in Fig.4 as function of the frequency.

The differential scheme is used to analyze scattering and attenuation in this

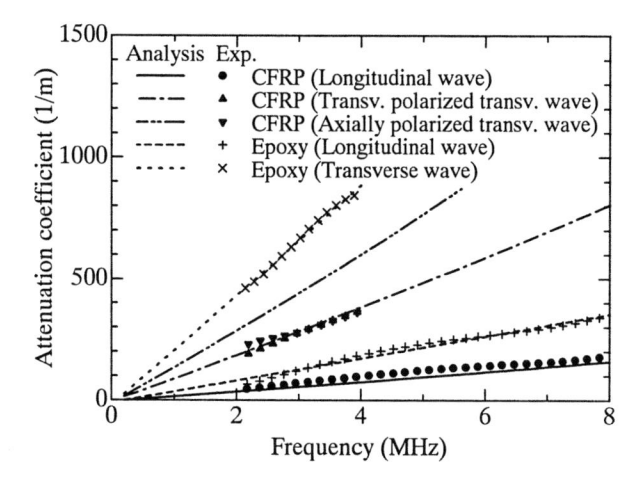

Figure 4. Attenuation spectra of unidirectional carbon/epoxy composite for fiber volume fraction of 60%.

composite since the fiber volume fraction is not small. The computed attenuation coefficients are shown in Fig.4 for the three wave modes. It can be found that the measured and computed attenuation spectra agree quite favorably, not only qualitatively but also in a quantitative sense, for the longitudinal and the transversely polarized transverse waves. It is to be noted, however, that some quantitative discrepancy lies between the measurement and the prediction for the case of axially polarized transverse wave. It is seen for this mode that the measured attenuation spectrum is lower than that computed by the analysis. In fact, the measured transverse attenuation coefficients for the two polarization directions are more or less indistinguishable in Fig.4. Computationally, the attenuation of the axially polarized transverse wave is predicted to be higher than that of the transversely polarized case due to larger mismatch between the acoustic properties of the fiber and the matrix and the resulting significant wave scattering. In strongly scattering cases such as the axially polarized transverse wave, the validity of the present differential scheme should be checked from a wider viewpoint. Presently, the implementation of the attenuation model into a certain self-consist scattering scheme is under development.

The attenuation coefficients of unidirectional CFRP with different fiber volume fractions are compared for the three modes, from both theoretical and experimental points of view. Figure 5 shows the computational and experimental results for some representative frequencies. The measurements were made for ten unidirectional CFRP samples ranging between 50 and 60 % of fiber volume fraction. In spite of some scatter of the measured values that are considered to be due not only to instrumentation errors but also to inherent sample variations, the measurements show a trend that the attenuation of CFRP decreases as the fiber fraction increases.

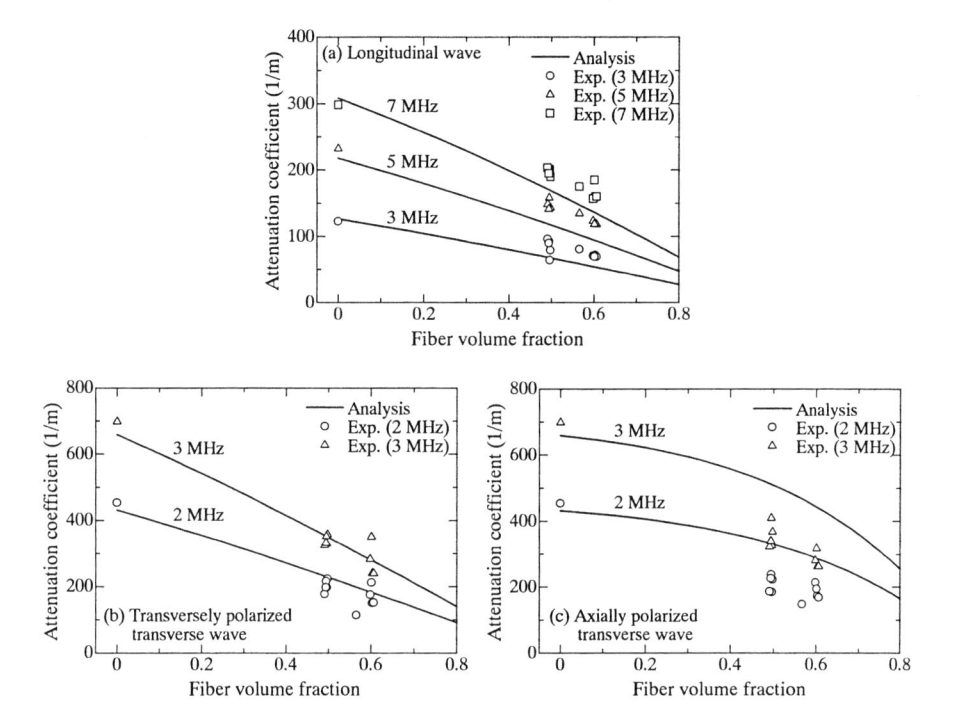

Figure 5. Attenuation coefficients of unidirectional carbon/epoxy composites for different fiber fractions and frequencies.

This trend is certainly reproduced by the computation for a wider range of the fiber volume fraction.

5. CONCLUSION

Wave scattering and attenuation characteristics of polymer-based composite materials have been examined from theoretical and experimental points of view. The analysis primarily rests on the independent scattering framework, and the energy loss of the plane wave in the composite is estimated by the scattering loss and the viscoelastic absorption. For non-dilute volume concentration of inclusions, the differential scheme is employed so as to account for the influence of the neighboring inclusions on the scattering response of a generic inclusion in an indirect and approximate manner. As a result, first the effect of the matrix viscoelasticity on the scattering cross-section of a spherical particle has been illustrated, which exhibits more or less linear dependence on the frequency in contrast to classical Rayleigh scattering that predicts the fourth-order frequency dependence. Numerical predictions of the attenuation have been demonstrated for the glass-particle-reinforced epoxy composite and the unidirectional carbon-fiber-reinforced epoxy composite. The theoretical predictions and the measurements have shown fair agreement from a qualitative point of view, and to substantial extent from a quantitative point of view, too. The present theoretical model is expected to offer a convenient basis to interpret the attenuation measurements for polymer-based composites and to seek for optimum means to evaluate or monitor these composites and composite structures.

6. REFERENCES

1. Kline, R.A. (1992) *Nondestructive Characterization of Composite Media*, Technomic, Lancaster.
2. Hosten, B., Deschamps, M. and Tittmann B.R. (1987) Inhomogeneous wave generation and propagation in lossy anisotropic solids: application to the characterization of viscoelastic composite materials, *Journal of the Acoustical Society of America* **82** 1763-1770.
3. Jeong, H. (1997) Effects of voids on the mechanical strength and ultrasonic attenuation of laminated composites, *Journal of Composite Materials* **31** 276-292.
4. Jayet, Y., Gaertner, R., Guy, P., Vassoille, R. and Zellouf, D. (2000) Application of ultrasonic spectroscopy for hydrolytic damage detection in GRFC: correlations with mechanical tests and microscopic observations, *Journal of Composite Materials* **34** 1356-1368.
5. Sayers, C.M. and Smith, R.L. (1983) Ultrasonic velocity and attenuation in an epoxy matrix containing lead inclusions, *Journal of Physics, D: Applied Physics* **16** 1189-1194.
6. Sabina, F.J. and Willis, J.R. (1988) A simple self-consistent analysis of wave propagation in particulate composites, *Wave Motion* **10** 127-142.
7. Varadan, V.K., Ma, Y. and Varadan, V.V. (1989) Scattering and attenuation of elastic waves in random media, in *Scattering and Attenuation of Seismic Waves*, Part II, edited by Wu, R.-S. and Aki, K., Birkhauser, Basel, pp.577-603.
8. Yang, R.-B. and Mal, A.K. (1994) Multiple scattering of elastic waves in a fiber-reinforced composite, *Journal of the Mechanics and Physics of Solids* **42** 1945-1968.
9. Beltzer, A.I. and Brauner, N. (1987) The dynamic response of random composites by a causal differential method, *Mechanics of Materials* **6** 337-345.
10. Biwa, S., Watanabe Y. and Ohno, N. (2001) Modelling of ultrasonic attenuation in unidirectional fiber reinforced plastics, in *Nondestructive Characterization of Materials X*, edited by Green, R.E., Jr., Kishi, T., Saito, T., Takeda, N. and Djordjevic, B.B., Elsevier, Oxford, pp.223-230.
11. Truell, R., Elbaum, C. and Chick, B.B. (1969) *Ultrasonic Methods in Solid State Physics*, Academic Press, New York.
12. Biwa, S., Watanabe Y. and Ohno, N. (2001) Attenuation of ultrasonic waves in unidirectional composites: analysis for shear waves by a differential scheme, *Materials Science Research International, Special Technical Publication* **2** 146-150.
13. Kinra, V.K., Petraitis, M.S. and Datta, S.K. (1980) Ultrasonic wave propagation in a random particulate composite, *International Journal of Solids and Structures* **16** 301-312.
14. Datta, S.K., Ledbetter, H.M. and Kriz, R.D. (1984) Calculated elastic constants of composites containing anisotropic fibers, *International Journal of Solids and Structures* **20** 429-438.
15. Pao, Y.-H. and Mow, C.-C. (1971) *Diffraction of Elastic Waves and Dynamic Stress Concentrations*, Crane Russak, New York.
16. Auld, B.A. (1973) *Acoustic Fields and Waves in Solids*, Vol.I, John-Wiley & Sons, New York.
17. Watanabe, Y., Biwa, S. and Ohno, N. (2002) Experimental investigation of ultrasonic attenuation behavior in carbon fiber reinforced epoxy composites, *Journal of the Society of Materials Science, Japan* **51** 451-457 (in Japanese).

DYNAMICS OF STRUCTURAL SYSTEMS WITH DEVICES DRIVEN BY FUZZY CONTROLLERS

Understanding fuzzy rules.

FABIO CASCIATI
*Department of Structural Mechanics, University of Pavia,
Via Ferrata 1, 27100, Pavia, Italy
E-mail: Fabio@dipmec.unipv.it*

ROBERTO ROSSI
*Department of Electronics, University of Pavia,
Via Ferrata 1, 27100, Pavia, Italy
E-mail: Roberto@ele.unipv.it*

1. INTRODUCTION

A stand-alone fuzzy-logic board has been designed by the authors and their colleagues at the Teaching Material and Structure Testing Laboratory of the University of Pavia [1][2]. It was implemented and realized by an innovative fuzzy micro-controller [3][4]. This system is suited to the implementation of adaptive fuzzy-logic strategy [5].

Indeed the design of fuzzy controllers [6][7][8] requires to select the parameters which define the membership functions in such a way that a desired level of performance is maintained. However, uncertain environmental conditions can make the controller inadequate. An adaptive controller is able to tune automatically the basic fuzzy controller. The adaptation law should pursue those values of the parameters for which stability and tracking convergence are guaranteed. The adopted fuzzy chip [9] has some programming features that make it very well suited to the implementation of an adaptive fuzzy-logic control system. In fact, more than one control algorithm can be designed by standard neuro-fuzzy algorithms [7] and stored in the chip memory.

Each algorithm can be optimized for a particular condition of the structure state. Therefore, as the parameters of the structure change, the best-suited control algorithm is automatically selected.

This task is accomplished by an acquisition system that monitors the parameters of the structure and chooses the optimal algorithm based on the actual values of some system variables [5][6].

The main goal of this paper is to show the relation between the heuristic design of the fuzzy controller and classical dynamics. Attention is focused on the special non linear character which fuzzy controllers possess.

2. THE REFERENCE CASE STUDY

2.1 The structural system

The specimen system is a three-story frame which is conveniently modeled by three degrees of freedom in series. Adding braces at each story, the actual number of degrees of freedom of the structure can be changed into two or one as in the case of Figure 1.

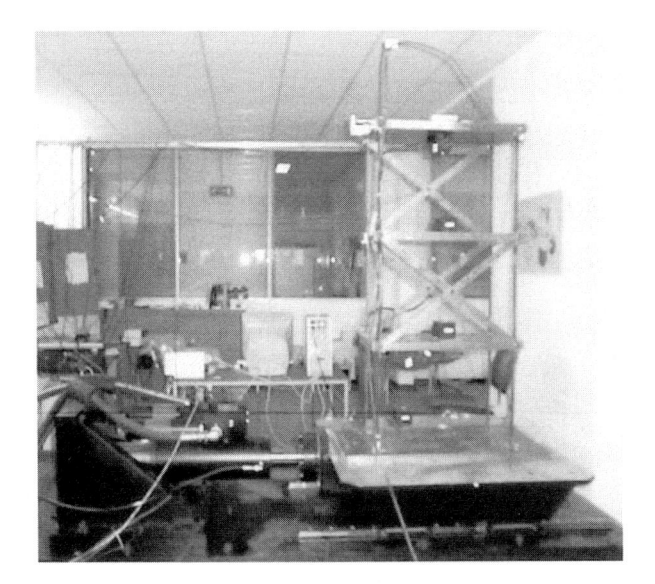

Figure 1 - Shaking table and steel structural prototype.

An active mass located on the top story is driven to control the structure. Accelerometers are positioned at every floor, which enables one to reconstruct velocities and displacements to monitor the structure response and to feed the controller with adequately chosen input variables

In the state space, one introduces the state variables x, the measured variables y, the force vector u and the structural matrices A, B, C and D and the mathematical model is written:

$$\dot{x} = Ax + Bu$$
$$y = Cx + Du$$
$$u_c = -k(y)y$$

where u_c is the control force, i.e. the component of vector u with the controller output. The vector $k(y)$ associates its value to the detected feedback.

2.2 The fuzzy controller

The fuzzy controller adopted in this paper was designed for the un-braced three story system in reference [3]. Figure 2 and Table 1 show the fuzzy sets and the fuzzy rules used.

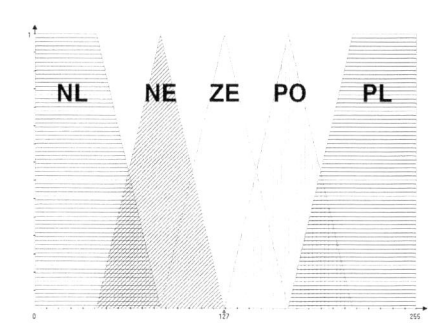

Figure 2 - Fuzzy sets for the input variables v_2 and v_3 over a (0, 255) axis.

Table 1 - Fuzzy rules for the controller used throughout this paper.

		v_2				
		NL	**NE**	**ZE**	**PO**	**PL**
v_3	**NL**	ZE	NE	XNL	NE	XNL
	NE	NE	ZE	NE	NL	NE
	ZE	XPL	PO	ZE	NE	XNL
	PO	PO	PL	PO	ZE	PO
	PL	XPL	PO	XPL	PO	ZE

It results from an original LQG (Linear Quadratic Gaussian) regulator designed and implemented in the laboratory environment in [9]. The LQG scheme was using four feedback parameters: the story accelerations a_1, a_2 and a_3 (with index ordered from the bottom to the top) and the acceleration a_d of the mass of the active mass damper mounted on the top of the frame:

$$u_c = -\{k\}^T \begin{Bmatrix} a_1 \\ a_2 \\ a_3 \\ a_d \end{Bmatrix}$$

with k independent on the feedback, leading to a linear dependency but also to the inability to account for the actuator saturation.

Moving from LQG to fuzzy control, the authors and their co-workers met the problem of conditioning the signals of the accelerometers. They clearly represent the expected structural response only when adequate antialiasing and filtering are introduced. By contrast a direct use of the accelerometer signals, quite disturbed and noisy, resulted in abnormal spikes in the command driving the active mass damper engine. This inconvenience is presently on the way to be solved by directly programming the appropriate filters in the faster new generation of microprocessor by which fuzzy controllers are implemented.

In the early stage of the study, however, it was more convenient to move toward an integrated (by circuits added to each accelerometers) signal. As a result the control rules were conceived on the story velocities v_2 (intermediate story) and v_3 (top story) while the signals coming from the other three accelerometers did not result significant enough:

$$u_c = -g\{k(w_2 v_2, w_3 v_3)\}^T \begin{Bmatrix} v_2 \\ v_3 \end{Bmatrix}$$

According to the previous equation, the rules in Table 1 must be completed with three further degrees of freedom: the weights w_2 and w_3 to be given to the two velocities and the gain g by which the controller output must be multiplied. One first assigns the weight to the top story velocity which has a dominant character since it is active in the first two modes of the structural system. Then, the weight of the second velocity results in the present example nearly one third of the previous one, while its influence has a different sign: otherwise the highest mode is excited by the actuator and this soon results in a failure of the controlled system.

3. DISCUSSING THE FUZZY RULES

It is worth noting that there are two ways toward a wrong controller:
1. the first consists in selecting the wrong sign for the predominant feedback parameters: the forces the actuator introduces add to the external forces resulting in an increasing amplification of the response;
2. the second consists in selecting the remaining parameters in such a way that the highest modes are not contrasted. The response shows an increasing influence of the highest mode leading to a different form of instability.

The numerical experiments carried out are conceived to support this mechanical translation of the rules arisen from the mere application of fuzzy logic.
Figure 3 collects the response to a sine excitation on the first mode obtained for different values of the gain g, namely 1, 1.5 and 2.. They are all built from the same values of the two weights w, with a fixed ratio between them. Increasing the gain to 3 the controller becomes instable. The role of the gain parameter will be discussed in the second next subsection.

3.1. The role of the weights

The axis on which the membership functions are built spans over the range (0, 255) due to the 8-bit architecture of the commercial chip adopted [9]. On the other hand, the digital value of the voltage coming from the integration block connected to each accelerometer is in the range (0, 2047) (due to the different 12-bit architecture of the computer) which corresponds to a physical range of (0 V, 5 V).
Assume that the maximum value achieved by the velocity signal v_3 be 0.1 V. The digital 11-bit value w_3v_3+1024 is what is actually sent to the fuzzy controller. The constant value 1024 sets the mean value to 2.5 V and guarantees that the resulting voltage always lies in the range (0 V, 5 V). For example, a weight of 5600 results in the value 1024+560=1584 to be sent to the fuzzy controller. In the 8-bit domain of the fuzzy controller this value corresponds to 1584/8=198, which is located at one half of the positive axis of the membership functions. But a weight of 2800 would correspond to ¼ of the positive axis. The controller will associate different reactions of the actuator to the same velocity depending on this weight w_3, and the dependency is also non linear.
A reduction of this weight (as well as a reduction of the gain g) would make stable controllers unstable, just by reducing the value of the unbeneficial control force. Similarly a controller with equal values of the weights of the two velocities could work for low values of the weights (or of the gain g).
A beneficial controller requires, as said above, w_3/w_2 fixed at –3. A more negative value of w_2, as well as a less negative one, would still be able to control the first mode but the resulting controller fails in recovering from the influence of higher modes, as arising from a broader frequency spectrum of the external excitation, from

the actuator action or from any spillover situation. The last one mainly applies to very negative values of w_2: it is well known that the problem with the spillover relies not in itself but in detecting its effects and accounting for them in the control law. On the other side, a less negative value of w_2 does not offer defense against the higher modes which could be excited by the actuator itself (see Figures 4 and 5).

Such a controller is also able to deal with an excitation on the second mode but fails (i.e. become instable) when the excitation is on the third mode. In this case a controller using as feedback just v_2 (with the same weight as before) should be adopted as replacement of the previous controller (see Figure 5). This remark gives relevance to the adaptive solutions pursued in [4] and [5].

3.2 The role of the gain g

For given values of the weights w_2 and w_3, Figure 3 emphasized how a higher damping initially means a higher global damping, i.e. a faster achievement of a satisfactory performance. However, the nonlinear character of the controller results in an inversion of trend, which for some values of *g* means less efficiency and, possibly, loss of stability due to the disruptive action of the highest mode. This happens, for instance, when *g*=3.
Figures 6 and 7 show that after a period of unsatisfactory response with gain 3, a reduction of the gain to 2 (Figure 6) is unable to re-stabilize the response, while a gain of 1.5 (Figure 7) is able to do it. It seems reasonable to regard the first case as a situation of conditioned stability, while the second case is of unconditioned stability.

Now, it is well known that one of the main drawbacks of fuzzy controllers is the inability to prove their stability in a mathematical manner [12], but, once again, adaptive control [11] can help to overcome the problem. Figure 8, in fact, shows that a controller with gain 2 is able to recover from the unstable situation created by the early use of $g = 3$, provided a preliminary use of the controller with $w_3 = 0$ is adopted until the highest mode effect is cleaned.

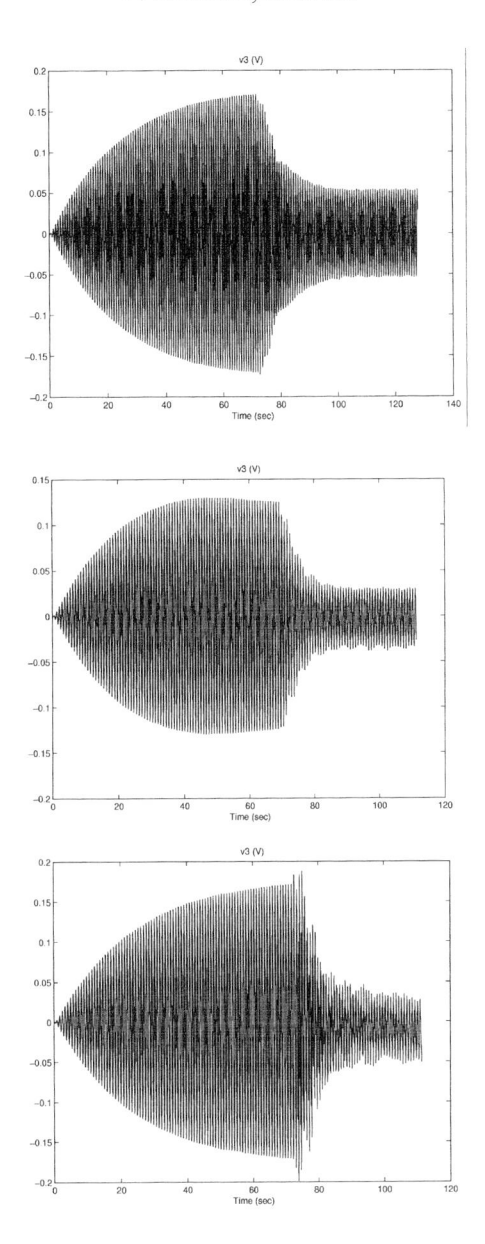

Figure 3 – Numerical response under a sine excitation with energy in the first mode. Optimal ratio w_3/w_2 for different values of the gain: a) 1 stable; b) 1.5 stable; c) 2: stable.

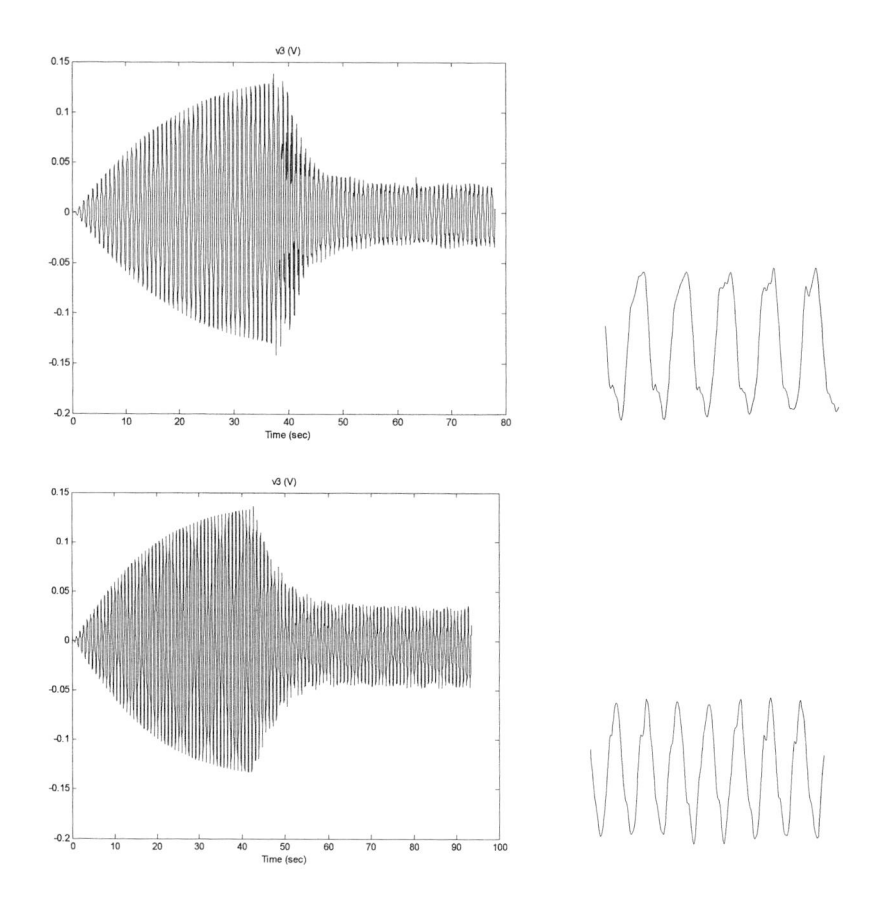

Figure4 – Sensitivity to the weight of the second velocity: w_2 =3400 (top) and 900 (bottom). Excitation on the first mode.

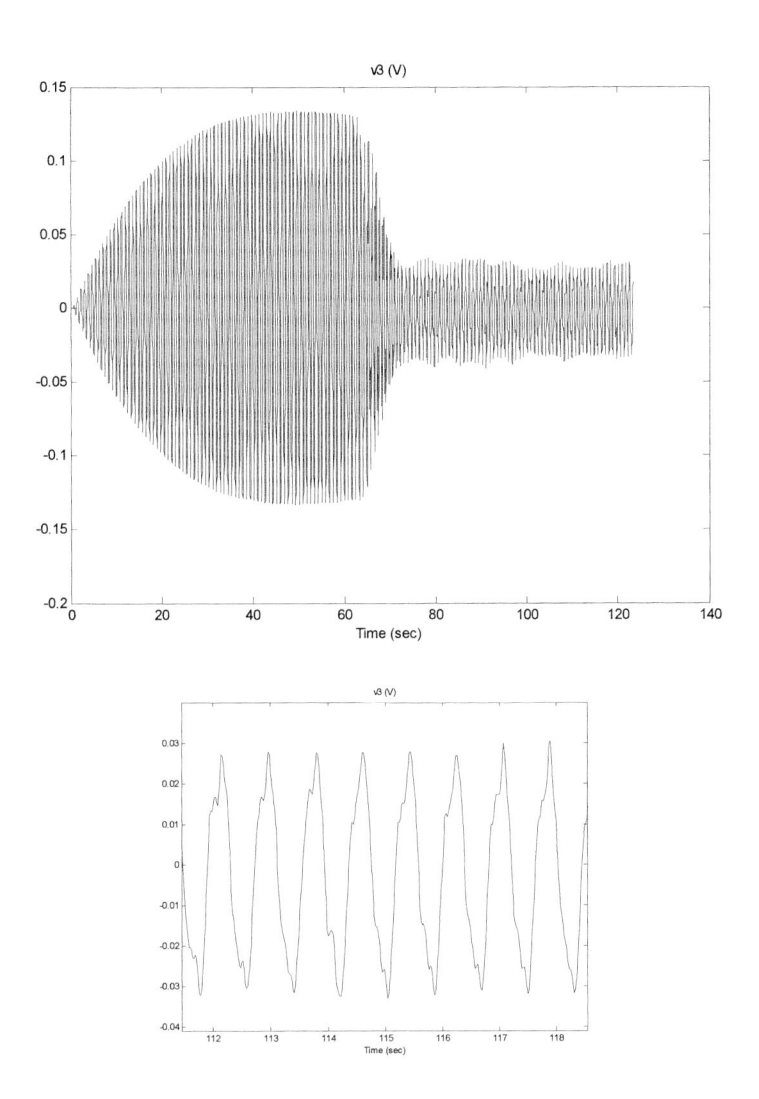

Figure 5 – Sensitivity to the weight of the second velocity: $w_2=5600$.. Excitation on the first mode.

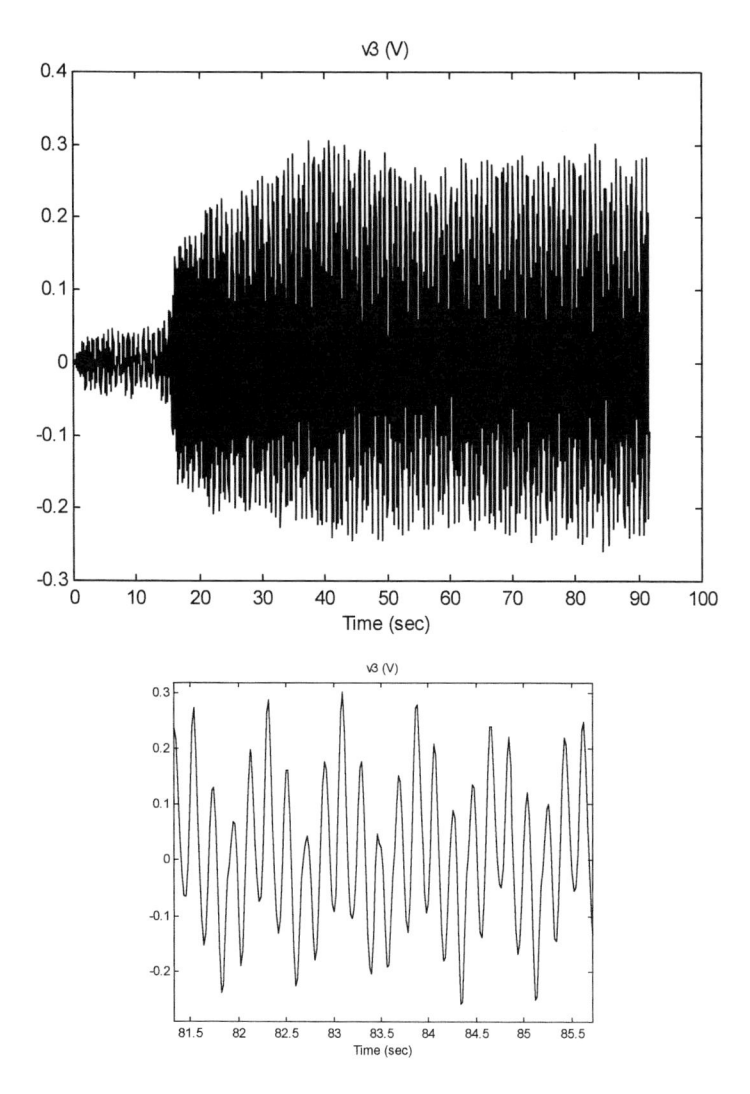

Figure 6 – Missed recover with gain 2 from the instable situation created by an early use of g=3. The second figure is a zoomed version of the first one.

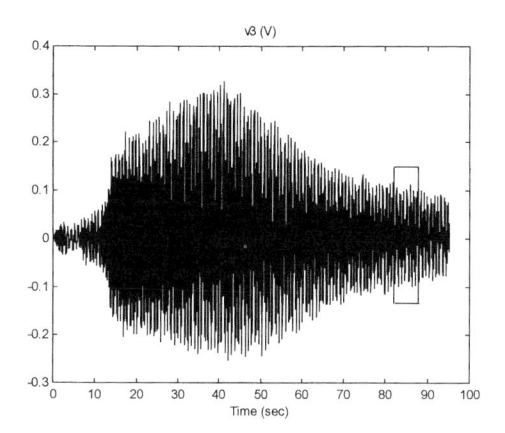

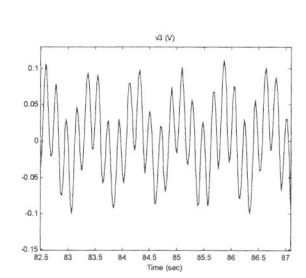

Figure 7 – Recover with gain 1.5 from the instable situation created by an early use of g=3. The second figure is a zoomed version of the first one.

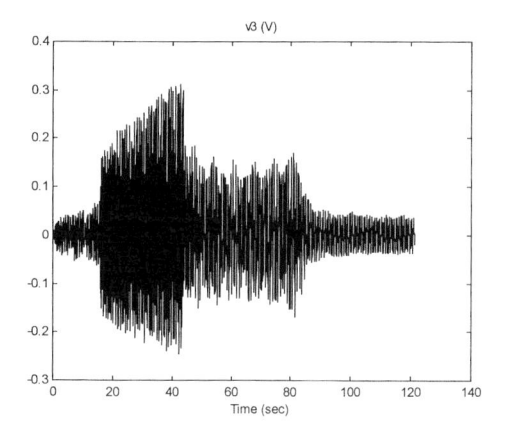

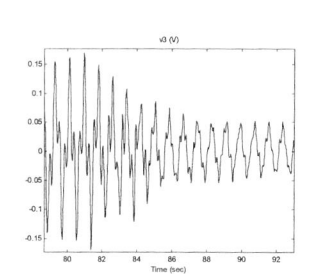

Figure 8 – Recover with gain 2 from the instable situation created by an early use of g=3. It is made possible by an intermediate use of the controller with w_3 =0 in order to clean the effects of the highest mode. The second figure is a zoomed version of the first one.

5. CONCLUSIONS

In this paper the authors describe the mechanical feature of a fuzzy controller implemented in a commercial micro-controller. The algorithm parameters are discussed in detail. Numerical results are produced to support the translation of heuristic rules into mechanical rationale.

6. ACKNOWLEDGEMENTS

The authors carried out the research activity summarized in this paper within the Italian Space Agency (ASI) program, which sees Professor Franco Bernelli-Zazzera, of the Polytechnic of Milan, as its national coordinator. The 2001 grant, with contract number I/R/187/01, is here acknowledged.

REFERENCES

1. Casciati F., Faravelli L. and Yao T. , (1996)., Control of Nonlinear Structures Using the Fuzzy Control Approach. *Non Linear Dynamics*, 11.
2. Casciati F., Faravelli L. and Torelli G. (1999)., A fuzzy Chip Controller for NonLinear Vibrations. *Nonlinear Dynamics* 20.
3. Faravelli L., Rossi R., (2000), Fuzzy Chip Controller Implementation, in F. Casciati and G. Magonette (eds.), *Proc. 3rd International Workshop on Structural Control – Structural Control for Civil and Infrastructure Engineering*, World Scientific, Singapore.
4. Faravelli L., Rossi R., (2002), Adaptive Fuzzy Control: Theory versus Implementation, *Journal of Structural Control*, 9 (1), 59-73.
5. Faravelli L., Rossi R., G. Torelli (2002), Numerical Testing of a Programmable Micro-controller with Fuzzy and Adaptive Features, submitted for publication in *SIMPRA Journal*
6. Passino K.M. and Yurkovich S. (1998)., *Fuzzy control*, Addison Wesley Longman Inc.
7. Jang J.S.R., Sun C.T. and Mizutani E. (1997), *Neuro-Fuzzy and Soft Computing*, Prentice Hall Inc.
8. Battaini M., Casciati F. and Faravelli L. (1998)., Fuzzy Control of Structural Vibration. An active Mass System Driven by a Fuzzy controller, *Earthquake Engineering and Structural Dynamics*, 27.
9. ST-ELECTRONICS, (2000), User manual.
10. Battaini M., (1994), Sistemi strutturali controllati: progettazione e affidabilità (in Italian), Ph.D. Thesis, Department of Structural Mechanics, University of Pavia.
11. Faravelli L., Rossi R., (2002), The Adaptive Character of a Fuzzy Chip Controller, *Proceedings 3rd. World Conference on Structural Control*, John Wiley & Sons, Chichester, UK
12. Casciati F., (1997), Checking the Stability of a Fuzzy Controller for Nonlinear Structures, *Microcomputers in Civil Engineering*, 12, pp. 205-215

MODEL REDUCTION FOR COMPLEX ADAPTIVE STRUCTURES

WOOSUK CHANG AND VASUNDARA V. VARADAN
Center for the Engineering of Electronic & Acoustic Materials, 212 EES
The Pennsylvania State University, University Park, PA 16802
E-mail: vvvesm@engr.psu.edu

1. INTRODUCTION

A complex adaptive structure for the purposes of this paper is treated as a structure with embedded piezoelectric sensors and actuators that are connected through a control loop so that the structure can adaptively or optimally respond to an external disturbance [1-4]. Such structures, also called smart structures, present special challenges in analytical and numerical modeling. The piezoelectric devices involve coupled electric and elastodynamic fields, the devices are multiple in number, the size of the structure is typically much larger in size relative to the sensors and actuators, and in real applications the geometry of the structure is complex. The response of such a structure to an external excitation and the response of the embedded sensors can only be simulated numerically. The finite element method has been successfully used to solve such problems, but the size of the resulting matrices becomes very large even for simple geometries [5-9]. To give an example, to obtain the transient response of a simple clamped plate with five pairs of sensors and actuators it is necessary to retain the first 50 structural modes and this results in an 800×800 matrix. In order to then interface such a numerical model with a control algorithm such as an H_∞ robust controller including modeling and device uncertainties, may challenge even the fastest computers to provide the required actuator excitations for real time controlled response of the adaptive structure. The objective of this paper is to address this issue and present a technique to condense or reduce the system model while still retaining the essential dynamical features of the smart structure [7].

The need for model reduction or condensation methods have broader relevance in a number of applications where one is interested in the real time response of large, complex systems. In the pre-computer era, it was necessary to resort to approximations and simplify the system model so that the resulting partial differential equations or integral equations were amenable to an analytical solution. The features of the system that were thus neglected were not necessarily irrelevant but rather dictated by the available analytical methods. With computational speeds increasing according to Moore's law, every detail of a complex system can be modeled leading to very large computational models and an overabundance of data about the system that now presents a challenge in terms of information overload. Thus robust model reduction techniques may be needed for a wide array of problems ranging from civil structures, air, land and sea vehicles, ecology and climate, biology and traffic patterns.

The plan for this paper is more modest and is confined to presenting a new technique for condensing the size of the plant or system model used to build a H_∞ controller incorporating uncertainties. The method is applied to a clamped plate with five pairs of collocated piezoelectric sensors and actuators. The finite element method as formulated by Allik and Hughes [10] for elastic/piezoelectric structures is used and the first 50 modes of the smart panel are computed. The second order modal representation is transformed into a first order state space form using previously defined mapping methods [9]. A prescription for finding the Modal Hankel Singular Values (MHSV) of the is presented. A user specified threshold value of the MHSV is used to select the most observable/controllable modes of the system. The results for the reduced model using MHSV, the full model and a reduced model using HSV in the state space representation are compared to show the effectiveness of the proposed model reduction scheme.

2. SYSTEM MODEL AND MODAL EQUATIONS

The finite element formalism proposed by Allik and Hughes [10] for piezoelectric materials is used to obtain the dynamic response of a panel with surface bonded piezoelectric patches [9]. For the sake of brevity, only the final equations are presented here. Following [10], the discretized finite element matrix equations of motion can be written as

$$[M_{uu}]\ddot{u} + [K_{uu}]u + [K_{u\phi}]\Phi = \{F\}$$
$$[K_{u\phi}^T]u + [K_{\phi\phi}]\Phi = \{Q\}$$
(1)

where M is the mass matrix, K_{uu} is the elastic stiffness matrix, $K_{\phi u}$ is the electro-mechanical coupling matrix, $K_{\phi\phi}$ is the dielectric matrix, **u** is the displacement vector of nodal values, Φ is set of electric potential nodal values, **F** is the applied force and Q are the nodal values of free electric charges. It is understood that linear piezoelectric constitutive equations are satisfied only in the piezoelectric regions and Hooke's law is obeyed in the purely elastic regions of the smart panel. Further the charge vector Q is nonzero only on the electrodes of the piezoelectric devices. The above matrix equations are written in partitioned form to reflect coupling between the elastic and the electric fields.

Twenty node brick elements are used to model the piezoelectric devices, 9 node shell elements are used to model the elastic plate and the interface elements between the device and the plate are modeled using 13 node transition elements [8]. This approach

considerably reduces the size of the model and also alleviates the numerical stiffening that would be caused by using brick elements to model the plate.
Equation (1) can be condensed to write the sensor potential in terms of the sensor displacement as

$$\begin{Bmatrix} u \\ \phi \end{Bmatrix} = \begin{bmatrix} I \\ -[K_{\phi\phi}]^{-1}[K_{u\phi}]^T \end{bmatrix} \{u\}. \tag{2}$$

After assembling all element matrices, and substituting the foregoing transformation matrix into Eq. (1), the system dynamic equation is written as

$$[M_{uu}]\ddot{u} + [K^*]u = \{F\} + \{F_c\} \tag{3}$$

where

$$[K^*] = [K_{uu}] - [K_{u\phi}][K_{\phi\phi}]^{-1}[K_{u\phi}]^T$$

$$\{F_c\} = -[K_{u\phi}][K_{\phi\phi}]^{-1}\{Q\}$$

and, $\{F\}$ and $\{F_c\}$ are called the mechanical force and the control feedback force, respectively.

The system equation of motion presented in Eq. (3) is not suitable for system analysis and control, because the number of degrees of freedom of the system is usually too large. Therefore, the equation has to be transformed into a set of properly chosen modal coordinates with smaller and more manageable number of degrees of freedom. The physical coordinate u is represented by

$$u = [U]q \tag{4}$$

where $[U]$ is the modal matrix, and q are referred to as modal coordinates. The modal matrix $[U]$ is simply a square matrix in which the columns corresponding to eigenvectors of system, satisfying the eigenvalue problem of Eq. (3). Some of the piezoelectric patches act as actuators and others as sensors, and it is therefore necessary to partition the electrical potential Φ to extract actuator and sensor contributions. By using transformation matrices T_a and T_s on the vectors of the voltages at the actuators and at the sensors [7], we obtain

$$[I]\ddot{q} + [2\varsigma\omega]\dot{q} + [\omega^2]q = [U]^T[K_d]d - [U]^T[K_{u\phi}][K_{\phi\phi}]^{-1}[T_a]\Phi_a$$
$$\Phi_s = -[T_s][K_{\phi\phi}]^{-1}[K_{u\phi}^T][U]q \tag{5}$$

where   Where ω_i, ζ_i are natural frequency and the modal damping ratio for the ith mode; T_a is an $m \times n$ actuator location matrix for the actuator degrees of freedom; and T_s is an $n \times m$ sensor location matrix for the sensor degrees of freedom. Here m is the number of total electrical potential degrees of freedom and n is the number of sensors and actuators. The columns of T_a matrix consist of the piezoelectric capacitance values of actuators corresponding to the actuator degrees of freedom, while the rows of T_s matrix contain unit values at the sensor degrees of freedom. The control forces are provided by n number of actuators and the sensor voltages are measured by n number of sensors.

3. STATE SPACE MODEL AND CONTROL

The governing equation can be written in state space form to provide a standard mathematical basis for control study [9]. The linear time-invariant equations of motion which include the effects of piezoelectric control forces and external disturbance force are expressed as follows

$$\dot{x} = \begin{Bmatrix} \dot{q} \\ \ddot{q} \end{Bmatrix} = \mathbf{A}\,x + \mathbf{B}_a \Phi_a + \mathbf{B}_d d, \qquad \Phi_s = \mathbf{C}\,x \tag{6}$$

where $\mathbf{A}$, $\mathbf{B}_d$, $\mathbf{B}_a$ and $\mathbf{C}$ are the corresponding state, disturbance, control and sensor matrices:

$$\mathbf{A} = \begin{bmatrix} 0 & I \\ -[\omega^2] & [2\zeta\omega] \end{bmatrix}, \qquad \mathbf{B}_a = \begin{bmatrix} 0 \\ -[U]^T [K_{u\phi}][K_{\phi\phi}]^{-1}[T_a] \end{bmatrix},$$

$$\mathbf{C} = \begin{bmatrix} 0 & -[T_s][K_{\phi\phi}]^{-1}[K_{u\phi}]^T [U] \end{bmatrix}, \qquad \mathbf{B}_d = \begin{bmatrix} 0 \\ [U]^T [K_d] \end{bmatrix} \tag{7}$$

Each resonance mode consists of two state variables, modal coordinate and its derivative. State space representation of a mode can be expressed as follow

$$\begin{Bmatrix} \dot{q}_i \\ \ddot{q}_i \end{Bmatrix} = \begin{bmatrix} 0 & 1 \\ -\omega_{ni}^2 & -2\xi_i \omega_{ni} \end{bmatrix} \begin{Bmatrix} q_i \\ \dot{q}_i \end{Bmatrix} + \begin{bmatrix} 0 & 0 \\ b_{di} & b_{ai} \end{bmatrix} \begin{Bmatrix} d \\ \Phi_a \end{Bmatrix}$$

$$\Phi_s = \begin{bmatrix} c_{di} & c_{vi} \end{bmatrix} \begin{Bmatrix} q_i \\ \dot{q}_i \end{Bmatrix} \tag{8}$$

where b_{di} and b_{ai} are ith row vectors of the disturbance matrix and input matrix respectively, and c_{di} and c_{vi} are the ith column vectors of the output or sensor matrix. The transfer functions $\mathbf{P}_a(s)$ and $\mathbf{P}_d(s)$, which map the control actuation Φ_a to sensors Φ_s and disturbance d to sensor Φ_s, can be defined in the Laplace s-domain with the parameters of the state space models. Thus

$$\Phi_s = \mathbf{P}_a(s)\Phi_a; \quad \Phi_s = \mathbf{P}_d(s)d \tag{9}$$

The transfer functions are defined as shown below and the abbreviated cross mark in the matrix as defined below is used for compactness.

$$\mathbf{P}_a(s) = \mathbf{C}(s\mathbf{I} - \mathbf{A})^{-1}\mathbf{B}_a = \left[\begin{array}{c|c} \mathbf{A} & \mathbf{B}_a \\ \hline \mathbf{C} & 0 \end{array}\right] \tag{10a}$$

$$\mathbf{P}_d(s) = \mathbf{C}(s\mathbf{I} - \mathbf{A})^{-1}\mathbf{B}_d = \left[\begin{array}{c|c} \mathbf{A} & \mathbf{B}_d \\ \hline \mathbf{C} & 0 \end{array}\right] \tag{10b}$$

It is clear that $\mathbf{P}_a(s)$ is related to the controllability of a state and $\mathbf{P}_d(s)$ is related to the observability of a state. Thus it appears desirable to identify those modes that are most observable and hence most controllable. This idea is sued below to identify the most important modes to reduce the size of the system model.

We may further define the state space system transfer matrix for the structure by combining the observability and controllability transfer matrices as follows:

$$\mathbf{P}(s) = \left[\begin{array}{c|c} \mathbf{P}_{ds}(s) & \mathbf{P}_{as}(s) \\ \hline \mathbf{P}_{ds}(s) & \mathbf{P}_{as}(s) \end{array}\right] = \left[\begin{array}{c|c|c} \mathbf{A} & \mathbf{B}_d & \mathbf{B}_a \\ \hline \mathbf{C} & 0 & 0 \\ \hline \mathbf{C} & 0 & 0 \end{array}\right] \tag{11}$$

4. MODEL REDUCTION

The balanced model reduction proposed by Moore [11] and Pernebo and Silverman [12] is developed to eliminate state variables and corresponding parameters which have low controllability and observability. This method utilizes Gramians to measure the controllability Gramian Ω_l and observability Gramian Ξ_l, which are the roots of the Lyapnov equations for controllability and observability respectively [13]

$$controllability: \mathbf{A}_i \Omega_i + \Omega_i \mathbf{A}_i^T + \mathbf{B}_i \mathbf{B}_i^T = 0$$
$$observability: \mathbf{A}_i^T \Xi_i + \Xi_i \mathbf{A}_i + \mathbf{C}_i^T \mathbf{C}_i = 0$$
(12)

These Gramians can be balanced with proper mathematical manipulations.

A measure of the balanced controllability and observability of each state can be realized by defining the Hankel Singular Value (HSV) of the Gramian. For efficient reduction of a flexible structure model, it is preferable to measure the balanced controllability and observability of each modal coordinate rather than that of each state in a state space representation. Hence, in this study, the balanced model reduction is applied to each mode and a unique HSV is derived for each resonance mode. This value is defined as a Modal Hankel Singular Value (MHSV). A threshold MHSV is adopted by the control designer to deselect resonance modes which have a lower MHSV. The modes with the highest HSV are the modes which are more observable and controllable. This process deselects insignificant modes and renders the reduced model more tractable for the design of a robust controller which can even be implemented in an analog fashion. The MHSV of the ith mode is obtained by taking the scalar product of Ω_I and Ξ_I. The MHSV of ith mode σ_i^M is defined as the largest eigenvalue of the Gramian of the balanced controllability and observability, which is simply the eigenvalue matrix of $\Omega_i \cdot \Xi_i^t$.

$$\sigma_i^M = \max\{\sigma_{i_1}, \sigma_{i_2}\} \text{ where } \Omega_i \Xi_i \mathbf{T}_i = \mathbf{T}_i \begin{bmatrix} \sigma_{i_1}^2 & \\ & \sigma_{i_2}^2 \end{bmatrix}$$
(13)

where $\mathbf{T}_i$ is a eigenvector matrix.

A threshold value of the MHSV is defined by the designer based on prior physical knowledge of the behavior of the system. Reduced-order transfer functions, defined as $\mathbf{P}_{ar}(s)$ and $\mathbf{P}_{dr}(s)$ are obtained by eliminating those modes with values of MHSV below the threshold. in Eq. (10a) and Eq. (10b). Model reduction with modal coordinates is simply the elimination of modes with low MHSV from the original state space model. Suppose that each modal state space component in Eq. (10a) is evaluated by MHSV and rearranged with high and low MHSVs as follows:

$$P^r(s) = \begin{bmatrix} 0 & 1 & & & & & & & & \\ \xi_1 \omega_1^{high} & (\omega_1^{high})^2 & & & & \cdots & & & b_{1d}^{high} & b_{1d}^{high} \\ & & 0 & 1 & & & & & & \\ & & \xi_2 \omega_2 & (\omega_2)^2 & & & & & b_{2d} & b_{2v} \\ & \vdots & & & \ddots & & & & & \\ & & & & & 0 & 1 & & & \\ & & & & & \xi_n \omega_n^{cutoff} & (\omega_n^{cutoff})^2 & & & \\ c_{1d}^{high} & c_{1v}^{high} & c_{2d} & c_{2v} & & c_{nd}^{cutoff} & c_{nv}^{cutoff} & & b_{nd}^{cutoff} & b_{nv}^{cutoff} \end{bmatrix}$$
(14)

where the subscript '*high*' indicates the modes with high MHSVs, and '*cutoff*' for the modes with the threshold MHSV chosen to deselect insignificant modes. The superscript '*r*' denotes that this is the reduced order transfer matrix. We notice that in the modal representation, the reduced transfer matrix is diagonally dominant.

The infinity norm quantifying the reduction is as follows

$$\left\|\mathbf{P}(s) - \mathbf{P}^r(s)\right\|_\infty = \left\|\mathbf{P}_{low}(s)\right\|_\infty \quad (15)$$

where $\mathbf{P}_{low}(s)$ denotes the model that consists of the low MHSV modes and is truncated from the original model. The model reduction error percentage is then defined as

$$\textit{Model reduction error} = \frac{\left\|\mathbf{P}_{low}(s)\right\|_\infty}{\left\|\mathbf{P}(s)\right\|_\infty} \times 100, \quad (16)$$

The reduced order transfer matrix $\mathbf{P}^r$ can also be written in the state space representation by finding the HSV of the product of the controllability and observability Gramian in an exactly similar fashion using the state space matrices.

$$\mathbf{P}^r(s) = \begin{bmatrix} a_{11}^{high} & a_{12}^{high} & \cdots & a_{1n}^{high} & b_1^{high} \\ a_{21} & a_{22} & & a_{2n} & b_2 \\ \vdots & & \ddots & & \\ a_{11}^{cutoff} & a_{12}^{cutoff} & & a_{1n}^{cutoff} & b_n^{cutoff} \\ c_1^{high} & c_2 & & c_n^{cutoff} & 0 \end{bmatrix} \quad (17)$$

In contrast to eq.(13), the reduced order transfer matrix in state space is a very full matrix.

5. COMPARISON OF MODEL REDUCTION RESULTS USING HSV AND MHSV

An aluminum panel with five patch piezoceramic sensors and actuators, shown in figure 1, is used to demonstrate the efficacy of the proposed model reduction using MHSV. Acoustic pressure loading on the top face of the panel is assumed as the external disturbance. All sides of the panel boundary are assumed fixed. Five sensors on the upper side sense the panel vibration and the 5 actuators on the lower side suppress the vibration. A 2nd order differential equation of motion is derived in terms of modal coordinates using finite element analysis as given in eq.(5) above. The external disturbance $\mathbf{B}_d$ is assumed to be uniformly loading the panel. So the dimensions of the state space parameters are given as

$$\mathbf{A}:(100\times100), \ \mathbf{B}_d:(100\times1), \ \mathbf{B}_a:(100\times5) \text{ and } \mathbf{C}:(5\times100). \quad (18)$$

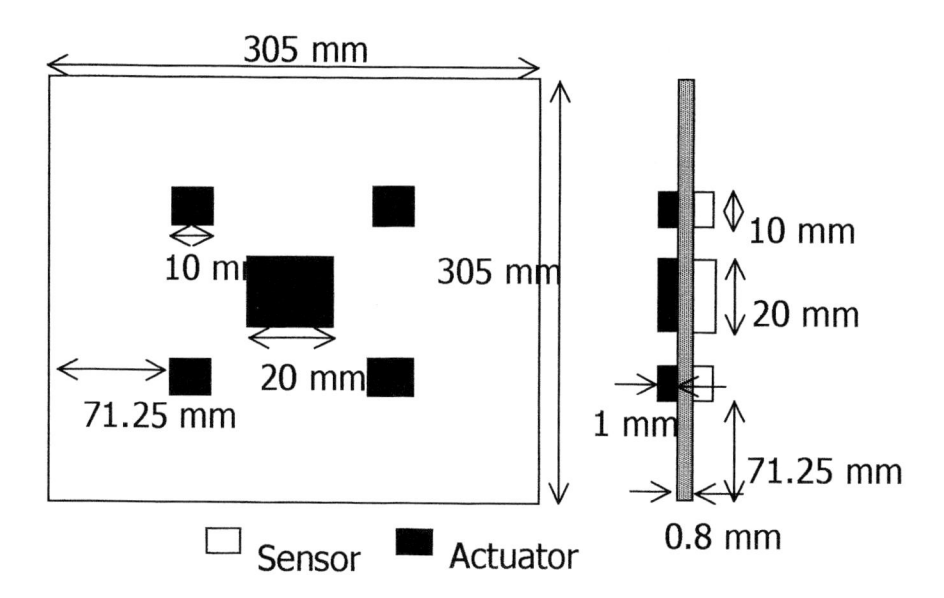

Figure 1: A clamped Aluminum plate with five pairs of surface mounted piezoelectric sensor and actuator patches.

The Gramians of the observability and controllability transfer matrices were obtained solving the Lyapunov equations in (11) from both the modal and state space transfer matrices. In the modal representation, the first 50 modes were retained resulting in 101x101 block diagonal transfer matrices. In the state space computation, the size of the matrix retained for computation of the Gramians was the same as that used for solving the first 50 eigenmodes. The MHSV and HSV are determined for each case, a threshold established and modes with HSV and MHSV below this value are deselected. Figure 2 shows the MHSVs of the 1st to 50th modes. The reduced model $\mathbf{P}^r(s)$ is defined by collecting the modes with MHSV higher than 1.3, resulting in the selection of the 1st, 5th, 8th, 20th, 31st and 38th modes, while the rest are ignored. The frequency of the 50th mode is about 1.3 kHz and that of the 38th mode (the highest selected mode) is 0.8 kHz which is significantly lower than 1.3 kHz.

The maximum singular value of the transfer matrix for observability and controllability are computed for 3 cases – (1) retaining all 50 modes, which is called the full or untruncated model; (2) the truncated model retaining the 7 modes with the highest MHSV; (3) the truncated model from the state space representation. These are compared in Figs. (2-4). We may interpret this value to represent the most observable mode and the most controllable mode or state and hence of significant importance for control of the adaptive structure. The $\mathbf{H}_\infty$ norm of the 50 modes model is 7.09 and that of the truncated mode is 2.37. The dimension of the state space model has reduced from 100 to 14 and resulted in a truncation error of 33%. We notice that there is excellent agreement between the reduced model using MHSV and the full model and the MHSV result departs modestly from the full result only after a frequency of 1 kHz. This to be expected since the modeled frequency range using MHSV is lower than this. The reduced model using HSV however departs widely from the full model and there is a very large model reduction error.

At least for the chosen example, we have demonstrated that model reduction using MHSV gives excellent results and is a very simple approach to reducing the size of a system model and still retain the dominant behavior of the system. The numerical procedure for obtaining the MHSV is stable and accurate because only block diagonal matrices are involved in the modal representation. The size of the matrix is also extremely small at each step since we are working on only one mode at a time. Since the mode shapes have a physical interpretation and the associated eigenfrequencies indicate what frequencies are included in the reduced model, we have a sense of the frequency range in which the reduced model can be used.

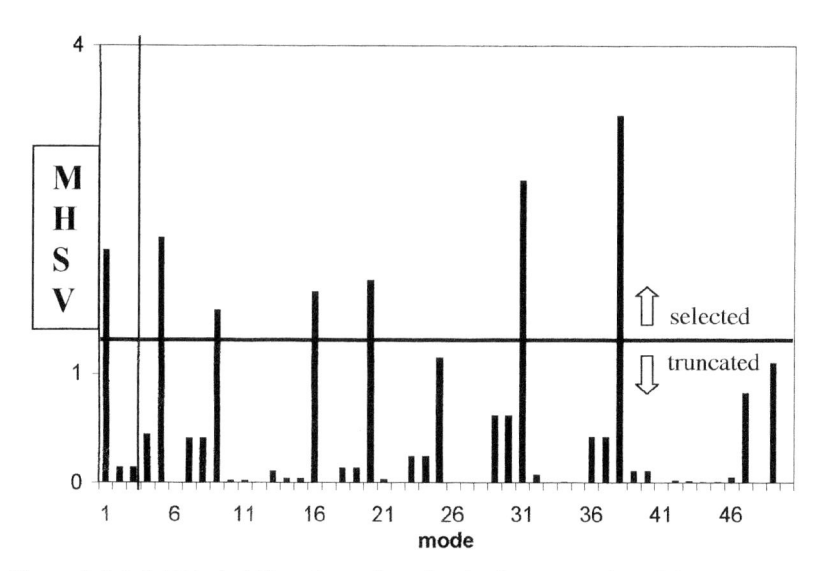

Figure 2: Modal Hankel Singular Values for the first 50 modes of the smart panel shown in figure 1

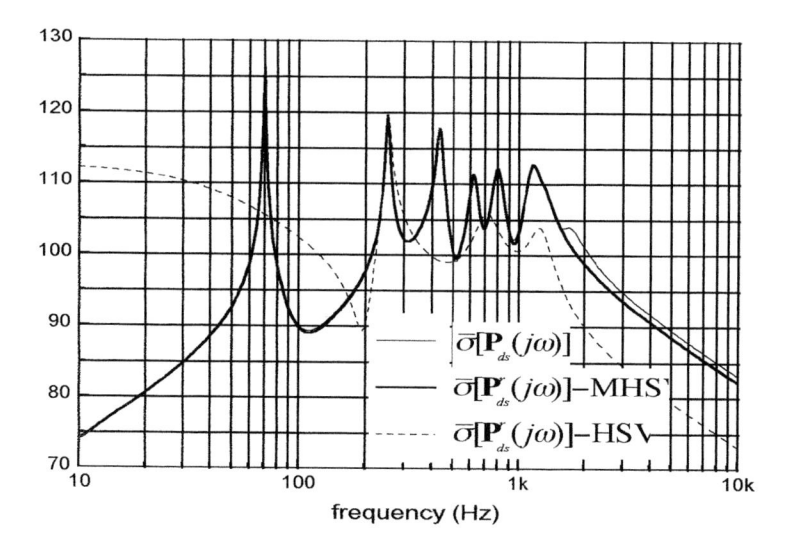

Figure 3: Maximum singular value of the observability transfer function for the full model (——); reduced model using MHSV (▬); reduced model using state space HSV (-----).

6. CONCLUSION

We have presented a simple but physically based approach to reducing the size of a system model using the concept of Modal Hankel Singular Values (MHSV). The transfer matrix computed with the reduced model agrees well with that calculated using the full model. It is proposed that this approach may be of use not only for designing real time controllers for adaptive structures, but also for several other applications such as real time animation, for developing new algorithms for inverse problems and also for interfacing structural codes with other system level codes in order to model an entire system. The proposed procedure will be more widely accepted if bounds can be mathematically derived for the errors that result from reduction scheme.

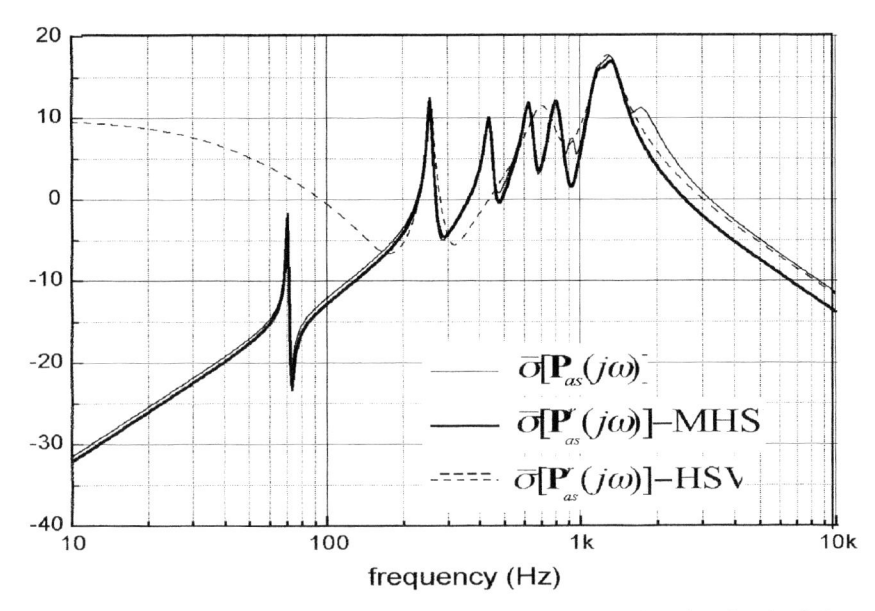

Figure 4: Maximum singular value of the controllability transfer function for the full model (——); reduced model using MHSV (——); reduced model using state space HSV (-----).

7. REFERENCES

1. Crawley, E. F. and Luis, D. J., 1987, "Use of Piezoelectric Actuator as Elements of Intelligent Structures", *AIAA Journal*, **25**, pp. 1373-1385.
2. Fuller R. C., 1990, "Active control of sound transmission/radiation from elastic plates by vibration inputs: I analysis," *Journal of Vibration and Acoustics*, pp. 1-15.
3. Hong, S-Y, Varadan, V. V. and Varadan, V. K., 1991, "Experiments on active vibration control of a thin plate using disc type piezoelectric actuators and sensors," *Proceeding of the ADPA/ASME/SPIE Conference on Active Materials and Adaptive Structures*, pp. 707-711
4. Bao, X., Varadan, V. V. and Varadan, V. K., 1995, "Active control of sound transmission through a plate using a piezoelectric actuator and sensor," *Smart Materials and Structures*, Vol. 4, pp. 231-239.
5. Koko, T. S., Orisamolu, I. R., Smith, M. J., and Akpan, U. O., 1997, "Finite-Element-Based Design Tool for Smart Composite Structures", *Proceedings of SPIE - Mathematics and Control in Smart Structures*, **3039**, San Diego, CA, USA, pp. 125-134.

6. Lim, Y. H., Varadan, V.V., and Varadan, V. K., 1997, "Closed loop finite element modeling of active structural damping in the frequency domain," *Smart Materials and Structures*, vol. 6, no. 2, pp. 161-168.
7. W. Chang, S. Gopinathan, V.V. Varadan, V.K. Varadan, "Design of Robust Vibration Controller for a Smart Panel Using Finite Element Model", <u>ASME J. Vibration</u> and Acoustics, Vol. 124, pp. 265-276, 2002.
8. J-W. Kim, V. V. Varadan and V. K. Varadan, "Finite Element Modeling of Structures Including Piezoelectric Active Devices", <u>International Journal of Numerical Methods in Engineering</u>, Volume 40, pp. 817-832, 1997.
9. Lim, Y-H., Gopinathan, S. V., Varadan, V. V. and Varadan, V.K., 1999, "Finite element simulation of smart structures using an optimal output feedback controller for vibration and noise control," *Journal of Smart Materials and Structures*, Vol. 8, pp. 324-337.
10. Allik, H. and Hughes, T. J. R., "Finite Element Method for Piezoelectric Vibration", *Int. J. for Numer. Meth. Engng.*, **2**, pp. 151-157.
11. Moore, B. C. 1981, "Principal component analysis in linear systems: controllability, observability, and model reduction," *IEEE Trans. Auto. Contr.*, Vol. 26, No. 2
12. Pernebo, L. and L. M. Silverman 1982, "Model reduction via balanced state space representation," *IEEE Tans. Automat. Contr.*, vol. AC-27, No. 2, pp. 382-387
13. Zhou, K., J. C. Doyle and K. Glover, 1995, "Robust and optimal control," Prentice Hall, New Jersey 07458

TRANSIENT ANALYSIS OF SMART STRUCTURES USING A COUPLED PIEZOELECTRIC-MECHANICAL THEORY

ADITI CHATTOPADHYAY, ROBERT P. THORNBURGH and
ANINDYA GHOSHAL
Department of Mechanical and Aerospace Engineering
Arizona State University
Tempe, AZ 85287-6106
Email: aditi@asu.edu

1. INTRODUCTION

Analysis of smart structures using piezoelectric materials (PZT) as either sensors or actuators has traditionally been performed using uncoupled models [1,2]. Uncoupled models make the assumption that the electric field within the piezoelectric material is constant and proportional to the ratio of electrode voltage to PZT thickness. Having made this assumption, the strain induced by an actuator is modeled with a single uncoupled equation and the charge output of a sensor is described by another uncoupled equation. This method is relatively simple, but it has its limitations. The mechanical and electric response of a piezoelectric device is in reality described by a pair of coupled equations [3] and cannot be accurately modeled if treated independently. It is necessary to simultaneously solve for both the electric response as well as the mechanical response regardless of whether the PZT is being used as a sensor or actuator. Also, the uncoupled model is not capable of taking into consideration any electrical circuitry connected to the piezoelectric device. This has been recognized in some specific applications and coupled equations have been used to model passive damping circuits [4,5] and develop self-sensing actuators [6]. Only recently have the coupled equations been simultaneously used for general analysis of adaptive structures [7-9].

In general the errors that result from using uncoupled models, as opposed to coupled ones, are relatively moderate. However, there are some cases in which very large differences exist when using the two approaches. One such case is for high frequency vibrations in adaptive structures. The objective of this work is to demonstrate the importance of proper modeling methods when analyzing high frequency vibration in smart composite structures and to show a comparison between the results predicted by uncoupled and coupled approaches.

2. COUPLED PIEZOELECTRIC-MECHANICAL FORMULATION

A recently developed two-way coupled piezoelectric-mechanical theory [9] is used to model composite plates with piezoelectric actuators. The construction of a model for smart composite laminates starts with the formulation of the constitutive relations. Traditionally these are expressed as a function of the components of strain (ε_{ij}) and electric field (E_i) as follows

$$\sigma_{ij} = c^{E}_{ijkl}\varepsilon_{kl} - e_{kij}E_k \qquad (1)$$

$$D_i = e_{ikl}\varepsilon_{kl} + \chi^{S}_{ik}E_k \qquad (2)$$

where σ_{ij} and D_i are the components of the mechanical stress and the electrical displacement, and c^{E}_{ijkl}, e_{ijk}, and χ^{S}_{ij} are the elastic, piezoelectric, and dielectric permitivity constants, respectively. It should be noted that the elastic constants used correspond to the zero electric field values (PZT is shorted out) and the dielectric permitivities correspond the zero strain values (clamped). These equations are traditionally used due to the ease with which piezoelectric materials can be modeled as either actuators or sensors. Most formulations make the assumption that the electric field is constant through the thickness of the material. However, if the strain is not constant through the thickness of the piezoelectric material, such as in the case of bending or transverse shear, then this method results in electric displacement varying through the thickness. This also implies differing amounts of charge on the upper and lower electrodes, which is a violation of the conservation of charge principle. This has been resolved by making the electric potential, and in turn the electric field, higher order functions of the through the thickness coordinate to match the displacement and strain fields in the structure. However, such an approach leads to additional degrees of freedom to describe the electric potential. Another drawback of such an approach is that the resulting system matrices in finite element implementation are not symmetric. This results in a sizable increase in the computational effort required to solve the system of equations.

To address these issues a different approach is used in which Eqs. (1) and (2) are reformulated in terms of the mechanical strain and electric displacement as follows

$$\sigma_{ij} = c^{D}_{ijkl}\varepsilon_{kl} - h_{kij}D_k \qquad (3)$$

$$E_i = -h_{ikl}\varepsilon_{kl} + \beta^{S}_{ik}D_k \qquad (4)$$

where c^{D}_{ijkl}, e_{ijk}, and β^{S}_{ij} are the open circuit elastic and zero strain dielectric constants, respectively. The coefficient h_{ijk} now represents the coupling between the strain and the electric displacement. Using this formulation, the electric displacement can be taken as constant through the thickness of the PZT, thus ensuring conservation of charge on each of the electrodes.

The equations of motion can be formulated using a variational approach and Hamilton's Principle [3]. The variational principle between times t_o and t, for the piezoelectric body of volume V can be written as follows

$$\delta\Pi = 0 = \int_{t_o}^{t} \int_{V} \left[\delta\left(\frac{1}{2}\rho\dot{\mathbf{u}}^T\dot{\mathbf{u}}\right) - \delta\mathbf{H}(\varepsilon, \mathbf{D}) \right] dVdt + \int_{t_o}^{t} \delta\mathbf{W} dt \qquad (5)$$

where the first term represents the kinetic energy, the second term the electric enthalpy and $\delta\mathbf{W}$ is the total virtual work done on the structure. The terms **u** and ρ refer to the mechanical displacement and density, respectively. The electric enthalpy (**H**) and the work done by body forces (**f**$_B$), surface tractions (**f**$_S$) and electrical potential (ϕ) applied to the surface of the piezoelectric material can be expressed by

$$\mathbf{H}(\varepsilon, \mathbf{D}) = \frac{1}{2}\varepsilon^T \mathbf{C}^D \varepsilon - \varepsilon^T \mathbf{h} \mathbf{D} + \frac{1}{2} \mathbf{D}^T \boldsymbol{\beta}^S \mathbf{D} \qquad (6)$$

$$\delta\mathbf{W} = \int_V \delta\mathbf{u}^T \mathbf{f}_B dV + \int_S \delta\mathbf{u}^T \mathbf{f}_S dS + \int_S \delta\mathbf{D}^T \phi dS \qquad (7)$$

Equations (5-7) provide the equations of motion for the piezoelectric body. To solve them, assumptions must be made concerning the nature of the mechanical strain and the electrical displacement. First, it is assumed that the piezoelectric material is oriented with its polarization axis normal to the plane of the plate and that the PZT has electrodes covering its upper and lower surfaces. This is the usual geometry for transversely operating piezoelectric actuators and sensors bonded to or embedded in plate structures. For this case, the electric displacement becomes zero along the two in-plane directions. The out-of-plane electric displacement can then be discretized over the surface of the piezoelectric device using finite elements.

A refined higher order laminate theory [10] is used to model the mechanical displacement field. The laminate is assumed to be a plate structure composed of an arbitrary number of orthotropic lamina arranged with varying orientations. The refined higher order theory assumes a parabolic distribution of transverse shear strain, thus providing accurate estimation of transverse shear stresses for moderately thick constructions with little increase in computational effort. The theory starts with a general third order displacement field and is simplified by imposing the stress free boundary conditions on the free surfaces. Since the laminate is orthotropic, this implies that the transverse shear strains are zero. The refined displacement field now takes the following form

$$\mathbf{u}_1 = u + z\left(\psi_x - \frac{\partial w}{\partial x}\right) - \frac{4z^3}{3h^2}\psi_x \qquad (8a)$$

$$\mathbf{u}_2 = v + z\left(\psi_y - \frac{\partial w}{\partial y}\right) - \frac{4z^3}{3h^2}\psi_y \qquad (8b)$$

$$\mathbf{u}_3 = w \qquad (8c)$$

where u, v, and w are the displacements of the midplane and ψ_x and ψ_y are the rotations of the normal at $z=0$ about the -y and -x axes, respectively. Note that u, v, w, ψ_x and ψ_y are all functions of the x and y coordinates only. The variable z

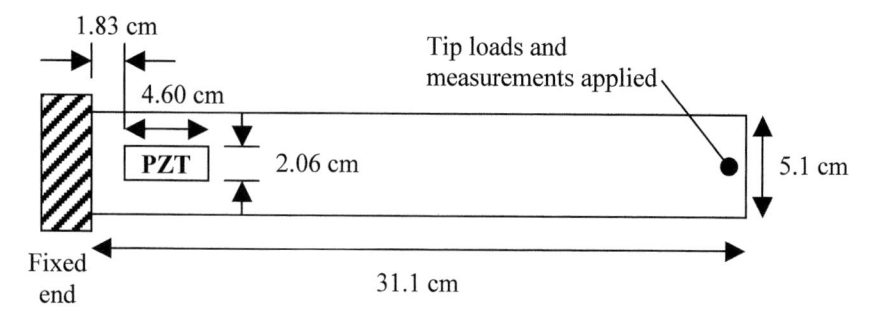

Figure 1. Cantilever plate layout.

represents the location with respect to the midplane of the plate, and h is the total plate thickness.

By using the above equations and the finite element method, the governing equations can be written in matrix form as follows

$$\begin{bmatrix} \mathbf{M}_u & 0 \\ 0 & 0 \end{bmatrix} \begin{Bmatrix} \ddot{\mathbf{u}}_e \\ \ddot{\mathbf{D}} \end{Bmatrix} + \begin{bmatrix} \mathbf{C}_u & 0 \\ 0 & 0 \end{bmatrix} \begin{Bmatrix} \dot{\mathbf{u}}_e \\ \dot{\mathbf{D}} \end{Bmatrix} + \begin{bmatrix} \mathbf{K}_{uu} & \mathbf{K}_{uD} \\ \mathbf{K}_{Du} & \mathbf{K}_{DD} \end{bmatrix} \begin{Bmatrix} \mathbf{u}_e \\ \mathbf{D} \end{Bmatrix} = \begin{Bmatrix} \mathbf{F}_u \\ \mathbf{F}_D \end{Bmatrix} \quad (9)$$

where $\mathbf{u}_e$ is the nodal displacements, $\mathbf{D}$ is the vector of the PZT nodal electrical displacements. The matrix $\mathbf{M}_u$ is the structural mass matrix. and $\mathbf{C}_u$ is the structural damping matrix. The matrix $\mathbf{K}_{uu}$ is the mechanical stiffness matrix, $\mathbf{K}_{DD}$ is the electrical stiffness matrix, and $\mathbf{K}_{uD}$ and $\mathbf{K}_{Du}$ are the stiffness matrices due to piezoelectric-mechanical coupling. The vectors $\mathbf{F}_u$ and $\mathbf{F}_D$ are the force vectors due to mechanical loading and applied voltages. To incorporate structural damping into the equations, a structural damping matrix $\mathbf{C}_u$ is added. The nature of the damping matrix can be chosen to meet the needs of the user.

The absence of any electrical inertia or damping terms in Eq. (9) is a result of only considering the mechanical aspects of the smart structure. When considering an integrated smart structural system as a whole, additional terms must be added for electrical components in the system. For a simple LRC circuit, the equations of motion for any electrical system attached to the smart structure can be formulated in the following form

$$\mathbf{M}_q \ddot{\mathbf{q}}_e + \mathbf{C}_q \dot{\mathbf{q}}_e + \mathbf{K}_q \mathbf{q}_e = \mathbf{F}_q \quad (10)$$

Since the electric displacement has been discretized, it is necessary to use the following relation between the electric displacement and the total PZT charge

$$q_i = \left(\int_S \mathbf{N}_q dS \right) \mathbf{D}_{ie} \quad \text{or} \quad \mathbf{q} = \mathbf{A}_q \mathbf{D}_e \quad (11)$$

Combining Eqs. (9) and (10), the resulting coupled electrical-mechanical system equations are obtained

$$\begin{bmatrix} M_u & 0 \\ 0 & A_q^T M_q A_q \end{bmatrix} \begin{Bmatrix} \ddot{u}_e \\ \ddot{D}_e \end{Bmatrix} + \begin{bmatrix} C_u & 0 \\ 0 & A_q^T C_q A_q \end{bmatrix} \begin{Bmatrix} \dot{u}_e \\ \dot{D}_e \end{Bmatrix} + \begin{bmatrix} K_{uu} & K_{uD} \\ K_{Du} & K_{DD} + A_q^T K_q A_q \end{bmatrix} \begin{Bmatrix} u_e \\ D_e \end{Bmatrix} = \begin{Bmatrix} F_u \\ F_D \end{Bmatrix} \quad (12)$$

To model a particular sensor configuration the electrical circuit must be modeled or appropriate electrical boundary conditions must be applied. If sensor charge flow is being measured, then the voltage is specified as zero. If sensor voltage is being measured, then the net charge flow, Eq. (11), is specified as zero.

The nonlinear transient analysis is conducted using Newmark-beta method with Newton-Raphson (N-R) iteration [11-12]. This results in a time integration method that can be iterated at each time step to provide accurate prediction of the transient response of the system.

3. RESULTS

The developed model is used to calculate the response of a composite plate with surface bonded actuators subjected to impulse loading. The objective is to demonstrate the nature and magnitude of errors that exist when simpler approaches are used. First, a comparison is made between the response predicted by the refined higher order laminate theory and the classical plate theory. Then the difference between the coupled piezoelectric-mechanical model and the uncoupled model is examined.

The plate under consideration is clamped at one end with a cantilevered section

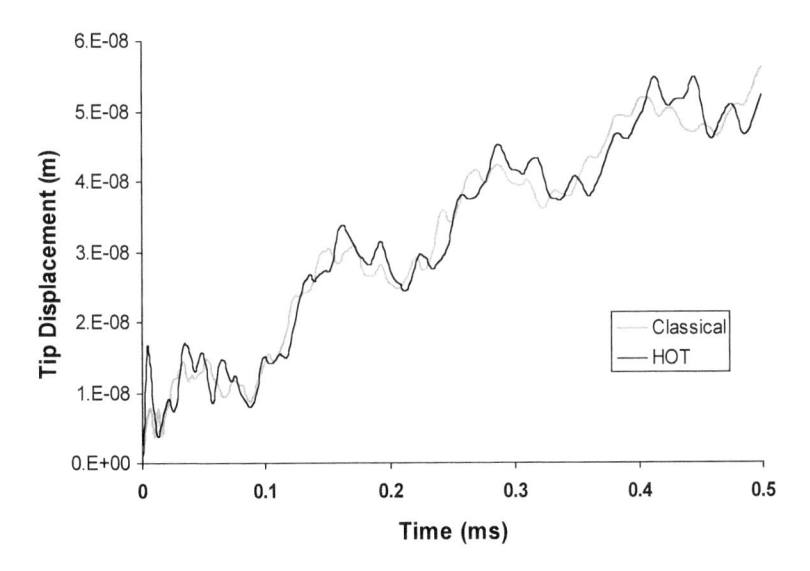

Figure 2. Tip displacement for $[0°, 90°]_{4s}$ laminate ($L/h=142$) under $1\mu s$ impulse loading.

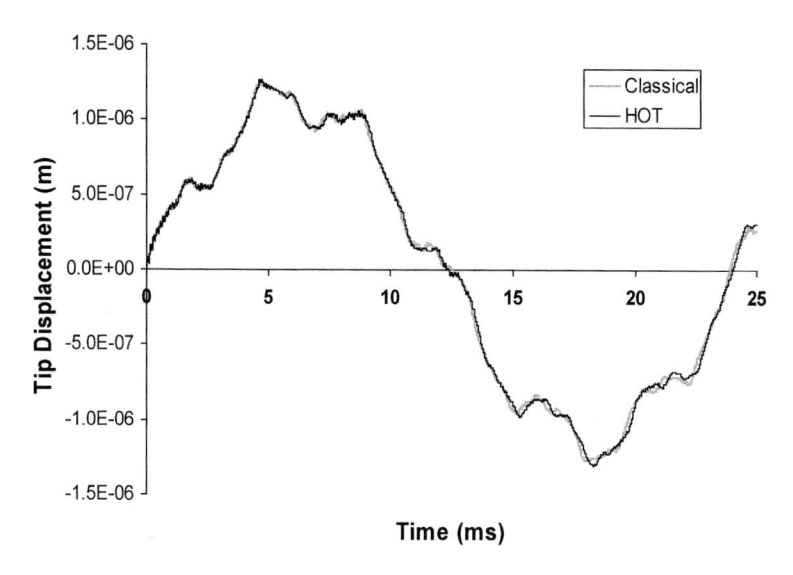

Figure 3. Tip displacement for $[0°,90°]_{4s}$ laminate (L/h=142) under 5μs impulse loading.

31.1cm long by 5.1cm wide. The plate is a graphite-epoxy laminate with sixteen plies of 0.137mm ply thickness. A variety of ply stacking sequences are considered, including cross-ply as well as balanced and unbalanced angle-ply lay-ups. A piezoelectric patch is bonded to the upper surface of the plate as shown in Figure 1. The patch is made of PZT-5H with a thickness of 0.25mm.

First, a comparison is made between the refined higher order laminate theory and the classical plate theory. A laminate stacking sequence of $[0°,90°]_{4s}$ is considered first, with the PZT left open circuited. This laminate has a ratio of plate length to thickness of 142, making this a relatively thin plate. The plate is modeled with the refined higher order laminate theory, which includes the effects of transverse shear, as well as with the classical plate theory, which neglect transverse shear. All other aspects of the modeling are identical in both cases. The plate is subjected to a 5-Newton, 1μs impulse point load at the tip. The analysis is performed for both cases with a 1μs time step. Figure 2 shows the resulting tip displacement calculated by the two models. The difference between the higher order theory and the classical theory is modest, but noticeable during the short time interval analyzed, even for this thin laminate. By including transverse shear stress, the refined higher order theory results in a model with lower natural frequencies than those predicted by the classical plate theory. The effects are not only more significant for thicker laminates, but also for the higher frequency modes. This difference in frequencies can be seen in Figure 2.

Next, the same plate is analyzed under a 1-Newton, 5μs impulse point load at the tip. The time step for the transient analysis is 5μs, so that a longer time segment could be examined. The tip displacement for the higher order and classical plate theories is shown in Figure 3 and the sensor output is shown in Figure 4. When a longer time step is used the differences between the two theories are much less

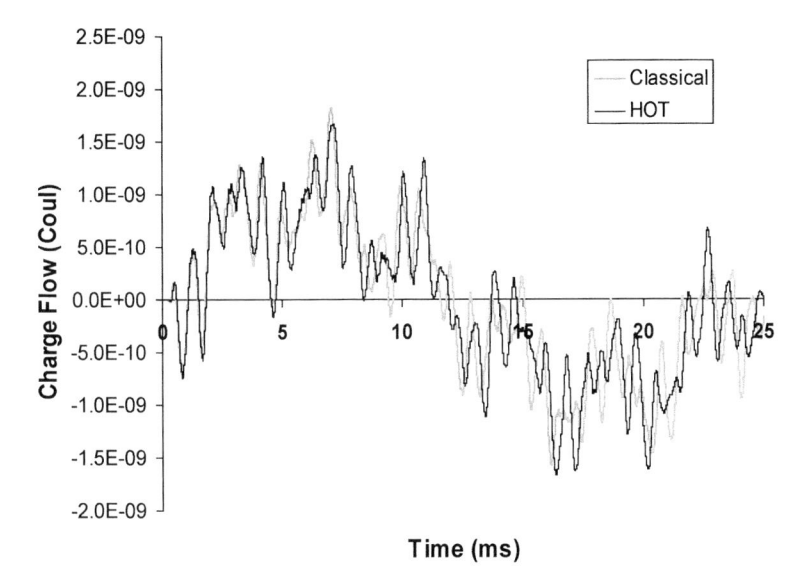

Figure 4. Sensor output for $[0°,90°]_{4s}$ laminate (L/h=142) under 5µs impulse loading.

significant for the thin plate, because the natural frequencies associated with the lower order modes predicted by the two plate theories are quite similar. However, it can be seen that the differences in the higher frequencies affect sensor output to a greater degree than tip displacement. Although the shape of the response is very similar, the magnitude of the signal differs significantly. This is because at high frequencies the output of the coupled piezoelectric-mechanical theory is very sensitive to changes in the strain field from the waves that move across the sensor.

An increase in plate thickness increases the difference in the response predicted by the higher order theory and the classical plate theory. This is shown in Figure 5, which depicts the tip response for a $[0°_5,90°_5]_{4s}$ laminate. The ratio of plate length to thickness is 28.4 for this case, making this a moderately thick plate. The impulse length and time step used in this case are both 1µs. By increasing plate thickness differences are observed even in the low frequency modes predicted by the two plate theories, resulting in sizable differences in plate deflections.

Next a comparison is made between the coupled and the uncoupled piezoelectric models using the higher order theory. The plate with $[0°,90°]_{4s}$ stacking sequence is considered. The plate is subjected to a 1-Newton, 5µs impulse point load at the tip and a 5µs time step is used for the transient analysis. The PZT is open circuited and voltage between the electrodes is measured for sensor output. The response is calculated using both the coupled theory presented in this work as well as the traditional uncoupled approach, which assumes that the electric field is constant value equal to the ratio of voltage to thickness. Only slight differences between the two models are observed when the tip displacements are compared. This is due to the relatively small size of the PZT patch and its limited contribution to the overall plate stiffness. This difference would be expected to be larger for thicker PZT

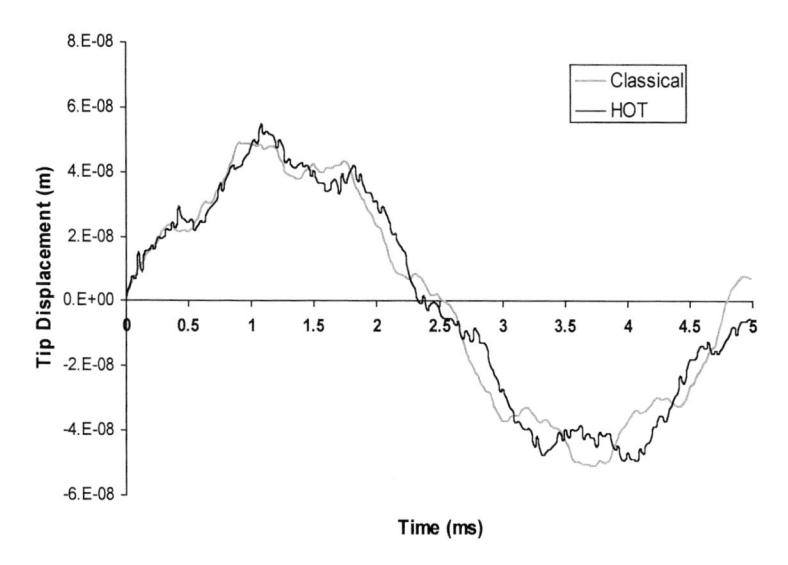

Figure 5. Tip displacement for $[0°_5, 90°_5]_{4s}$ laminate (L/h=28.4) under 1μs impulse loading.

patches. Figure 6 shows the voltage output of the sensor for both cases. Here it can be seen that the coupled theory predicts dramatically different results from the uncoupled approach, even though the displacements are shown to be similar. Figure 7 presents the output if the charge flow from the PZT is instead measured. The time delay can be seen between the impulse and the deformation wave reaching the sensor and again the coupled theory predictions greatly differ from the uncoupled approach.

The reason for the large differences in the electrical response is due to the fact that the uncoupled theory assumes that the electric field is constant over the entire area of the PZT. During impulse loading high frequency bending waves travel across the length of the plate. As these high frequency waves move across the PZT, the piezoelectric material is subjected to areas of local compression and tension. As a result, the electric displacement is positive in some local areas and negative in others. The net charge output is thereby reduced since charge merely flows from one region of the patch to another. When the PZT is open circuited, the charge flow within the PZT still occurs making the voltage output very sensitive to local strain. During impulse loading there exists a continuously changing strain gradient across the PZT patch, which leads to the difference in results predicted by the coupled and uncoupled models.

4. CONCLUSIONS

A new approach has been developed for modeling the transient response of composite laminates with piezoelectric sensors and actuators. The mathematical model uses a coupled piezoelectric-mechanical theory that accurately captures both electrical and mechanical characteristics of adaptive structures. Parametric studies

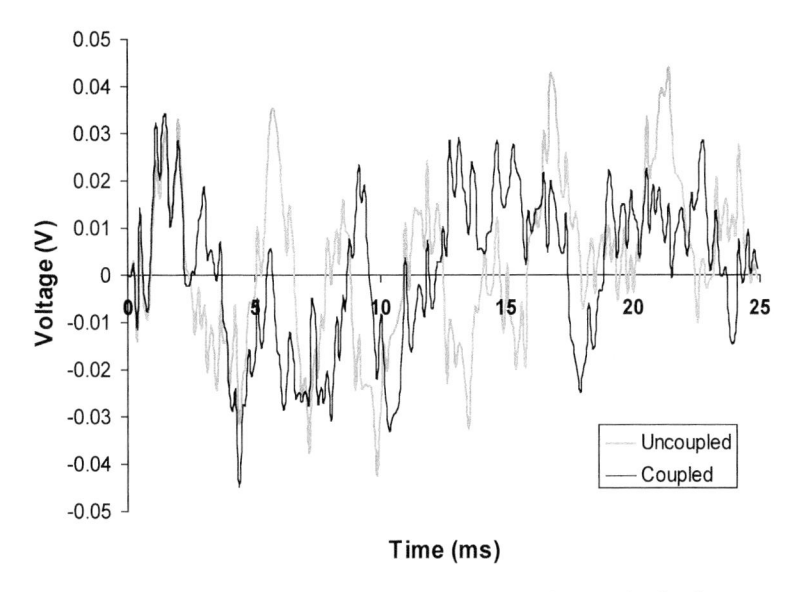

Figure 6. Sensor output for $[0°,90°]_{4s}$ laminate under impulse loading.

were performed to assess the difference in results predicted between the refined higher order laminate theory and the classical plate theory, as well as between the coupled and uncoupled piezoelectric models. The following conclusions were made from the present study

1. For thin plates, the differences in mechanical displacements predicted by the higher order and the classical plate theories are significant only when the time step is sufficiently small to capture very high frequency modes. The differences in sensor output are more significant in all cases.
2. For thicker plates, significant differences existed in both mechanical displacements and sensor output predicted by the higher order and the classical plate theories during all analyses.
3. Although only moderate differences are observed in the predictions of mechanical displacements, larger differences are observed for both charge and voltage measurements, predicted by the coupled and uncoupled models.
4. The assumption of constant electric field over the entire PZT area in the uncoupled theory leads to inaccuracies in modeling the effect of high frequency vibrations that create both areas of local compression and local tension within the PZT patch.

5. ACKNOWLEDGEMENT

This research was supported by the Air Force Office of Scientific Research, Technical monitor, Daniel Segalman.

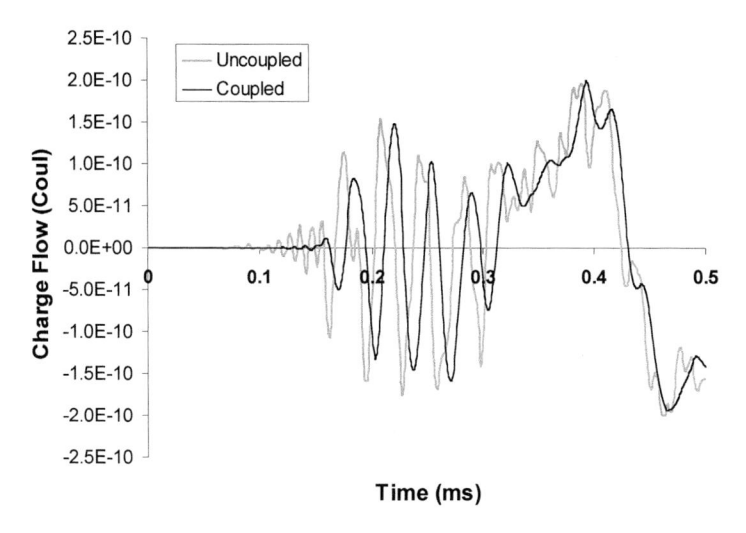

Figure 7. Sensor charge flow for $[0°,90°]_{4s}$ laminate under impulse loading.

6. REFERENCES

1. Crawley, E. F., "Use of Piezoelectric Actuators as Elements of Intelligent Structures," *AIAA Journal*, Vol. 25, No. 10, 1987, pp. 1373-1385.
2. Detwiler, D. T., Shen, M. H. and Venkayya, V. B., "Finite Element Analysis of Laminated Composite Structures Containing Distributed Piezoelectric Actuators and Sensors," *Finite Elements in Analysis and Design*, Vol. 20, No. 2, 1995, pp. 87-100.
3. Tiersten, H. F., 1967, "Hamilton's Principle for Linear Piezoelectric Media," *IEEE Proceedings*, Vol. 55, No. 8, pp. 1523-4.
4. Hagood, N. W. and Von Flotow, A., "Damping of Structural Vibrations With Piezoelectric Materials and Passive Electrical Networks," *Journal of Sound and Vibration*, Vol. 146, No. 2, 1991, pp. 243-68.
5. Wu, S. Y., "Piezoelectric Shunts with a Parallel R-L Circuit for Structural Damping and Vibration Control," *Proceedings of the International Society for Optical Engineering*, Vol. 2720, 1996, pp. 259-69.
6. Anderson, E. H. and Hagood, N. W., "Simultaneous Piezoelectric Sensing/Actuation: Analysis and Application to Controlled Structures," *Journal of Sound and Vibration*, Vol. 174, No. 5, 1994, pp. 617-639.
7. Mitchell, J. A. and Reddy, J. N., "A Refined Hybrid Plate Theory for Composite Laminates with Piezoelectric Laminae," *International Journal of Solids Structure*, Vol. 32, No. 16, 1995, pp. 2345-2367.
8. Chattopadhyay, A., Li, J. and Gu, H., "A Coupled Thermo-Piezoelectric-Mechanical Model For Smart Composite Laminates," *AIAA Journal*, Vol. 37, No. 12, 1999, pp. 1633-1638.
9. Thornburgh, R. P. and Chattopadhyay, A., 2001, "Electrical-Mechanical Coupling Effects on the Dynamic Response of Smart Composite Structures," *Proceedings of the International Society for Optical Engineering*, Vol. 4327, pp. 413-424.
10. Reddy, J. N., 1984, "A Simple Higher-Order Theory for Laminated Composite Plates," *Journal of Applied Mechanics*, Vol. 51, pp. 745-5.
11. Bathe, K. J., "Finite Element Procedures," Prentice Hall, Englewood Cliffs, NJ, 1996.
12. Argyris:J & Mlejnek, H. P. "Dynamics of Structures", Text on Computational Mechanics, V5, Elsevier Sci. Publishers, 1991.

DYNAMIC BEHAVIOR OF SHAPE MEMORY ALLOY STRUCTURAL DEVICES: NUMERICAL AND EXPERIMENTAL INVESTIGATION

Martensite versus austenite in SMA behaviour.

LUCIA FARAVELLI AND SARA CASCIATI
*Department of Structural Mechanics, University of Pavia,
Via Ferrata 1, 27100, Pavia, Italy
E-mail: lucia@dipmec.unipv.it*

1. INTRODUCTION

The first author was recently engaged in an editorial effort [1] having the modeling and applications of shape memory alloys (SMA) as subject. This activity followed a wide band cooperation period with specialty teams, working in Russia, France and Spain, and their associates [2]. The second author is taking care of an attempt to implement SMA devices in the seismic retrofitting of an ancient monument: the Luxor Memnon Colossi, in Egypt [3][4].

Most of the collected literature spans over the exploitation of shape memory alloys in their austenite form, mainly in wires in order to avoid any form of material deterioration. Their hyper-elastic hysteresis cycles result in energy dissipation which can usefully be adopted in devices of passive structural control [1]. Nevertheless, when dealing with large size structures, massif components could be required and the martensite phase of shape memory alloys possesses properties, as the residual strain recovery by heating, which are fascinating in view of innovative technical applications. In this area a wide experimental campaign was conducted to confirm the asymmetry in tension compression, the variability of the Young modulus after loading-unloading cycles and the potential risk of precipitates with no shape memory effect [5].

A further aspect, which is the core of the developments summarized in the first part of this paper, is the possible devastating consequence of the normal temperature increase under repeated deformation cycles. Indeed when dealing with shape memory alloys in the martensitic phase, one achieves soon temperatures which are very close to the martensite-austenite transformation temperature (Figure 1).

The second part of the paper pursues the modeling of the material constitutive law. The ultimate goal is the design of devices for passive structural control, able either to re-center themselves thanks to the shape memory effect, in the martensite phase, or to sew independent structural blocks without preventing from relative motions, in the austenite phase.

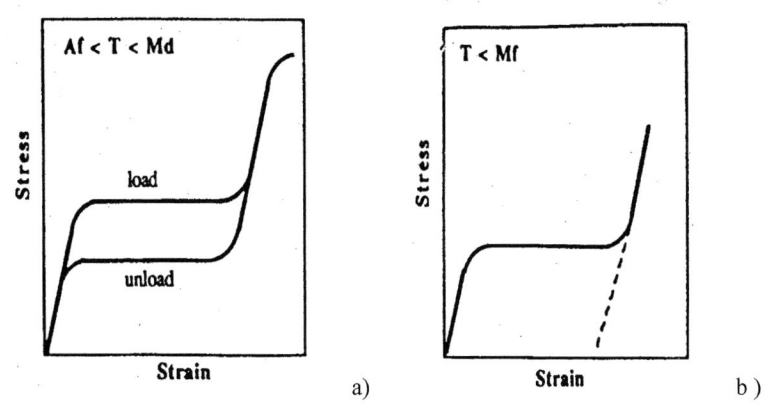

Figure 1. a) Pseudoelastic stress-strain relationship; b) Martensite stress-strain relationship.

2. THE TESTING ENVIRONMENT

Some specimens (Figure 2) of Nickel-Titanium alloy, collected in the United States, were tested. This special alloy is in its martensite phase at the environment temperature. When dealing with such material, machining the original bars to create the standard specimens to be tested can produce significant troubles, both in terms of failures during the material working and of local alterations of the material properties.

For this purpose the adoption of digitally controlled machines is recommended together with the adoption of miniature tool able to removing minimal parts of material during a single cycle. The tested specimens were obtained by a machine with manual control and this resulted in loosing almost 20% of the specimens.

Figure 2. The specimen forms tested in the experimental campaign.

Standard specimens of minimum diameter 5 mm and cylindrical specimens (i.e. without surface alterations) of diameter 13 mm (see Figure 2) were both tested in similar conditions in order to check potential consequences, on the material properties, of the machining process.

The tests summarized in this paper were all obtained by cyclic torsion loading and unloading, in order to avoid a direct discussion of the asymmetry in tension and compression that NiTi alloys show.

The adequacy of the experimental environment is checked by testing the two classes of specimens in angle control.. The results are collected in Figures 3, 4 and 5. The first two figures report results obtained for the 5mm specimen: hysteresis cycles for different maximum angles at a very low speed (0.5 degree per second) and for an angle of 22° repeated 10 times at the speed of 240 deg/sec.

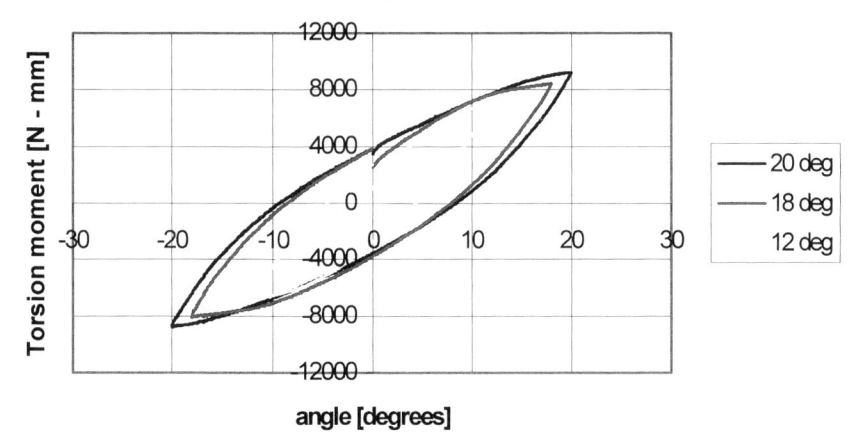

Figure 3 – 5 mm specimen: hysteresis cycles for different values of the maximum angle control.

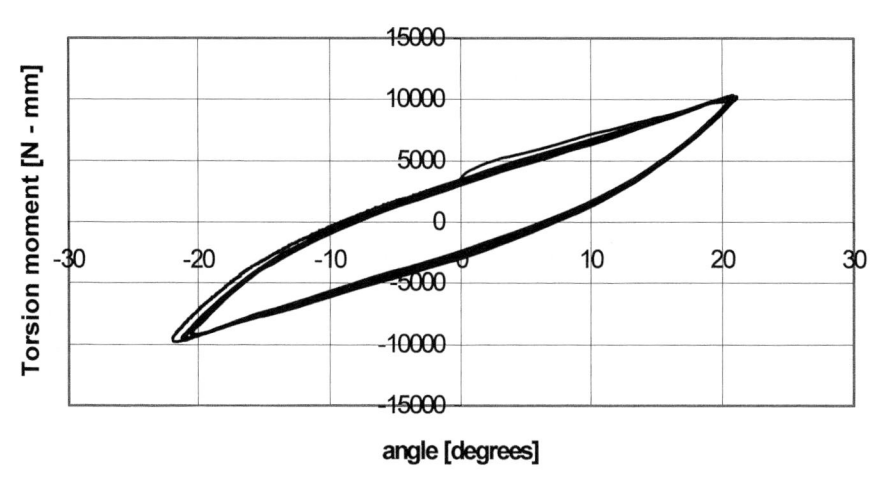

Figure 4 – 5mm specimen: 10 cycles at 22 degree of maximum angle with speed 240 deg/sec.

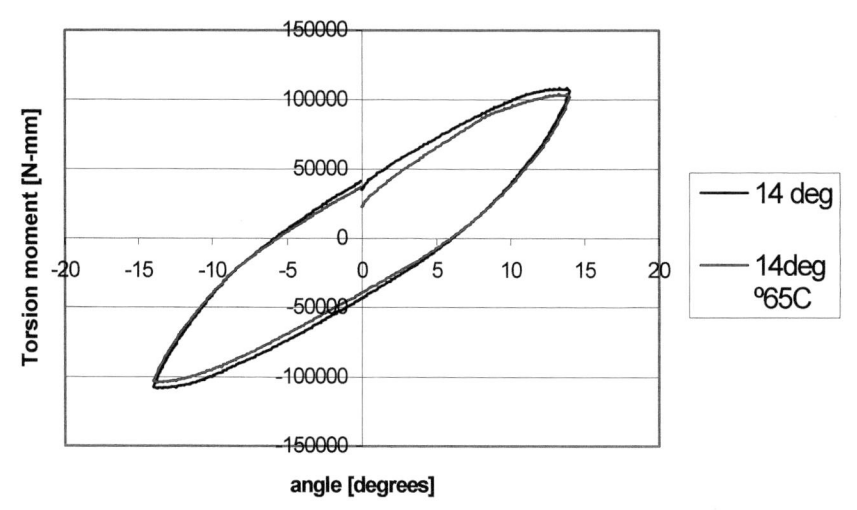

Figure 5 – 13mm specimen: slow tests (0.5 degree per sec) with maximum angle of 14° for environment (30 °C) temperature and 65°C.

Figure 5 gives evidence of the different order of magnitude required for the torsion moment in testing the 13mm specimen. Comparing the initial stiffness in the tests of Figures 3 and 5 one finds 1400 Nmm against 65000 Nmm which gives exactly the ratio between the forth power of the specimen radii as suggested by De Saint Venant theory. Figure 5 also shows a decrease of stiffness as the temperature increases (as expected for standard materials): the maximum moment decreases to 104200 Nmm from 108450 Nmm.

3. TEMPERATURE EFFECT

The next step is to increase the number (500) of cycles (i.e. to increase the dissipated energy) as well as the angular velocity of the torsion actuator, so that the heat dissipation capacity of the environment is reduced. Figure 6 shows how the specimen temperature increases (becoming almost stationary after one half of the test) and the consequent modification of the cycle: a higher value of the final moment is required. It also results in a slimmer and slimmer hysteresis cycle as the temperature increases. One is still far from the transformation temperature of martensite into austenite, but locally this can occurs and mechanically evidence is found in the cycle modification.

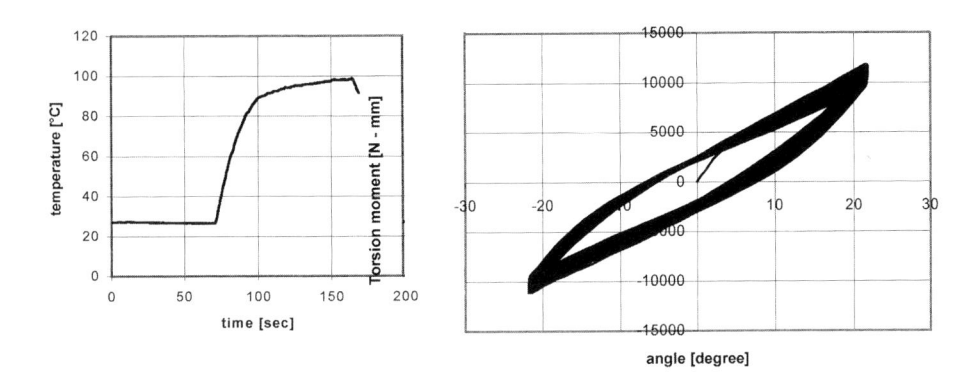

Figure 6 – Temperature increase and consequent hysteresis modification during a 500 cycles test at 200 deg/sec on the 5mm specimen.

Moving to the bar of larger diameter, the larger volume undergoing hysteresis is able to dissipate a higher amount of energy and, of consequence, to produce an even more significant increase of temperature. Figure 7 collects the temperature time histories recorded during different tests. The first of them is conducted in conditions quite similar to those of Figure 6. The result in terms of diagram angle-torsion moment is given in Figure 8. Once again the value of torsion moment required to achieved the assigned value of maximum angle increases as the temperature increases, while the hystersis cycles become slimmer and slimmer.

Subsequent tests pursued a greater increase of the temperature during the test. For this purpose both the number of cycles (to 1000) and the maximum angle (to 18°) were increased. The resulting temperature time history in Figure 7b) shows the important feature that there is no longer an equilibrium between the heat produced in energy dissipation and the environment absorbing capacity. By contrast the temperature continues to increase in a rather linear way.

a)

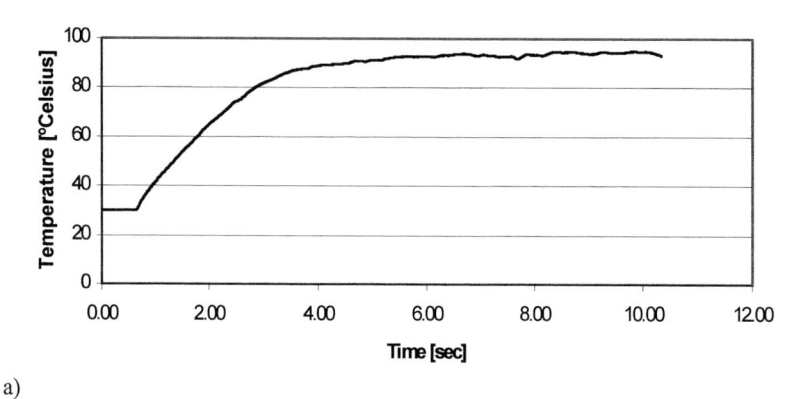

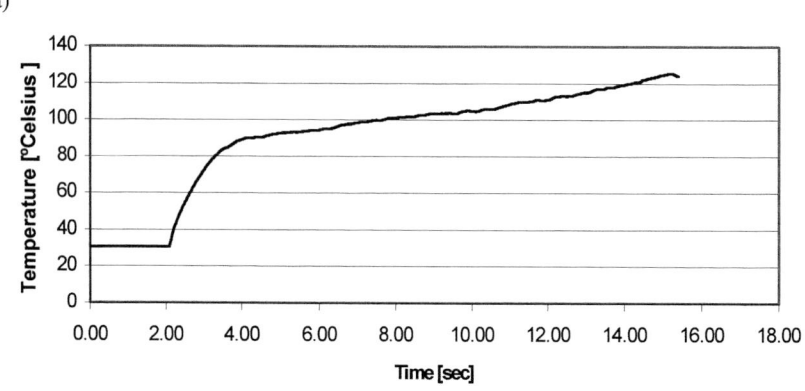

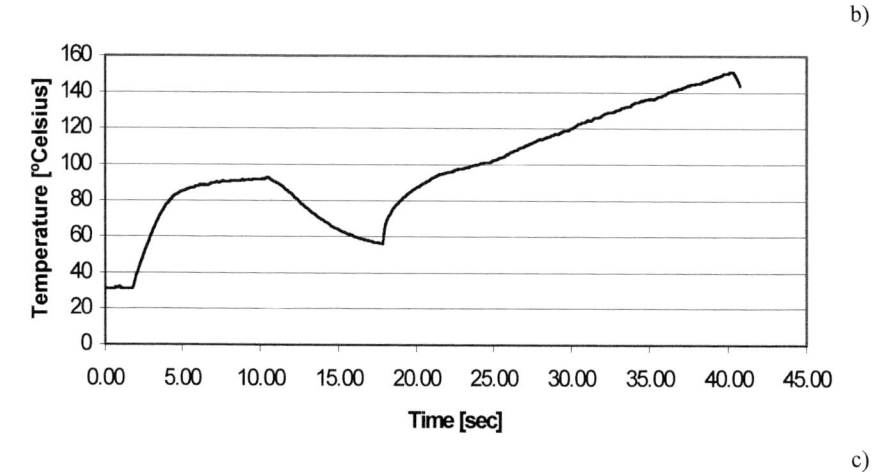

b)

c)

Figure 7 – Temperature time histories during a) a 500 cycles at 200 degrees per second with maximum angle of 14°; b) a sequence of 500 cycles at 200 degrees per second with maximum angle of 18° and c) 1000 cycles, after pre-heating, at.18° and 300 deg/sec.

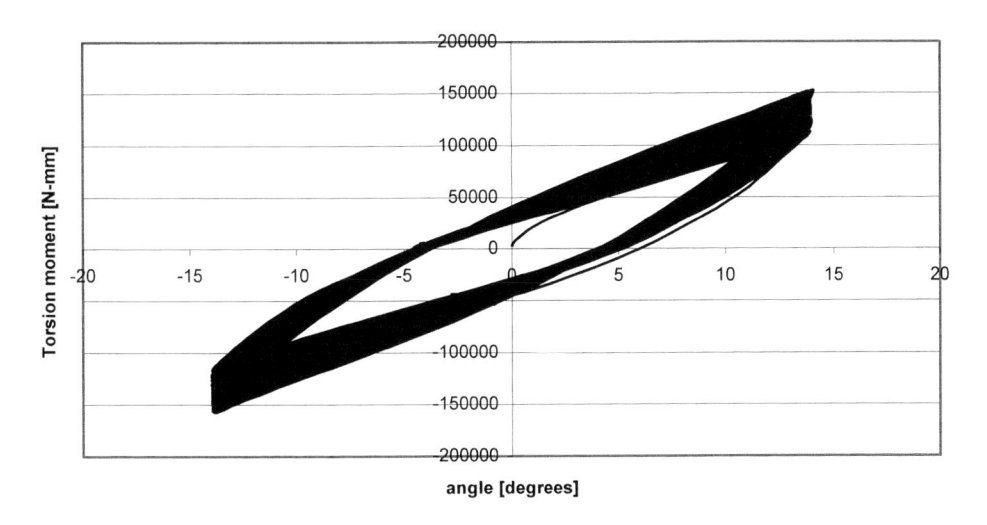

Figure 8 – 500 cycles with 14° maximum angle: temperature as in Figure 7 a).

A further test was therefore conceived: the 1000 cycles of the previous test are first doubled. But this sequence of 2000 cycles starts without having, as before, the environment temperature as initial condition. The new initial condition is a pre-heated situation generated by a preliminary sequence of cycles.
During this test, the temperature increases up to 150° C, i.e. well above the martensite austenite transformation temperature estimated by the DSC (Differential Scanning Calorimeter) test.

The tests illustrated in this section emphasize some features which seem quite significant in view of conceiving a mathematical model of the SMA response:

- the macroscopic specimen moment-angle diagram depends on the temperature;
- this dependence shows a sort of saturation, provided that such an equilibrium temperature is well below the austenite-martensite transformation temperature;
- when the dissipated energy associated to the hysteresis cycles does not allow such a saturation, there is a temperature threshold after which the temperature increase shows a linear dependency on time.

4. RETROFITTING BY SMA DEVICES

A possible seismic retrofitting of the Southern Memnon Colossus, located in Luxor, Egypt (Figure 9a), was conceived following the Italian expertise collected in masonry bell-tower rehabilitations [1]. Shape memory alloy wires were used to sew the structure in such applications. The result was to limit, but not to eliminate, the benefic relative motion of the masonry blocks.

In the Memnon Colossus, the blocks resulting from cracking must be sewed one with the other in order to avoid local overturning which could results in a global collapse. Shape memory alloys, in their austenitic phase (Figure 1a)), offers the unique possibility of sewing two blocks, by applying them a pre-tension, without preventing the blocks from relative motions.

In the case of the Colossus non-apparent devices can be added by strengthening the basement below the soil level. The material constitutive law between stress the σ and the strain ε along the hysteresis loop is, for given temperature and for given stress σ_0 at the beginning of the step [1][4]:

$$\sigma = \sigma_0 + E_1 \ \varepsilon + \delta' \ E_2 \ \varepsilon \quad \text{for } \dot{\varepsilon} > 0$$

$$\sigma = \sigma_0 + E_1 \ \varepsilon + \delta'' \ E_2 \ \varepsilon \quad \text{for } \dot{\varepsilon} < 0$$

where, for a NiTi alloy, $E_1 = 60000$ N/mm^2; $E_1/E_2 = 8$, strain at the elastic limit 0.01 and thickness of the hysteresis given by $\Delta\sigma = 600$ N/mm^2. The parameters δ' and δ'' governs the hysteresis loop. On the basis of this constitutive law, given a pre-tension stress, the wire undergoes plastic deformation with practically no modification in the stress value. Some comparative analysis between NiTi and Cu-based alloys are however in progress toward an optimization of the resulting device.

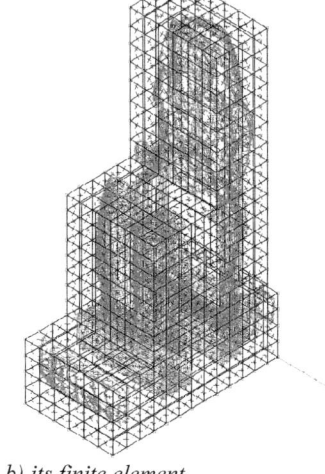

Figure 9 – a) The cracked monument under investigation and b) its finite element discretization.

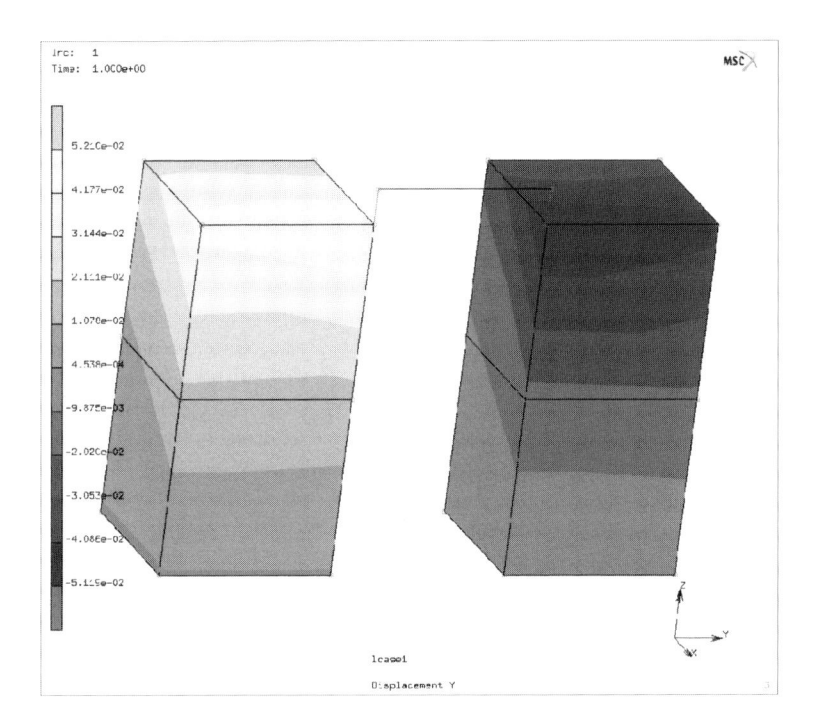

Figure 10 – The shape memory alloy link between the two blocks introduces a pre-tensioning which makes the two part collaborating under horizontal actions, but the hyperelastic hysteretic behavior of the alloy allows the two blocks to undergo relative displacements.

Therefore, within the project CHIME, funded by the European Union, one has:

- produced a finite element discretization of the cracked monument (Figure 10) suitable for structural analysis within the general purpose code MARC [6][7];
- written a consistent material user subroutine able to introduce the shape memory alloy behavior depicted above;
- planned a series of numerical simulations aiming at studying the effect of the temperature on the benefit associated with the introduction of shape memory alloy devices.

Figure 10 shows how the shape memory alloy wire is working in a simplified structural scheme, while the global analyses of the retrofitted Colossus are in progress.

5. CONCLUSIONS

Several aspects of the dynamic response of shape memory alloy structural components have been investigated at the authors' laboratory. As reported in [5], they cover: (i) the symmetry of the response under cyclic loading in tension-compression and torsion; (ii) the dependence of the response on the strain rate; (iii) the ability to recover strain; and, mainly, (iv) the risk of moving from martensite to austenite due to the heating generated by the energy dissipation during hysteretic cycles of loading-unloading. The last aspect received special attention within this paper where the results of a further experimental campaign are illustrated.

The application of this material, in its austenitic phase, to the retrofitting of cracked monolithic monuments is also discussed. The policy of filling the cracks with mortar or other suitable plastic materials cannot be regarded as a suitable repair. Indeed, in this way one would loose the possibility of relative motions between blocks which is a convenient way of dissipating energy under dynamic excitation.

The implementation of the material model, as resulting from the testing campaign, in a user-subroutine written for a selected general purpose finite element code, provides eventually a design tool for such a device.

ACKNOWLEDGEMENT

This paper reports some results of an experimental research activity carried out within the national research program COFIN'01 with Professor Fabrizio Davì acting as national coordinator. The authors acknowledge the cooperation in this activity of Mr. Danilo Miozzari.

The monument retrofitting study is part of a research activity supported by the European Project CHIME (ICA3-1999-00006), with Professor F. Casciati as coordinator

REFERENCES

[1] Auricchio F., Faravelli L., Magonette G. and Torra V. (eds.), 2001, *Shape Memory Alloy. Advances in Modelling and Applications*, CIMNE, Barcelona, Spain.

[2] Torra V. (ed.), 2001, *Proceedings of the Workshop Trends on Shape Memory Behavior. The guaranteed long time SMA*, CIRG-DFA-UPC, Barcelona, Spain.

[3] Casciati, F., El Attar, A., Casciati, S., 2001, "Conceiving Semiactive Control Devices for Large Size Monolithic Monuments", in Liu S.C. (ed.), Smart Systems for Bridges, Structures and Highways, Proceedings of SPIE , Newport Beach, USA, p. 212-217.

[4] Casciati, S, 2001, *"Analisi di pericolosità, fragilità sismica ed ipotesi di adeguamento per uno dei Colossi di Memnone"* (in Italian), Master Thesis, Dept. of Structural Mechanics, University of Pavia, Italy.

[5] Faravelli, L., 2002, Experimental Approach to the Dynamic Behavior of SMA in Their Martensitic Phase, in F. Casciati (ed.), *Proceedings 3rd World Conference on Structural Control*, John Wiley & Sons, Chichester, UK

[6] MARC, 1994, *Program User Manual*.

[7] Casciati F., Faravelli L., 1997, Coupling SMA and Steel in Seismic Control Devices, *Proc. Of Saint-Venant Symposium*, ECPC, Paris, 355-362.

OVERALL DESIGN AND SIMULATION OF SMART STRUCTURES

ULRICH GABBERT, HEINZ KÖPPE, FALKO SEEGER,
TAMARA NESTOROVIĆ TRAJKOV

*Institut für Mechanik, Otto-von-Guericke-Universität Magdeburg,
Universitätsplatz 2, 39106 Magdeburg, Germany
E-Mail: ulrich.gabbert@mb.uni-magdeburg.de*

1. INTRODUCTION

Increasing engineering activities in the development and industrial application of piezoelectric smart structures require effective and reliable simulation methods and design tools [2], [9], [14]. In our opinion, the finite element method (FEM) is an excellent approach to develop a tool, which meets the relevant engineering requirements. Recently, an overall simulation and design tool for piezoelectric controlled smart structures has been developed, containing both a suitable controller design and the appropriate actuator/sensor placement. To model complex engineering structures, a library of multi-field finite elements has been developed with coupled electric and mechanical degrees of freedom for 1D, 2D and 3D continua as well as for thin-walled layered structures [4].

The quality of a smart structure decisively depends on the amount, the shape and the distribution of the active material across the passive structure. The optimal design of the actuator/sensor configuration is a very complex problem, which has not yet been fully solved to date. First, we developed an approach to describe the positions of the actuators and/or sensors by means of discrete (0-1) variables, and to include the parameters of the control law as continuous variables. However, long and time-consuming computational procedures are required to determine an optimal number and position of actuators by solving a discrete-continuous optimization problem. Hence, this paper discusses alternative concepts based on controllability and observability indices.

The paper mainly focusses on the relationship between the controller design and the multi-physical structure consisting of the base passive structure with actuators, sensors and also the control electronics. For designing a general-type controller the well-known Matlab/Simulink software package is used, which is coupled with the finite element tool via a data interface. However, the controller may be also included in the finite element tool to test it under realistic conditions.

As a benchmark example the active vibration suppression of an excited plate structure is discussed in detail to demonstrate the applicability of our finite element based overall design and simulation software. Thereafter, two industrial applications are mentioned, where our overall approach was used to design and analyze structures.

2. FINITE ELEMENT ANALYSIS OF SMART STRUCTURES

The semi-discrete form of the equations of motion of any finite element can be derived by using the well established approximation method of displacements and electric potential and following the standard finite element procedure (see [1], [5], [7]). This approach has been used to develop a comprehensive library of multi-field finite elements (1D, 2D, 3D elements, thick and thin layered composite shell elements, etc.) and suitable numerical methods to simulate the static and dynamic structural behavior of smart structures. [1]. For thin shell applications curved multilayer shell elements developed on the basis of the classical *SemiLoof* elements [8],[12] are perfectly suited. Following the Kirchhoff-Love hypothesis three different approaches were developed including electromechanical coupling effects within these thin shell elements: i) the electric influence is included in terms of distributed forces and moments, ii) the difference between the electric potential at the top and the bottom of each active layer was considered as an additional element degree of freedom (poling in normal direction), and iii) each element node has as many additional electric degrees of freedom as there are active layers in the composite (in-plane electric poling) [4],[12]. As revealed by test results, these elements are much more accurate and efficient in thin shell applications compared with 3D elements. In cases where temperature effects also play a role, a solution to the fully coupled thermo-electro-mechanical three-field problem is required, including heat generation caused by electric resistance. As turned out, an iterative strategy is the most efficient computational technique to include the temperature effect as two sets of equations, i.e. (i) the heat conduction equations and (ii) the electromechanical equations are solved separately for each time step, while considering the coupling terms as force vectors on the right hand side of the corresponding equations [7].

3. CONTROL OF SMART STRUCTURES

If a given structure is approximated by any number of finite elements considering also controller influences, the semi-discrete form of the equations of motion is obtained as

$$\mathbf{M}\ddot{\mathbf{q}} + \mathbf{D}_d\dot{\mathbf{q}} + \mathbf{K}\mathbf{q} = \mathbf{F} = \mathbf{F}_E + \mathbf{F}_C = \overline{\mathbf{E}}\mathbf{f}(t) + \overline{\mathbf{B}}\mathbf{u}(t), \tag{1}$$

where **M**, **D**$_d$ and **K** are the mass matrix, the damping matrix and the stiffness matrix, respectively, **f**(*t*) is the vector of external disturbances, and **u**(*t*) is the vector of the controller influence on the structure. The total load vector **F** is divided into the vector of the external forces $\mathbf{F}_E = \overline{\mathbf{E}}\mathbf{f}(t)$ and the vector of the control forces $\mathbf{F}_C = \overline{\mathbf{B}}\mathbf{u}(t)$, where the forces are generalized quantities including also electric charges or electric potentials. The matrices $\overline{\mathbf{E}}$ and $\overline{\mathbf{B}}$ describe the positions of the forces and the control parameters in the finite element structure, respectively.

In terms of the control theory, the state space form of the equations of motion is more convenient. The generalized displacements and the velocities are chosen as state variables. With the state vector $\mathbf{z}^T = \begin{bmatrix} \mathbf{q}^T & \dot{\mathbf{q}}^T \end{bmatrix}$ the equation (1) can be written as

$$\dot{\mathbf{z}} = \begin{bmatrix} 0 & \mathbf{I} \\ -\mathbf{M}^{-1}\mathbf{K} & -\mathbf{M}^{-1}\mathbf{D} \end{bmatrix} \mathbf{z} + \begin{bmatrix} 0 \\ \mathbf{M}^{-1}\overline{\mathbf{B}} \end{bmatrix} \mathbf{u}(t) + \begin{bmatrix} 0 \\ \mathbf{M}^{-1}\overline{\mathbf{E}} \end{bmatrix} \mathbf{f}(t) = \mathbf{A}\mathbf{z}(t) + \mathbf{B}\mathbf{u}(t) + \mathbf{E}\mathbf{f}(t). \tag{2}$$

In smart structures some of the active materials are used as sensors, which are able to sense components of the state vector, e.g. the electric potential; however, a variety of other signals, such as exciting forces, displacements and accelerations can be also measured in order to characterize the state of the structure. Based on these data the measurement equation can be written in a general form as

$$\mathbf{y} = \mathbf{C}\mathbf{z} + \mathbf{D}\mathbf{u}(t) + \mathbf{F}\mathbf{f}(t), \tag{3}$$

where **C**, **D**, **F** are mapping matrices describing the relations between the measured quantities of the vectors **z**, **u** and **f** with respect to the measurement vector **y**.

3.1. Controllability/observability index

The efficiency of a smart structure decisively depends on the amount and distribution of active materials in the passive structure as well as on the controller design. Of course, the actuator and sensor locations are critical for the controllability and the observability of a controlled structure and exert a major influence on the efficiency of the control system and the required control effort to satisfy a given design criterion. The placement of actuators is one of the main problems in designing adaptive structures. Difficulties arise from the relationship between the position of the actuators and the control algorithm used. On the other hand, the position and the mode of operation of the actuators also influence the dynamic behavior of the structure itself, which needs to be considered. In static cases the situation is much simpler, and in particular in truss structures the problem of actuator placement is well investigated [11]. In the distributed control of continua (e.g. plate and shell structures attached with piezoelectric wafers) the estimation of an optimal actuator and sensor shape as well as their placement is a very complex problem and has not yet been fully solved. Over the past few years two Ph.D. theses have been written in the research group of the authors of this paper; these theses focused on the automatic design of smart structures while considering the actuator/sensor placement and the controller design, and to an extent also the structural design [15], [11]. A combined structure/control design optimization problem was formulated, where discrete (0,1) variables describe the actuator/sensor positions and continuous variables describe the controller parameters. The non-linear discrete-continuous programming problem was solved iteratively by employing gradient-based methods for the continuous subproblem and an explicit enumeration for the discrete subproblem [6], [10]. However, the computational effort of such a method is quite high and the software developed was suited only to solve very simple truss structures, frames and beam structures. Hence, simpler techniques were investigated to design complex smart structures with a good overall control behavior in a given frequency domain. Obviously, the placement of actuators and sensors must guarantee the controllability and observability of a structure. To evaluate a given structure, the linear eigenvalue problem $(\mathbf{K} - \lambda_k \mathbf{M})\boldsymbol{\varphi}_k = \mathbf{0}$ has to be solved first, which results in the ($n \times r$) modal matrix $\boldsymbol{\Phi} = [\boldsymbol{\varphi}_1 \mid \boldsymbol{\varphi}_2 \mid \cdots \mid \boldsymbol{\varphi}_r]$ and the ($r \times r$) spectral matrix $\boldsymbol{\Lambda} = diag(\lambda_k)$, where $\boldsymbol{\Phi}$ is ortho-normalized with $\boldsymbol{\Phi}^T \mathbf{M} \boldsymbol{\Phi} = \mathbf{I}$ and $\boldsymbol{\Phi}^T \mathbf{K} \boldsymbol{\Phi} = \boldsymbol{\Lambda}$. Inserting the modal coordinates $\mathbf{q} = \boldsymbol{\Phi} \mathbf{x}$ into Eq. (1) the modal truncated system of ($r \times r$) differential equations is obtained as

$$\ddot{\mathbf{x}} + \boldsymbol{\Delta} \dot{\mathbf{x}} + \boldsymbol{\Lambda} \mathbf{x} = \boldsymbol{\Phi}^T \overline{\mathbf{E}} \mathbf{f}(t) + \boldsymbol{\Phi}^T \overline{\mathbf{B}} \mathbf{u}(t), \tag{4}$$

where $\boldsymbol{\Delta}$ is the modal damping matrix. The controllability index of the kth natural mode (see [13]) is given by

$$\mu_k = \boldsymbol{\varphi}_k^T \overline{\mathbf{B}} \, \overline{\mathbf{B}}^T \boldsymbol{\varphi}_k. \tag{5}$$

A system is completely controllable if $\mu_k > 0$, $\forall k$, where μ_k should be as high as possible. From a practical point of view only some of the modes need to be selected for control purposes. Similarly, also an observability index can be calculated. These indices can only be evaluated if actuators have been placed and the stiffness matrix and the mass matrix as well as the electric degrees of freedom of the active structural components are included in the finite element model.

From a computational point of view a simulation tool is required for the design process to perform efficient calculations of the controlled behavior of a structure, where it should be possible to automatically place and shift actuators or sensors across the base structure without any restrictions. This can be achieved by a remeshing technique [12]. If an active patch is to be placed at any position and in any orientation on the base structure, the meshing algorithm removes the mesh in the surrounding of the active part and fits a new adapted mesh at this position. The new automatic meshing technique was implemented in our finite element package for smart structures [12] and facilitates shifting the actuator and sensor patches continuously across the structure, with the mesh automatically adapted to the new positions.

Designing a structure, we start with the assumption that the specification of the structure itself is known, including the objective of the controlled behavior, external disturbances, the frequency range, etc. Then, the amount and the positions of the required actuators and sensors are estimated by using the controllability and observability indices Eq. (5). Based on the results of an eigenvalue analysis at each structural point, the modal strains, and consequently, the modal electric voltage are calculated yielding the controllability index $\mu_k(\mathbf{x}_P)$ of the kth mode at the structural point position $\mathbf{x}_P$. The best positions $\mathbf{x}_P$ to control the first r eigenmodes are the positions with the highest overall controllability index $\mu(\mathbf{x}_P) = \prod_{k=1}^{r} \mu_k(\mathbf{x}_P)$. In the second step these preselected positions are modified through a sensitivity design process taking into account the placed actuators and sensors. Finally, the controlled structural behavior is examined and may be used for further improving the actuator and sensor distribution on the basis of an overall simulation model including the controller design. To this end, our finite element simulation software needs to be linked with controller design tools.

3.2. Coupling of finite element analysis and control

To facilitate a numerical simulation of smart structures within the finite element frame, an overall finite element model which comprises the passive structure, the active sensor and actuator elements as well as an appropriate controller is required. Today, comprehensive design tools, such as Matlab/Simulink, are available to support the design process. To exchange data and information between the finite element model and the controller design tool, a general data exchange interface is required. We have coupled our finite element software COSAR with MatLab/Simulink by means of such bi-directional data interface [5].

Large-scale finite element models are not suitable for controller design purposes. Hence, a suitable method is needed to reduce the number of *dof's* of the finite element model. The modal truncation method turned out to be an appropriate approach.

To design the controller, the matrices **A, B, E, C, D, F** of the state space model are transferred to Matlab/Simulink via the data interface. The controller can be directly incorporated in a dSPACE system, enabling the designer to work in a *hardware-in-the-loop* configuration to test and modify the designed controller in real experiments. Before such experiments are performed, the structural behavior may be also tested on a virtual computer model of the structure, obtained from the original finite element model. Subsequently, the controller can be retransferred to the finite element software via the data interface. In an *LTI* system the designed controller matrix **R** is used to generate the actuator signal as $\mathbf{u}(t) = \mathbf{Rz}(t)$. Alternatively, the controller may be also directly incorporated in our finite element software tool as a C-code subroutine resulting from Matlab/Simulink.

3.3. Discrete-time controller design with additional dynamics

A state-space model of the controlled plant, obtained from the finite element analysis and modal reduction can be used for designing the controller. The starting point for the controller design is a state-space model, i.e. Eqs. (2), (3), assuming the presence of a disturbance vector **f** in the state and output equations in a general case. The controller design is performed to suppress disturbance-induced vibrations, and should include an *a priori* knowledge about the occurring disturbance type.

A discrete-time controller based on the optimal LQ tracking system with additional dynamics shows a good behavior. Additional dynamics features are provided to compensate for the presence of disturbances and ensure tracking of the reference input with the prescribed magnitude and frequency for suppressing vibrations. Such controller with additional dynamics features serves controlling purposes if the reference input to be tracked and the disturbance acting upon the structure can be described by a rational discrete function. This condition is fulfilled by a sine function used as a disturbance model.

The additional dynamics features are determined in a state-space form on the basis of disturbance and/or reference input poles λ_i. For a sinusoidal disturbance or a reference input with the radian frequency $\omega_0=2\pi f_0$, the poles are $\lambda_i=\pm j\omega_0$, whereas for a step reference input (or disturbance) the pole equals zero. The fact that the disturbance and reference input poles appear in the closed-loop system as poles of the compensator, provides robustness to disturbances and model inaccuracies in the steady-state. With defined additional dynamics poles both the reference input and disturbances are considered, i.e. a closed-loop tracking system, which is able to reject certain disturbances (defined by additional dynamics features), is also able to track the reference input described by the same additional dynamics features, and vice versa.

The feedback gain matrix **L** of the optimal *LQ* regulator is calculated on the basis of a design model such that the feedback law $\mathbf{u}[k]= -\mathbf{L}\mathbf{x}_d[k]$ minimizes the performance index

$$J = \frac{1}{2}\sum_{k=0}^{\infty}(\mathbf{x}_d[k]^T \mathbf{Q}\mathbf{x}_d[k] + \mathbf{u}[k]^T \mathbf{R}\mathbf{u}[k]) \tag{6}$$

subject to the constraint equation (2) in the discrete-time form, where **Q** and **R** are symmetric, positive-definite matrices. Thereafter, the feedback gain matrix **L** is partitioned into the submatrices $\mathbf{L}_1$ and $\mathbf{L}_2$ corresponding to the plant and the additional dynamics, respectively.

The following chapter presents the design of a smart plate structure. This test structure is one of the reference examples to evaluate the theoretical and numerical techniques developed.

4. OVERALL DESIGN AND SIMULATION - EXAMPLES

4.1. Design of a smart plate structure

The objective of this example is to demonstrate the active reduction of structural vibrations under varying excitation conditions. Various piezoelectric patches serving as sensors and/or actuators were glued to the top of the clamped plate structure (see Fig. 1). The arrangement of these actuator/sensor pairs was determined by employing Eq. (5) to yield a good overall damping behavior under the first seven bending eigenmodes in the controller design. A configuration of the controller design with four actuators and four sensors is shown in Fig. 1. The dimensions of the plate are 900x600x1.75 mm. The active elements consist of the piezoceramic material PIC151, 0.3 mm thick, 50 mm in length and 32 mm in width.

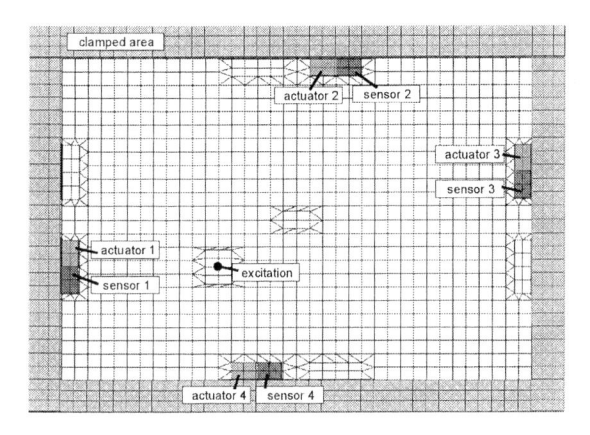

Figure 1: Plate with sensor and actuator configuration (eigenfrequencies [Hz]: 29.70 / 45.03 / 71.63 / 73.7 / 88.39 / 108.95 / 108.95 / 113.86 /.

For analytical purposes it is assumed that there are two different time-dependent forces. First, an excitation consisting of any superposition of four harmonics in the frequency range from 29.7 Hz to 73.74 Hz is investigated and second, a Dirac impulse excitation is studied. However, only the results of the sine-type excitations are reported here.

First, we calculated the eigenvalues and eigenmodes of the passive plate structure following the finite element method. To this end, we used a finite element mesh consisting of 1490 *SemiLoof* shell type elements. On the basis of the calculated eigenmodes we developed the state space model for the controller design in Matlab/Simulink by performing a modal reduction of the finite element model. The data required for the state space model were exported into the Matlab/Simulink software system via the data exchange interface.

First, an *LQR* controller (co) was designed and tested following a common procedure and thereafter a controller (cwdm) was used while taking into account the excitation or disturbance input, referred to as *Digital Tracking System Design* [16], [17]. Although such disturbances are usually unknown, there are cases in which their characteristics or statistical features are known and can be included.

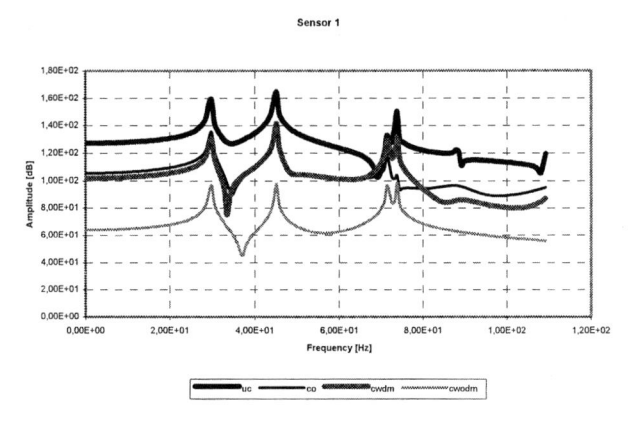

Fig. 2 Frequency response of sensor 1

If the disturbances are known functions, such as constants, ramps, exponentials or sinusoidal sequences with unknown amplitudes, the disturbance model can be considered as a part of the plant model. Under the assumption that vibrations due to the sinusoidal disturbance of any of the seven given frequencies should be suppressed, this additional information can be used for the controller design in a third test case (cwodm). With this *a priori* knowledge the controller was designed with additional dynamics features including the conjugate complex poles $\pm j(2\pi f_i)$ of the seven sinusoidal disturbances. After replicating additional dynamics four times (corresponding to four outputs) and forming a cascade combination with the plant model, the design model of the 70th order is obtained. The weighting matrices **Q** and **R** are of the 70th and the fourth order respectively. Employing different combinations of the weighting matrices yields different controller designs applicable to specific sine-type disturbances with frequencies corresponding to the seven eigenfrequencies of interest as well as to disturbances consisting of any superposition of the harmonics in the frequency range of interest. The investigations presented have been verified by the simulation results so far. Experimental verifications of the plate structure employing dSPACE system are under progress as well.

4.2 Brief description of two industrial applications

The methodology described above has been successfully applied to designing smart structures in the engineering practice. The following sections describe two examples.

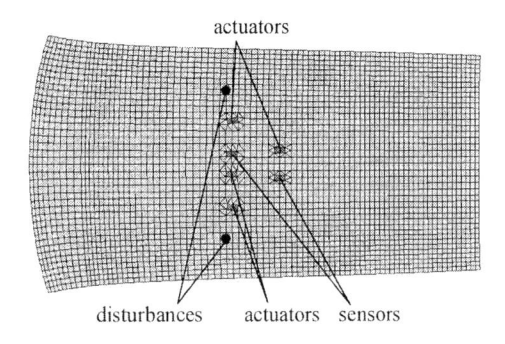

Figure 3: Actuator/sensor positions at a car roof

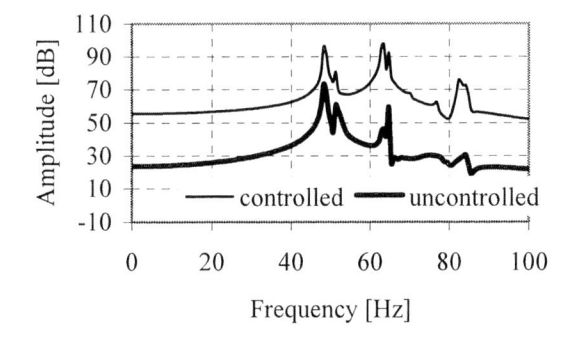

Figure 4: Frequency response to a sinusoidal excitation

The first example refers to a car roof, which can be characterized as a curved shell type structure. The task is to perform computational investigations to develop an optimal actuator/sensor design, which results in an active reduction of noise and vibration originating from the car roof. The final objective is to improve the driver's and passengers' comfort. This task, which is solved in close cooperation with VOLKSWAGEN (VW), is one of the engineering test cases under *Leitprojekt Adaptronik* – a research project financially supported by the German Federal Ministry of Education and Research. A series of actuators is to be mounted to the inner surface of the car roof to reduce the vibrations by means of a suitable control system. Fig. 3 shows the actuator and sensor distribution in one of the test cases, which were calculated based on the observability and controllability indices Eq. (5). Fig. 4 depicts the frequency response functions of the controlled and uncontrolled behavior of the car roof excited with a sinusoidal force [12].

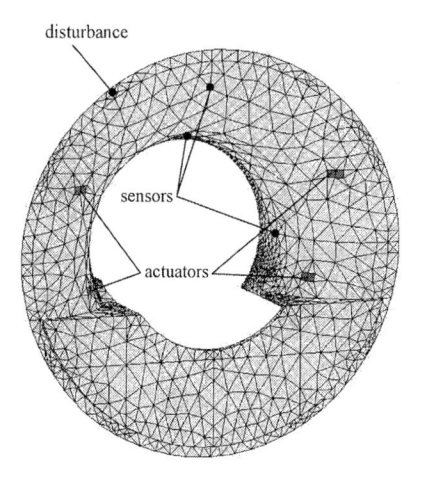

Figure 5: The funnel – a part of a MRT

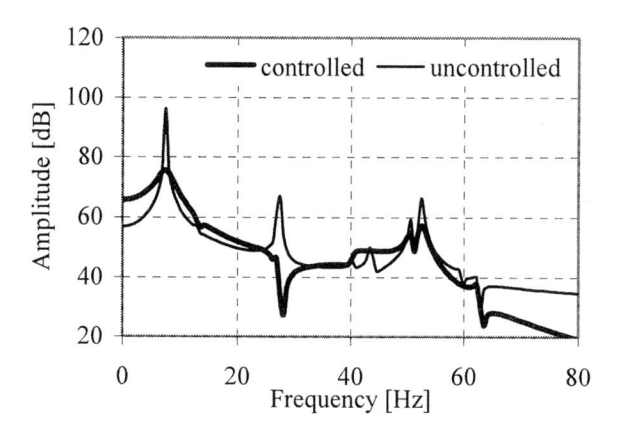

Figure 6: Frequency response to a sinusoidal excitation

The second example consists of another thin-walled structure, which is part of magnetic resonance tomographs (MRT) made by SIEMENS. This test case is also part of the *Leitprojekt Adaptronik* project. The structure we are investigating is a funnel-shaped component made of a synthetic material. Its shape is rather complex with a variable thickness. The tests are aimed at reducing the noise for patients undergoing a magnetic resonance tomograph examination. For the test purposes only four actuators made of piezoelectric ceramics are arranged on one side of the machine. For a simpler verification of the results by experimental investigations the funnel is not fixed by any boundary conditions influencing the frequency range of interest. Fig. 5 presents the positions of actuator and sensor as well as points at which the structure is excited. The actuator and sensor positions were calculated again on the basis of controllability and observability indices. For test purposes several different excitations were investigated, where in each of the test cases an optimal *LQR* controller was designed on the basis of a reduced state space model of the funnel. In Fig. 6 the results of a sinusoidal force excitation without and with control are shown.

5. CONCLUSIONS

The paper presents a general finite element based overall simulation and design tool for piezoelectric controlled smart structures. This tool can be used for (i) simulating both static and dynamic structural problems, (ii) designing controllers using Matlab/Simulink on the basis of a general bilateral data exchange interface between our finite element analysis software COSAR [3] and the controller design tool Matlab/Simulink, and (iii) simulating the controlled structural behavior.

A simple controllability index was developed and successfully tested to determine the arrangement of actuators and sensors at engineering smart structures. This design can be improved by a sensitivity approach taking into account e.g. the transfer behavior of the structure. As a benchmark example to test such new developments a simply supported plate structure is proposed where an analytical solution is available to verify the finite element based numerical optimization techniques [18].

Large-scale finite element models can not be used for controller design purposes. To this end, an appropriate model reduction is required. We consider modal truncation the best method for the controller design based on finite element models. We used the well-established Matlab/Simulink software package for designing and testing our controllers as it provides a wide range of special design tools and algorithms.

A reference example yielding an active vibration suppression of an excited plate structure was discussed to demonstrate the applicability of our finite element based overall design and simulation software. Two different *LQR* controllers were investigated - (a) a common *LQR* controller, and (b) a *Digital Tracking System Design* controller responding also to the excitation or disturbance input. Finally, two industrial applications were briefly described, where our overall approach was followed to design and to analyze the structures.

ACKNOWLEDGEMENT

This work is a part of the *Leitprojekt ADAPTRONIK* supported by the German Federal Ministry of Education and Research. This support is gratefully acknowledged.

REFERENCES

1. Berger, H., Gabbert, U., Köppe, H., Seeger, F. (2000): Finite Element Analysis and Design of Piezoelectric Controlled Smart Structures. *Journal of Theoretical and Applied Mechanics*, 3, 38, pp. 475-498.
2. Chee, C. Y., Tong, L., Steven, G. P. (1998): A Review on the Modelling of Piezoelectric Sensors and Actuators Incorporated in Intelligent Structures, *J. of Intelligent Material Systems and Structures*, Vol. 9, pp. 3-19.
3. COSAR - General Purpose Finite Element Package: *Manual*, FEMCOS GmbH Magdeburg (see also: http://www.femcos.de).
4. Gabbert, U., Köppe, H., Fuchs, K., Seeger, F. (2000): Modeling of Smart Composites Controlled by Thin Piezoelectric Fibers, in Varadan, V.V. (Ed.): *Mathematics and Control in Smart Structures*, SPIE Proceedings Series, Vol. 3984, pp. 2-11.
5. Gabbert, U., Köppe, H., Seeger, F. (2001): Overall design of actively controlled smart structures by the finite element method, SPIE Proceedings Series, Vol. 4326, 2001, 113-122
6. Gabbert, U., Weber, Ch.-T. (1999): Optimization of Piezoelectric Material Distribution in Smart Structures, in Varadan,V.V. (Ed.): *Mathematics and Control in Smart Structures*, SPIE Proceedings Series, Vol. 3667, pp. 13-22.
7. Görnandt, A., Gabbert, U. (2002): Finite Element Analysis of Thermopiezoelectric Smart Structures. *Acta Mechanica*, 154, 129-140 (2002)
8. Irons, B. M. (1976): The Semiloof shell element, in Ashwell D. G. and Gallagher R. H. (Eds.): *Finite Elements for Thin Shells and Curved Members*, J. Wiley, London.
9. Janocha, H. (1999): *Adaptronics and Smart Structures - Basics, Materials, Design, and Applications*, Springer Berlin.
10. Schulz, I., Gabbert, U. (2000): Automatic design of smart structures using frequency domain methods, *Journal of Structural and Multidisciplinary Optimization* (submitted).
11. Schulz, I. (2001): *Automatischer Entwurf adaptiver mechanischer Systeme im Frequenzbereich*, Dissertation Uni. Magdeburg, VDI-Fortschritt-Berichte, Reihe 11 (Schwingungstechnik), VDI-Verlag, Düsseldorf, 2001.
12. Seeger, F., Gabbert, U., Köppe, H., Fuchs, K. (2002): Analysis and Design of Thin-Walled Smart Structures in Industrial Applications, SPIE's 9th Annual International Symposium on *Smart Structures and Materials*, Conference on *Industrial and Commercial Applications of Smart Structures Technologies*, 17-21 March 2002, San Diego, CA, paper No. 4698-38.
13. Sepulveda, A.E., Schmidt, L.A. (1991): Optimal Placement of Actuators and Sensors in Control-augmented Structural Optimization, *Int. J. for Num. Meth. in Eng.*, Vol. 32, 1165-1187.
14. Tzou, H.-S., Guran, A. (Eds.), (1998): *Structronic Systems: Smart Structures, Devices and Systems*, Part 1: *Materials and Structures*, Part 2: *Systems and Control*, World Scientific.
15. Weber,Ch.-T. (1998): *Ein Beitrag zur optimalen Positionierung von Aktoren in adaptiven mechanischen Strukturen*. Dissertation Uni. Magdeburg, VDI-Fortschritt-Berichte, Reihe 11 (Schwingungstechnik), Nr. 265,VDI-Verlag, Düsseldorf.
16. Santina, M., Stubberud, A., Hostetter, G. (1994): *Digital Control System Design*, Saunders College Publishing, a Harcourt Brace College Publisher.
17. Föllinger, O. (1994): *Regelungstechnik - Einführung in die Methoden und ihre Anwendung*, Hüthig GmbH Heidelberg.
18. Seeger, F. (2002): Optimal placement of distributed actuators and sensors in a controlled system for a elastic plate, *PAMM - Proceedings in Applied Mathematics and Mechanics*, GAMM Conference Augsburg 2002, Volume 2 (2002) Number 1 – 3 (submitted)

BIO-MIMETIC SMART MICROSTRUCTURES

Attachment Devices in Insects as a Possible Source for Technical Design

STANISLAV N. GORB
Biological Microtribology Group, Max-Planck-Institute of Developmental Biology, Spemannstr. 35, D-72076, Tübingen, Germany, stas.gorb@tuebingen.mpg.de

Evolutionary Biomaterials Group, Max-Planck-Institute of Metals Research, Heisenbergstr. 3, D-70569, Stuttgart, Germany

1. FUNCTIONAL SOLUTIONS FROM BIOLOGY

The theory of *inventive problem solving* in engineering is based on three main principles: (1) all patents are based on about 40 inventive principles; (2) all technology trends are predictable; (3) important inventions come from outside the industry within which they are applied. During evolution, nature has constantly been called upon to act as an engineer in solving technical problems [1]. Organisms have evolved an immense diversity of shapes and structures. Many functional solutions are based on a variety of ingenious structural solutions. There is numerous biomechanical systems in insects adapted for attachment of parts of the body to each other or for attaching the organism to a substrate [2; 3]. Understanding these systems is of major scientific interest, since we can learn about their use as structural elements and their biological role and function. This knowledge is also highly relevant for technical applications [4].

2. SURFACE MICROSTRUCTURES ADAPTED FOR ATTACHMENT FUNCTION

One of the greatest challenges for today's engineering science is miniaturization. During their evolution, insects and other animals have solved many problems related to extremely small body size. There are three main areas, where nature's solutions of attachment problems may be applied: (1) precise mechanics, (2) gluing technology, and (3) material science of surface-active composite materials. Possible innovations may also appear at the boundaries of the named areas.

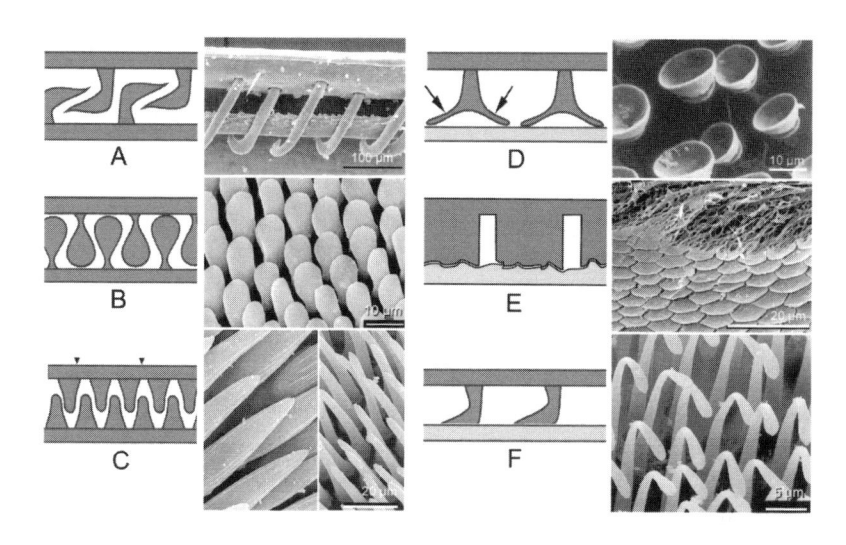

Figure 1. Attachment microstructures of the insect body based on different principles. Left panels show diagrams explaining principles of attachment in structures shown in right panels. A. Hooks of the wing locking mechanism of the sawfly Cimbex femoratus. *B. Lock of the head in the damselfly* Pyrrhosoma nymphula. *C. Friction-active structures of the head-arresting mechanism of the dragonfly* Aeshna mixta. *D. Microsuckers of the first legs in the males of the beetle* Dytiscus marginatus. *E. Soft pads of the grasshopper* Tettigonia viridissima. *F. Hairy pads with anisotropic terminal elements on legs of the beetle* Rhagonycha fulva. *Dark-gray is an animal body; light gray indicates a substrate.*

The presence of fields of specialized *protuberances* or *outgrowths* on an insect body may result in an increase of the frictional and/or adhesive forces in the region of contact with the corresponding surface or with a variety of surfaces (Figure 1). We studied a broad variety of mechanical systems of the insect cuticle adapted for attachment [3]. Systems, such as *leg pads, head-arresting systems, wing-to-body locking devices,* and *intersegmental frictional areas* of leg articulations contain surfaces evolved to fix parts of the body to each other, or to attach an animal to the substratum. Attachment provided by these systems is fast, precise and reversible.

For various cases of contact pairs in biology, anti-friction systems always have a predefined pair of surfaces, whereas among friction systems there are some that deal with predefined surfaces and others, in which one surface remains unpredictable [5]. The first type of friction system occurs, for example in wing-locking devices and head-arresting systems and is called *probabilistic fasteners*. The second type is mainly represented by insect attachment pads of two alternative designs: hairy and smooth [6]. Relationship between surface patterns and/or mechanical properties of materials of contact pairs results in two main working principles of frictional devices: mechanical interlocking and maximization of the contact area. In the present paper, we give a short overview of the functional design of two main groups of friction-based attachment devices in insects: probabilistic fasteners and smooth attachment pads.

3. PROBABILISTIC FASTENERS

Probabilistic fasteners are composed of two functionally corresponding surfaces covered with cuticular outgrowths, such as *setae, acanthae* or *microtrichia* [7]. These fasteners are called probabilistic [2], because the interlocking takes place without precise positioning of both surfaces. In this case, attachment is based on the use of the surface profile and mechanical properties of materials. The single outgrowths, which are called *elements*, should certainly not be designed as hooks in a fashion similar to those of *Velcro fasteners*. The mechanism of attachment in such systems also differs from the hook principle. Probabilistic fasteners with parabolic elements have been described in head-arresting systems [8], intersegmental fixators of leg joints [9], and wing-to-body locking devices [10; 11]. The most studied examples of the systems are wing-locking mechanisms in beetles and the head arrester in dragonflies. An example of functional significance of such a pair of surfaces in the mechanical system of *the chain of mobile segments* is given below.

3.1. Frictional Systems in Insect Joints

In insects, the leg is a tube made of a *cuticle*, a natural composite material. It contains muscles that control parts of the tube, called *segments*. The material properties of various parts of the tube are completely different. The segments are, more or less, hard and stable [12]. The *intersegmental membranous area* is flexible and/or elastic [13]. The most fascinating aspect is that the tube is not interrupted between the segments, but consists of a continuum of various material properties. The design of insect joints varies from simple articulations with an endless number of rotation axes, to monoaxial joints with a singe rotation axis, and to complex structures with several precisely defined rotation axes. Usually, profiles of contacting surfaces of both segments are smooth within the joint. This results in decreasing friction within the joint during segment movements. However, many insect joints contain structured surfaces that are covered by cuticular protuberances or depressions (Fig. 2 D-F). Such surfaces exist not only in leg joints, but also in joints between segments of antenna, or at the base of large mobile spines.

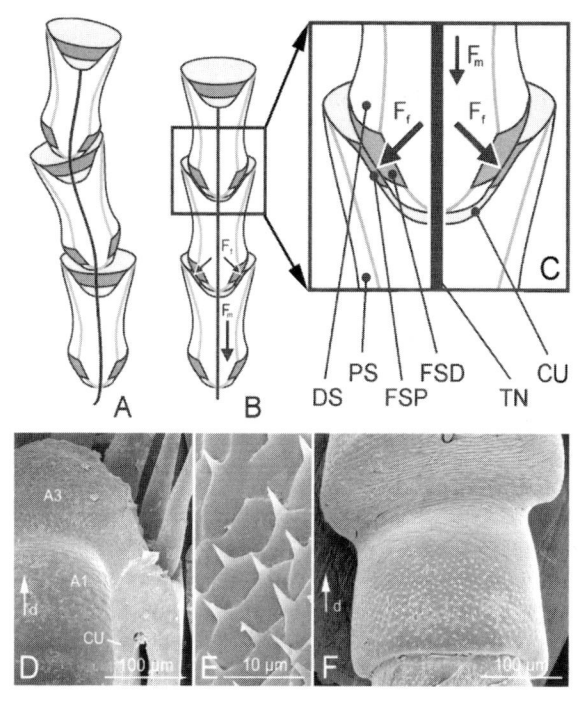

Figure 2. Multisegmental chain equipped with frictional areas. A. Mobile chain-like structure of two joints. B-C. After application of muscular force to the tendon running through the chain, the structure becomes stiff due to the interlocking of frictional surfaces at both ends of each segment. D-F. Examples of link-fixing systems in leg joints of the beetle Melolontha melolontha *(Scarabaeidae). D-E. Distal end of the segment, joint cup. F. Proximal end of the segment. A1, A3, rough areas covered by outgrowths; CU , joint cup; d, distal direction; DS, distal segment; F_f, frictional force; F_m, muscular force; FSD, frictional surfaces at the distal end; FSP, frictional surfaces at the proximal end; TN, tendon.*

The role of such frictional surfaces in insect joints can be explained with the example of the *tarsal chain* of the leg (Fig. 2 A). Neighboring segments of the tarsal chain lack muscles. They are connected to each other through a series of joints that resemble a telescopic structure, where the *proximal end* (located closer to the body) of the segment fits into the cup-like *distal end* (located far from the body) of the previous segment. The whole tarsus is driven by a single muscle, usually located elsewhere, and connected with the distal part of the tarsus (*pretarsus*) through a long tendon. When the muscle contracts, the entire tarsus becomes stiff due to joint-interlocking (Fig. 2 B). The structures responsible for this mechanism are fields of outgrowths, located on the cup surface at the distal end of the proximal segment (Figs. 2 D, E) and at the proximal end of the next distal segment (Fig. 2. F). Behavior of an ideal patterned pair of surfaces under unidirectional load is considered below with the aid of an artificial model.

3.2. Artificial Model of Probabilistic Fasteners

Force measurements on the artificial system, composed of two surfaces covered by *parabolic pins* (Fig. 3), show that the attachment force is strongly dependent on the load force (Fig. 3 B, C). At small loads, the increase of attachment is very slow, whereas rapid increase of attachment was detected at higher loads. At very high loads, a saturation of the attachment force was revealed. A simple explanation of the attachment principle follows: with an increasing load, elements of both surfaces slide into gaps of the corresponding part. This results in an increase of lateral loading forces acting on elements. High lateral forces lead to an increase of friction between single sliding elements.

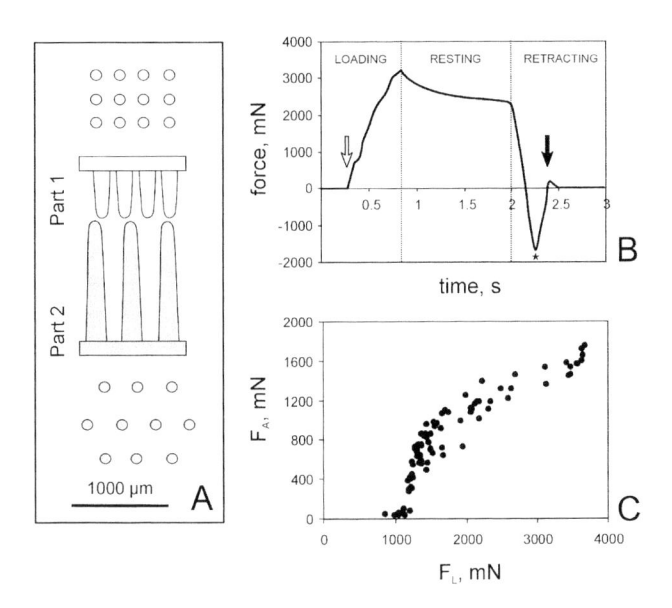

Figure 3. Results of force measurements on the dry system of artificial probabilistic fasteners. A. The model system used in the force measurements; circles above and below the scheme indicate distribution of single elements on each counterpart. B. An example of the force-time curve consisting of three main parts: loading, resting and retracting. White arrow indicates contact initiation; black arrow indicates contact breakage; asterisk indicates the beginning of element sliding. C. Dependence of the attachment force (F_A) on the load force (F_L).

The main feature of such a system is the existence of a critical compressive force needed to "interlock" the frictional fastener. After overcoming this critical value, the attachment force increases with the loading force. The attachment force has the same order of magnitude as the loading force needed to achieve interlocking. The attachment force is, however, always lower than the loading force and is of the same order of magnitude as the elastic force needed to deflect the fastener elastically in the horizontal direction to a distance equal to the diameter of the element tips. This feature can be used as an experimental test of the frictional nature of a fastener.

A theoretical model of probabilistic fasteners with parabolic elements shows that dependence of the attachment force on the loading force is sensitive to the shape of the element [14]. For example, in the case of cone-shaped elements, the attachment force is linearly proportional to the loading force and no critical interlocking force exists. The stronger the convexity of the basic curve of the rotating body of the element, the higher the critical interlocking force.

4. ATTACHMENT PADS WITH ANISOTROPIC FRICTION

4.1. Smooth Attachment Pads in Insects

Attachment pads of legs are structures which are able to attach to various surfaces. Such a property makes them interesting from a technical point of view. In their evolution, insects have developed two distinctly different mechanisms to attach themselves to a variety of substrates: *smooth pads* or *hairy surfaces*. Due to the surface structuring (hairy systems) or flexibility of the material (smooth systems) of the attachment structures, both mechanisms can maximize the possible contact area with the substrate, regardless of its macro- and microsculpture. Smooth pads bear a number of interesting properties, such as smooth appearance, frictional anisotropy, and rather high adhesion. Therefore structural and functional principles of the pads may be used for mimicking artificial materials and surfaces with similar properties.

4.2. Frictional Properties of the Grasshopper Pad

Frictional behavior of the pad of the grasshopper *Tettigonia viridissima* (Orthoptera, Tettigoniidae) was obtained by oscillating the sample over a distance of 10 μm along an x-axis (distal-proximal) in both directions (Fig. 4 A-C) [15]. The experiments revealed that the static friction during proximal movement was larger compared with distal movement (Fig. 4 D). The dependence of the friction force on load is given in Fig. 4 E. The ultrastructural study shows that the inner architecture of pads provides mechanical stability and, simultaneously, extreme flexibility [16]. This allows the pad material to adapt to different substrate roughness, which is unpredictable for mobile insects.

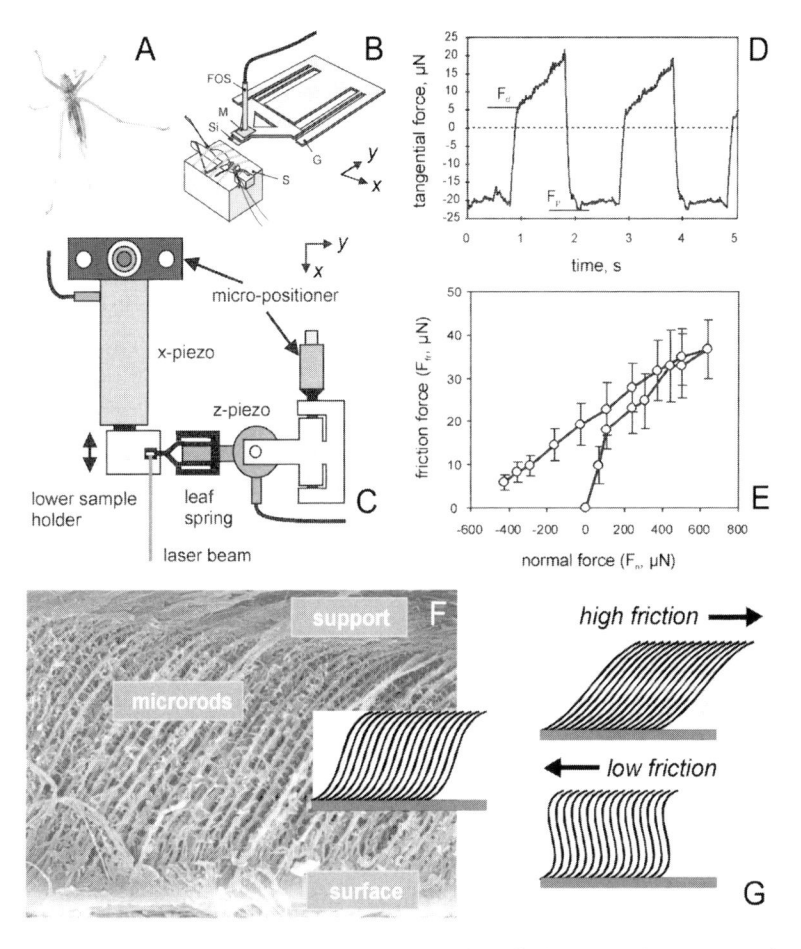

Figure 4. Friction of the attachment pads of the grasshopper Tettigonia viridissima. *A. Animal holding onto a vertical glass surface. B-C. Force-tester, in which the oscillatory motion is provided by an x-piezo. The pad is attached to the x-piezo. A silicon plate, attached to a glass spring, served as an upper sample. A laser beam, reflected by a mirror attached to the spring, was used to detect deflection of the spring. In the z-direction, a z-piezo is attached to adjust the normal force. D. Friction behavior of the pad in different directions. F_d, friction force to the distal direction of the pad; F_p, friction force to the proximal direction of the pad. E. Friction force vs normal force. The lower part of the curve was obtained with an increasing normal force, the upper part was recorded with a decreasing normal force after full contact between the pad and silicon surface was reached. F-G. Architecture of the pad material and schematic explanation of the frictional anisotropy.*

4.3. Material Structure

The cuticle of smooth pads consists of a natural friction-active material with a specific inner structure. Tiny *microrods* or *filaments* (0.1 μm in diameter) are located just under the *epicuticle* of the pad (Fig. 4 F). These filaments are branches of thicker microrods (1 μm in diameter), located more deeply in the cuticle [17;18]. It has been shown that these threads can change their shape under loads [16].

The key property of smooth attachment devices is deformability and the viscoelastic properties of the material. Profile changes of the surface of grasshopper pads and of the microrod orientation were visualized by means of scanning electron microscopy followed by freezing-substitution experiments [16]. The results show that the flexible pad material deforms replicating the substrate profile down to the micrometer roughness. The pad material showed both elastic and viscous behavior under loads. Elastic modulus of the pad is rather low (27.2 kPa).

Low stiffness of the material aids in surface replication and in an increase of the area of real contact between the pad and the underlying substrate. Microrod orientation in a distal direction is reflected in the frictional anisotropy of the material, because sloped microrods can more easily be recruited by the sliding movement in the direction opposite to the microrod slope (Fig. 4 F, G).

5. CAST TECHNIQUE AS THE FIRST STEP TOWARDS MICROFABRICATION TECHNOLOGY

Since many micro electro-mechanical devices (MEMS) become smaller and smaller, all systems known from the existing macroscale devices have to be miniaturized in different ways. Assembling of parts of the MEMS can be achieved by gluing parts together, but sometimes releasable attachment fasteners are also required. An engineering approach, applied after detailed studies on the natural system, would be most promising. However, engineers can also copy the surface shape of a variety of scales and materials using available technologies of chemistry and rapid prototyping. Both approaches could run parallel for some time and possibly converge later.

An overview of modern technologies which may be applied for prototyping of diverse surface microsculpture, is given elsewhere [19]. A simple technique using a two-component fluid silicon can be applied to produce negative surface casts of a living surface at room temperature. Positive casts can then be obtained in an epoxy resin. Fluid silicon, applied to a biological surface, spreads very quickly over the sample and fills surface irregularities (Fig. 5 A-B). After hardening (2-5 min), the biological sample can be easily removed without any damage because of the high elasticity of the hardened wax (Fig. 5 C-D). Then the wax casts are filled with Spurr's resin [15] or other low viscosity epoxy resin, which becomes relatively stiff after polymerization at 60-70°C for 24 h, and again can be removed from the «negative» without damage (Fig. 5 E-F). The «negative» replica can be re-used again to produce a number of resin «positives». This method provided good results with relatively long cuticle outgrowths (10-30 µm), as well as with small structures (0.02 µm) (Fig. 5 G-I).

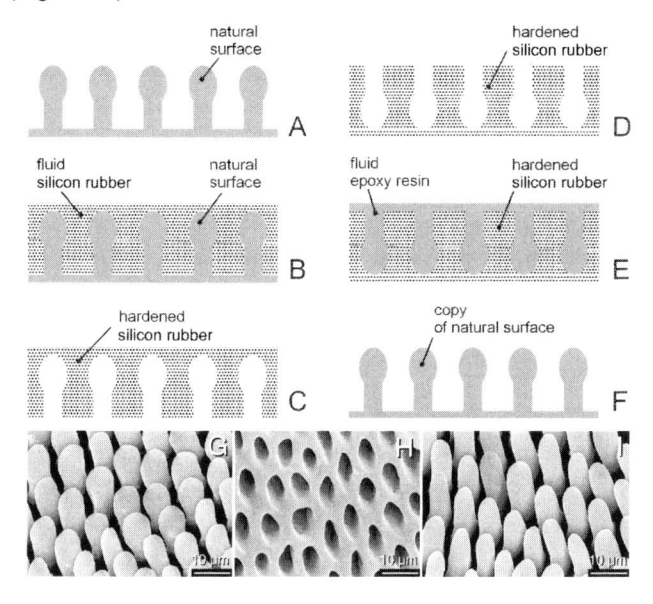

Figure 5. Dental wax cast method for surface replicas. A-F. Diagram of the method (explanations in text). G. Ventral microtrichia field of the head-arresting system of a damselfly, Pyrrhosoma nymphula, *air dried. H. «Negative» replica of the surface shown in G. I. «Positive» replica of the surface shown in G made by using the «negative» shown in H.*

By varying the resin composition, polymerized material can demonstrate a variety of material properties from very hard to quite soft. By using samples with the same shape and different mechanical properties of material, dependence of attachment forces on material properties at a given surface profile can be tested.

7. ACKNOWLEDGEMENTS

Victoria Kastner kindly provided linguistic corrections. This work is supported by the Federal Ministry of Education, Science and Technology, Germany to SNG (Project BioFuture 0311851).

8. REFERENCES

1. Nachtigall, W. (1998) *Bionik. Grundlagen und Beispiele für Ingenieure und Naturwissenschaftler*, Springer, Berlin et al.

2. Nachtigall, W. (1974) *Biological mechanisms of attachment*, Springer-Verlag, Berlin, Heidelberg, New York.

3. Gorb, S.N. (2001) *Attachment devices of insect cuticle*, Kluwer Academic Publishers, Dordrecht, Boston, London.

4. Scherge, M. and Gorb, S.N. (2000) Using biological principles to design MEMS, *J. Micromech. Microeng.* **10** 359-364.

5. Scherge, M. and Gorb, S.N. (2001) *Biological micro- and nanotribology*, Springer, Berlin et al.

6. Beutel, R. and Gorb, S.N. (2001) Ultrastructure of attachment specializations of hexapods (Arthropoda): evolutionary patterns inferred from a revised ordinal phylogeny, *Journal of Zoological Systematics and Evolutionary Research* **39** 177-207.

7. Richards, A.G. and Richards, P.A. (1979) The cuticular protuberances of insects, *Int. J. Insect Morphol. Embryol.* **8** 143-157.

8. Gorb, S.N. (1999) Evolution of the dragonfly head-arresting system, *Proc. Roy. Soc. London B* **266** 525-535.

9. Gorb, S.N. (1996) Design of insect unguitractor apparatus, *J.Morphol.* **230** 219-230.

10. Schrott, A. (1986) Vergleichende Morphologie und Ultrastruktur des Cenchrus-Dornenfeldapparates bei Pflanzenwespen (Insecta: Hymenoptera, Symphyta), *Berichte Naturwiss. Med. Ver. Innsbruck* **73** 159-168.

11. Gorb, S.N. (1998) Frictional surfaces of the elytra to body arresting mechanism in tenebrionid beetles (Coleoptera: Tenebrionidae): design of co-opted fields of microtrichia and cuticle ultrastructure, *Int. J. Insect Morphol. Embryol.* **27** 205-225.

12. Hepburn, H.R. and Chandler, H.D. (1976) Material properties of arthropod cuticles: the arthrodial membranes, *J. Comp. Physiol. A* **109** 177-198.

13. Hackman, R.H. and Goldberg, M. (1987) Comparative study of some expanding arthropod cuticles: the relation between composition, structure and function, *J. Insect Physiol.* **33** 39-50.

14. Gorb, S.N. and Popov, V.L. (2002) Probabilistic fasteners with parabolic elements: biological system, artificial model and theoretical considerations, *Philosophical Transactions: Mathematical, Physical and Engineering Sciences* **360** 221-226.

15. Gorb, S.N. and Scherge, M. (2000) Biological microtribology: anisotropy in frictional forces of orthopteran attachment pads reflects the ultrastructure of a highly deformable material, *Proc. Roy. Soc. London B* **267** 1239-1244.

16. Gorb, S.N., Jiao, Y. and Scherge, M. (2000) Ultrastructural architecture and mechanical properties of attachment pads in *Tettigonia viridissima* (Orthoptera Tettigoniidae), *J. Comp. Physiol. A* **186** 821-831.

17. Kendall, U.D. (1970) The anatomy of the tarsi of *Schistocerca gregaria* Forskål, *Z. Zellforsch.* **109** 112-137.

18. Henning, B. (1974) Morphologie und Histologie der Tarsen von *Tettigonia viridissima* L. (Orthoptera, Ensifera), *Z. Morphol. Tiere* **79** 323-342.

19. Whitesides, G.M. and Love, J.C. (2002) Die große Kunst klein zu bauen, *Spektrum der Wissenschaft Spezial* **N2** 14-23.

STRESS-FOCUSING EFFECTS IN A SPHERICAL INCLUSION EMBEDDED IN AN INFINITE MEDIUM CAUSED BY INSTANTANEOUS PHASE TRANSFORMATION

TOSHIAKI HATA
Faculty of Education, Shizuoka University,
Shizuoka City, Shizuoka 422-8529 Japan
E-mail: eithata@ipc.shizuoka.ac.jp

1. INTRODUCTION

Recently, transformation toughening of ceramics has attracted considerable attention in several works [1]. The mechanism in the toughening of ceramics is the stress-induced phase transformation of a Zirconia particle [2], which is accompanied by volumetric expansion under cooling process. Due to this expansion, the composite material consisting of Zirconia particles within a brittle matrix becomes more resistant to thermal fracture. While in the dynamic state the mechanism in the toughening of ceramics is not well understood. Therefore the interaction between a thermal shock and a possible phase transformation in the transformation-toughened ceramics with Zirconia particles should be studied. In this paper a phenomenological model is proposed to describe the situation, which involves the interaction between a thermal shock and a dynamic inhomogeneity with a stress-induced martensitic transformation in a spherical particle of Zirconia embedded in an infinite elastic matrix.

When an infinite elastic medium with a spherical inclusion of Zirconia is suddenly subjected to a uniform temperature fall through the temperature of phase transformation, stress waves occur at the surface of the spherical inclusion the moment thermal impact is applied. The stress wave in an inclusion proceeds radially inward to the center of the inclusion. The wave may accumulate at the center and may show the stress-focusing effects [3], even though the initial thermal stress should be relatively small. Stress waves, which develop following rapid uniform cooling of spherical inclusions, display the stress-focusing effects as they proceed radially towards the center in this geometry.

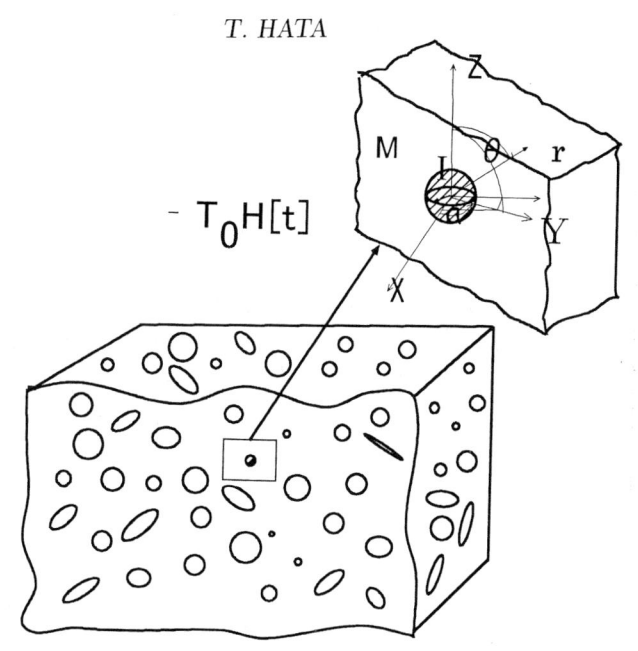

Figure 1. *Coordinate system of a spherical inclusion embedded in an infinite elastic medium.*

As for the study of these stress-focusing effects, Hata [4] solved, in an exact manner, the thermal stress-focusing effect in a spherical inclusion embedded in an infinite medium by using the ray integrals. In this paper, following the ray methods, we clarify the interaction between the thermal stress-focusing effect and the phase transformation in a spherical inclusion of Zirconia caused by the instantaneous cooling. It should be noted that the mechanism in the toughening of ceramics in the steady state does not hold in the dynamic state.

2. FORMULATION OF PROBLEM

The geometry of the problem is shown in Figure 1. The medium and the inhomogeneity are denoted M and I, respectively. Consider an infinite isotropic elastic medium of M containing a spherical inclusion of I with an eigenstrain (or transformation strain) $e_{ij}^*(\in \Omega)$. The coefficients of thermal expansion are α_I and α_M for an inclusion and the medium, respectively. The governing equations for an inclusion are [5]

$$\left. \begin{array}{l} \sigma_{ij,j}^I = \rho_{0I} \ddot{u}_i^I \\ \sigma_{ij}^I = C_{ijkl}^I (e_{kl}^I - e_{kl}^* - \delta_{kl}\alpha_I T) \\ e_{kl}^I = (u_{k,l}^I + u_{l,k}^I)/2 \end{array} \right\} \quad (1)$$

where ρ_{0I} is the mass density and C_{ijkl}^I is the elastic tensor of the inclusion as follows;

$$C_{ijkl}^I = \mu_I(\delta_{ik}\delta_{jl} + \delta_{il}\delta_{jk}) + \lambda_I \delta_{ij}\delta_{kl} \quad (2)$$

The following eigenstrain $e_{ij}^*(\mathbf{x}, t)$ in an inclusion is considered as

$$\begin{aligned} e_{kl}^*(\mathbf{x}, t) &= e_{kl}^*(\mathbf{x}) f(t) \\ e_{kl}^*(\mathbf{x}) &= \begin{cases} e_{kl}^* & \mathbf{x} \in \Omega \\ 0 & \mathbf{x} \in (I - \Omega) \end{cases} \end{aligned} \qquad (3)$$

The formulation of the problem in the infinite medium is

$$\begin{aligned} \sigma_{ij,j}^M &= \rho_{0M} \ddot{u}_i^M \\ \sigma_{ij}^M &= C_{ijkl}^M (e_{kl}^M - \delta_{kl} \alpha_M T) \\ e_{kl}^M &= (u_{k,l}^M + u_{l,k}^M)/2 \end{aligned} \qquad (4)$$

For the medium with a spherical inclusion as in Figure 1, the boundary conditions on the interface of $r = a$ are

$$\sigma_r^I = \sigma_r^M, \qquad u_r^I = u_r^M \qquad (5)$$

and the additional condition is that the displacement of the infinite medium at infinity is $u_r^M = 0$. The medium with a spherical inclusion is at rest prior to time $t = 0$ and the initial conditions of displacement are

$$\begin{aligned} u_r^I(r, t) &= u_r^I(r, t)_{,t} = 0 \\ u_r^M(r, t) &= u_r^M(r, t)_{,t} = 0. \end{aligned} \qquad (6)$$

2.1. Formulation of Stress Problem in an Isotropic Spherical Inclusion

The elastic medium with an inclusion is initially heated beyond the temperature of phase transformation. The elastic medium is at rest prior to time $t = 0$ and assumed to be in the stress-free state. For $t > 0$ the medium is cooled by the constant temperature T_0 from the reference state. The temperature distribution is assumed to have the following form

$$T(r, t) = -T_0 H(t) \qquad (7)$$

where $H(t)$ is the Heaviside step function.
The associated thermal stresses of Eq.(7) in a spherical inclusion are

$$\sigma_{rI}^T = \sigma_{\theta I}^T = \sigma_{\varphi I}^T = \frac{\alpha_I E_I T_0 H(t)}{1 - 2\nu_I} \qquad (8)$$

The stress-strain relations of Eq.(1) take the forms as

$$\begin{aligned} \sigma_{rI} &= 2\mu_I(e_r - e_r^*) + \lambda_I(e - e^*) - \beta_{IT} T \\ \sigma_{\theta I} &= 2\mu_I(e_\theta - e_\theta^*) + \lambda_I(e - e^*) - \beta_{IT} T \\ \sigma_{\varphi I} &= 2\mu_I(e_\varphi - e_\varphi^*) + \lambda_I(e - e^*) - \beta_{IT} T \end{aligned} \qquad (9)$$

where $e = e_r + e_\theta + e_\varphi$, $\sigma_{\theta I} = \sigma_{\varphi I}$, and $\beta_{IT} = \alpha_I(3\lambda_I + 2\mu_I)$.
The strain-displacement relations of Eq.(1) are

$$e_r = \frac{\partial u_I}{\partial r}, \quad e_\theta = e_\varphi = \frac{u_I}{r} \tag{10}$$

The equation of motion of Eq.(1) is given by

$$\frac{\partial \sigma_{rI}}{\partial r} + \frac{2(\sigma_{rI} - \sigma_{\theta I})}{r} = \rho_{0I} \frac{\partial^2 u_I}{\partial t^2} \tag{11}$$

Therefore the displacement equation of motion is obtained as

$$\frac{\partial^2 u_I}{\partial r^2} + \frac{2}{r}\frac{\partial u_I}{\partial r} - \frac{2u_I}{r^2} - \frac{1}{c_I^2}\frac{\partial^2 u_I}{\partial t^2} = \frac{(1+\nu_I)}{(1-\nu_I)}\alpha_I \frac{\partial T}{\partial r}$$
$$+ \frac{1-2\nu_I}{1-\nu_I}\{\frac{\partial e_r^*}{\partial r} + \frac{\nu_I}{1-2\nu_I}\frac{\partial e^*}{\partial r} + \frac{2(e_r^* - e_\theta^*)}{r}\} \tag{12}$$

where c_I is the dilatational wave speed, which is denoted as
$c_I = \sqrt{(\lambda_I + 2\mu_I)/\rho_{0I}}$.
In the analysis the eigenstrains of phase transformation are given by

$$\begin{aligned} e_r^* = e_\theta^* = e_\varphi^* = \epsilon_0 f(t) & \quad (0 < r \le r_0) \\ e_r^* = e_\theta^* = e_\varphi^* = 0 & \quad (r_0 < r \le a) \end{aligned} \tag{13}$$

If $f(t) = H(t)$ and $r_0 = a$, the associated eigenstresses of Eq.(13) in a spherical inclusion are

$$\sigma_{rI}^p = \sigma_{\theta I}^p = \sigma_{\varphi I}^p = -\frac{E_I \epsilon_0}{1 - 2\nu_I} H(t) \tag{14}$$

Here, when we introduce the displacement potential function Φ_I^S of homogeneous equation defined as

$$u_I = \frac{\partial \Phi_I^S}{\partial r}, \tag{15}$$

the function Φ_I^S should satisfy the equation

$$\frac{\partial^2 \Phi_I^S}{\partial r^2} + \frac{2}{r}\frac{\partial \Phi_I^S}{\partial r} - \frac{1}{c_I^2}\frac{\partial^2 \Phi_I^S}{\partial t^2} = 0 \quad . \tag{16}$$

Applying the Laplace transform to Eq.(16), we can find the homogeneous solution as follows;

$$\bar{\Phi}_I^S(r,p) = C_1 \left\{ h_0^{(1)}\left(ipr/c_I\right) + h_0^{(2)}\left(ipr/c_I\right) \right\} \tag{17}$$

where $h_0^{(1)}(z) = -ie^{iz}/z$ and $h_0^{(2)}(z) = ie^{-iz}/z$.

We find the displacement and the associated radial stress in a spherical inclusion as

$$\bar{u}_I^S(\rho, p) = C_1 \left\{ h_0^{(1)}(ip^*\rho)_{,\rho} + h_0^{(2)}(ip^*\rho)_{,\rho} \right\}/a$$

$$\bar{\sigma}_{rI}^S(\rho, p) = C_1 \left\{ -\frac{4\mu_I}{\rho}(ip^*) h_0^{(1)\prime}(ip^*\rho) + \rho_{0I} p^2 h_0^{(1)}(ip^*\rho) \right\} \quad (18)$$

$$+ C_1 \left\{ -\frac{4\mu_I}{\rho}(ip^*) h_0^{(2)\prime}(ip^*\rho) + \rho_{0I} p^2 h_0^{(2)}(ip^*\rho) \right\}$$

where $p^* = pa/c_I$ and $\rho = r/a$.

2.2. Formulation of Stress Problem in an Infinite Medium

The elastic medium with an elastic inclusion is at rest prior to time $t = 0$ and, for $t > 0$, the medium is cooled from the reference state to the uniform temperature $-T_0$. The corresponding thermal stresses of Eq.(7) for the elastic infinite medium are

$$\sigma_{rM}^T = \sigma_{\theta M}^T = \sigma_{\varphi M}^T = \frac{\alpha_M E_M T_0 H(t)}{1 - 2\nu_M} \quad (19)$$

These stresses do not satisfy the boundary conditions on the interface between the spherical inclusion and the infinite medium. In order to satisfy the boundary conditions, we should solve an ordinary dynamic stress problem.

For the medium with a spherical inclusion, the boundary conditions on the interface of $r = a$ are given by Eq.(5) and the additional condition is that the displacement at infinity should be $u_M = 0$. The medium with a spherical inclusion is at rest prior to time $t = 0$ and the initial conditions of displacement are given by Eq.(6).

Upon introduction of the displacement potential ϕ_M^S of Eq.(15), the equation of motion may be expressed as

$$\frac{\partial^2 \phi_M^S}{\partial r^2} + \frac{2}{r}\frac{\partial \phi_M^S}{\partial r} = \frac{1}{c_M^2}\frac{\partial^2 \phi_M^S}{\partial t^2} \quad (20)$$

where c_M is the dilatational wave speed, which is denoted as $c_M = \sqrt{(\lambda_M + 2\mu_M)/\rho_{0M}}$.

Applying the Laplace transform to Eq.(20) and solving the transformed equations, we find that the displacement potential for the infinite medium is given by

$$\bar{\phi}_M^S(r, p) = C_2 h_0^{(1)}(i\frac{p}{c_M}r). \quad (21)$$

Differentiating $\bar{\phi}_M^S(r, p)$ of Eq.(21) by r, we obtain the displacement of the infinite medium as

$$\bar{u}_M(r, p) = C_2 h_0^{(1)}(i\frac{p}{c_M}r)_{,r} \quad (22)$$

The associated radial stress is given by

$$\sigma_{rM}^{\bar{S}} = C_2 \left\{ -\frac{i4\mu_M p^* l_1}{\rho} h_0^{(1)'}(ip^* l_1 \rho) + \rho_{0M} p^2 h_0^{(1)}(ip^* l_1 \rho) \right\} \quad (23)$$

where

$$p^* = \frac{pa}{c_I}, \qquad \rho = \frac{r}{a}, \qquad l_1 = \frac{c_I}{c_M}. \quad (24)$$

The unknown constants C_1 and C_2 may be determined from the boundary conditions of Eq.(5) as follows;

$$C_1(c'_{11} + c'_{12}) + \bar{\sigma}_{rI}^T + \bar{\sigma}_{rI}^P = C_2 c'_{13} + \bar{\sigma}_{rM}^T$$

$$C_1 c_{21} - \alpha_I T_0 a + \epsilon_0 a = C_2 ip^* l_1 h_0^{(1)'}(ip^* l_1)/a - \alpha_M T_0 a \quad (25)$$

where

$$\begin{aligned}
c'_{11} &= -i4\mu_I p^* h_0^{(1)'}(ip^*) + \rho_{0I} p^2 h_0^{(1)}(ip^*) \\
c'_{12} &= -i4\mu_I p^* h_0^{(2)'}(ip^*) + \rho_{0I} p^2 h_0^{(2)}(ip^*) \\
c'_{13} &= -i4\mu_M p^* l_1 h_0^{(1)'}(ip^* l_1) + \rho_{0M} p^2 h_0^{(1)}(ip^* l_1) \\
c_{21} &= ip^* \{h_0^{(1)'}(ip^*) + h_0^{(2)'}(ip^*)\}/a
\end{aligned} \quad (26)$$

Then Eq.(25) yields

$$C_1 = \frac{g_2(p)}{c_{11} + c_{12}} \qquad C_2 = \frac{g_3(p)}{c_{11} + c_{12}} \quad (27)$$

where

$$\begin{aligned}
g_1(p) &= \frac{c'_{13}}{il_1 p^* h_0^{(1)'}(il_1 p^*)} \\
g_2(p) &= \bar{\sigma}_{rM}^T - \bar{\sigma}_{rI}^T - \bar{\sigma}_{rI}^P + g_1(p)\{T_0(\alpha_I - \alpha_M)a - \epsilon_0 a\}/p \\
g_3(p) &= \frac{(c'_{11} + c'_{12})\{T_0(\alpha_M - \alpha_I)a + \epsilon_0 a\}/p + c_{21}(\bar{\sigma}_{rM}^T - \bar{\sigma}_{rI}^T - \bar{\sigma}_{rI}^P)}{il_1 p^* h_0^{(1)'}(il_1 p^*)} \\
c_{11} &= c'_{11} - g_1(p)(ip^*) h_0^{(1)'}(ip^*) \\
c_{12} &= c'_{12} - g_1(p)(ip^*) h_0^{(2)'}(ip^*)
\end{aligned} \quad (28)$$

3. STRESS-FOCUSING EFFECT IN A SPHERICAL INCLUSION BY RAY INTEGRALS

In order to analyze the wave propagation in a spherical inclusion, we apply the ray theory to Eq.(17). Substituting Eq.(27) into Eq.(17), we obtain

$$\overline{\Phi_I^S} = \frac{g_2(p)}{c_{11} + c_{12}} \left\{ h_0^{(1)}(ip^* \rho) + h_0^{(2)}(ip^* \rho) \right\} \quad (29)$$

Here, we introduce the reflection coefficient R, which is defined as

$$R = -\frac{c_{11}}{c_{12}} \tag{30}$$

By introducing the formula

$$\frac{1}{1-R} = 1 + R + R^2 + \ldots \qquad (|R| \leq 1) \tag{31}$$

and rewriting Eq.(29) in the form of Eq.(31), we obtain the displacement potential for a spherical inclusion by using the formulas of $h_0^{(1)}(z) = -ie^{iz}/z$ and $h_0^{(2)}(z) = ie^{-iz}/z$ as

$$\overline{\Phi_I^S}(\rho, p) = \sum_{j=0}^{\infty} \bar{\Psi}_j^S(\rho, p) \tag{32}$$

where

$$\begin{aligned}
\bar{\Psi}_0^S(\rho, p) &= -e^{p^*(\rho-1)} \frac{l_1^2 u^{IM}(\lambda_M + 2\mu_M) p^{*2} + (l_1 p^* + 1)(4\mu_M u^{IM} + \sigma^{IM})}{f_1(p)\rho} \\
\bar{\Psi}_1^S(\rho, p) &= e^{-p^*(\rho+1)} \frac{l_1^2 u^{IM}(\lambda_M + 2\mu_M) p^{*2} + (l_1 p^* + 1)(4\mu_M u^{IM} + \sigma^{IM})}{f_1(p)\rho} \\
\bar{\Psi}_j^S(\rho, p) &= R(p) \bar{\Psi}_{j-2}^S(\rho, p) \qquad (j = 2, 3, 4, \ldots)
\end{aligned} \tag{33}$$

The notations u^{IM}, σ^{IM}, and function $f_1(p)$ in Eq.(33) are given by

$$\begin{aligned}
u^{IM} &= T_0(\alpha_M - \alpha_I)a + \epsilon_0 a \\
\sigma^{IM} &= \bar{\sigma}_{rI}^T + \bar{\sigma}_{rI}^P - \bar{\sigma}_{rM}^T \\
f_1(p) &= l_1(\lambda_I + l_1\lambda_M + 2\mu_I + 2l_1\mu_M) p^{*3} \\
&+ \{\lambda_I + 2\mu_I - l_1(l_1\lambda_M + 4\mu_I + 2l_1\mu_M - 4\mu_M)\} p^{*2} \\
&+ 4(l_1 - 1)(\mu_I - \mu_M) p^* + 4(\mu_I - \mu_M)
\end{aligned} \tag{34}$$

Since the inverse Laplace transforms of Eq.(33) are easily obtained by using the inverse Laplace formulas, the high order terms of $\Psi_j^S(\rho, t)$ are obtained through the convolution integral. Therefore the displacement potential for a spherical inclusion is obtained from Eq.(32) as

$$\Phi_I^S(\rho, t) = \sum_{j=0}^{\infty} \Psi_j^S(\rho, t) \tag{35}$$

Finally, the total displacement and stresses in a spherical inclusion are

$$\left.\begin{aligned}
u_I(\rho, t) &= u_I^S(\rho, t) - \alpha_I T_0 a\rho + \epsilon_0 a\rho \\
\sigma_{rI}(\rho, t) &= \sigma_{rI}^S(\rho, t) + \sigma_{rI}^T(\rho, t) + \sigma_{rI}^P(\rho, t) \\
\sigma_{\theta I}(\rho, t) &= \sigma_{\theta I}^S(\rho, t) + \sigma_{\theta I}^T(\rho, t) + \sigma_{\theta I}^P(\rho, t)
\end{aligned}\right\} \tag{36}$$

4. WAVE PROPAGATION IN AN INFINITE MEDIUM BY RAY INTEGRALS

In order to analyze the wave propagation in the infinite medium, we apply the ray theory to Eq.(21). Substituting Eq.(27) into Eq.(21), we obtain

$$\overline{\phi_M^S}(\rho,p) = \frac{g_3(p)}{c_{11}+c_{12}} h_0^{(1)}(il_1 p^* \rho) \tag{37}$$

Here, introducing the reflection coefficient R and rewriting Eq.(37) in the form of Eq.(31), we obtain the displacement potential for the infinite elastic medium as

$$\overline{\phi_M^S}(\rho,p) = \frac{g_3(p)}{c_{12}}[1 + R + R^2 + \ldots]h_0^{(1)}(il_1 p^* \rho)$$
$$= \sum_{j=0}^{\infty} \bar{\varphi}_j^S(\rho,p) \tag{38}$$

where

$$\bar{\varphi}_0^S(\rho,p) = -e^{-l_1 p^*(\rho-1)} \frac{u^{IM}(\lambda_I + 2\mu_I)p^{*2} + (p^* + 1)(4\mu_I u^{IM} + \sigma^{IM})}{f_1(p)\rho}$$
$$\bar{\varphi}_1^S(\rho,p) = e^{p^*(-l_1\rho + l_1 - 2)} \frac{u^{IM}(\lambda_I + 2\mu_I)p^{*2} + (p^* + 1)(4\mu_I u^{IM} + \sigma^{IM})}{f_1(p)\rho}$$
$$\bar{\varphi}_j^S(\rho,p) = R(p)\bar{\varphi}_{j-2}^S(\rho,p) \quad (j = 2,3,4,\ldots) \tag{39}$$

Since the inverse Laplace transforms of Eq.(39) are easily obtained by using the inverse Laplace formulas, the high order terms of $\varphi_j^S(\rho,t)$ are obtained through the convolution integral. Therefore the displacement potential for an infinite medium is derived from Eq.(38) as

$$\phi_M^S(\rho,t) = \sum_{j=0}^{\infty} \varphi_j^S(\rho,t) \tag{40}$$

Finally, the total displacement and stresses in the infinite medium are

$$\left.\begin{array}{l} u_M(\rho,t) = u_M^S(\rho,t) - \alpha_M T_0 a \rho \\ \sigma_{rM}(\rho,t) = \sigma_{rM}^S(\rho,t) + \sigma_{rM}^T(\rho,t) \\ \sigma_{\theta M}(\rho,t) = \sigma_{\theta M}^S(\rho,t) + \sigma_{\theta M}^T(\rho,t) \end{array}\right\} \tag{41}$$

5. NUMERICAL RESULTS AND DISCUSSION

To show the mechanism in the toughening of ceramics subjected suddenly to a constant temperature fall, we performed numerical calculations by using the

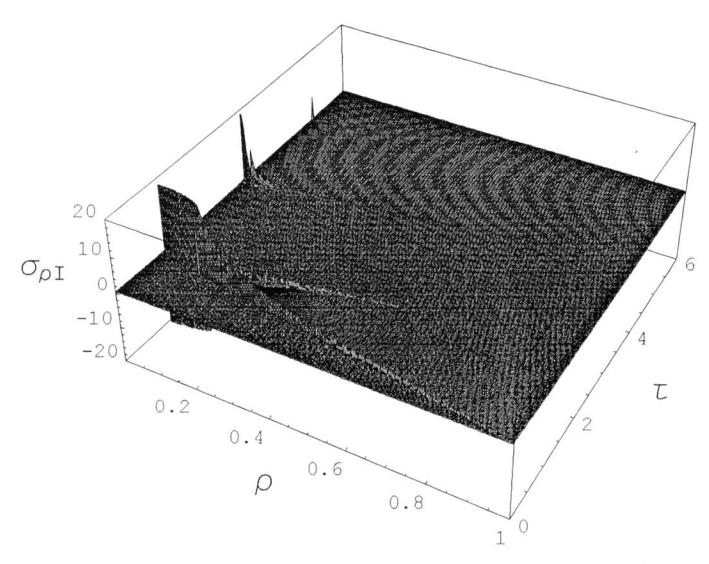

Figure 2. The behavior shape of the radial stress $\sigma_{\rho I}$ in an inclusion with phase transformation.

material constants [6] of a $Z_r O_2$ spherical inclusion embedded in the Al_2O_3 medium as

$$\frac{E_M}{E_I} = 1.857, \quad \frac{\alpha_M}{\alpha_I} = 0.667, \quad \frac{c_I}{c_M} = 0.632,$$
$$\nu_I = 0.3, \quad \nu_M = 0.25, \quad \frac{\epsilon_0}{\alpha_I T_0} = 0.9 \tag{42}$$

The results of the numerical evaluation of stress variation are illustrated in Figures 2 to 3. In the figures we use the following nondimensional variables as

$$\sigma_{\rho I} = \frac{\sigma_{rI}}{\rho_{0I} c_I^2 \alpha_I T_0}, \quad \tau = \frac{c_I t}{a} \tag{43}$$

The behavior shape of radial stress with phase transformation as a function of time is illustrated in Figure 2. The behavior shape of radial stress without phase transformation as a function of time is illustrated in Figure 3. In these figures we can observe that the waves reflected from the interface accumulate at the center of a spherical inclusion and give rise to very large stress magnitudes, even though the initial thermal stresses should be relatively small. The stress-focusing effects in a spherical inclusion with phase transformation are similar to those in an inclusion without phase transformation, whereas the positive peak stresses in an inclusion with phase transformation are much higher than those in a Zirconia inclusion without phase transformation.

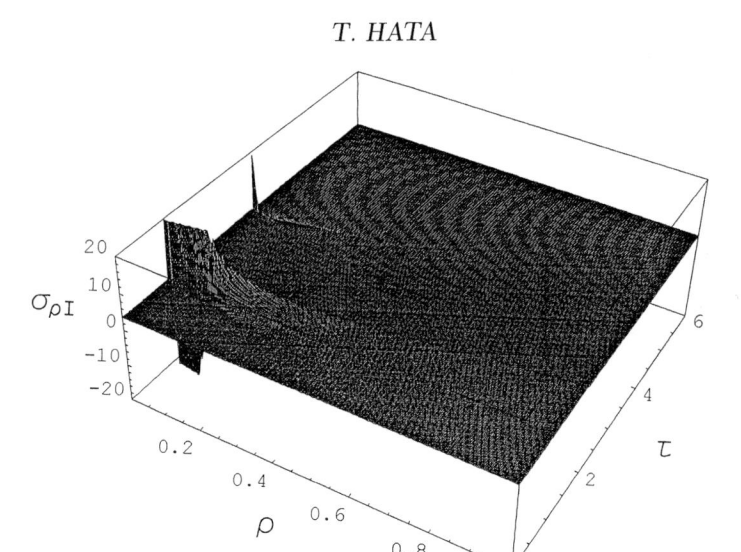

Figure 3. The behavior shape of the radial stress $\sigma_{\rho I}$ in an inclusion without phase transformation.

6. CONCLUSION

The major accomplishment of this study has been in gaining a better understanding of the stress-focusing effect in an instantaneously cooled Zirconia spherical inclusion with phase transformation embedded in an infinite ceramics matrix. In the paper, we demonstrate that the stress-focusing effects in a spherical inclusion play an important role. Therefore it should be noted that the stress-induced mechanism of phase transformation in the toughening of the Aluminum Oxide ceramics with a Zirconia inclusion in the steady state does not hold in the dynamic state.

7. REFERENCES

1. Mikata, Y.and Nemat-Nasser, S.(1990) Elastic Field Due to a Dynamically Transforming Spherical Inclusion, *Journal of Applied Mechanics*, **57** 845-849.
2. Garvie, R.C., Hannink,R.H. and Pascoe,R.T.(1975) Ceramic Steel, *Nature*, **258** 703-704.
3. Hata, T.(1997) Thermal Stress-Focusing Effect Following Rapid Uniform Heating of Spheres and Long Cylindrical Rods, *Journal of Thermal Stresses*, **20** 819-852.
4. Hata, T.(1999) Analysis of the Thermal Stress-Focusing Effect in a Spherical Inclusion Embedded in an Infinite Elastic Medium, *JSME International Journal*, **42** 176-182.
5. Mura, T.(1987) *Micromechanics of Defects in Solids*, Martinus Nijhoff Publishers, Dordrecht, The Netherlands, second edition.
6. Lange, F.F.(1982) Transformation toughening Part 4 Fabrication, Fracture Toughness and Strength of $Al_2O_3 - Z_rO_2$, *J. Materials Science*, **17** 247-254.

FREE LARGE VIBRATIONS OF BUCKLED LAMINATED PLATES

RUDOLF HEUER
Civil Engineering Department, Vienna University of Technology,
Wiedner Hauptstraße 8-10/E201, A-1040 Vienna, Austria
E-mail: rh@allmech9.tuwien.ac.at

Dedicated to Prof. Dr. Dr.h.c. Franz Ziegler on the occasion of his 65th birthday

1. INTRODUCTION

Fundamental works concerning the vibrations of pre-buckled structures are dating back to Mettler [1] and Eisley [2]. Nonlinear forced oscillations of buckled beams are studied in [3] and detailed finite element analysis of buckled plate vibrations is performed in Yang and Han[4]. Internal resonance phenomena are treated in Tien and Namachchivaya [5], [6], and nonlinear random vibrations of thermally buckled skew plates are analyzed by Heuer et al. [7].

In the present contribution large amplitude natural flexural vibrations of plates about a buckled reference configuration are investigated. In particular, thermally buckled polygonal plates made of multiple transversely isotropic layers are considered. Thermally stressed isotropic plates exhibiting large amplitude natural flexural vibrations about the plane reference configuration have been studied in Heuer et al. [8]. In case of hard hinged supports of the straight boundary segments of skew or even more generally shaped polygonal plates, a multi-modal approach combined with the Galerkin procedure gives a finite nonlinearly coupled set of ordinary differential equations of the Duffing type. In the following article that formulation is extended to the (unsymmetric) free vibrations about the thermally buckled plate position. The latter is assumed to be associated to a (postcritical) spatial distribution of cross-sectional mean temperature. In the special case of layered plates with physical properties symmetrically disposed about the middle surface, a correspondence to moderately thick homogeneous plates is found.

The analysis starts with the equations of motion according to the dynamic version of the von Kármán plate theory [9], modified by Mindlin´s kinematic hypothesis [10] in order to include shear deformation. Berger´s approximation [11] is applied which is known to be a reliable simplification if the in-plane displacements are constrained on the boundary. Following the lines of [8], a multimodal approach, where the eigenfunctions of the corresponding linear plate problem are selected as the space variables, gives a finite nonlinearly coupled set of ordinary differential equations containing both quadratic and cubic nonlinearities. In a single-term approximation, a closed-form solution is found in terms of Jacobian elliptic functions which is independent of the special polygonal plate geometry. For an evaluation of the real-time spectrum of the nonlinear natural fundamental frequency from this unifying similarity solution, only the Dirichlet-Helmholtz-eigenvalue of the corresponding plate must be known. The influence of multiple degrees-of-freedom on the fundamental frequency is studied numerically by means of the computer program MAPLE.

2. THE INITIAL-BOUNDARY VALUE PROBLEM FOR LARGE FREE FLEXURAL VIBRATIONS OF SHEAR-DEFORMABLE THERMALLY BUCKLED PLATES

Free moderately large flexural vibrations of buckled laminated plates of constant thickness h are investigated. We assumed that the buckled plate configuration is achieved through a constantly distributed in-plane force; in particular a thermal pre-stress is considered.

For laminates which are composed of N symmetrically arranged transversal isotropic layers, the global dynamic plate behavior is characterized by the following stiffness and mass parameters evaluated by homogenization procedure, see e.g. [12]:

$$D = \sum_{k=1}^{N} \int_{z_{k-1}}^{z_k} \frac{E_k}{(1-v_k^2)} dz, \quad v = \frac{1}{K} \sum_{k=1}^{N} K_k v_k, \quad \mu = \sum_{k=1}^{N} \int_{z_{k-1}}^{z_k} \rho_k dz, \quad (1)$$

$$K = \sum_{k=1}^{N} K_k = \sum_{k=1}^{N} \int_{z_{k-1}}^{z_k} \frac{E_k}{(1-v_k^2)} z^2 dz, \quad \frac{1}{S} = \kappa^2 \sum_{k=1}^{N} \int_{z_{k-1}}^{z_k} G_{ck} dz. \quad (2)$$

The stiffness coefficients D, K contain Young's modulus and Poisson's ratio of the individual layers k, κ^2 denotes a common shear factor, G_{ck} takes into account the effects of transverse isotropy, and ρ_k stands for the mass density of the k-th layer.

In case of a layerwise theory for sandwich plates made of three moderately thick layers, the global coefficients of Eq. (2) have to be replaced by, see [13] for a detailed derivation:

$$K = \varepsilon \frac{E_2 h_2^3}{12(1-v_2^2)}, \quad \varepsilon = \frac{G_2}{G_1} \frac{(1-v_2)}{(1-v_1)} \left(1-\tilde{t}_2^2\right) + \tilde{t}_2^3, \quad \tilde{t}_2 = \frac{t_2}{h}, \quad (3)$$

$$\frac{1}{s} = \frac{G_2 h}{\chi}, \quad \chi = \frac{(1-\nu_2)}{(1-\nu_1)} \left[\frac{G_1}{G_2} \frac{3}{2} \left(\tilde{t}_2 - \tilde{t}_2^3 \right) + \left(1 - \frac{3}{2}\tilde{t}_2 + \frac{1}{2} \tilde{t}_2^3 \right) \right] + \tilde{t}_2^3 . \tag{4}$$

The mid-plane of the plate contains an orthogonal coordinate system x, y, and z is the coordinate perpendicular to the middle plane. u, v, w denote the components of the displacement at the middle plane. Then, according to the equations suggested by von Kármán [9], the mid-plane strains due to moderately large deflections w are

$$e_x = u_{,x} + \frac{1}{2} w_{,x}^2, \quad e_y = v_{,y} + \frac{1}{2} w_{,y}^2, \quad e_{xy} = u_{,y} + v_{,x} + w_{,x} w_{,y} . \tag{5}$$

Furthermore, the effect of transverse shear is included by means of Mindlin's [10] kinematic hypotheses. Thus, the total strains at any point in the plate become

$$\begin{aligned} \varepsilon_x &= e_x + z\, \psi_{x,x}, \quad \varepsilon_y = e_y + z\, \psi_{y,y}, \\ \gamma_{xy} &= e_{xy} + z(\psi_{x,y} + \psi_{y,x}), \quad \gamma_{xz} = w_{,x} + \psi_x, \quad \gamma_{yz} = w_{,y} + \psi_y, \end{aligned} \tag{6}$$

where ψ_x and ψ_y denote the cross-sectional rotations.

The strain energy of a thermally pre-stressed plate under consideration can be decomposed into the four following parts: The membrane energy,

$$U_m = \frac{1}{2} D \int_A \left[I_e^2 - 2(1-\nu) II_e \right] dA, \quad I_e = e_x + e_y, \quad II_e = e_x e_y - \frac{1}{4} e_{xy}^2 , \tag{7}$$

and the bending energy

$$U_b = \frac{K}{2} \int_A \left\{ \left[\psi_{x,x}^2 + \psi_{y,y}^2 + \frac{1}{2} (\psi_{x,y} + \psi_{y,x})^2 \right] + \nu \left[2\psi_{x,x} \psi_{y,y} - \frac{1}{2} (\psi_{x,y} + \psi_{y,x})^2 \right] \right\} dA . \tag{8}$$

Considering plates with immovable edges, Berger's [11] assumption is applied, which states that the second invariant of the middle surface strain tensor, II_e, can be neglected in equation (7), i.e.

$$II_e = 0 \;\Rightarrow\; U_m \cong \frac{1}{2} D \int_A I_e^2 \, dA . \tag{9}$$

Shear deformation is expressed by the part

$$U_s = \frac{1}{2s} \int_A \left[(w_{,x} + \psi_x)^2 + (w_{,y} + \psi_y)^2 \right] dA , \qquad (10)$$

and finally, the effect of thermal pre-stress is included by means of

$$U_\theta = -(1+\nu) D \int_A n_\theta^* I_e \, dA , \quad n_\theta^* = \frac{1}{D(1+\nu)} \sum_{k=1}^{N} \int_{z_{k-1}}^{z_k} \frac{E_k \alpha_k \theta}{(1-\nu_k)} dz . \qquad (11)$$

Neglecting in-plane and rotatory inertia, the kinetic energy of the flexural vibrations is approximated by

$$T = \frac{1}{2} \int_A \dot{w}^2 \mu \, dA , \qquad (12)$$

where a dot stands for the time derivative.

By virtue of Hamilton's principle the equations of motion of the nonlinear plate problem can be derived. After eliminating the cross-sectional rotations, a single fourth-order differential equation for the plate midplane deflection $w(x, y)$ is obtained:

$$K(1+ns) \Delta\Delta w - n \Delta w - K\mu s \Delta \ddot{w} + \mu \ddot{w} = 0 . \qquad (13)$$

Berger's hydrostatic tensile in-plane force n is time-variant but it is constant within the plate domain,

$$n = D\left[I_e - (1+\nu) n_\theta^* \right] = const. \qquad (14)$$

Considering plates with immovable edges, a procedure developed by Wah [14] for the isothermal plate problem can be extended to thermally stressed plates, that renders

$$n = -\frac{D}{2A} \int_A w \Delta w \, dA - D \bar{n}_\theta^*, \quad \bar{n}_\theta^* = \frac{(1+\nu)}{A} \int_A n_\theta^* dA . \qquad (15)$$

Analyzing structures with hard hinged supports the corresponding boundary conditions for the case of polygonal geometry can be expressed by, compare [15],

$$\Gamma: w = 0, \quad \Delta w = 0. \tag{16}$$

3. DISCRETIZATION PROCEDURE

Applying a multi-modal approach, the lateral plate deflection is sought in the separable form,

$$w(x,t) = \sum_j c_j \left[q_{jb}^* + q_j^*(t) \right] w_j^*(x), \tag{17}$$

with initial conditions

$$q_j^*(t=0) = 1, \quad \dot{q}_j^*(t=0) = 0. \tag{18}$$

The coefficients q_{jb}^* determine the buckled position of the plate, and the generalized coordinates $q_j^*(t)$ give the vibrations about that pre-buckled plate. The spatial distribution of the deflection is expressed by the orthogonal set of eigenfunctions of the linearized flat plate problem, $w_j^*(x)$, and the multipliers c_j carry the dimension of length. The superscript (*) stands for non-dimensional quantities. In [15] it has been shown that eigenfunctions of those shear deformable plates with boundary conditions according to Eq. (16) are governed by a set of second-order Helmholtz differential equations with homogeneous Dirichlet's boundary conditions,

$$\Delta w_j^* + \alpha_j w_j^* = 0, \quad j = 1, 2...N, \tag{19}$$

$$\Gamma: w_j^* = 0. \tag{20}$$

α_j is the j-th eigenvalue of an effectively prestressed membrane of the same planform and it is related to the j-th linear natural frequency of the plate under consideration through (the thermal loading is absent)

$$\omega_{jL0} = \left[K\alpha_j^2/\rho h(1+Ks\alpha_j) \right]^{1/2}. \tag{21}$$

Substituting expression (17) together with Eq. (19) into Eqs. (13) and (14), and going through the Galerkin's procedure, we obtain the following coupled set of nonlinear ordinary differential equations for the non-dimensional amplitudes:

$$\ddot{q}_j^* + \gamma_{\theta j}^* \omega_{jL0}^2 \left(q_{jb}^* + q_j^*\right) + \frac{D\alpha_j}{2\mu}\left(q_{jb}^* + q_j^*\right) \sum_{k=1}^{N} \alpha_k^* \beta_k^* c_k^2 \left(q_{kb}^* + q_k^*\right)^2 = 0, \quad (22)$$

where

$$\gamma_{\theta j}^* = \left[1 - \bar{n}_\theta^* / \bar{n}_{\theta cj}^*\right], \; \bar{n}_{\theta cj}^* = \alpha_j h^2 / \left[1 + Ks\alpha_j\right], \; \beta_j^* = \frac{1}{A}\int_A w_j^{*\,2} dA. \quad (23)$$

4. UNIFYING SIMILARITY SOLUTION OF THE NONLINEAR NATURAL FREQUENCIES

Transformation to the eigen-time scale,

$$t^* = \omega_{1L0} t, \; \omega_{1L0} = \left[K\alpha_1^2 / \rho h \left(1 + s^*\right)\right]^{1/2}, \quad (24)$$

gives the nondimensional set of coupled ordinary differential equations containing both quadratic and cubic nonlinearities,

$$q_j^{*\,\prime\prime} + \gamma_{\theta j}^* \alpha_j^{*\,2} \frac{(1+s^*)}{(1+s^*\alpha_j^*)}\left(q_{jb}^* + q_j^*\right)$$

$$+ \delta^*(1+s^*)\alpha_j^*\left(q_{jb}^* + q_j^*\right)\sum_{k=1}^{N}\alpha_k^* \beta_k^* c_k^{*\,2}\left(q_{kb}^* + q_k^*\right)^2 = 0, \quad (25)$$

with

$$s^* = Ks\alpha_1, \; \alpha_j^* = \frac{\alpha_j}{\alpha_1}, \; c_j^* = \left(\frac{c_j}{h}\right), \; \gamma_{\theta j}^* = 1 + \left(\gamma_{\theta 1}^* - 1\right)\frac{(1+\alpha_j^* s^*)}{\alpha_j^*(1+s^*)}, \; \delta^* = \frac{Dh^2}{2K}, \quad (26)$$

and $()^\prime$ denotes the derivative with respect to t^*.

The thermally buckled position has to be determinated from the nonlinear algebraic system of equations,

$$\gamma_{\theta j}^* \alpha_j^{*2} \frac{(1+s^*)}{(1+s^*\alpha_j^*)} q_{jb}^* + \delta^*(1+s^*) \alpha_j^* q_{jb}^* \sum_{k=1}^{N} \alpha_k^* \beta_k^* c_k^{*2} q_{kb}^{*2} = 0 . \quad (27)$$

In the special case where the buckling configuration is proportional to the basic eigenmode of the linearized plate problem, it follows

$$q_{1b}^* = \pm \frac{1}{c_1^*} \sqrt{-\frac{\gamma_{\theta 1}^*}{\delta^*(1+s^*)\beta_1^*}} , \quad (\gamma_{\theta 1}^* < 0), \quad q_{kb}^* = 0 \ldots k > 1 , \quad (28)$$

and

$$j=1: \quad q_1^{*''} - 2\gamma_{\theta 1}^* q_1^* + 3\delta^*(1+s^*)\beta_1^* c_1^{*2} q_{1b}^* q_1^{*2} + \delta^*(1+s^*)\beta_1^* c_1^{*2} q_1^{*3}$$

$$+ \delta^*(1+s^*)\left(q_{1b}^* + q_1^*\right) \sum_{k=2}^{N} \alpha_k^* \beta_k^* c_k^{*2} q_k^{*2} = 0, \quad (29)$$

$j \geq 2$:

$$q_j^{*''} + \gamma_{\theta j}^* \alpha_j^{*2} \frac{(1+s^*)}{(1+s^*\alpha_j^*)} q_j^* + \delta^*(1+s^*) q_j^* \sum_{k=1}^{N} \alpha_k^* \beta_k^* c_k^{*2} \left(\delta_{1k} q_{kb}^* + q_k^*\right)^2 = 0. \quad (30)$$

5. CLOSED-FORM SOLUTION OF SINGLE-TERM APPROXIMATION

In a single-term approximation Eq. (25) reduces to

$$q^{*''} + \gamma_\theta^* \left(q_b^* + q^*\right) + \delta^*(1+s^*)\left(q_b^* + q^*\right)^3 \beta^* c^{*2} = 0 , \quad (31)$$

with the static buckled position according to Eq. (28). Thus, the initial value problem reads

$$q^{*''} + C_1 q^* + C_2 q^{*2} + C_3 q^{*3} = 0, \quad q^*(t^* = 0) = 1, \quad q^{*'}(t^* = 0) = 0, \quad (32)$$

where

$$C_1 = -2\overset{*}{\gamma_\theta}, \quad C_2 = c^*\sqrt{-9\delta^*(1+s^*)\beta^*\overset{*}{\gamma_\theta}}, \quad C_3 = \delta^*(1+s^*)\beta^* c^{*2}. \qquad (33)$$

Following Weigand [16], a closed-form solution for that initial value problem can be formulated by means of the Jacobian elliptic function. The period of this cosinus amplitudinis function gives the nonlinear natural frequency parameter in a unifying manner,

$$\frac{\overset{*}{\omega_N}}{2\pi} = \frac{1}{\overset{*}{T_N}} = \frac{m\sqrt{C_3/2}}{4\,K(k^2)}. \qquad (34)$$

K denotes the complete elliptic integral of the first kind. Its modulus k and the factor m are calculated from the roots of the characteristic equation,

$$C_1 + \frac{2}{3}C_2 + \frac{1}{2}C_3 - C_1 z^2 - \frac{2}{3}C_2 z^3 - \frac{1}{2}C_3 z^4 = 0 \;\Rightarrow\; z_1, \ldots, z_4, \qquad (35)$$

$$m' = \sqrt{(z_1 - z_3)(z_2 - z_4)}, \quad m'' = \begin{cases} \sqrt{(z_1 - z_2)(z_3 - z_4)} & \ldots C_3 < 0, \\ \sqrt{(z_1 - z_4)(z_2 - z_3)} & \ldots C_3 > 0, \end{cases} \qquad (36)$$

$$m = \frac{m' + m''}{2}, \quad k = \frac{m' - m''}{m' + m''}. \qquad (37)$$

Finally the real-time nonlinear natural frequency becomes

$$\omega_N = \omega_{1L0}\,\overset{*}{\omega_N}. \qquad (38)$$

6. NUMERICAL STUDIES

In the first numerical example a single-term approximation is considered for plates with $\beta^* = 0.5$ according to Eq. (31). That geometry factor holds e.g. holds for a plate strip or an right-angled isosceles triangular plate. A shear coefficient of $s^* = 0.02$ is taken into account and a thermal pre-stress of $\overset{*}{\gamma_\theta} = -1.5$ is chosen. Fig. 1 shows the influence of the initial amplitude c^* on the frequency parameter.

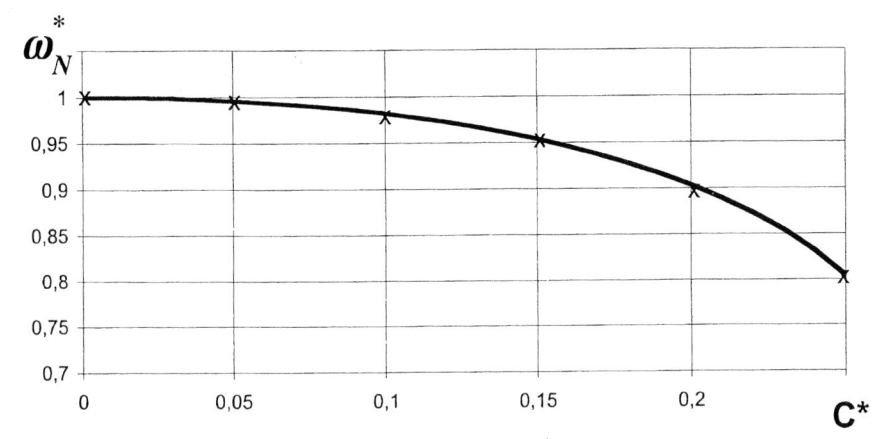

Figure 1. Nondimensional frequency parameter ω_N^* according to a single-term approximation. Plate parameters: $\beta^* = 0.5$ (e.g. plate strip or right-angled isosceles triangular plate), $\gamma_\theta^* = -1.5$, $s^* = 0.02$.

——— ... *closed form solution;* **X** ... *numerical solution*

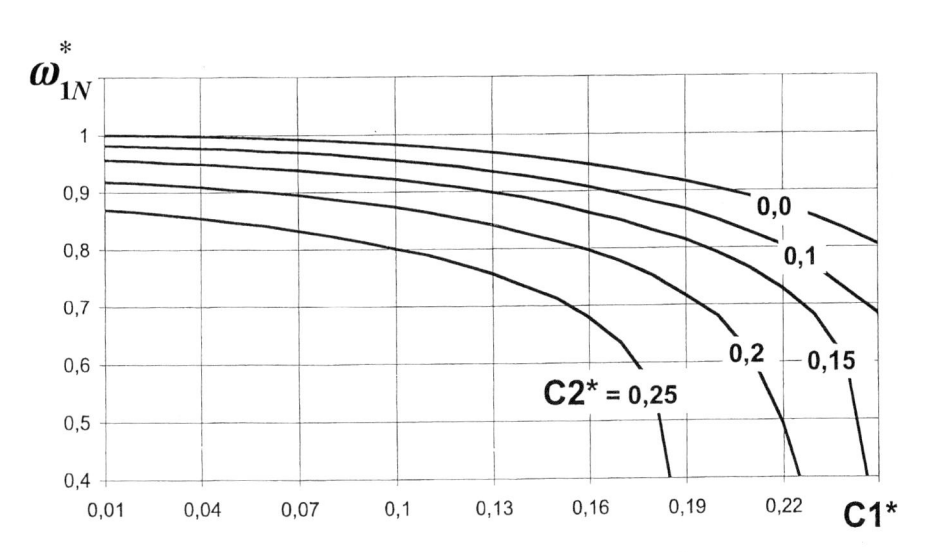

Figure 2. Nondimensional fundamental frequency parameter ω_{1N}^* of the nonlinear natural frequency in the parameter space according to a two-term approximation. Plates parameters: $\beta_1^* = \beta_2^* = 0.5$, $\alpha_2^* = 4$ (e.g. plate strip), $\gamma_\theta^* = -1.5$, $s^* = 0.02$; numerical solution for various c_1^* and c_2^*

$\omega_N^*(c^*)$ exhibits a nonlinear soft spring behavior of the buckled plate. The analytical function according to Eq. (34) is compared to computational results for selected values of c^* evaluated by the computer program MATLAB.

The second example, Fig. 2, studies the nonlinear influence of the first two eigenmodes, represented by c_1^* and c_2^* on the fundamental frequency parameter $\omega_{1N}^* = \omega_{1N} / \omega_{1L0}$ of a plate strip, where $\beta_1^* = \beta_2^* = 0.5$ and $\alpha_2^* = 4$. These results again are numerically determined by MAPLE.

7. REFERENCES

1. Mettler, E. (1962) Dynamic Buckling, in *Handbook of Engineering Mechanics*, edited by W. Flügge, McGraw-Hill, New York, pp.62-1 – 62-11.
2. Eisley, J.G. (1964) Large amplitude vibration of buckled beams and rectangular plates, *AIAA-Journal* **2** 2207-2209.
3. Tseng, W.-Y. and Dugundji, J. (1971) Nonlinear vibrations of a buckled beam under harmonic excitation, *Journal of Applied Mechanics* **38** 467-476.
4. Yang, T.Y. and Han, A.D. (1983) Buckled plate vibrations and large amplitude vibrations using high-order triangular elements, *AIAA-Journal* **21** 758-766.
5. Tien, W.-M., Namachchivaya, N.S. and Bajaj, A.K. (1994) Non-linear dynamics of shallow arch under periodic excitation – I. 1:2 Internal resonance, *Int. J. Non-Linear Mechanics* **29** 349-366.
6. Tien, W.-M., Namachchivaya, N.S. and Malhotra, N. (1994) Non-linear dynamics of shallow arch under periodic excitation – II. 1:1 Internal resonance, *Int. J. Non-Linear Mechanics* **29** 367-386.
7. Heuer, R., Irschik, H. and Ziegler, F. (1993) Nonlinear random vibrations of thermally buckled skew plates, *Probabilistic Engineering Mechanics* **8** 265-271.
8. Heuer, R., Irschik, H. and Ziegler, F. (1990) Multi-modal approach for large natural flexural vibrations of thermally stressed plates, *Nonlinear Dynamics* **1** 449-458.
9. von Kármán, Th. (1910) Festigkeitsprobleme im Maschinenbau, *Encyklopädie der mathematischen Wissenschaften* **IV**, Teubner, Leipzig.
10. Mindlin, R.D. (1951) Influence of rotatory inertia and shear on flexural motions of isotropic, elastic plates, *Journal of Applied Mechanics* **18** 31-38.
11. Berger, H.M. (1955) A new approach to the analysis of large deflection of plates, *Journal of Applied Mechanics* **22** 465-472.
12. Jones, R.M. and Klein, S. (1968) Equivalence between single-layered and certain multilayered shells, *AIAA Journal* **6** 2295-2300.
13. Heuer, R., Irschik, H. (1991) Erzwungene Schwingungen elastischer Sandwichplatten mit dicken Deckschichten, *ZAMM* **71** T86-T88.
14. Wah, T. (1963) Large amplitude flexural vibrations of rectangular plates, *International Journal of Mechanical Sciences* **5** 425-438.
15. Irschik, H. (1985) Membrane-type eigenmotions of Mindlin plates, *Acta Mechanica* **55** 1-20.
16. Weigand, A. (1941) Die Berechnung freier nichtlinearer Schwingungen mit Hilfe der elliptischer Funktionen, *Forschung auf dem Gebiete des Ingenieurwesens* **12**, VDI-Verlag, 264-284.

HIGH-PERFORMANCE IMPACT ABSORBING MATERIALS - THE CONCEPT, DESIGN TOOLS AND APPLICATIONS

JAN HOLNICKI-SZULC, PIOTR PAWLOWSKI
Institute of Fundamental Technological Research,
Swietokrzyska 21, 00-049 Warsaw, Poland
holnicki@ippt.gov.pl

1. INTRODUCTION

Motivation for the undertaken research is to respond to requirements for high impact energy absorption e.g. in the following cases: i) structures exposed for risk of extreme blast, ii) light, thin wall tanks with high impact protection, iii) vehicles with high crashworthiness, iv) protective barriers, etc. Typically, the suggested solutions focus on the design of passive energy absorbing systems. These systems are frequently based on the aluminium and/or steel honeycomb packages characterised by a high ratio of specific energy absorption. However high is the energy absorption capacity of such elements they still remain highly redundant structural members, which do not carry any load in an actual operation of a given structure. In addition, passive energy absorbers are designed to work effectively in pre-defined impact scenarios. For example, the frontal honeycomb cushions are very effective during a symmetric axial crash of colliding cars but are completely useless in other types of crash loading. Consequently, distinct and sometimes completely independent systems must be developed for specific collision scenarios.

In contrast to the standard passive systems the proposed approach focuses on *active adaptation* of energy absorbing structures (equipped with sensor system detecting impact in advance and controllable semi-active dissipaters, so called *structural fuses*) with highest ability of adaptation to extremal overloading. The quasistatic formulation of this problem allows developing effective numerical tools necessary for farther considerations concerning dynamic problem of optimal design for the best structural response (see [2]). The structures with the highest impact absorption properties can be designed in this way. The proposed optimal design method combines sensitivity analysis with redesign process, allowing stress limits control in structural fuses. So-called Virtual Distortion Method, (VDM, see [1]), leading to analytical formulas for gradient calculations, has bee used in numerically efficient algorithm. Another approach to similarly formulated problem is presented in [3].

2. THE CONCEPT OF ADAPTIVE MULTIFOLDING MICROSTRUCUTRE

The objective of this paper is to propose new concept of adaptive micro-structure with high strain energy absorbing characteristics. Let us discuss the truss-like microstructure (similar to honeycomb layout) shown in Fig.1 equipped with specially designed micro-fuses (Fig.2a) with stacked thin washers made of SMA (Shape Memory Alloys) as controllable stickers (Fig.2b.) The micro-structure response to external pressure strongly depends on the yield stress levels applied to micro-fuses and these levels can be controlled activating proper numbers of SMA micro-washers in each sticker.

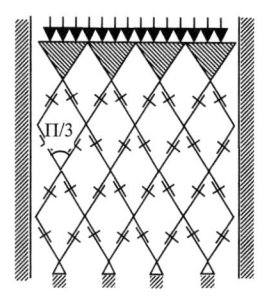

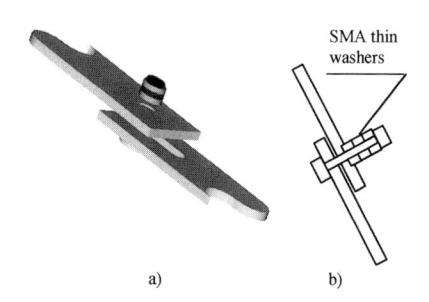

Figure 1. Truss-like micro-structure

Figure 2. Controllable micro-stickers

To analyse the performance of the proposed micro-structure, let us follow the response of the model shown in Fig.3, corresponding to behaviour of one row of the discussed hypothetic smart material.

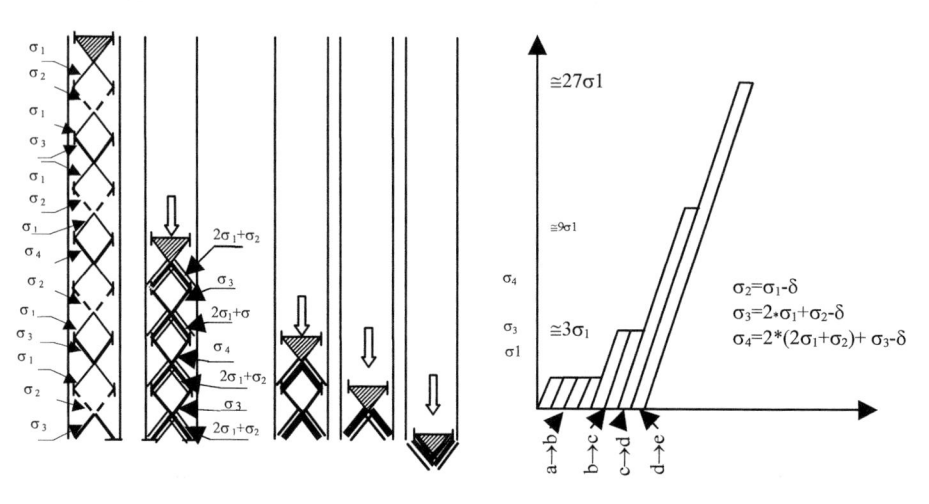

$\sigma_2 = \sigma_1 - \delta$
$\sigma_3 = 2 \cdot \sigma_1 + \sigma_2 - \delta$
$\sigma_4 = 2*(2\sigma_1 + \sigma_2) + \sigma_3 - \delta$

Figure 3. Model of Adaptive Multifolding Microstructure (MFM)

Assuming idealised truss structure model (Fig.3a) composed of idealised elasto-plastic members with the shown yield stress levels (realised through properly activated stickers), the sequence of its collapse stages is shown in Figs 3b, 3c and 3d, respectively. The corresponding effect of energy dissipation (Fig.3e) is 343% higher than for the same kind of micro-structure, made of the same material volume and with homogeneously distributed yield stress levels.

The crucial point to get the additional value of energy dissipation (due to synergy of repetitive use of dissipaters) is to pre-design the optimal distribution of yield stress levels in all stickers, triggering desired sequence of local collapses. Let us call the discussed adaptive micro-structure the *Adaptive Multifolding Microstructure* (MFM). The piece-wise linear constitutive model of MFM shown in Fig.4 (applicable in computational simulations) can be proposed. Cyclically loading-unloading adaptive members will have their characteristics with high hysteresis (Fig.4b). Additionally, fictitious members (dotted lines in Fig.4a) with piece-wise linear locking properties (Fig.4c) are proposed to model variable contact problem in loading scenario. The numerical model for simulation of MFM performance will require taking into account both: physical and geometrical non-linearities.

3. NUMERICAL SIMULATION

It is necessary to simulate numerically MFM performance to design desired yield stress levels in all stickers. To this end, let us introduce notation of strains and stresses (cf.[1]) as superposition of linear structural response ε_i^L and σ_i^L, respectively, to external load p and the component caused by *virtual distortions* β_j^0 modelling real, *plastic-like* distortions in adaptive members (the set B_σ of elements) and *locking-like* distortions in fictitious members (the set B_ε of elements) simulating variable contact conditions in loading process (cf. Fig.4).

$$\varepsilon_i = \varepsilon_i^L + \sum_j D_{ij} \beta_j^0 \qquad \sigma_i = \sigma_i^L + \sum_j E_i (D_{ij} - \delta_{ij}) \beta_j^0 \qquad (1)$$

where virtual distortions β_j^0 has to satisfy the following conditions:

$$\sigma_i - \sigma_i^* = \gamma_i E_i (\varepsilon_i - \varepsilon_i^*) \qquad (2)$$

and D_{ij} denote strain caused in member *i* by the unit virtual distortion $\beta_j^0 = 1$ generated in member *j*. Assume that for adaptive elements (i ∈ B_σ) γ_i is a small positive value modelling behaviour close to ideal plasticity while for fictitious elements (i ∈ B_ε) γ_i takes large negative values modelling behaviour close to locking material. $\sigma_i^* = E_i \cdot \varepsilon_i^*$ denotes yield stress level for adaptive member while ε_i^* (equal to –1 in our case) denotes locking level for the very flexible fictitious members ($E_i \cong 0$)

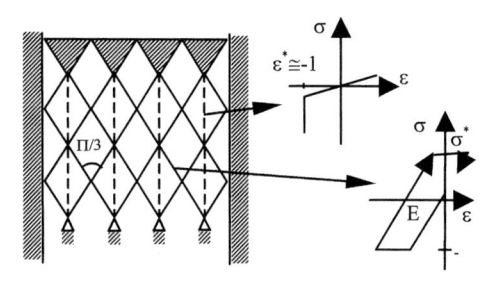

Figure 4. Constitutive model of microstructure

Substituting (1) to (2), the following set of equations determining virtual distortions modelling MFM response to external load can be obtained.

$$[(1-\gamma_i)D_{ij} - \delta_{ij}]\beta_j^0 = -(1-\gamma_i)(\varepsilon_i^L - \varepsilon_i^*) \qquad (3)$$

The above description is valid for geometrically linear problem. In our case, however, due to large deformations and necessity of sequential modification of the global stiffness matrix, the incremental approach has to be applied. Then, the set of equations (3) should be modified as follows:

$$[(1-\gamma_i)D_{ij} - \delta_{ij}]\Delta\beta_j^0 = -(1-\gamma_i)\left(\varepsilon_i' + \Delta\varepsilon_i^L \frac{l_i}{l_{0i}} - \varepsilon_i^*\right) \qquad (4)$$

where l_i and D_{ij} are determined for the actual geometric configuration, ε_i' denotes final strains determined for previous load increment and l_{0i} denotes the initial length of the member. The VDM based algorithm for simulation of MFM (with determined yield stress levels σ_i^*) non-linear response to external load is shown in Table 1.

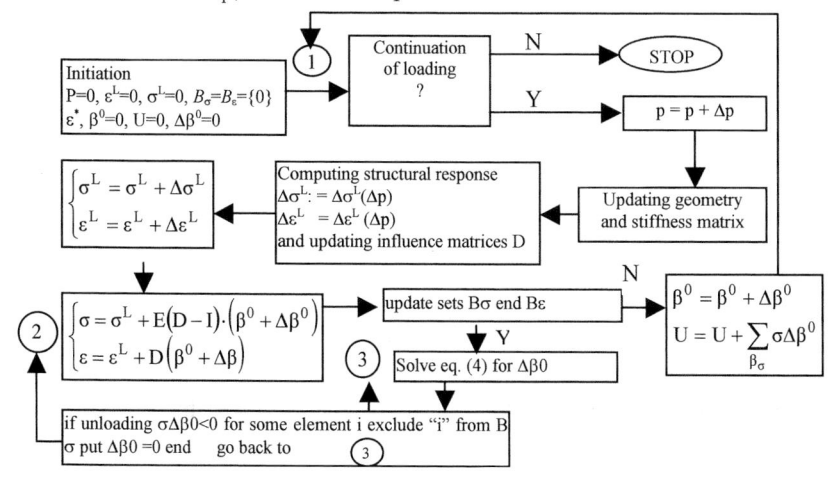

Table 1. VDM based algorithm for simulation of MFM non-linear response

4. OPTIMAL CONTROL

The VDM based non-linear analysis described above allows simulating performance of the MFM micro-structure with determined stress levels triggering plastic-like behaviour of micro-stickers. However, in order to improve MFM response adapting to particular load, a control strategy should be proposed, where triggering stress levels σ_i^* are control parameters.

The problem can be formulated as follows:
for given load maximise the plastic-like energy dissipation:

$$\max U^0 = \sum_i \sigma_i \Delta \beta_i \qquad (5)$$

subject to following constraints

$$\left|\beta_i^0\right| \leq \beta^u, \quad \sigma^* \leq \overline{\sigma} \qquad (6)$$

where σ_i is coupled with strains and the control parameters trough relations (1) and (2). The solution of this qasi-static problem will allow maximally smooth load reception. Analogous formulation can be applied also to the fully dynamic problem with the accumulated energy dissipation as the objective function. The solution of the above static problem exists if the external load intensity is not higher then the maximal safe load level.

The procedure to determine this maximal load level still safe for the adaptive micro-structure can be proposed when MFM with initially determined triggering stresses ($\sigma_i = \overline{\sigma}$) is not able to sustain the applied load with assumed constraints $|\beta^0{}_i| \leq \beta^u$ imposed on plastic distortions. Then, the algorithm (Table 2) of adaptation (mostly lowering) of the control parameters σ_i^* can be applied, where maximisation of the energy dissipation U has been chosen as the control strategy.

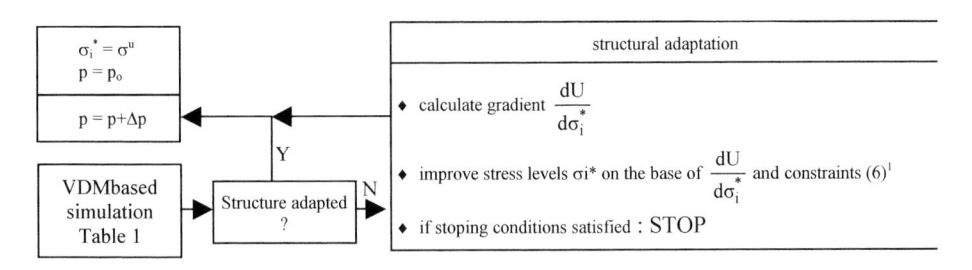

$\sigma_i^* = \sigma^u$ $p = p_o$	structural adaptation
$p = p + \Delta p$	• calculate gradient $\dfrac{dU}{d\sigma_i^*}$
VDMbased simulation Table 1 → Structure adapted ?	• improve stress levels σi^* on the base of $\dfrac{dU}{d\sigma_i^*}$ and constraints (6)[1] • if stoping conditions satisfied : STOP

Table 2. The algorithm searching for maximal load level, safe for adaptive MFM

The gradient based procedure of MFM adaptation can be driven by the following, analytically determined formulas, obtained from equations (4):

$$[(1-\gamma_i)D_{ij} - \delta_{ij}]\frac{\partial \Delta \beta_j^0}{\partial \varepsilon_k^*} = -(1-\gamma_i)\left(\frac{\partial \varepsilon_i'}{\partial \varepsilon_k^*} - \delta_{ik}\right) \qquad (6)$$

where gradient $\dfrac{\partial \varepsilon_i'}{\partial \varepsilon_k^*}$ was calculated for the previous load level.

Having the following relation for actual strains:

$$\varepsilon_i = \varepsilon_i' + \Delta \varepsilon_i^L + \sum_j D_{ij} \Delta \beta_j^0 \qquad (7)$$

where ε_i' has been determined for the previous load level, the corresponding gradiend relations can be provided:

$$\frac{\partial \varepsilon_i}{\partial \varepsilon_k^*} = \frac{\partial \varepsilon_i'}{\partial \varepsilon_k^*} + \sum_j D_{ij} \frac{\partial \beta_j^0}{\partial \varepsilon_k^*} \qquad (8)$$

Analogously

$$\frac{\partial \sigma_i}{\partial \varepsilon_k^*} = \frac{\partial \sigma_i'}{\partial \varepsilon_k^*} + E_i \sum_j (D_{ij} - \delta_{ij}) \frac{\partial \beta_j^0}{\partial \varepsilon_k^*} \qquad (9)$$

Finally, gradient of the objective function (5) can be calculated as follows:

$$\frac{\partial U}{\partial \varepsilon_k^*} = \sum_i \left(\frac{\partial \sigma_i}{\partial \varepsilon_k^*} \Delta \beta_i^0 + \sigma_i \frac{\partial \Delta \beta_i^0}{\partial \varepsilon_k^*}\right) \qquad (10)$$

where $\Delta \beta_i^0$ is determined by (4), $\dfrac{\partial \Delta \beta_i^0}{\partial \varepsilon_k^*}$ is determined by (6) end $\dfrac{\partial \sigma_i}{\partial \varepsilon_k^*}$ is determined by (9).

Following the non-linear incremental analysis of MFM response (for fixed σ_i^* parameters) described in Table 1, the solution of equations (4) for each load level is needed. With small extra cost (modifying right hand side vectors) derivatives $\dfrac{\partial \Delta \beta_i^0}{\partial \varepsilon_k^*}$ can be determined (cf. 6) end stored. Afterwards the gradients (10) can be also step by step computed and cumulated. Finally having global structural response and the gradient (10) value, the decision about modification of control parameters σ_i^* can be taken.

5. NUMERICAL EXAMPLE

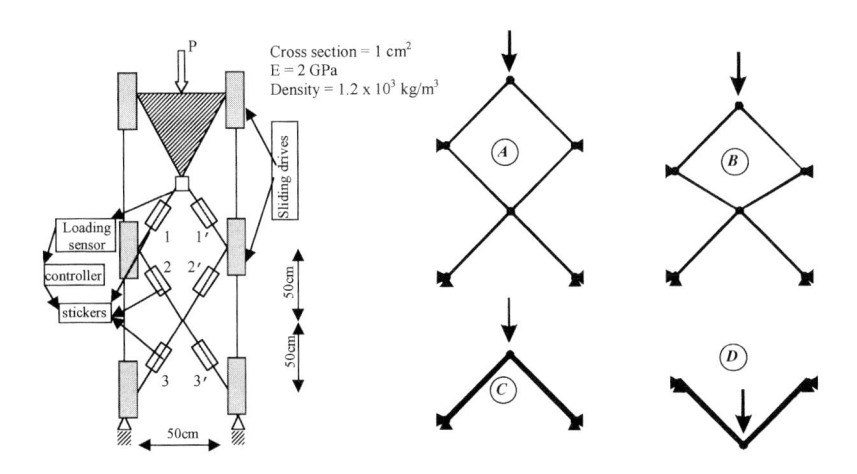

Figure 5. MFM demonstrator set-up and desired multifolding sequence

A numerical model of the MFM demonstrator set-up (Fig.5) has been created and tested numerically. The objective function (5) distribution as the quasi-static structural response to external load P=30kN, for two control parameters (selected systematically): σ_1^* - describing the yield stresses for elements 1, 1', 3, 3' and σ_2^* - describing the yield stresses for elements 2, 2' is shown in Fig. 6. The evolution of stresses, strains and plastic distortions for characteristic elements No.1 and 2, corresponding to the optimal control parameters σ_1^*=60 MPa and σ_2^*=40 MPa (cf.fig.6) are shown in Figs 7 and 8, respectively.

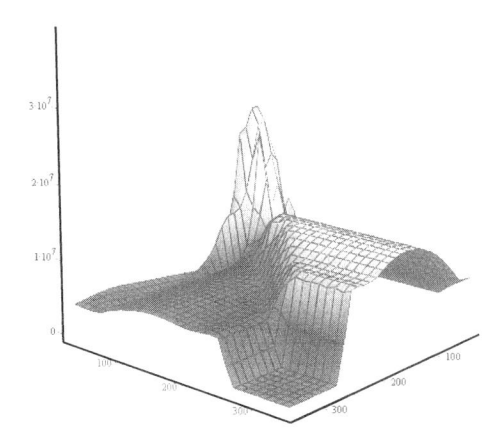

Figure 6. The energy dissipation for various yield stress values

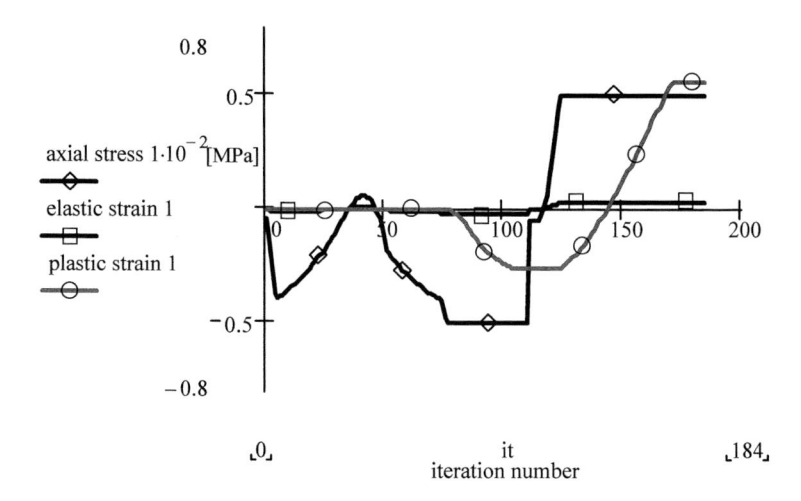

Figure 7. The evolution of stress, strain and plastic distortion for element No.1

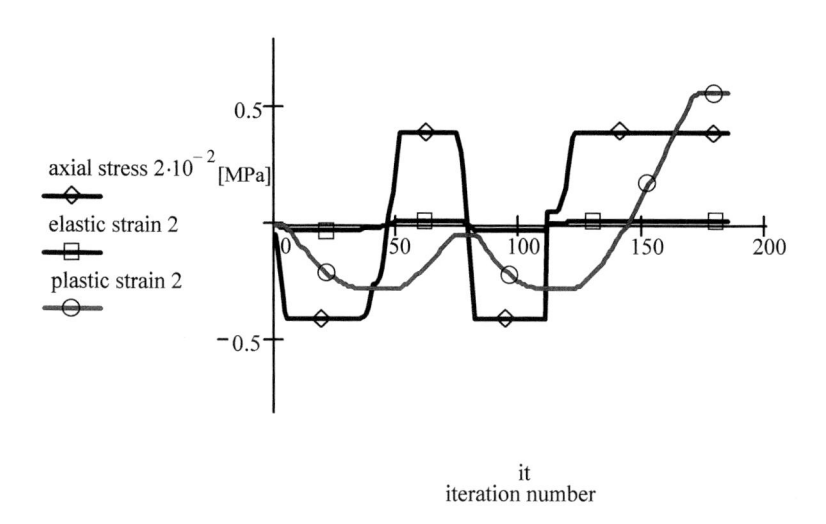

Figure 8. The evolution of stress, strain and plastic distortion for element No.2

Fully dynamic response of the considered model (with plastic stress limits identical to the above optimal quasi-static solution) is presented in Figs 10 and 11 while the evolution of kinetic, elastic and plastic (dissipated) strain energy for the whole analysed structure is exposed in Fig.9.

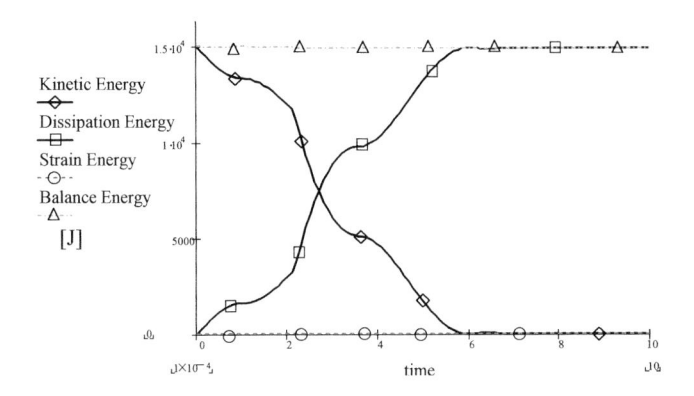

Figure 9. Evolution of energy components for dynamic response

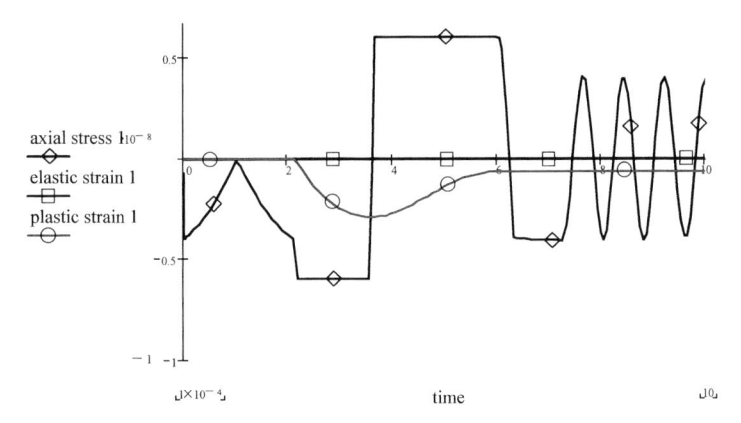

Figure 10. The evolution of stress, strain and plastic distortion for element No.1

6. CONCLUSIONS

It has been demonstrated effectiveness of the yield stress level adaptation to applied load on the intensity of strain energy dissipation. If the structure can be decomposed into elements with own microstructure inside, the above approach can be applicable on the macro-structural as well as micro-structural level.
The following general methodology in design of adaptive MFM can be proposed.

- design topological pattern of MRF for variety of all expected extreme loadings
- determine optimal yield stress level distribution (quasi-static approach on macro- structural level, without the multifolding effect) for each extreme loading

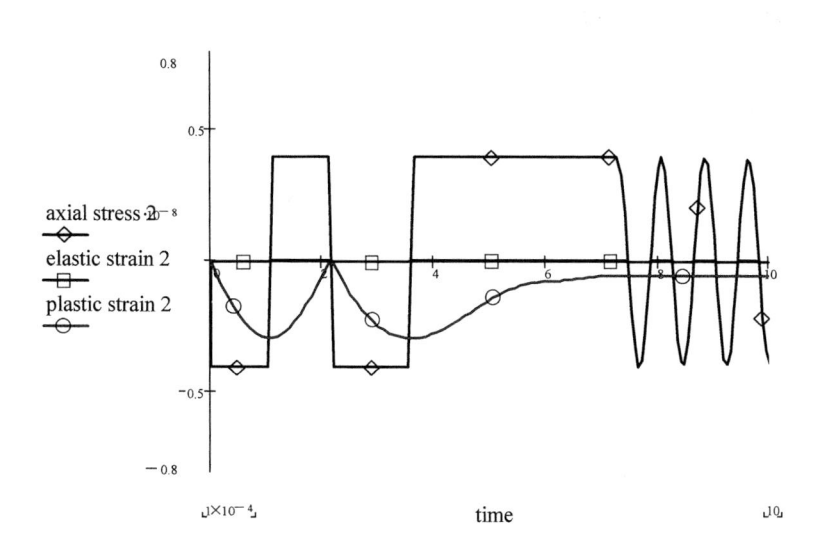

Figure 11. The evolution of stress, strain and plastic distortion for element No.2

- determine optimal yield stress level distribution (quasi-static approach on micro-structural level, including the multifolding effect) for each extreme loading
- simulate fully dynamic response of adaptive MFM for each extreme loading
- apply in real time the pre-computed control strategy as the response for detected (through a sensor system) impact.

ACKNOWLEDGEMENT

This work was supported by the grant No. KBN7T07A02516 from the Institute of Fundamental Technological Research, funded by the National Research Committee and presents a part of the Ph.D. thesis of the second author, supervised by the first author.

REFERENCES

1. Holnicki-Szulc, J., Gierliński, J.T.: *Structural Analysis, Design and Control by the VDM Method*, J.Wiley & Sons, Chichester, 1995.
2. Knap, L., Holnicki-Szulc, J.: *Optimal Design of Adaptive Structures for the Best Crash-Worthiness*, Proc. 3rd World Congress on Structural and Multidisciplinary Optimisation, Buffalo, May 1999.
3. Holnicki-Szulc, J., Mackiewicz, A., Kołakowski, P.: *Design of Adaptive Structures for Improved Load Capacity*, AIAA Journal vol.36, No.3 March 1998.

MYSEL'S FORMULA FOR SMALL VIBRATIONS SUPERIMPOSED UPON LARGE STATIC DEFORMATIONS OF PIEZOELASTIC BODIES*

HANS IRSCHIK AND UWE PICHLER
Division of Technical Mechanics, University of Linz
A-4040 Linz-Auhof
Email: hans.irschik@jku.at, uwe.pichler@jku.at

*) Dedicated to Professor Dr. Dr. h.c. Franz Ziegler, Technical University Vienna, Austria, on the occasion of his 65th birthday

1. INTRODUCTION

Maysel´s formula was originally developed for the linear static theory of thermoelasticity, [1]. It renders the thermoelastic displacement by a convenient volume integration, using isothermal influence functions as the kernels of the integrals. Maysel's formula was brought to the knowledge of a wider audience through the book on thermoelasticity by W. Nowacki [2]. Later, Nowacki presented an important extension of this formula to the dynamic problem of piezo-thermoelasticity, [3]. Nowacki's extension makes use of Green's functions of the coupled piezo-thermoelastic problem. The proof of Maysel's original formula presented by Parkus [4] demonstrates that the simple form of Maysel's original formulation can be retained also in the case of coupling between temperature and elastic deformation, because the latter coupling needs not to be addressed in the proof. The value of Maysel´s original formula thus lies in the fact that known solutions of an auxiliary isothermal force problem can be utilized for presenting a formal solution of the linear coupled thermoelastic problem. The coupling to the thermal field can be often neglected in practical applications, particularly in the case of quasi-static motions.

Several technically important applications and extensions of Maysel's formula were presented by F. Ziegler and his co-workers, see the review articles [5], [6], [7] and [8], and the literature cited therein. In the following, we only mention some papers directly attributed to Maysel's formula. Various static structural applications concerning beams, plates, shells and plane problems were considered in the state-of-the-art review by Ziegler and Irschik [9]. An extension considering auxiliary problems with different boundary conditions was also presented in [9], opening Maysel's formula to the computational field of Boundary Element Methods. It was moreover pointed out in Ref. [9] that Maysel's formula may be used for eigenstrain loading in general. The name eigenstrain originally was created by Mura [10] for non-compatible strains such as thermal expansion strains and plastic misfit strains. An extension of Maysel's formula to the dynamic eigenstrain problem was given by Irschik, Fotiu and Ziegler [11]. The application of Maysel's formula to dynamic problems of small strain elasto-plasticity is documented in the review articles [5]–[8]. The analogy between electric field and temperature was utilized by Irschik and Ziegler [12] and Irschik, Krommer and Ziegler [13] in order to apply Maysel's formula to small strain vibrations of piezoelastic structures. These formulations gave raise to exact solutions of the problem of shape control of small strain linear displacements by eigenstrain actuation, see Irschik and Ziegler [14] for the static case, Irschik and Pichler [15] for the problem of linear vibrations, and Irschik [16] for a review. A first account of an extension of Maysel's formula to small vibrations superimposed upon large static deformations was presented in the review article by Irschik and Ziegler [5]. Maysel's formula of thermoelasticity was extended to static deformations of anisotropic thermoelastic materials at finite strain by Irschik, Pichler et al. [17].

The present paper is concerned with an extension of Maysel's formula to small vibrations superimposed upon large static deformations of piezoelastic bodies. Our presentation rests upon the comprehensive expositions on the non-linear theory of continuum mechanics of electromagnetic solids to be found in the literature, particularly on the books by Parkus [18] and Maugin [19]. We first derive a reciprocity relation for large deformations superimposed upon an arbitrarily strained static intermediate configuration in the presence of ponderomotive forces. We thereby use a valuable formulation of piezoelasticity with a symmetric second Piola-Kirchhoff tensor introduced by Toupin [20]. We then specialize the reciprocity theorem to the case of small dynamic piezoelastic deformations superimposed upon large deformations. A special choice of the dummy problem in the reciprocity relation eventually leads to extensions of Graffi's theorem and Maysel's formula of the linear theory of thermoelasticity. Since we deal with the non-linear theory, the common assumption of linear piezoelasticity to neglect the ponderomotive forces, see Kamlah [21] for a justification, is performed only in the last step of our derivation.

2. A GENERAL RECIPROCITY RELATION

2.1. Fundamental Equations in Material Formulation

The local form of the equation of balance of momentum reads in the material formulation of continuum electromechanics:

$$\rho_0 \boldsymbol{a} = DIV \, \boldsymbol{P} + \boldsymbol{b}_0 + \boldsymbol{b}_0^{em}. \tag{1}$$

Here, ρ_0 denotes the density in the reference configuration, u is the displacement vector to the actual configuration, a denotes the acceleration, DIV stands for the divergence with respect to the place X in the reference configuration, P is the first Piola-Kirchhoff stress tensor, and b_0 is the imposed body force per unit volume in the reference configuration. The material electromagnetic or ponderomotive volume force is denoted as b_0^{em}. Cauchy's classical fundamental theorem on stresses at the reference surface of the body ∂B_0 with normal vector n_0 is replaced by

$$P n_0 = t_0 + t_0^{em}, \qquad (2)$$

where t_0 is the Lagrange surface traction, and t_0^{em} is the ponderomotive traction in the material description. The first Piola-Kirchoff stress tensor P is connected to the second Piola-Kirchoff tensor S by the transformation

$$P = FS, \qquad (3)$$

where F denotes the deformation gradient. In the mechanics of non-polar continua, the equation of balance of angular momentum asserts that S is symmetric, $S = S^T$, the subscript T denoting the transpose of a tensor. In electromechanics, however, the skew tensor

$$S - S^T = W^{em} \qquad (4)$$

generally can not be assumed to vanish. This is due to the presence of a couple stress tensor and a surface couple density, both of electromagnetic origin.

The above sketched presence of the ponderomotive forces and electromagnetic couples, linking the balance equations of mechanics explicitly to the Maxwell equations of electrodynamics, makes the theory highly complex. For some thorough presentations, we particularly refer to the books by Parkus [18] and Maugin [19]. The problem is somewhat simplified in the present case of piezoelectric bodies, which can be assumed to be non-magnetizable and non-conducting. Nevertheless, b_0^{em}, t_0^{em} and W^{em} generally do not vanish in the latter case, too.

2.2. Introduction of a Static Intermediate State

We first introduce an intermediate configuration of the body, indicated by the subscript (i) in the following. We assume that this configuration, called the intermediate state, exhibits large strains with respect to the unstrained reference configuration. The intermediate state is assumed to represent a static equilibrium position. Hence, all of the vectors and tensors identified by the subscript (i) are understood as being time-invariant. A static state additionally is designated by a superposed hat subsequently. The increments from the intermediate state to the actual state are assumed to be due to a dynamic motion, and they are designated by the index (+). Since we refer both, the intermediate and the actual state, to the same reference configuration, we have the following additive decomposition in the above equations:

$$\begin{aligned} &\boldsymbol{u} = \hat{\boldsymbol{u}}_i + \boldsymbol{u}_+, \quad \boldsymbol{F} = \hat{\boldsymbol{F}}_i + \boldsymbol{F}_+, \\ &\boldsymbol{b}_0 = \hat{\boldsymbol{b}}_{0i} + \boldsymbol{b}_{0+}, \quad \boldsymbol{t}_0 = \hat{\boldsymbol{t}}_{0i} + \boldsymbol{t}_{0+}, \\ &\boldsymbol{b}_0^{em} = \hat{\boldsymbol{b}}_{0i}^{em} + \boldsymbol{b}_{0+}^{em}, \quad \boldsymbol{t}_0^{em} = \hat{\boldsymbol{t}}_{0i}^{em} + \boldsymbol{t}_{0+}^{em}, \\ &\boldsymbol{P} = \hat{\boldsymbol{P}}_i + \boldsymbol{P}_+, \quad \boldsymbol{S} = \hat{\boldsymbol{S}}_i + \boldsymbol{S}_+. \end{aligned} \quad (5)$$

The increments in Eq.(5) need not to be small. Note that Eq.(3) yields

$$\boldsymbol{P}_+ = \hat{\boldsymbol{F}}_i \boldsymbol{S}_+ + \boldsymbol{F}_+ \hat{\boldsymbol{S}}_i + \boldsymbol{F}_+ \boldsymbol{S}_+ . \quad (6)$$

Due to the additive decomposition given in Eqs.(5), the local equation of balance of momentum, Eq.(1), and Cauchy's fundamental theorem on stresses, Eq.(2), do also apply for the increments. We now perform a scalar multiplication of the corresponding incremental version of the equation of balance of momentum by a smooth test vector field $\tilde{\boldsymbol{u}}$. We then integrate over the volume of the body in the reference configuration, B_0. From the divergence theorem, we have

$$\int_{B_0} DIV \, \boldsymbol{P}_+ \cdot \tilde{\boldsymbol{u}} \, dV = \int_{\partial B_0} (\boldsymbol{P}_+^T \tilde{\boldsymbol{u}}) \cdot \boldsymbol{n}_0 \, dS - \int_{B_0} \boldsymbol{P}_+ \cdot \tilde{\boldsymbol{H}} \, dV , \quad (7)$$

where $\tilde{\boldsymbol{H}} = GRAD \, \tilde{\boldsymbol{u}}$. We use the tensorial notation introduced by Gurtin [22], a dot denoting the inner product of two vectors or second order tensors, respectively. With Eq.(2), the result of the volume integration can be written as

$$\int_{B_0} \rho_0 \boldsymbol{a}_+ * \tilde{\boldsymbol{u}} \, dV = \int_{\partial B_0} (\boldsymbol{t}_{0+} + \boldsymbol{t}_{0+}^{em}) * \tilde{\boldsymbol{u}} \, dS + \int_{B_0} (\boldsymbol{b}_{0+} + \boldsymbol{b}_{0+}^{em}) * \tilde{\boldsymbol{u}} \, dV \\ - \int_{B_0} \boldsymbol{P}_+ * \tilde{\boldsymbol{H}} \, dV , \quad (8)$$

where the asterisk (*) indicates that we have applied an additional convolution. This means that the first term in a product is to be taken at some time τ, and the second one at a shifted time $(t-\tau)$, the instant $\tau=0$ being referred to the intermediate state. One then integrates over the time interval under consideration, see Gurtin [22] for the properties of convolution integrals.

2.3. General Reciprocity Relation for Actual and Dummy Increments

In order to obtain a reciprocity relation, we assume the test vector field $\tilde{u}$ to represent a dynamic virtual displacement from the static intermediate state, produced by a dummy system of forces and electromagnetic fields. The same procedure as for the derivation of Eq.(8) then is applied to this dummy problem, using the actual increment u_+ as test function. With an obvious notation, this yields

$$\int_{B_0} \rho_0 \tilde{a} * u_+ \, dV = \int_{\partial B_0} (\tilde{t}_0 + \tilde{t}_0^{em}) * u_+ \, dS + \int_{B_0} (\tilde{b}_0 + \tilde{b}_0^{em}) * u_+ \, dV - \int_{B_0} \tilde{P} * H_+ \, dV. \tag{9}$$

Recall that both, the intermediate and the actual and virtual increments in Eqs.(8) and (9) may represent large deformations. The intermediate state is contained in these relations only implicitly. Since the intermediate state is time-invariant, however, we may assume that all of the initial conditions are homogeneous. Integration by parts then yields

$$a_+ * \tilde{u} = u_+ * \tilde{a}. \tag{10}$$

Eqs.(8) and (9) thus can be combined to the following reciprocity relation concerning the virtual work of the increments of the imposed and ponderomotive forces and the virtual work of the incremental first Piola-Kirchhoff stress:

$$\int_{\partial B_0} (t_{0+} + t_{0+}^{em}) * \tilde{u} \, dS + \int_{B_0} (b_{0+} + b_{0+}^{em}) * \tilde{u} \, dV - \int_{B_0} P_+ * \tilde{H} \, dV$$
$$= \int_{\partial B_0} (\tilde{t}_0 + \tilde{t}_0^{em}) * u_+ \, dS + \int_{B_0} (\tilde{b}_0 + \tilde{b}_0^{em}) * u_+ \, dV - \int_{B_0} \tilde{P} * H_+ \, dV. \tag{11}$$

This reciprocity relation holds for large incremental deformations, and irrespective of the form of the constitutive relations of the problem in hand. Neglecting the ponderomotive forces and identifying the intermediate state with the unstrained reference configuration, $\hat{F}_i = I$, the dynamic relation Eq.(11) in the static limit leads to Hill's theorem [23] for large static deformations.

2.4. Reciprocity Relation in Case of Elastic Dielectrics

As is discussed in the book by Maugin [19], a convenient reformulation of Eqs.(1)–(4) is possible that reveals the simple structure of the classical theory of continuum mechanics with a symmetric stress tensor. In the case of non-magnetizable elastic dielectrics, this formulation was introduced by Toupin [20]. Motivated by this work, we replace the first Piola-Kirchhoff stress tensor in Eq.(1) by

$$\boldsymbol{P}^e = \boldsymbol{F}\boldsymbol{S}^e, \tag{12}$$

such that the new second Piola-Kirchhoff stress tensor $\boldsymbol{S}^e$ is symmetric:

$$\boldsymbol{S}^e - \boldsymbol{S}^{eT} = \boldsymbol{0}. \tag{13}$$

This re-formulation requires to replace the genuine ponderomotive body forces and surface tractions by counterparts that are responsible for the symmetry of $\boldsymbol{S}^e$. Let the newly introduced ponderomotive forces be denoted by $\boldsymbol{b}_0^e$, and the newly introduced ponderomotive surface tractions by $\boldsymbol{t}_0^e$. These new type of ponderomotive forces are not expressed explicitly in Ref.[19], but they can be computed from the formulas given there. Then Eq.(11) reads

$$\int_{\partial B_0} (\boldsymbol{t}_{0+} + \boldsymbol{t}_{0+}^e) * \tilde{\boldsymbol{u}}\, dS + \int_{B_0} (\boldsymbol{b}_{0+} + \boldsymbol{b}_{0+}^e) * \tilde{\boldsymbol{u}}\, dV - \int_{B_0} \boldsymbol{P}_+^e * \tilde{\boldsymbol{H}}\, dV$$
$$= \int_{\partial B_0} (\tilde{\boldsymbol{t}}_0 + \tilde{\boldsymbol{t}}_0^e) * \boldsymbol{u}_+\, dS + \int_{B_0} (\tilde{\boldsymbol{b}}_0 + \tilde{\boldsymbol{b}}_0^e) * \boldsymbol{u}_+\, dV - \int_{B_0} \tilde{\boldsymbol{P}}^e * \boldsymbol{H}_+\, dV. \tag{14}$$

3. SMALL DYNAMIC DISPLACEMENTS SUPERIMPOSED UPON LARGE STATIC DEFORMATIONS

3.1. Linearization of the Incremental State

When the increments superimposed upon the intermediate state are small, they represent the linearizations of the respective entities about the intermediate state in the direction of the incremental displacements. For some concise discussions of this linearization technique, we refer to the expositions of Stein and Barthold [24] and Bonet and Wood [25]. In the present case, Eqs.(6) and (12) yield immediately the linearized increments of the first Piola-Kirchhoff stress as

$$\boldsymbol{P}_+^e = \boldsymbol{H}_+ \hat{\boldsymbol{S}}_i^e + \hat{\boldsymbol{F}}_i \boldsymbol{S}_+^e, \tag{15}$$

$$\tilde{P}^e = \tilde{H}\,\hat{S}^e_i + \hat{F}_i\,\tilde{S}^e\ . \tag{16}$$

The linearizations of the Green strain tensor are given by

$$E_+ = \tfrac{1}{2}(H^T_+\hat{F}_i + \hat{F}^T_i H_+)\ , \tag{17}$$

$$\tilde{E} = \tfrac{1}{2}(\tilde{H}^T \hat{F}_i + \hat{F}^T_i \tilde{H})\ . \tag{18}$$

3.2. Virtual Work of the First Piola-Kirchhoff Stress

The virtual workings of the increments of the first Piola-Kirchhoff stress done on the small superimposed deformations in Eq.(14) become

$$P^e_+ * \tilde{H} = H_+\,\hat{S}^e_i * \tilde{H} + \hat{F}_i\,S^e_+ * \tilde{H}\ , \tag{19}$$

$$\tilde{P}^e * H_+ = \tilde{H}\,\hat{S}^e_i * H_+ + \hat{F}_i\,\tilde{S}^e * H_+\ . \tag{20}$$

see Eqs.(15) and (16). Taking into account the following properties, valid for every second-order tensor A, B and D,

$$A \cdot (B\,D) = (A\,D^T) \cdot B = (B^T\,A) \cdot D\ , \tag{21}$$

$$A * B = B * A\ , \tag{22}$$

and utilizing the symmetry of the second Piola-Kirchhoff stress S^e, Eq.(14), it is found that

$$H_+\,\hat{S}^e_i * \tilde{H} = \tilde{H}\,\hat{S}^e_i * H_+ \tag{23}$$

in Eqs.(19) and (20). Furthermore, for every second-order tensor A and a symmetric second-order tensor S there is

$$S * A = S * \tfrac{1}{2}(A + A^T)\ . \tag{24}$$

It thus follows that

$$\hat{F}_i\,S^e_+ * \tilde{H} = S^e_+ * \tilde{H}^T \hat{F}_i = S^e_+ * \tilde{E}\ , \tag{25}$$

$$\hat{F}_i \check{S}^e * H_+ = \check{S}^e * H_+^T \hat{F}_i = \check{S}^e * E_+ , \qquad (26)$$

see Eqs.(19) and (20). Hence, in case of small displacements superimposed upon the intermediate state, the reciprocity theorem of Eq. (14) takes on the form

$$\int_{\partial B_0} (t_{0+} + t_{0+}^e) * \tilde{u} \, dS + \int_{B_0} (b_{0+} + b_{0+}^e) * \tilde{u} \, dV - \int_{B_0} S_+^e * \tilde{E} \, dV$$
$$= \int_{\partial B_0} (\tilde{t}_0 + \tilde{t}_0^e) * u_+ \, dS + \int_{B_0} (\tilde{b}_0 + \tilde{b}_0^e) * u_+ \, dV - \int_{B_0} \check{S}^e * E_+ \, dV . \qquad (27)$$

3.3. Incremental Constitutive Relations

In short, the second Piola-Kirchhoff stress in a non-linear piezoelastic body can be taken as a function of the Green strain, of the material electric field vector and of temperature, compare Maugin [19] for a similar formulation. In the case of small deformations superimposed upon the intermediate state, by linearization about the intermediate configuration we thus obtain the following constitutive relations for the increments

$$S_+^e = C_i [E_+] + M_+ , \qquad (28)$$

where C_i is the fourth-order tensor of elastic coefficients. The second order stress actuation tensor M_+ gathers the influence of the electric and thermal fields. When we consider small increments ε_+ of the electric field and small increments of temperature θ_+, then M_+ can be taken as a linear mapping in the form

$$M_+ = M_{\varepsilon i} \varepsilon_+ + M_{\theta i} \theta_+ . \qquad (29)$$

In Eq. (29), $M_{\varepsilon i}$ denotes the third order tensor of piezoelectric constants, and $M_{\theta i}$ is the second order stress-actuation tensor due to thermal expansion. For the case of small deformations from the unstrained reference configuration, $\hat{F}_i = I$, see e.g. Kamlah [21] and Carlson [26]. It is important to note that the tensors of material parameters C_i, $M_{\varepsilon i}$ and $M_{\varepsilon i}$ in the present case do depend on the Green strain, the electric field and the temperature in the intermediate state, and therefore are to be considered as non-homogeneous functions of the place in the reference configuration. In a fully coupled theory, the influence of the deformation upon ε_+ and θ_+ is taken into account in the electromechanical and thermodynamic field equations, as well as in the ponderomotive forces.

For the dummy problem, we write the counterpart of Eq.(28) as

$$\tilde{S}^e = \tilde{C}_i [\tilde{E}] + \tilde{M}. \tag{30}$$

Here, we have taken the liberty of choosing a tensor of elastic coefficients $\tilde{C}_i$ and a stress-actuation tensor $\tilde{M}$ that differ from the tensors C_i and M_+, respectively. The difference in $\tilde{M}$ follows simply by considering different increments of electric field and temperature in the dummy problem. We furthermore are allowed to use a different tensor of elastic coefficients $\tilde{C}_i$, since the above derivations only require the deformation gradient $\hat{F}_i$ and the second Piola-Kirchhoff stress $\hat{S}_i^e$ to be common in the intermediate state. These latter two static fields can be arrived in a non-unique manner, assigning different constitutive behaviour, forces and electromagnetic fields to the body in the reference configuration.

Eqs.(28) and (30) now are used to reformulate the virtual workings of the second Piola-Kirchhoff tensors in Eq. (27). Noting that for every second-order tensor A and B and a fourth-order tensor C there is

$$C[A] * B = C^T [B] * A, \tag{31}$$

we obtain the relations

$$S_+^e * \tilde{E} = C_i [E_+] * \tilde{E} + M_+ * \tilde{E}, \tag{32}$$

$$\tilde{S}^e * E_+ = \tilde{C}_i^T [E_+] * \tilde{E} + \tilde{M} * E_+. \tag{33}$$

4. AN EXTENSION OF GRAFFI'S THEOREM

The following result rests upon the assumption that we are allowed to chose the dummy tensor of elastic coefficients in Eq.(33) as

$$\tilde{C}_i^T = C_i. \tag{34}$$

When the stress is derivable from a potential function, the symmetry of the tensors of the elastic coefficients follows from a standard argument. The same tensor of elastic coefficients, $\tilde{C}_i = C_i$, then has to be used in the actual and the dummy problem.

Inserting Eqs. (32)–(34) into the reciprocity relation of Eq.(27) yields an extension of Graffi's dynamic theorem of the linear thermoelastic theory to the case of small dynamic piezoelastic displacements superimposed upon large static deformations,

$$\int_{\partial B_0} (t_{0+} + t^e_{0+}) * \tilde{u} \, dS + \int_{B_0} (b_{0+} + b^e_{0+}) * \tilde{u} \, dV - \int_{B_0} M_+ * \tilde{E} \, dV$$

$$= \int_{\partial B_0} (\tilde{t}_0 + \tilde{t}^e_0) * u_+ \, dS + \int_{B_0} (\tilde{b}_0 + \tilde{b}^e_0) * u_+ \, dV - \int_{B_0} \tilde{M} * E_+ \, dV . \tag{35}$$

For the linear theory of thermoelasticity, in which the intermediate state corresponds to the reference configuration, $\hat{F}_i = I$, see Carlson [26].

5. AN EXTENSION OF MAYSEL'S FORMULA

We now choose a restricted dummy problem, namely a problem in which the piezoelectric and thermal effects are neglected, such that

$$\tilde{t}^e_0 = 0, \quad \tilde{b}^e_0 = 0, \quad \tilde{M} = 0 . \tag{36}$$

This special selection of a dummy incremental problem does not restrict the validity of the above formulations, and it could have been introduced in all of the above relations from the onset. When we furthermore adopt the common assumption in engineering applications of piezoelectricity to neglect the actual ponderomotive forces,

$$t^e_0 = 0, \quad b^e_0 = 0 , \tag{37}$$

then we arrive at an equation which is of the type of Maysel's integration method:

$$\int_{\partial B_0} \tilde{t}_0 * u_+ \, dS + \int_{B_0} \tilde{b}_0 * u_+ \, dV = -\int_{B_0} M_+ * \tilde{E} \, dV$$

$$+ \int_{\partial B_0} t_{0+} * \tilde{u} \, dS + \int_{B_0} b_{0+} * \tilde{u} \, dV . \tag{38}$$

In the sense of Maysel's original formulation [1], see [9] and Ziegler [27], we finally study a problem in which the incremental body forces b_{0+} are absent, and in which the surface tractions t_{0+} vanish at those parts of the boundary, on which the dummy displacements do not vanish. We thus obtain the following simple relation

$$\int_{\partial B_0} \tilde{t}_0 * u_+ \, dS + \int_{B_0} \tilde{b}_0 * u_+ \, dV = -\int_{B_0} M_+ * \tilde{E} \, dV . \tag{39}$$

Particularly, take as the dummy loading a single impulsive unit force, applied at the place $\bar{X}$ at the time-instant $\bar{t}$ in the direction of some unit vector e_k in the reference configuration. Then the following expression for the dynamic displacement increment u_+ in the direction of the force is obtained:

$$1\, e_k \cdot u_+(\bar{X}, t-\bar{t}) = -\int_0^t \int_{B_0} \tilde{E}_k(X, \bar{X}; \tau-\bar{t}) \cdot M_+(X, \tau)\, dV(X)\, d\tau \,. \tag{40}$$

For the case of small dynamic displacements from the unstrained reference configuration, $\hat{F}_i = I$, see Irschik, Fotiu and Ziegler [11]. For the case of small static displacements superimposed upon large static deformations, see Irschik, Pichler et al. [17]. We note that the formulations given in Eqs. (38)–(40) do not mean that we are restricted to an uncoupled problem. Only the ponderomotive forces have been neglected, however the influence of the mechanical deformation upon ε_+ and θ_+ can be retained in the electromechanical and thermodynamic field equations. When we are allowed to neglect the latter coupling also, Eq.(40) can be used to compute directly the incremental displacements u_+, given the solution of the dummy problem, $\tilde{E}_k$. When the cited coupling is present, Eq.(40) nevertheless holds and describes the formal structure of the solution of the problem in hand.

6. CONCLUSIONS

It has been shown how Maysel's formula of the linear theory of thermoelasticity extends to the case of small piezoelectrically induced vibrations superimposed upon large static deformations of hyperelastic bodies. Special emphasis has been laid upon discussing under which assumptions Maysel's classical formula for small vibrations about an unstrained reference configuration does apply directly.

6. ACKNOWLEDGEMENT

The present paper has been performed in the framework of the K+ Linz Center of Competence in Mechatronics (LCM). Support of the LCM by the Austrian K+ program and by the Government of Upper Austria is gratefully acknowledged. We especially acknowledge the participation of Siemens Cooperate Technology (Munich, Germany, Group of Professor Dr. Hans Meixner) in the LCM. The latter industrial cooperation has motivated to undertake the above investigations.

7. REFERENCES

1 Maysel, V.M. (1941) *A generalization of the Betti-Maxwell theorem to the case of thermal stresses and some of its applications*, Dokl. Akad. Sci. USSR **30**, 115-118.

2. Nowacki, W. (1962) *Thermoelasticity*, Pergamon Press, Oxford.
3. Nowacki, W. (1975) *Dynamic problems of Thermoelasticity*, Noordhoff, Leyden.
4. Parkus, H. (1976) *Thermoelasticity*, 2nd ed., Springer-Verlag, Wien-New York.
5. Irschik H. and Ziegler F. (1988) *Dynamics of linear elastic structures with selfstress: A unified treatment for linear and nonlinear problems*, Zeitschrift für Angewandte Mathematik und Mechanik (ZAMM) **68**, 199-205.
6. Irschik, H. and Ziegler, F. (1995) *Dynamic processes in structural thermo-viscoplasticity*, Applied Mechanics Reviews **48**, 301-315.
7. Ziegler, F. and Irschik H. (1998) *Nonlinear structural vibrations and waves: A multiple field concept*, in *Modelling and Simulation based Engineering*, edited by Atluri, S.N., ODonughue, P.E., Proc. of the International Conference on Computational Engineering Science, Vol. 1, Tech. Science Press.
8. Ziegler, F. (2002) *The multiple field concept of structural analyses: with an outlook on deformation control*, in *Proc. Fifth World Congress on Computational Mechanics*, edited by Mang, H.A. and Rammerstorfer, F.G., Technical University of Vienna, Austria.
9. Ziegler, F. and Irschik H. (1987) *Thermal stress analysis based on Maysel's formula*, in *Thermal Stresses II*, edited by Hetnarski, R. B., North-Holland, Amsterdam.
10. Mura T. (1991), *Micromechanics of Defects in Solids*, 2nd ed., Kluwer, Dordrecht.
11. Irschik, H., Fotiu, P. Ziegler, F. (1993) *Extension of Maysel's Formula to the Dynamic Eigenstrain Problem*, J. Mechanical Behavior of Materials **5**, 59-66.
12. Irschik, H. and Ziegler F. (1996) *Maysel's formula generalized for piezoelectric vibrations: Application to thin shells of revolution*, AIAA Journal **34**,
13. Irschik H., Krommer M. and Ziegler F. (1997) *Dynamic Green's Function Method Applied to Vibrations of Piezoelectric Shells*, in *Proc. 1st Int. Conference on Control of Oscillations and Chaos (COC'97)* Vol.3, edited by Chernousko, F.L., and Fradkov, A.L., IEEE, Piscataway, NJ.
14. Irschik, H. and Ziegler, F. (2001) *Eigenstrain without Stress and Static Shape Control of Structures*, AIAA-Journal **39**, 1985-1999.
15. Irschik, H., Pichler, U. (2001) *Dynamic Shape Control of Solids and Structures by Thermal Expansion Strains*, Journal of Thermal Stresses **24**, 565-576.
16. Irschik, H. (2002) *A review on static and dynamic shape control of structures by piezoelectric actuation*. Engineering Structures **24**, 5-11.
17. Irschik, H., Pichler, U., Gerstmayr, J. H. and Holl, H. J. (2001) *Maysel's formula of thermoelasticity extended to anisotropic materials at finite strain*, Int. J. Solids and Struct. **38**, 9479-9492.
18. Parkus, H. (1972) *Magneto-thermoelasticity*, Springer-Verlag, Wien-New York.
19. Maugin, G.A. (1988) *Continuum Mechanics of Electromagnetic Solids*, North-Holland, Amsterdam.
20. Toupin, R.A. (1963) *A Dynamical Theory of Dielectrics*, Int. J. Eng. Sciences **1**, 101-126.
21. Kamlah, M. (2001) *Ferroelectric and ferroelastic piezoceramics modelling of electromechanical hysteresis phenomena*, Continuum Mech. Thermodyn. **13**, 219-268.
22. Gurtin, M. E. (1972) *The linear theory of elasticity*, in *Handbuch der Physik, Vol. VIa/2*, edited by Flügge, S., Springer-Verlag, Berlin.
23. Hill, R. (1957) *On uniqueness and stability in the theory of finite elastic strain*. J. Mechanics and Physics of Solids **5**.
24. Stein, E. and Barthold, F. J. (1996) *Elastizitätstheorie*, in *Der Ingenieurbau*, edited by Mehlhorn, G., Ernst und Sohn, Berlin.
25. Bonet, J. and Wood R. D. (1997) *Nonlinear Continuum Mechanics for Finite Element Analysis*. Cambridge University Press, Cambridge.
26. Carlson, D.E. (1972) *Linear Thermoelasticity*, in *Handbuch der Physik, Band VIa/2*, edited by Flügge, S., Springer-Verlag, Berlin.
27. Ziegler, F. (1998) *Mechanics of Solids and Fluids*, 2nd ed., 2nd printing. Springer-Verlag, NewYork, Vienna.

THE ANALYSIS OF TRANSIENT THERMAL STRESSES IN PIEZOTHERMOELASTIC SEMI-INFINITE BODY WITH AN EDGE CRACK

M. ISHIHARA*, O. P. NIRAULA** and N. NODA*
*Department of Mechanical Engineering, Shizuoka University
3-5-1, Johoku, Hamamatsu, Shizuoka, 432-8561, JAPAN
E-mail: tmmishi@ipc.shizuoka.ac.jp
**Physics Department,Tribhuvan University Amrit Campus
G. P. O. Box 6108, Kathmandu, NEPAL

1. INTRODUCTION

Piezoelectric ceramics have been attracted much attention as one of important elements of smart structures in recent years. They are often subjected to severe temperature environment and, due to their brittleness, tend to develop cracks under thermal load. Therefore, evaluation of thermal effect on the cracks in a piezoelectric body is very important and the problems of the cracks within the body are treated by many researchers. Lu, Tan and Liew [1] studied a two-dimensional piezoelectric material with an elliptic cavity under a uniform heat flow. Shen and Kuang [2] presented interface cracks in bi-piezothermoelastic media and the interaction with a point heat source. Wang and Noda [3] discussed fracture of a piezoelectric material strip under steady thermal load, and investigated thermally induced fracture of a smart functionally graded composite structure [4].

As often observed, piezoelectric ceramics develop cracks from their surface due to severe thermal load. Therefore, Niraula and Noda treated a problem of the crack breaking at the surface of piezothermoelastic semi-infinite body [5] and a strip [6] under steady thermal load. However, cracks often propagate under a sudden change of thermal load. Therefore, the problem of cracks under transient thermal load is important and Wang and Noda [7] discussed a surface thermal shock fracture of a semi-infinite piezoelectric medium under one-dimensional temperature change.

In this paper, we aim at the solution needed to evaluate the propagation of an surface-breaking crack in a piezothermoelastic body due to suddenly changing thermal load and to seek the methods to arrest the propagation. Our analytical model to accomplish this aim is a two-dimensional piezothermoelastic semi-infinite body with an edge crack which is subjected to the sudden themperature change as an unavoidable thermal environment and is subjected to the electrical displacement to compensate the effect of the thermal environment. We use the potential method suggested by Ashida, Tauchert and Noda [8] to obtain the piezothermoelastic field. The Laplace transform with respect to the time variable is introduced to treat the transient field. The problem is reduced to a problem of a singular integral equation. The thermal stress intensity factor (SIF) of the crack is obtained by solving the equation by the collocation method.

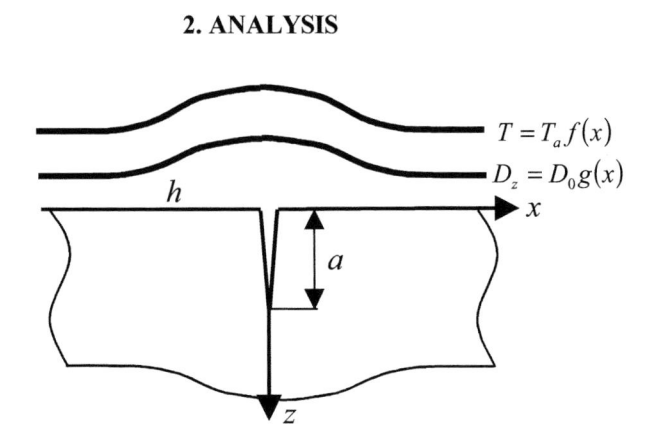

2. ANALYSIS

Figure 1. Analytical model.

The analytical model is a two-dimensional piezothermoelastic semi-infinite body with an edge crack of length a, as shown in Figure 1. From the initial time $t=0$, the surface $z=0$ is subjected to the temperature change $T_a f(x)$. It is assumed to be an unavoidable thermal environment which would intensify the thermoelastic field at the crack tip. Furthermore, the surface $z=0$ is also subjected to the electric displacement $D_0 g(x)$ in z-direction in order to compensate the effect of the thermal environment. Let $f(x)$ and $D_0 g(x)$ be expressed as $f(x) = \int_0^\infty f^*(\alpha)\cos\alpha x\, d\alpha$ and $D_0 g(x) = \int_0^\infty D_1(\alpha)\cos\alpha x\, d\alpha$, respectively.

The governing equation and the boundary and initial conditions for the temperature distribution $T(x,z,t)$ are as follows:

$$\frac{1}{\kappa}\frac{\partial T}{\partial t} = \frac{\partial^2 T}{\partial x^2} + \lambda^2 \frac{\partial^2 T}{\partial z^2}, \quad \left(\lambda^2 = \frac{\lambda_z}{\lambda_x}\right) \tag{1}$$

$$T = T_a f(x) \text{ on } z = 0; \quad T = 0 \text{ at } t = 0 \tag{2}$$

where λ_x and λ_z denote the thermal conductivities in the x- and z-directions, respectively. The solutions of Eqs. (1) and (2) is obtained in the Laplace domain with respect to t as

$$\overline{T} = \int_0^\infty \frac{T_a f^*(\alpha)}{p} e^{-sz} \cos\alpha x\, d\alpha, \quad \left(s = \sqrt{p + \kappa\alpha^2}/(\lambda\sqrt{\kappa})\right) \tag{3}$$

where the function with an overbar denotes the Laplace transform of the function and p denotes the parameter of the Laplace transform hereafter.

The constitutive equations for the two-dimensional piezothermoelastic field are expressed as

$$\left.\begin{array}{l}\sigma_{xx} = c_{11}\varepsilon_{xx} + c_{13}\varepsilon_{zz} - e_1 E_z - \beta_1 T;\ \sigma_{zz} = c_{13}\varepsilon_{xx} + c_{33}\varepsilon_{zz} - e_3 E_z - \beta_3 T \\ \sigma_{zx} = c_{55}\varepsilon_{zx} - e_5 E_x \\ D_x = e_5\varepsilon_{zx} + \eta_1 E_x;\ D_z = e_1\varepsilon_{xx} + e_3\varepsilon_{zz} + \eta_3 E_z + p_3 T \\ \left(\varepsilon_{xx} = u_{x,x};\ \varepsilon_{zz} = u_{z,z};\ \varepsilon_{xz} = u_{x,z} + u_{z,x};\ E_x = -\Phi_{,x};\ E_z = -\Phi_{,z}\right)\end{array}\right\} \tag{4}$$

where σ_{xx}, σ_{zz} and σ_{zx} denote stresses; ε_{xx}, ε_{zz} and ε_{zx} denote strains; D_x and D_z denote electric displacements; E_x and E_z denote electric fields; u_x and u_z denote displacements; Φ denotes electric potential. The equations of equilibrium for stresses and electric displacements are

$$\sigma_{xx,x} + \sigma_{zx,z} = 0;\ \sigma_{zx,x} + \sigma_{zz,z} = 0;\ D_{x,x} + D_{z,z} = 0 \tag{5}$$

The displacements and the electric potential satisfying Eqs. (4) and (5) can be expressed as

$$u_x = \left(\phi_1 + \phi_2 + \sum_{i=1}^3 m_i \psi_i\right)_{,x};\ u_z = \left(k_1\phi_1 + j\phi_2 + \sum_{i=1}^3 n_i \psi_i\right)_{,z};\ \Phi = \chi + \sum_{i=1}^3 l_i \psi_{i,z} \tag{6}$$

where ϕ_1 is the piezothremoelastic potential function which is a particular solution of

$$\left(\frac{\partial^2}{\partial x^2} + \mu_1 \frac{\partial^2}{\partial z^2}\right)\left(\frac{\partial^2}{\partial x^2} + \mu_2 \frac{\partial^2}{\partial z^2}\right)\left(\frac{\partial^2}{\partial x^2} + \mu_3 \frac{\partial^2}{\partial z^2}\right)\phi_1 = \delta_1\left(\frac{\partial^2}{\partial x^2} + \gamma_1 \frac{\partial^2}{\partial z^2}\right)\left(\frac{\partial^2}{\partial x^2} + \gamma_2 \frac{\partial^2}{\partial z^2}\right)T$$

$$...(7)$$

ϕ_2 and χ are particular solutions of equations:

$$\phi_{2,zz} = (\phi_{1,xx} + v_1\phi_{1,zz} - \delta_1 T)/\xi_1 \; ; \; \chi_{,z} = (\phi_{2,xx} + v_2\phi_{2,zz} - \delta_2 T)/\xi_2 \tag{8}$$

and $\psi_i \, (i=1,2,3)$ is the piezoelastic potential function satisfying

$$\psi_{i,xx} + \mu_i \psi_{i,zz} = 0 \; (i=1,2,3) \tag{9}$$

In Eqs. (6) to (9), l_i, m_i, n_i, k_1, j, δ_i, γ_i, μ_i, v_i and ξ_i are the constants determined by the material properties in Eq. (4) and their definitions are given in [8]. The mechanical and electrical boundary conditions are given as follows:

$$\left.\begin{array}{ll}
\sigma_{zx} = 0 \text{ at } z=0, 0 \le x \le \infty \; ; & \sigma_{zz} = 0 \text{ at } z=0, 0 \le x \le \infty \\
\sigma_{xz} = 0 \text{ at } x=0, 0 \le z \le \infty \; ; & \sigma_{xx} = 0 \text{ at } x=0, 0 \le z \le a \\
u_x = 0 \text{ at } x=0, a \le z \le \infty \; ; & \sigma_{ij} = \text{value at } \sqrt{x^2+z^2} \to \infty \\
D_z = D_0 g(x) \text{ on } z=0 \; ; & D_x = 0 \text{ at } x=0, 0 \le z \le \infty \\
D_z = \text{value at } z \to \infty &
\end{array}\right\} \tag{10}$$

The potential functions in the Laplace domain, $\overline{\phi}_1$, $\overline{\phi}_2$ and $\overline{\chi}$, are obtained as

$$\{\overline{\phi}_1, \overline{\phi}_2, \overline{\chi}_1\} = \int_0^\infty \{g_1(\alpha), g_2(\alpha), g_3(\alpha)\} e^{-sz} \cos\alpha x \, d\alpha \tag{11}$$

and the general solution of Eq. (9) in the Laplace domain is obtained as

$$\overline{\psi}_i = \int_0^\infty \overline{A_i(\alpha)} e^{-\alpha z/\sqrt{\mu_i}} \cos\alpha x \, d\alpha + \int_0^\infty \overline{B_i(q)} e^{-q\sqrt{\mu_i}x} \cos qz \, dq \; (i=1,2,3) \tag{12}$$

where

$$\left.\begin{array}{l}
g_1(\alpha) = \dfrac{(\gamma_1 s^2 - \alpha^2)(\gamma_2 s^2 - \alpha^2)}{(\mu_1 s^2 - \alpha^2)(\mu_2 s^2 - \alpha^2)(\mu_3 s^2 - \alpha^2)} \delta_1 T_a f^*(\alpha)/p \\[2mm]
g_2(\alpha) = \dfrac{(v_1 s^2 - \alpha^2)g_1(\alpha) - \delta_1 T_a f^*(\alpha)/p}{\xi_1 s^2} \\[2mm]
g_3(\alpha) = -\dfrac{(v_2 s^2 - \alpha^2)g_2(\alpha) - \delta_2 T_a f^*(\alpha)/p}{\xi_2 s}
\end{array}\right\} \tag{13}$$

and $\overline{A_i(\alpha)}$ and $\overline{B_i(q)}$ are the unknown functions to be determined by the mechanical and electrical boundary conditions. By using Eqs. (4), (6), (11) through (13) and the Laplace inversion, u_x, u_z, σ_{xx}, σ_{zz}, σ_{xz}, D_x and D_z, are obtained as, for example,

$$\sigma_{xx} = \int_0^\infty \left[-\sum_{i=1}^3 \frac{\alpha^2}{\mu_i} (c_{11} m_i \mu_i - c_{13} n_i - e_1 l_i) A_i(\alpha) e^{-\alpha z/\sqrt{\mu_i}} + T_a f^*(\alpha) S_{xx}(\alpha, z, t) \right] \cos \alpha x \, d\alpha$$

$$+ \int_0^\infty \sum_{i=1}^3 q^2 (c_{11} m_i \mu_i - c_{13} n_i - e_1 l_i) B_i(q) e^{-q\sqrt{\mu_i} x} \cos qz \, dq \qquad (14)$$

In Eq. (14), the term including $S_{xx}(\alpha, z, t)$ comes from ϕ_1, ϕ_2 and χ. The definition of $S_{xx}(\alpha, z, t)$ and those of $S_{zz}(\alpha, z, t)$, $S_{zx}(\alpha, z, t)$ and $D_z(\alpha, z, t)$ obtained in the same manner for σ_{zz}, σ_{zx} and D_z, respectively, are as follows:

$$S_{xx}(\alpha, z, t) =$$

$$\delta_1 \sum_{i=1}^3 \left\{ F_{2,i}(\alpha, z, t) \frac{C_i}{\mu_i} \left[(\mu_i c_{11} - k_1 c_{13}) + (\mu_i c_{11} - j c_{13}) \frac{v_1 - \mu_i}{\xi_1} - \frac{e_1}{\xi_1 \xi_2} (v_1 - \mu_i)(v_2 - \mu_i) \right] \right\}$$

$$+ F_1(\alpha, z, t) \left\{ \frac{e_1 \delta_2}{\xi_2} + \beta_1 - \delta_1 \sum_{i=1}^3 \frac{C_i}{\mu_i} \left[c_{13}\left(k_1 + j \frac{v_1 - \mu_i}{\xi_1}\right) + \frac{e_1 v_2}{\xi_1 \xi_2}(v_1 - \mu_i) \right] \right\}$$

$$S_{zz}(\alpha, z, t) =$$

$$\delta_1 \sum_{i=1}^3 \left\{ F_{2,i}(\alpha, z, t) \frac{C_i}{\mu_i} \left[(\mu_i c_{13} - k_1 c_{33}) + (\mu_i c_{13} - j c_{33}) \frac{v_1 - \mu_i}{\xi_1} - \frac{e_3}{\xi_1 \xi_2} (v_1 - \mu_i)(v_2 - \mu_i) \right] \right\}$$

$$+ F_1(\alpha, z, t) \left\{ \frac{e_3 \delta_2}{\xi_2} + \beta_3 - \delta_1 \sum_{i=1}^3 \frac{C_i}{\mu_i} \left[c_{33}\left(k_1 + j \frac{v_1 - \mu_i}{\xi_1}\right) + \frac{e_3 v_2}{\xi_1 \xi_2}(v_1 - \mu_i) \right] \right\}$$

$$S_{zx}(\alpha, z, t) =$$

$$- \left\langle \delta_1 \sum_{i=1}^3 F_{4,i}(\alpha, z, t) \frac{C_i}{\mu_i} \left\{ c_{55}\left[(1+k_1) + (1+j)\frac{v_1 - \mu_i}{\xi_1}\right] + \frac{e_5}{\xi_1 \xi_2}(v_1 - \mu_i)(v_2 - \mu_i) \right\} \right.$$

$$\left. + F_3(\alpha, z, t) \left\| -\frac{\delta_2 e_5}{\xi_2} + \delta_1 \sum_{i=1}^3 \frac{C_i}{\mu_i} \left\{ c_{55}\left[(1+k_1) + (1+j)\frac{v_1 - \mu_i}{\xi_1}\right] + \frac{e_5 v_2}{\xi_1 \xi_2}(v_1 - \mu_i) \right\} \right\| \right\rangle$$

$$DZ(\alpha, z, t) =$$

$$\delta_1 \sum_{i=1}^3 \left\{ F_{2,i}(\alpha, z, t) \frac{C_i}{\mu_i} \left[(\mu_i e_1 - k_1 e_3) + (\mu_i e_1 - j e_3) \frac{v_1 - \mu_i}{\xi_1} + \frac{\eta_3}{\xi_1 \xi_2} (v_1 - \mu_i)(v_2 - \mu_i) \right] \right\}$$

$$+ F_1(\alpha, z, t) \left\{ -\frac{\eta_3 \delta_2}{\xi_2} - p_3 - \delta_1 \sum_{i=1}^3 \frac{C_i}{\mu_i} \left[e_3\left(k_1 + j \frac{v_1 - \mu_i}{\xi_1}\right) - \frac{\eta_3 v_2}{\xi_1 \xi_2}(v_1 - \mu_i) \right] \right\} \qquad (15)$$

where

$$F_1(\alpha,z,t) = \frac{1}{2}\left[e^{\frac{\alpha z}{\lambda}}\operatorname{erfc}\left(\frac{z}{2\lambda\sqrt{\kappa t}}+\alpha\sqrt{\kappa t}\right)+e^{-\frac{\alpha z}{\lambda}}\operatorname{erfc}\left(\frac{z}{2\lambda\sqrt{\kappa t}}-\alpha\sqrt{\kappa t}\right)\right]$$

$$F_{2,i}(\alpha,z,t) = \frac{\lambda^2/\mu_i}{1-\lambda^2/\mu_i}\left[F_1(\alpha,z,t)-e^{-\alpha^2(1-\lambda^2/\mu_i)\kappa t}F_1(\lambda\alpha/\sqrt{\mu_i},z,t)\right]$$

$$F_3(\alpha,z,t) = -\frac{\lambda}{2}\left[e^{\frac{\alpha z}{\lambda}}\operatorname{erfc}\left(\frac{z}{2\lambda\sqrt{\kappa t}}+\alpha\sqrt{\kappa t}\right)-e^{-\frac{\alpha z}{\lambda}}\operatorname{erfc}\left(\frac{z}{2\lambda\sqrt{\kappa t}}-\alpha\sqrt{\kappa t}\right)\right]$$

$$F_{4,i}(\alpha,z,t) = \frac{\lambda^2/\mu_i}{1-\lambda^2/\mu_i}\left[F_3(\alpha,z,t)-\frac{1}{\lambda/\sqrt{\mu_i}}e^{-\alpha^2(1-\lambda^2/\mu_i)\kappa t}F_3(\lambda\alpha/\sqrt{\mu_i},z,t)\right]$$

(16)

and C_i satisfies

$$\begin{bmatrix} \mu_2\mu_3 & \mu_3\mu_1 & \mu_1\mu_2 \\ \mu_2+\mu_3 & \mu_3+\mu_1 & \mu_1+\mu_2 \\ 1 & 1 & 1 \end{bmatrix} \begin{Bmatrix} C_1 \\ C_2 \\ C_3 \end{Bmatrix} = \begin{Bmatrix} \gamma_1\gamma_2 \\ \gamma_1+\gamma_2 \\ 1 \end{Bmatrix} \quad (17)$$

In order to determine $A_i(\alpha)$ and $B_i(q)$, we introduce the function $\omega(z)$ defined as

$$\omega(z) = \begin{cases} u_{x,z}\big|_{x=0} = \int_0^\infty q^2 \sum_{i=1}^{3} m_i\sqrt{\mu_i}B_i(q)\sin qz\,dq\,; & 0 \leq z \leq a \\ 0\,; & z \geq a \end{cases} \quad (18)$$

Then, from Eqs. (10), $A_i(\alpha)$ and $B_i(q)$ are obtained as

$$\frac{\alpha^2}{\mu_i}A_i(\alpha) = A_i^*(\alpha) + \int_0^a A_i''(\alpha,z')\omega(z')dz'\,, \quad B_i(q) = \frac{2B_i^*}{\pi q^2\sqrt{\mu_i}}\int_0^a \omega(z')\sin qz'\,dz' \quad (19)$$

where $A_i^*(\alpha)$, $A_i''(\alpha,z')$ and B_i^* satisfy

$$\mathbf{M}\begin{Bmatrix} A_1^*(\alpha) \\ A_2^*(\alpha) \\ A_3^*(\alpha) \end{Bmatrix} = \begin{Bmatrix} T_a f^*(\alpha)S_{zx}(\alpha,0,t) \\ -T_a f^*(\alpha)S_{zz}(\alpha,0,t) \\ -D_1(\alpha)-T_a f^*(\alpha)DZ(\alpha,0,t) \end{Bmatrix} \quad (20)$$

$$\mathbf{M}\begin{Bmatrix} A_1''(\alpha,z') \\ A_1''(\alpha,z') \\ A_1''(\alpha,z') \end{Bmatrix} = \frac{2}{\pi}\begin{Bmatrix} 0 \\ \sum_{i=3}^{\infty}(c_{13}m_i\mu_i - c_{33}n_i - e_3l_i)(B_i^*/\mu_i)e^{-\alpha z/\sqrt{\mu_i}} \\ \sum_{i=3}^{\infty}(e_1m_i\mu_i - e_3n_i + \eta_3l_i)(B_i^*/\mu_i)e^{-\alpha z/\sqrt{\mu_i}} \end{Bmatrix} \quad (21)$$

and

$$\begin{bmatrix} (m_1+n_1)c_{55}+e_5l_1 & (m_2+n_2)c_{55}+e_5l_2 & (m_3+n_3)c_{55}+e_5l_3 \\ (m_1+n_1)e_5-\eta_1l_1 & (m_2+n_2)e_5-\eta_1l_2 & (m_3+n_3)e_5-\eta_1l_3 \\ m_1 & m_2 & m_3 \end{bmatrix}\begin{Bmatrix} B_1^* \\ B_2^* \\ B_3^* \end{Bmatrix} = \begin{Bmatrix} 0 \\ 0 \\ 1 \end{Bmatrix} \quad (22)$$

respectively, where

$$\mathbf{M} = \begin{bmatrix} [(m_1+n_1)c_{55}+e_5l_1]\sqrt{\mu_1} & [(m_2+n_2)c_{55}+e_5l_2]\sqrt{\mu_2} & [(m_3+n_3)c_{55}+e_5l_3]\sqrt{\mu_3} \\ c_{13}m_1\mu_1 - c_{33}n_1 - e_3l_1 & c_{13}m_2\mu_2 - c_{33}n_2 - e_3l_2 & c_{13}m_3\mu_3 - c_{33}n_3 - e_3l_3 \\ e_1m_1\mu_1 - e_3n_1 + \eta_3l_1 & e_1m_2\mu_2 - e_3n_2 + \eta_3l_2 & e_1m_3\mu_3 - e_3n_3 + \eta_3l_3 \end{bmatrix}$$
...(23)

By applying the condition for σ_{xx} in Eqs. (10), we finally obtain a singular integral equation for $\omega(z)$ as

$$\int_0^a \left[\frac{1}{z'-z} + R(z,z')\right]\omega(z')dz' = RR(z,t); \quad (0 < z < a) \quad (24)$$

where

$$\begin{aligned} R(z,z') &= \frac{1}{z+z'} - \frac{1}{CB}\sum_{i=1}^{3}(c_{11}m_i\mu_i - c_{13}n_i - e_1l_i)\int_0^{\infty} A_i''(\alpha,z')e^{-\alpha z/\sqrt{\mu_i}}d\alpha \\ RR(z,t) &= \frac{1}{CB}\int_0^{\infty}\left[\sum_{i=1}^{3}(c_{11}m_i\mu_i - c_{13}n_i - e_1l_i)A_i^*(\alpha) + T_a f^*(\alpha)S_{xx}(\alpha,z,t)\right]d\alpha \\ CB &= \sum_{i=1}^{3}(c_{11}m_i\mu_i - c_{13}n_i - e_1l_i)B_i^*/(\pi\sqrt{\mu_i}) \end{aligned} \quad (25)$$

By introducing

$$\omega(z) = \Psi(z)/\sqrt{a-z} \qquad (26)$$

the solution of Eq. (24) can be obtained using the collocation technique. Then, the thermal stress intensity factor at the crack tip is obtained as

$$K_I = -2\sqrt{2\pi}\,\Psi(a)CB \qquad (27)$$

3. NUMERICAL EXAMPLES

We consider the cadmium selenide material as an example and its material properties [9] are given as

$$\left.\begin{array}{l} c_{11} = 74.1[\text{GPa}],\ c_{13} = 39.3[\text{GPa}],\ c_{33} = 83.6[\text{Gpa}],\ c_{55} = 13.17[\text{GPa}] \\ \hat{a}_1 = 0.621[\text{MPa/K}],\ \hat{a}_3 = 0.551[\text{MPa/K}] \\ e_1 = -0.160[\text{C/m}^2],\ e_3 = 0.347[\text{C/m}^2],\ e_5 = -0.138[\text{C/m}^2] \\ \varsigma_1 = 8.25\times10^{-11}[\text{C}^2/\text{Nm}^2],\ \varsigma_3 = 9.02\times10^{-11}[\text{C}^2/\text{Nm}^2] \\ p_3 = -2.94\times10^{-6}[\text{C/m}^2\text{K}] \end{array}\right\} \qquad (28)$$

The thermal conductivity for cadmium selenide could not be found in the literature. Therefore, we assume $\lambda^2 = 0.25$. The distribution functions of the temperature and electric displacement applied to the surface $z = 0$ are given as

$$f(x) = H(b-|x|),\quad g(x) = H(c-|x|) \qquad (29)$$

where $H(\cdot)$ denotes the Heaviside's unit step function and b and c are a half of the width of the temperature and electric displacement, respectively. In order to illustrate the examples, we use non-dimensional quantities as follows:

$$\hat{b} \equiv \frac{b}{a},\ \hat{c} \equiv \frac{c}{a},\ \tau \equiv \frac{\kappa t}{a^2},\ \hat{K}_I \equiv \frac{K_I}{\sqrt{\pi a}\,\alpha_x Y_x T_a},\ \hat{D}_0 \equiv \frac{D_0}{\alpha_x T_a \sqrt{Y_x \eta_1}} \qquad (30)$$

where Y_x and α_x are the Young's modulus and the coefficient of linear thermal expansion defined, respectively, as

$$Y_x = \frac{c_{11}c_{33} - c_{13}^2}{c_{33}},\quad \alpha_x = \frac{c_{33}\beta_1 - c_{13}\beta_3}{c_{11}c_{33} - c_{13}^2} \qquad (31)$$

Figure 2 shows the variation of the non-dimensional SIF $\hat{K}_I$ with the non-dimensional time τ for various values of the non-dimensional electric displacement $\hat{D}_0$ $(\hat{b}=1)$. It is found that the non-dimensional SIF $\hat{K}_I$ increases as time and it tends to a constant value as time tends to infinity. It is also found that all the time the non-dimensional SIF $\hat{K}_I$ without applied electric displacement $\hat{D}_0=0$ is positive, which means that the thermal stress σ_{xx} around the crack tip is intensified in such manner as $r^{-1/2}$ for $r \to 0$ (r: distance from the crack tip) even when the surface $z=0$ is subjected to temperature rise. This fact derives from the assumption $\lambda^2 = 0.25$ which means that heat flows more in the x-direction than in the z-direction. From Figure 2, it is also found that the non-dimensional SIF $\hat{K}_I$ is increased by the positive electric displacement and it is reduced by the negative electric displacement. An important aspect obtained from this example is that the stress intensity factor by the thermal load on the surface which has a crack can be compensated by applying some proper electric displacement on the same surface.

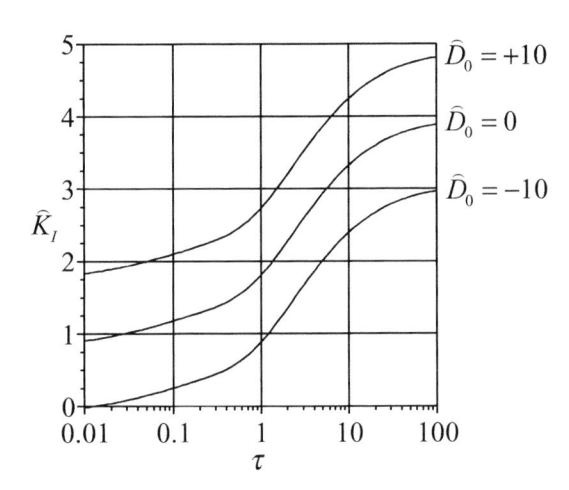

Figure 2. The non-dimensional stress intensity factor $\hat{K}_I$ with the non-dimensional time τ $(\hat{b}=1, \hat{c}=1)$.

4. CONCLUSION

We analyze the problem of the surface-breaking crack in a semi-infinite piezoelectric body subjected to the suddenly changing thermal load and to the electric displacement. The transient piezothermoelastic field is obtained by the potential method [8] and the Laplace transform. The stress intensity factor is obtained by the solution of a singular integral equation. Using numerical examples, it is found that the stress intensity factor due to the suddenly changing thermal load can be compensated by applying the electric displacement to the cracked surface.

REFERENCES

1. Lu, P., Tan, M.J. & Liew, K.M. (1998). Piezothermoelastic analysis of a piezoelectric material with an elliptic cavity under uniform heat flow. Archive of Applied Mechanics, 68, 719-733.
2. Shen, S. & Kuang, Z.-B. (1998). Interface crack in bi-piezothermoelastic media and the interaction. International Journal of Solids and Structures, 35, 3899-3915.
3. Wang, B.L. & Noda, N. (2001). Fracture of a piezoelectric material strip under steady thermal load. Journal of Thermal Stresses, 24, 281-299.
4. Wang, B.L. & Noda, N. (2001). Thermally induced fracture of a smart functionally graded composite structure. Theoretical and Applied Fracture Mechanics, 35, 93-109.
5. Niraula, O.P. & Noda, N. (2002). The analysis of thermal stresses in piezothermoelastic semi-infinite material with a crack. Archive of Applied Mechanics, 72, 119-126.
6. Niraula, O.P. & Noda, N. (2002). Thermal stress analysis of piezothermoelastic strip with an edge crack. Journal of Thermal Stresses, 25, 389-405.
7. Wang, B.L. & Noda, N. (2001). Thermally induced fracture of a smart functionally graded structure, Theoretical and Applied Fracture Mechanics, 35, 93-109.
8. Ashida, F., Tauchert, T.R. & Noda, N. (1994). A general solution technique for piezothermoelasticity of hexagonal solids of class 6mm in cartesian coordinates. ZAMM, 74, 87-95.
9. Berlincourt, D., Jaffe, H. & Shiozawa, L.R. (1963). Electroelastic Properties of the Sulfides, Selenides, and Tellurides of Zinc and Cadmium. Physical Review, 129, 1009-1017.

TRANSIENT DYNAMIC STRESSES AROUND A RECTANGULAR CRACK IN A NONHOMOGENEOUS LAYER BETWEEN TWO DISSIMILAR ELASTIC HALF-SPACES

SHOUETSU ITOU
Department of Mechanical Engineering, Kanagawa University
Rokkakubashi, Kanagawa-ku, Yokohama, 221-8686, Japan
E-mail: itous001@kanagawa-u.ac.jp

1. INTRODUCTION

For an interface crack, one of the physically acceptable solutions seems to be provided by Delale and Erdogan [1]. They considered that a crack does not exist in the interface between two dissimilar elastic half-planes but in the thin interfacial layer. They assumed that the elastic constants vary continuously across the interfacial layer within the range from the elastic constants of the upper half-plane to those of the lower half-plane. If a plastic material is reinforced by an aluminum fiber, a thin diffusion layer possibly connects the two materials mechanically. Thus, Delale and Erdogan's solutions are meaningful from an engineering viewpoint. The stress singularity at the crack tip is the same of the conventional solution for a homogeneous cracked elastic plate.

If composite materials joined by a cracked nonhomogeneous layer are loaded suddenly, it is necessary to clarify the transient dynamic stress intensity factors. Babaei and Lukasiewicz solved the transient dynamic problem for a crack in a nonhomogeneous layer between two dissimilar elastic half-planes [2]. The crack surfaces are loaded suddenly by anti-plane shear traction, and the dynamic stress intensity factor for Mode III loading is obtained. Later, Itou solved the corresponding Mode I solution [3].

In the previous paper, a transient problem has been solved for a rectangular crack in an adhesive layer between two dissimilar elastic half-spaces [4]. In the present paper, two dissimilar elastic half-spaces are connected with a thin nonhomogeneous layer. A sudden internal pressure is applied to the surfaces of a rectangular crack in the layer. The material constants in the layer are assumed to vary continuously within a range from those of the upper half-space to those of the lower half-space. In order to circumvent the difficulty associated with the cracked nonhomogeneous layer, the interfacial layer is divided into several homogeneous sub-layers that have different material properties. By letting the number of sub-layers approach infinity, the stresses and displacements can be obtained for the nonhomogeneous layer.

The problem is first formulated with the aid of Fourier and Laplace transforms, and then the mixed conditions with respect to stresses and displacements are reduced to dual integral equations in the Laplace domain. In order to solve the equations, the differences in the displacements of the crack are expanded into a double series of functions that become zero outside the crack. Unknown coefficients in the series are determined using the Schmidt method [5]. Stress intensity factors are defined in the Laplace domain and are inverted numerically in physical space using Miller and Guy's method [6].

2. FUNDAMENTAL EQUATIONS

A nonhomogeneous interfacial layer is sandwiched between two elastic half-spaces, as shown in Fig. 1. A rectangular crack is located at $x_2 = 0$ along the x_1-axis from $-a$ to a and along the x_3-axis from $-b$ to $+b$ with reference to the rectangular coordinate system (x_1, x_2, x_3). The cracked layer (A) is denoted by $(-H_B \leq x_2 \leq H_C)$. An upper half-space (C) and a lower half-space (B) are denoted by $(H_C \leq x_2)$ and $(x_2 \leq -H_B)$, respectively. The shear modulus, Poisson's ratio and the density of the layer (A) are represented by μ_A, ν_A and ρ_A. Those of the lower half-space (B) and upper half-space (C) are represented using the subscripts B and C, respectively. To solve the problem, layer (A) must be further divided into layer (A-1), denoted by $(0 \leq x_2 \leq H_C)$, and layer (A-2), denoted by $(-H_B \leq x_2 \leq 0)$. Elastic constants (μ_A, ρ_A, ν_A) most likely vary continuously with respect to x_2 in the interfacial layer, as shown in Fig. 2.

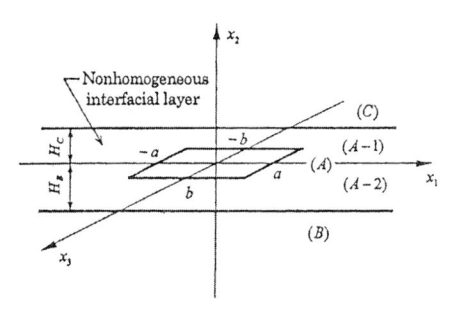

Figure 1. Rectangular crack and coordinate system.

The incident stresses that propagate through the upper half-space (C) can be expressed in the form:

$$\tau_{22C}^{(inc)} = pH(t + x_2/c_{LC}), \quad \tau_{23C}^{(inc)} = \tau_{31C}^{(inc)} = \tau_{12C}^{(inc)} = 0 \tag{1}$$

with

$$c_{LC} = [(\lambda_C + 2\mu_C)/\rho_C]^{1/2} \tag{2}$$

where p is a constant, $H(t)$ is the Heaviside unit step function, c_{LC} is the longitudinal wave velocity, and the subscript C indicates variables for the upper half-space (C). Variables having the superscript (inc) represent those for an incident stress field. If a wave is incident on the interfacial layer, it is reflected and

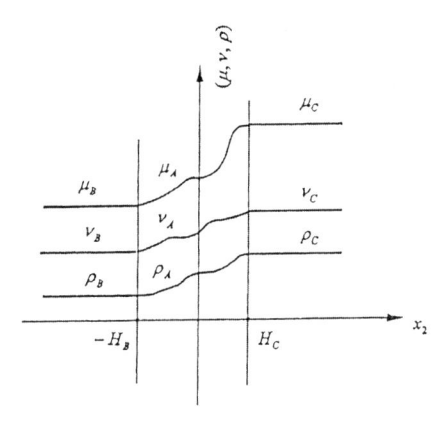

Figure 2. Variation of elastic constants in interfacial layer.

refracted at the interface $x_2 = H_C$, at the crack surfaces, and at the interface $x_2 = -H_B$, in a complicated manner. However, it is very likely that stress waves similar to Eq. (1) pass across the crack. Therefore, the boundary conditions for the problem to be studied are assumed to be:

$$\tau_{22A1}^{0} = -pH(t), \quad \tau_{12A1}^{0} = 0, \quad \tau_{23A1}^{0} = 0 \quad \text{for } x_2 = 0, \; |x_1| \le a, \; |x_3| \le b, \tag{3}$$

$$u_{1A2}^{0} = u_{1A1}^{0}, \quad u_{2A2}^{0} = u_{2A1}^{0}, \quad u_{3A2}^{0} = u_{3A1}^{0},$$
$$\text{for } x_2 = 0, \; (|x_1| \le a, \; b \le |x_3|), \; (a \le |x_1|, \; |x_3| \le \infty), \tag{4}$$

$$\tau_{22A2}^{0} = \tau_{22A1}^{0}, \quad \tau_{12A2}^{0} = \tau_{12A1}^{0}, \quad \tau_{23A2}^{0} = \tau_{23A1}^{0},$$
$$\text{for } x_2 = 0, \; |x_1| \le \infty, \; |x_3| \le \infty, \tag{5}$$

$\tau_{22C} = \tau_{22A1}^0$, $\tau_{12C} = \tau_{12A1}^0$, $\tau_{23C} = \tau_{23A1}^0$, $u_{1C} = u_{1A1}^0$, $u_{2C} = u_{2A1}^0$, $u_{3C} = u_{3A1}^0$,

for $x_2 = H_C$, $|x_1| \le \infty$, $|x_3| \le \infty$, (6)

$\tau_{22A2}^0 = \tau_{22B}$, $\tau_{12A2}^0 = \tau_{12B}$, $\tau_{23A2}^0 = \tau_{23B}$, $u_{1A2}^0 = u_{1B}$,

$u_{2A2}^0 = u_{2B}$, $u_{3A2}^0 = u_{3B}$, for $x_2 = -H_B$, $|x_1| \le \infty$, $|x_3| \le \infty$, (7)

where the superscript 0 refers to quantities evaluated at $x_2 = 0$. Variables with the subscripts A1, A2, B and C are those for layer (A-1), layer (A-2), lower half-space (B) and upper half-space (C), respectively.

3. DIVISION OF INTERFACIAL LAYER INTO SUB-LAYERS

The nonhomogeneous interfacial layer (A) is divided into several homogeneous sub-layers that have different material properties. The number of the sub-layers, l, should be odd rather than even. In order to illustrate the process by which the problem is solved, l is set to three. If, $l = 3$ the layer (A) is divided into four layers because the sub-layer that contains a crack is divided into two separate layers.

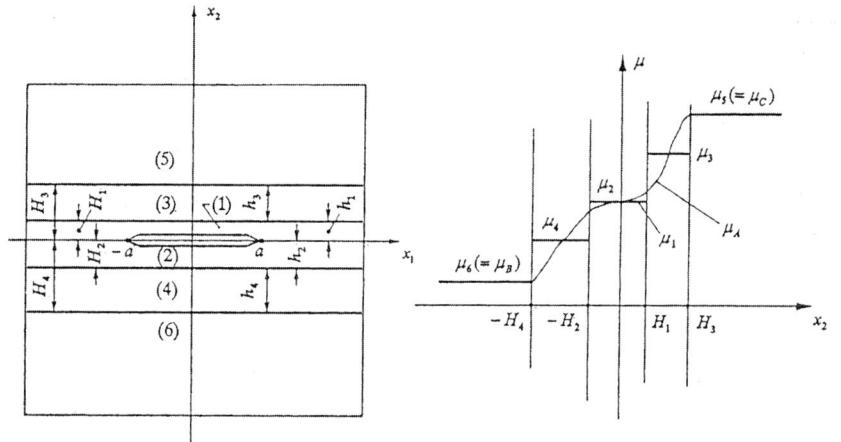

Figure 3. Interfacial layer replaced by three sub-layers.

Figure 4. Shear moduli in sub-layers used to represent interfacial layer.

More precisely, the interfacial layer (A) is divided into sub-layer (1) ($0 \le x_2 \le H_1$), sub-layer (2) ($-H_2 \le x_2 \le 0$), sub-layer (3) ($H_1 \le x_2 \le H_3$) and sub-layer (4) ($-H_4 \le x_2 \le -H_2$), as shown in Fig. 3, where h_1, h_2, h_3 and h_4 are the respective thicknesses of the layers. The upper half-space (C) and the lower half-space (B) are numbered by (5) and (6), respectively. For $l = 3$, the shear moduli $\mu_i (i = 1,2,3,4)$ for the four homogeneous layers are as shown in Fig. 4, and are given by the following equations:

$$\begin{aligned}
\mu_4 &= \mu_A \quad at \quad x_2 = -H_4 + h_4/2, \\
\mu_2 &= \mu_A \quad at \quad x_2 = -H_2 + (h_2 + h_1)/2, \\
\mu_1 &= \mu_2, \\
\mu_3 &= \mu_A \quad at \quad x_2 = H_1 + h_3/2.
\end{aligned} \qquad (8)$$

Poisson's ratios v_i and densities $\rho_i (i = 1,2,3,4)$ have relationships that are quite similar to those expressed in Eq. (8).

Boundary conditions (3) ~ (7) must be changed accordingly.

4. ANALYSIS

A Laplace transform pair is defined by the equations

$$f^*(s) = \int_0^\infty f(t)\exp(-st)dt \qquad (9.1)$$

$$f(t) = 1/(2\pi i)\int_{Br.} f^*(s)\exp(st)ds \qquad (9.2)$$

where the second integral is over the Bromwich path. In addition, the two-dimensional Fourier transforms are introduced by the equations

$$\bar{\bar{f}}(\xi,x_2,\varsigma) = \int_{-\infty}^{\infty}\int_{-\infty}^{\infty} f(x_1,x_2,x_3)\exp[i(\xi x_1 + \varsigma x_3)]dx_1 dx_3 \qquad (10.1)$$

$$f(x_1,x_2,x_3) = 1/(2\pi)^2 \int_{-\infty}^{\infty}\int_{-\infty}^{\infty} \bar{\bar{f}}(\xi,x_2,\varsigma)\exp[-i(\xi x_1 + \varsigma x_3)]d\xi d\varsigma. \qquad (10.2)$$

Following his previous paper [4], it can be seen that the boundary conditions are reduced to the equation

$$\sum_{m=1}^{\infty}\sum_{n=1}^{\infty} e_1^{mn} P_{mn}(x_1,x_3) + \sum_{m=1}^{\infty}\sum_{n=1}^{\infty} e_2^{mn} Q_{mn}(x_1,x_3)$$

$$+ \sum_{m=1}^{\infty}\sum_{n=1}^{\infty} e_3^{mn} R_{mn}(x_1,x_3) = -w_0(x_1,x_3)$$

$$\sum_{m=1}^{\infty}\sum_{n=1}^{\infty} e_1^{mn} S_{mn}(x_1,x_3) + \sum_{m=1}^{\infty}\sum_{n=1}^{\infty} e_2^{mn} U_{mn}(x_1,x_3)$$

$$+ \sum_{m=1}^{\infty} \sum_{n=1}^{\infty} e_3^{mn} V_{mn}(x_1, x_3) = 0$$

$$\sum_{m=1}^{\infty} \sum_{n=1}^{\infty} e_1^{mn} W_{mn}(x_1, x_3) + \sum_{m=1}^{\infty} \sum_{n=1}^{\infty} e_2^{mn} X_{mn}(x_1, x_3) \qquad (11)$$

$$+ \sum_{m=1}^{\infty} \sum_{n=1}^{\infty} e_3^{mn} Y_{mn}(x_1, x_3) = 0$$

$$\text{for} \quad 0 \leq x_1 \leq a, \quad 0 \leq x_3 \leq b,$$

where expressions of known functions $P_{mn}(x_1, x_3), \cdots, Y_{mn}(x_1, x_3)$ and $w_0(x_1, x_3)$ are omitted and e_1^{mn}, e_2^{mn} and e_3^{mn} are the unknown coefficients to be solved using the Schmidt method [5].

5. STRESS INTENSITY

In the same manner used in [4], the stress intensity factors in the Laplace transform domain can be easily defined by the forms:

$$K_1^{a*} = \lim_{x_1 \to a+} \tau_{22(1)}^{0*} [2\pi(x_1 - a)]^{1/2}$$

$$= \sum_{m=1}^{\infty} \sum_{n=1}^{\infty} e_1^{mn} (2m-1)(-1)^m / [\pi(\pi/a)^{1/2}]$$

$$\times q_1(\xi_L, 0) / \xi_L \times \cos[(2n-1) \sin^{-1}(x_3/b)],$$

$$K_2^{a*} = \lim_{x_1 \to a+} \tau_{12(1)}^{0*} [2\pi(x_1 - a)]^{1/2}$$

$$= \sum_{m=1}^{\infty} \sum_{n=1}^{\infty} e_2^{mn} (2m)(-1)^m / [\pi(\pi/a)^{1/2}]$$

$$\times q_8(\xi_L, 0) / \xi_L \times \cos[(2n-1) \sin^{-1}(x_3/b)],$$

$$K_3^{a*} = \lim_{x_1 \to a+} \tau_{23(1)}^{0*} [2\pi(x_1 - a)]^{1/2}$$

$$= \sum_{m=1}^{\infty} \sum_{n=1}^{\infty} e_3^{mn} (2m-1)(-1)^m / [\pi(\pi/a)^{1/2}] \qquad (12)$$

$$\times q_{15}(\xi_L, 0) / \xi_L \times \sin[(2n) \sin^{-1}(x_3/b)],$$

$$K_1^{b*} = \lim_{x_3 \to b+} \tau_{22(1)}^{0*} [2\pi(x_3 - b)]^{1/2}$$

$$= \sum_{m=1}^{\infty}\sum_{n=1}^{\infty} e_1^{mn}(2n-1)(-1)^n /[\pi(\pi/b)^{1/2}]$$
$$\times q_1(\xi_L,0)/\xi_L \times \cos[(2m-1)\sin^{-1}(x_1/a)],$$

$$K_2^{b*} = \lim_{x_3 \to b+} \tau_{12(1)}^{0*}[2\pi(x_3-b)]^{1/2}$$
$$= \sum_{m=1}^{\infty}\sum_{n=1}^{\infty} e_3^{mn}(2n)(-1)^n /[\pi(\pi/b)^{1/2}]$$
$$\times q_{15}(\xi_L,0)/\xi_L \times \cos[(2m-1)\sin^{-1}(x_1/a)],$$

$$K_3^{b*} = \lim_{x_3 \to b+} \tau_{23(1)}^{0*}[2\pi(x_3-b)]^{1/2}$$
$$= \sum_{m=1}^{\infty}\sum_{n=1}^{\infty} e_2^{mn}(2n-1)(-1)^n /[\pi(\pi/b)^{1/2}] \tag{13}$$
$$\times q_8(\xi_L,0)/\xi_L \times \sin[(2m)\sin^{-1}(x_1/a)].$$

where the expressions of known functions $q_i(\xi,\zeta)$ ($i=1, 8, 15$) are omitted and ξ_L is a large value of ξ.

In Sections 3, 4 and 5, stress intensity factors K_k^{a*} and K_k^{b*} ($k=1,2,3$) were solved only for $l=3$. However, solutions for $l=5, 7$ can also be obtained in a straightforward manner. The stress intensity factors are calculated numerically for $l=3, 5, 7$ and plotted with respect to $1/l$. The stress intensity factors in the cracked nonhomogeneous layer can be obtained only if the layer is replaced by an infinite number of sub-layers. If the numerical results are plotted with respect to $1/l$, the stress intensity factors for the interfacial layer, the material constants of which vary continuously with respect to x_2, can be obtained from the values at $(1/l) \to 0$. This process is explained in detail in Section 6.

The inverse Laplace transformations of the stress intensity factors are carried out by the numerical method described by Miller and Guy [6].

We have the relation between $g^*(s)$ and $g(t)$ as

$$\lim_{s \to 0}[sg^*(s)] = \lim_{t \to \infty} g(t). \tag{14}$$

The static results of the stress intensity factors in physical space can be obtained using Eq. (14).

6. NUMERICAL EXAMPLES AND RESULTS

The transient dynamic stress intensity factors are calculated for composite materials of an epoxy resin half-space and an aluminum half-space using the quadruple precision. The material constants are listed in Table 1. Ratio a/b is fixed at 1.0, namely, K_1^a, K_2^a and K_3^a are identical to K_1^b, K_2^b and K_3^b, respectively. It is also assumed that ratio H_C/H_B is equal to 1.0. In the interfacial layer (A), the material constants are assumed to vary linearly with respect to x_2. The nonhomogeneous layer is divided into equal sub-layers. If $l = 3$, then $h_3 = h_4 = (H_B + H_C)/3$, $h_1 = h_2 = (H_B + H_C)/6$.

Table 1. Material constants.

Constants	Aluminum	Epoxy Resin
μ (GPa)	26.9	0.889
ν	0.34	0.35
ρ ($\times 10^3$ kg/m^3)	2.7	1.25

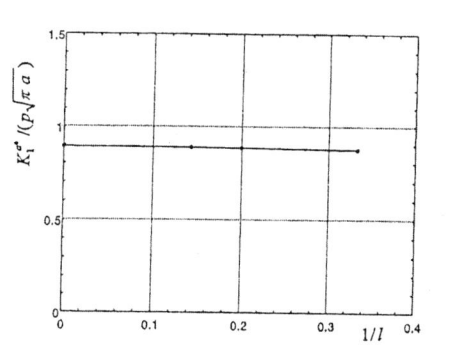

Figure 5. Curve of K_1^{a*} expressed by
$K_1^{a*} = a_1(1/l)^3 + a_2(1/l)^2 + a_3$ for H_B/a $(= H_C/a) = 0.5$, $(sa/c_{T1}) = 1.0$.

The stress intensity factors K_1^{a*} are calculated for H_B/a $(= H_C/a) = 0.5$, $(sa/c_{T1}) = 1.0$ and are then plotted with respect to $1/l$ in Fig. 5, where c_{T1} is the shear wave velocity. As the number of sub-layers, l, approaches infinity, K_1^{a*} rapidly approaches a correct value for a nonhomogeneous interfacial layer. Then, using the values for $(1/l) = 1/3$, $1/5$ and $1/7$ in Fig. 5, we approximate K_1^{a*} using the following equation:

$$K_1^{a*} = a_1(1/l)^3 + a_2(1/l)^2 + a_3 \tag{15}$$

where constants a_1, a_2 and a_3 can be easily determined. It is considered that the value of $dK_1^{a*}/d(1/l)$, which is the first order derivative of with respect to $(1/l)$, should be zero as the l approaches infinity. Thereby, K_1^{a*} can be approximated by Eq. (15). The correct value of K_1^{a*} for a cracked nonhomogeneous layer is given by a_3 in Eq. (15).

The stress intensity factors in physical space, inverted using Miller and Guy's method, are affected by the values of the parameters δ, β and N. However, if the value in physical space varies slowly with time, the numerical Laplace inversion can be performed easily, as described in a previous paper [7]. The present Laplace inversion is just such a case. All the stress intensity factors are inverted by setting $(\beta = 0.0,\ \delta = 0.2,\ N = 11)$ in Ref. [6].

In Fig. 6, values of the dynamic stress intensity factors K_1^a and K_2^a at $x_3/a = 0.0$ are plotted versus $c_{T1}t/a$ for $H_C/a = 1.0,\ 0.5$. Curves of K_3^a

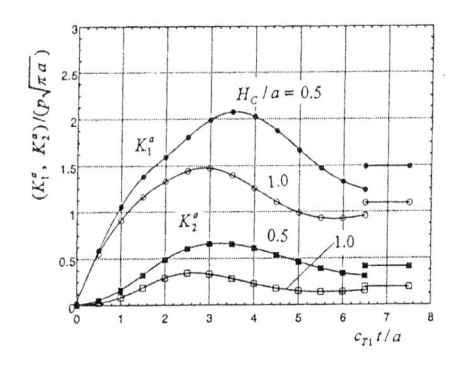

Figure 6. Dynamic stress intensity factors K_1^a and K_2^a at $x_3/a = 0.0$ plotted versus $c_{T1}t/a$ for $H_C = H_B$, $a = b$.

are not provided because they are zero. Static results are calculated using Eq. (14) and are indicated by straight lines in the figure.

7. DISCUSSION

Increasing ratio $K_1^{a(peak)}/K_1^{a(static)}$ is 1.389 $(= 2.073/1.492)$ for $H_C/a = 0.5$

with $K_1^{a(peak)}$ and $K_1^{a(static)}$ being the peak and static values of K_1^a. The increasing ratio is 1.346 $(=1.473/1.094)$ for $H_C/a=1.0$. Therefore, it is estimated that the increasing ratio is about 1.4 for K_1^a. In the same manner, it is verified that the ratio $K_2^{a(peak)}/K_2^{a(static)}$ is about 1.7.

If b/a approaches ∞, the dynamic stress intensity factors K_1^a and K_2^a at $x_3/a=0.0$ are coincident with $K_1^{plane\ strain}$ and $K_2^{plane\ strain}$ for the plane state of strain. The stress intensity factors K_1^a and K_2^a are only calculated for $a/b=1.0$. For an arbitrary a/b ratio, the values are probably smaller than $K_1^{plane\ strain}$ and $K_2^{plane\ strain}$, and these exist between those given in the present paper and those given in Ref. [3].

A peak value of the dynamic stress intensity factor $K_1^{a(peak)}/(p\sqrt{\pi a})$ around a square rectangular crack in an infinite medium is about 1.0 [8]. On the contrary, the value considerably increases as H_c/a ratio decreases for a crack in a non-homogeneous interfacial layer between an epoxy resin half-space and an aluminum half-space. Therefore, if an interfacial crack is detected in the non-homogeneous layer, it is of importance to carefully estimate the peak value of the dynamic stress intensity factors.

8. REFERENCES

1. Delale, F. and Erdogan, F. (1988) On the mechanical modeling of the interfacial region in bonded half-planes, ASME J. Appl. Mech. 55 317-324.
2. Babaei, R. and Lukasiewicz S. (1998) A dynamic response of a crack in a functionally graded material between two dissimilar half planes under anti-plane shear impact load, Eng. Frac. Mech. 60 479-487.
3. Itou, S. (2001) Transient dynamic stress intensity factors around a crack in a nonhomogeneous interfacial layer between two dissimilar elastic half-planes, Int. J. Solids and Structures 38 3631-3645.
4. Itou, S. (200x) Dynamic stress intensity factors around a rectangular crack in an elastic layer between two dissimilar elastic half-spaces under impact load, in contribution.
5. Morse, P. M. and Feshbach, H. (1958) Methods of Theoretical Physics, Vol. 1, McGraw-Hill, New York.
6. Miller, M. and Guy, T. (1966) Numerical inversion of the Laplace transform by use of Jacobi polynomials, SIAM J. Numer. Anal. 3 624-635.
7. Itou, S. (1983) Dynamic stress concentration around a circular hole in an infinite elastic strip, ASME J. Appl. Mech. 50 57-62.
8. Itou, S. (1980) Transient analysis of stress waves around a rectangular crack under impact load, ASME J. Appl. Mech. 47 958-959.

REMOTE SMART DAMAGE DETECTION VIA INTERNET WITH UNSUPERVISED STATISTICAL DIAGNOSIS

ATSUSHI IWASAKI and AKIRA TODOROKI
Department of Mechanical Sciences and engineering
Tokyo Institute of Technology
2-12-1,Ohokayama, Meguro-ku, Tokyo, 152-8552, Japan
E-mail: aiwasaki@ginza.mes.titech.ac.jp

TSUNEYA SUGIYA
DMW Corporation
3-27, Miyoshi-chou, Mishima-shi, Shizuoka, 441-8560, Japan

1. INTRODUCTION

Structural health monitoring system has many sensors mounted on, and evaluates damage state of the entire structure or structural components on real time. In order to prevent serious failures of civil structures such as bridge, gas piping, etc, the structural health monitoring system becomes noticeable technology recently. Especially in Japan, it is very significant to minimize seismic disaster, and to develop a system that diagnoses the condition of civil structure in short time at lower cost is an urgent subject. The present research proposes a new remote structural health monitoring system at the lowest cost.

Structural damage monitoring system requires a large number of sensors, and the structural integrity is monitored through the vast amount of measured data. For the structural health monitoring (SHM) system, distributed optical fiber strain sensors are generally employed as sensors. However, multiple kinds of sensors like speed counters, leakage sensors, gas sensors, intensity-based non-distributed fiber optic sensors and CCD cameras are required for practical SHM systems. In some cases, actuators may be necessary for closing safety valves or activating vibration exciters. If these sensors and actuators are mounted on the structures using conventional analog lead wires, the system requires a large number of bundles of analog lead wires, and this causes cumbersome handling of the bundles of lead wires. The bundles of lead wire may also cause significant increase of weight. In some cases, the bundles of the lead wires make it impractical to replace some structural components when it requires repairs or arrangements. The Internet is generally adopted to transfer digital data packets for computer networks such as e-mail, multimedia information or Web data. To transfer analog data of sensors via the Internet has already attempted in several cases[1,2]. Conventional cases employ PCs for data acquisition and transfer. The present study proposes new tiny tools for SHM via the Internet. One of the tools is a tiny size Smart terminal that has a network socket, a CPU, memory, a large capacity silicon disk tip, A/D converter, D/A converter and digital I/O ports. Linux operating system is adopted, and it has a web-server, mail-server and diagnostic methods for structural damage monitoring.

The present paper proposes a health monitoring system using this system and a new automatic damage diagnostic method for the system that does not require data of damaged state structure. The diagnostic method employs system identification from the measured data of the structure. And the damage is automatically diagnosed by testing the similarity of the identified system by statistical methods.

As an example of health monitoring system using this system, the present study deals delamination detection of a CFRP beam. Delaminations are detected from a change of a set of surface strain. As a result, the damage of the structure is successfully diagnosed with the method, automatically.

2. HEALTH MONITORING SYSTEM VIA THE INTERNET

For conventional applications that adopted Ethernet for data transfer, PCs, A/D converters and Ethernet cards were required. This requirement made the health monitoring via Ethernet cumbersome. As a solution for this problem, the present paper proposes a new tiny tool for structural health monitoring system via Internet.

2.1. Smart terminal for structural health monitoring

Smart terminal is a Linux based small computer that has a network socket, a CPU, memory, a large capacity silicon disk tip, A/D converter, D/A converter and digital I/O ports as hardware. And it has a web-server, mail-server and diagnostic methods for structural damage monitoring as software. In the health monitoring system, sensors are connected to the terminal and the terminal is connected to the Internet. Health monitoring of the structure is automatically performed by the built-in CPU, and result can be confirmed via web browser from a remote places.

The smart terminal, which we are developing now, has several A/D converter channels and D/A channels in a very small package like a lunch box. Since this new smart station has D/A converter, users can react on the basis of sensor information using these smart terminals. For example, when fracture of a gas pipeline is detected, the pipeline valves can be closed remotely. Using the smart terminals, we can have distributed sensor systems with conventional sensors.

Health monitoring system using this smart terminal is shown in following paragraph and automatic damage diagnostic technique is shown in next chapter.

Figure 1. Smart terminal

2.2. Health monitoring system using smart terminal

Using the smart terminal described above, it is possible to construct health monitoring system via Internet. Multiple kinds of conventional sensors such as optical sensors, thermometers, vibration sensors and speed counters are required for the structural health monitoring. If these sensors are mounted on the structures using conventional analog lead wires, the system requires a large number of bundles of analog lead wires. This causes many troubles as previously described. Using the smart terminal, these troublesome tasks can be neglected, and the only task we have to do is connecting the lead wires to the smart terminal. Since the smart terminal has a CPU, quasi-real-time measurements can be performed with installing a data transfer control program to avoid data collision. The smart terminal includes a web server; we can remotely confirm the diagnosis that the smart terminal sent us with some additional measurements trough the web services.

Using Ethernet for health monitoring gives quite a lot of advantages as follows.
(1) Since the Ethernet is a digital technology, this is very tough against noise. When the 10-base-F is adopted, electro magnetic noise can be neglected.
(2) Multiple sensors that are not distributed sensors can be connected in a cable.
(3) Replacements of sensors and actuators in service are very easy.
(4) Multiple kinds of sensors and actuators can be connected in a cable.
(5) Troubles of sensors and actuators do not affect network system.
(6) Changing network topology is possible by just changing plugs.
(7) Using dynamic routing and multiple network cables, trouble of a network can be automatically avoided.

3. UNSUPERVISED STATISTICAL DAMAGE DIAGNOSTIC METHOD USING SYSTEM IDENTIFICATION

This section addresses a new damage detecting method for health monitoring systems. For a usual health monitoring systems, relations between measured sensor data and damage location or size are indispensable to identify damages. These relations are derived from modeling of the entire structure or experiments. These works for obtaining the relations are very time consuming and require high computational and/or experimental cost.

The new statistical diagnostic method proposed in the present paper is a low cost simple system. The diagnostic method employs system identification using response surface(RS) and the damage is automatically diagnosed by testing the similarity of the RS by statistical methods.

The system does not require the relation between measured sensor data and damages. The method does not require a FEM model of the entire structure. This method diagnoses slight change of the relation between the measured sensor data.

3.1. System identification using response surface methodology

Response surface methodology[3] is use for the system identification in this method. Response surface methodology is employed for the process optimization in a quality-engineering field. Response surface methodology consists of a design of experiments to select the most suitable points for fitting the surfaces effectively and the least square method to regress response surfaces. Response surface is the approximation function that expresses the relationship between a response and predictors. Generally, a response surface is represented with the following formula.

$$y = f(x_1, x_2, \cdots, x_l) + \varepsilon \qquad (1)$$

Where x are predictors, y is a response, ε is a regression error and l is a number of predictors. In general, polynomials are used.

For simplification, let us consider the case that a response is approximated by quadratic polynomials of two predictors as followings.

$$y = \beta_0 + \sum_{j=1}^{l} \beta_j x_j + \sum_{j=1}^{l} \beta_{jj} x_j^2 + \sum_{i=1}^{l-1} \sum_{j=i+1}^{l} \beta_{ij} x_i x_j \qquad (2)$$

Where β is regression coefficient. If squares or interactions of the predictors x_j^2, $x_i x_j$ are replaced by new predictors $x_j (j > l)$, the formula (2) becomes the linear regression model as follows.

$$y = \beta_0 + \sum_{j=1}^{k} \beta_j x_j \qquad (3)$$

Where k is number of the predictors after the replacement. In terms of n observations, the equation (3) can be written in matrix form as follows.

$$Y = X\beta + \varepsilon \qquad (4)$$

$$X = \begin{bmatrix} 1 & x_{11} & x_{12} & \cdots & x_{1k} \\ 1 & x_{21} & x_{22} & \cdots & x_{2k} \\ \vdots & \vdots & \vdots & \ddots & \vdots \\ 1 & x_{n1} & x_{n2} & \cdots & x_{nk} \end{bmatrix}, Y = \begin{Bmatrix} y_1 \\ y_2 \\ \vdots \\ y_n \end{Bmatrix}, \beta = \begin{Bmatrix} \beta_0 \\ \beta_1 \\ \vdots \\ \beta_k \end{Bmatrix}, \varepsilon = \begin{Bmatrix} \varepsilon_1 \\ \varepsilon_2 \\ \vdots \\ \varepsilon_n \end{Bmatrix}$$

Unbiased estimator of β(**b**) is obtained using the least-square-method as follows.

$$\mathbf{b} = \left(\mathbf{X}^T \mathbf{X}\right)^{-1} \mathbf{X}^T \mathbf{Y} \qquad (5)$$

3.2. Similarity test of response surfaces using F-test

Let us consider we have two response surfaces that are created from two different sets of experiments.

$$\begin{aligned} \mathbf{Y}_1 &= \mathbf{X}_1 \boldsymbol{\beta}_1 + \boldsymbol{\varepsilon}_1 \\ \mathbf{Y}_2 &= \mathbf{X}_2 \boldsymbol{\beta}_2 + \boldsymbol{\varepsilon}_2 \end{aligned} \qquad (6)$$

Where the numbers of experiments for regression are n_1 and n_2 respectively. In order to investigate the similarity of the two response surfaces, a null hypothesis is introduced. The hypothetical definition is shown as follows.

$$H_0 : \boldsymbol{\beta}_1 = \boldsymbol{\beta}_2 \qquad (7)$$

Assuming that each error term (ε) is independent and has the same distribution in two sets of experiments. In this case, the F-statistic value F_0 is defined as follows.

$$F_0 = \frac{SSE_0 - SSE_{12}}{SSE_{12}} * \frac{n - 2p}{p} \qquad (8)$$

$$n = n_1 + n_2 \qquad SSE_{12} = SSE_1 + SSE_2$$

Where SSE represents residual sum of squares of a response surface. This F-statistic value F_0 follows F-distribution of degree of freedom(p, n-$2p$) under the null hypothesis. When both of the response surfaces are similar with each other F_0 becomes small value. Critical limit for the rejection of the hypothesis H_0 is defined as follows.

$$F_0 > F^\alpha{}_{p,n-2p} \tag{9}$$

Where α is significance level. The similarity of response surfaces is rejected when F_0 is larger than $F^\alpha{}_{p,n-2p}$.

3.3. Procedure of damage detection using system identification

Procedure of the new diagnostic method for damage detection is shown in Figure 2 and Figure 3. First, we perform system identification of a structure of intact state using response surface and create a response surface from the measured sensor data obtained from the initial state(Figure 2). The response surface is named the initial response surface. For example, data of a sensor is selected as a response and the data obtained from the adjacent sensors are selected as predictors. Of course, we can select natural frequencies obtained from vibration data instead of using the measured data directly. After the training process, damage monitoring process is started. During the monitoring process(Figure 3), a set of every sensor data is periodically obtained by cycling measurements several times. From the measured set

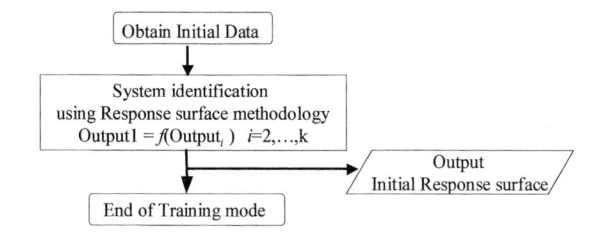

Figure 2. Flow of training process

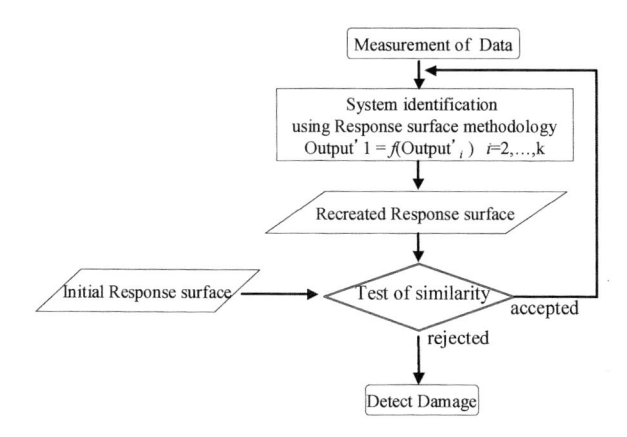

Figure 3. Flow of monitoring mode

of data, we perform system identification at the structure and a response surface is recreated. The response surface is named as recreated response surface here. The two response surfaces are compared with each other using a statistical similarity test with F-test. When the identified system of monitoring process (=recreated response surface) is discriminated from the identified system of initial state (=initial response surface), it means that relation between the sensor data is changed, and that can be concluded that something happens in the structure. Of course this does not always means damage initiation, but this can provide a low cost solution for structural health monitoring to decide the necessity of the precise investigation.

4. DAMAGE DETECTION OF CFRP BEAM USING CHANGE OF SURFACE STRAIN DISTRIBUTION

4.1. Specimen for experiment

As mentioned before, the new diagnostic method is applied to delamination monitoring of a CFRP beam, and the effectiveness of the method is experimentally investigated here. Conventional strain gages are employed as sensors in the present study. Specimen configuration of the present study is shown in Figure 4. The specimen is a CFRP cantilever beam with a thickness of 1.4mm and stacking sequence of the specimen is $[0_2/90_2]_s$.

Three conventional strain gages are mounted on the specimen surface as shown in Figure 4. Gage length of center gage is 10mm and the gage length of the side gages is 2mm.

A delamination crack is created under the center gage with a short beam shear test. Schematic illustration of the delamination is shown in Figure 5. Delamination occurs between $0°$ and $90°$ plies with matrix cracking in the middle $90°$ ply that is a trigger of the delamination initiation. Length of delamination (a) is defined as Figure 5.

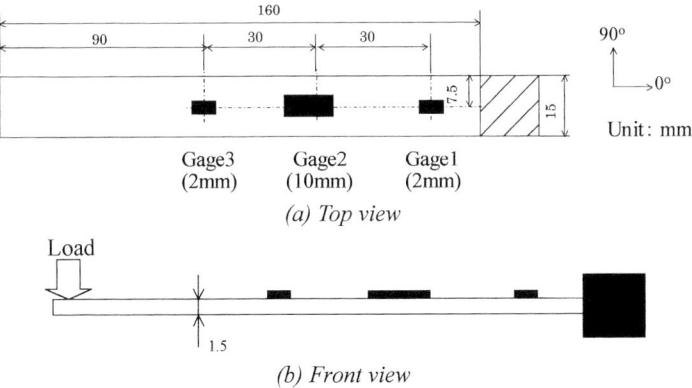

(a) Top view

(b) Front view

Figure 4. Specimen configuration

Delamination
gage
0°ply
90°ply
0°ply
Delamination size a Matrix cracking

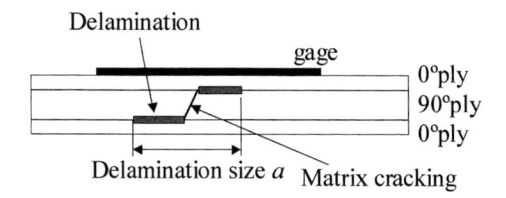

Figure 5. Schema of delamination

4.2. Damage detection method

In the present study, strain data of the middle gage #2 is the response, and the strain data of the gage #1 and #3 are predictors. Quadratic polynomials are employed for the creation of the response surface. The response surfaces are approximated as follows.

$$\varepsilon_2 = \beta_0 + \beta_1 \varepsilon_1^2 + \beta_2 \varepsilon_3^2 + \beta_3 \varepsilon_1 \varepsilon_3 + \beta_4 \varepsilon_1 + \beta_5 \varepsilon_3 \qquad (10)$$

Where ε_i represents the strain data of gage #i. In the experiments, right end of the specimen is held and excitation is applied to left end of specimen. A response surface of the equation (10) is regressed from the strain 30 data sets selected from time series of strain of gages at random.

Since the critical limit for the similarity test of response surfaces must be defined first, similarity tests between the response surface of the initial state and the response surface measured before creation of a delamination must be performed. 20000 tests were conducted to obtain distribution of the F_0 value of the intact state. Figure 6 shows frequency distribution of F_0 of the intact state. In the figure, the ordinate represents probability, and the horizontal axis shows range of F-value. When 99% confidence probability is adopted, the limit line is placed on 3.20 as shown in Figure 6.

4.3. Identification results

After the definition of the limit, delamination detections of the CFRP beams were conducted. 20000 tests of various delamination sizes were conducted. From the experimental difficulties to create delamination cracks of the exact length, the delamination length is quantized into four levels of spacing of 2mm. Table 1 shows the average value of F_0 statistics for each size level. F_0 increases uniformly according to the increase of the delamination length.

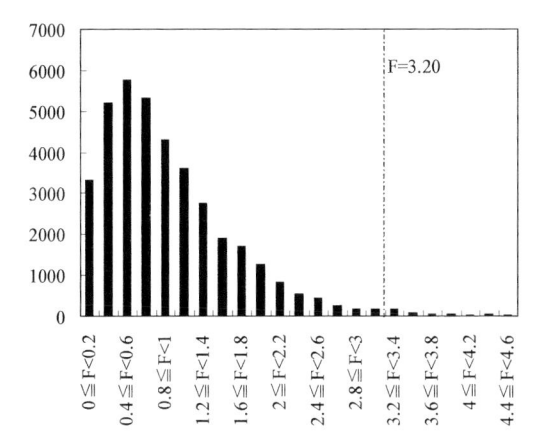

Figure 6. Frequency distribution of F_0

Table 1. Average of F_0 of each size region

Delamination Length	Average of F0
Intact	1.44
$0 < a \leqq 2$	2.46
$2 < a \leqq 4$	2.54
$4 < a \leqq 6$	2.90
$6 < a \leqq 8$	33.18
$8 < a \leqq 10$	52.38
$10 < a \leqq 12$	59.99

Table 2. Diagnostic accuracy of detection of delamination

Delamination Length	Reliability of Estimation
Intact	99.0%
$0 < a \leqq 2$	19.2%
$2 < a \leqq 4$	43.3%
$4 < a \leqq 6$	50.6%
$6 < a \leqq 8$	96.8%
$8 < a \leqq 10$	99.9%
$10 < a \leqq 12$	100.0%

Using the limit of 3.20 defined before, the performance tests of the diagnostic method are conducted to investigate the effectiveness of the method. Table 2 shows probability of diagnosis. For the in that case (delamination size is zero), the similarity tests of the two response surfaces passed by the performance of 99%. When the delamination crack is longer than 6mm, the similarity of the two response surfaces is rejected by the performance of 96.8%, which is almost perfect diagnosis for delamination existences. On the basis of the performance results, the new diagnostic method provides high performance at low cost. The new method does not require test of damaged state to define the limit between the intact state and damaged state.

5. CONCLUSIONS

By conducting similarity tests of the two response surfaces with F-test, the present paper describes the new damage diagnostic method without measurements of damaged state. As an example of the health monitoring system by this method, the present study deals monitoring of delamination detections of a composite beam. The delaminations are detected from the slight changes of measured strain data mounted on the specimen surface by using the new statistical tools. The new method employs response surfaces and F-Statistics to discriminate two response surfaces obtained from different sets of measured data, the results obtained are the follows.

(1) A new statistical diagnostic method is proposed. The method discriminates the similarity of the two response surfaces that are created two sets of measured data using the statistical F-test.
(2) Critical limit to discriminate the intact state from the damaged state can be defined only from the F_0-distribution of the intact state. This causes that the method does not require a large number of experiments of damaged state.
(3) The new method is applied to delamination detections using conventional strain gages. This method successfully detected delamination cracks larger than 6mm by the almost perfect performance.

5. REFERENCES

1. Felippa, C. A., K. C. Park and M. R. Justino Filho. 1998. "The construction of free-free flexibility matrices as generalized stiffness inverses", *Computers & Structures*, 68-4(1998), p.411-418
2. Kameyama,M., Y.Ogi and H.Fukunaga. 1999. "DAMAGE IDENTIFICATION OF LAMINATED PLATES USING VIBRATION DATA", Proc.6th Japan International SAMPE Symposium, (1999), p.987-990.
3. Okafor, A. C., K. Chandrashekhara and Y. P. Jiang. 1996. "Delamination prediction in composite beams with built-in piezoelectric devices using modal analysis and neural network," *Smart Mater. Struct.*, 5 (1996), p.338-347.
4. Keilers, C. H. Jr. and F. Chang. 1995. "Identifying Delamination in Composite Beams Using Built-in Piezoelectric: Part II - An Identification method." *J. Intell. Matls. Sys. & Struct.*, 6 (1995), p.664-672.
5. Myers,R.H. and D.C.Montgomery. 1995. "Response Surface Methodology: Process and Product optimization Using Designed Experiments", John Wiley & Sons. Inc, (1995)

SMART ACTUATION FROM COUPLING BETWEEN ACTIVE POLYMER GELS AND FIBROUS STRUCTURES

Can we learn from biological systems ?

GEORGE JERONIMIDIS
Centre for Biomimetics, Department of Engineering,
The University of Reading, Whiteknights, Reading RG6 2AY
E-mail:g.jeronimidis@reading.ac.uk

1. INTRODUCTION

Active polymer gels have been around now for nearly thirty years [1-2]. Their unique swelling and de-swelling characteristics in the present of solvents, combined with the possibility of triggering the volumetric expansion or contraction by a number of physical and chemical means, make them particularly interesting for actuation. Changes in temperature, pH or ionic strength are some of the "stimuli" which can be used to get to gels to respond by expanding or contracting in volume [3-6]. Electro-active polymer gels, responding to electrical fields, are particularly suited for actuation because of the ease of controlling the stimulus [7-9]. A great deal of work has already been done in characterising the physico-chemical aspects of their behaviour, quantifying the magnitude of the effects involved and in trying to develop actuators which mimic to some extent the behaviour of muscles [10-12]. By their very nature, actuators based on active polymer gels are better suited to applications where large displacements and low forces are useful, such as perhaps in the medical field. The large volumetric swelling capacity, often with swelling ratios of 10 or more, provides the means of achieving large deformations. The intrinsic low elastic modulus of a swollen gel, typically of the order of 10-100 kPa, limits very significantly the magnitude of the forces which can be generated [13-14]. However, there are in nature several examples of biological actuation besides muscular action, which can provide some solution to the problem of limited force generation. Turgid plant cells, for example, depend for their actuation function on mechanisms which are conceptually similar to those which govern the behaviour of active polymer gels but which are capable of generating large forces. They rely on partial confinement of water within a flexible fibrous container, the cell wall, and on a semi-permeable membrane, the lipid bi-layer, with a difference of chemical potential across the membrane leading to osmotic processes [15]. In this paper the potential across the membrane leading to osmotic processes [15]. In this paper the concept of integrating active polymer gels with fibrous structures will be explored, discussing the advantages and limitations of this approach. The potential benefits of such systems for large-displacement low-force or small-displacement large-force actuation will be illustrated using numerical simulations based on measured properties.

2. INTEGRATION OF ACTIVE POLYMER GELS AND FIBROUS STRUCTURES

2.1 Biological Actuation

The two basic mechanisms for mechanical actuation in biology are based on muscle contraction and on the generation of high turgor pressures inside plant cells. Muscle contraction depends on the specific molecular "ratchet" type interaction between actin and myosin molecules, leading to the sliding filaments model [16]. Turgor pressure inside plant cells depends on the differences in the water chemical potential across the biologically active membrane which can control the passage of ions and hence alter the osmotic potential and the internal pressure [15]. The anisotropy of muscle contraction is due primarily to the directionality of the sliding filaments, which occurs at the nano-scale. The anisotropy of actuation in plants manifests itself either at the level of a cell (micro-scale) via the shape of the cell and the orientation of cellulose fibres in it, or at the level of a tissue (macro-scale), via interactions between cells of different shapes, aspect ratios and cell wall thickness [17]. Although muscles and plant cell walls appear to have little in common as actuation devices there are similarities and common features which can inspire a biomimetic approach to the design of bespoke actuators based on harnessing the swelling potential of active polymer gels. Both systems share the following:

- Organised and extremely heterogeneous hierarchical structures capable of swelling by drawing in solvent;
- Swelling rate controlled by diffusion;
- Swelling potential partially confined by a fibrous structure made of flexible but high modulus fibres, collagen in muscle, cellulose in plant cells;
- Anisotropic swelling resulting from interactions between swelling element and confining structure;
- Direct conversion of chemical energy into mechanical energy.

2.2 Active Polymer Gel, Fibres and Actuation

Current actuation technology is based either on high modulus-low strain materials, such as piezoceramics and magnetostrictors, or on components, such as hydraulic, pneumatic or electromagnetic devices. The former are capable of working at high stresses but low strains, the latter can produce large strains or displacement but at comparatively low stresses. The only materials capable of delivering both high forces and large displacements are shape memory alloys but they require very high energy inputs and they are based on very dense materials. In fact, a study on the performance indices of mechanical actuators [18] shows that there is a gap between the high stress-low strain and the low-stress-high strain groups. By themselves, owing to their chemical and physical properties, active polymer gels fall typically in the low stress-high strain group, together with muscle. Free isotropic volumetric swelling can be very large indeed, with swelling ratios of 10-12, but is omni-directional. Differential swelling across a thin strip of active gel can induce bending but, owing to the very low elastic modulus of the material, the forces which can be generated are very limited. Anisotropic behaviour and increase in the force-generation potential of active polymer gels can be achieved if the free swelling potential of the gel is partially confined to extract mechanical work in a tailored fashion. There is an analogy with the free expansion of a gas which, unless confined and restricted by some "container", will not provide useful work [19-20].

Flexible fibrous structures are ideally suited to provide the coupling between the swelling potential of a gel and the conversion of chemical energy into mechanical work. The gel acts essentially as the "motor" and the fibrous structure confines the volumetric expansion, generating tensile stresses in the fibres and producing global deformations of the structure. The response can be tailored by the arrangement of fibres in the confining structure and can be designed to generate large forces at small displacements or, vice-versa, large displacements at low forces.

The fibrous structures needed to interact with the swelling gel perform two important functions: partial confinement of the gel which is not allowed to expand up to its "free" equilibrium value of swelling ratio and conversion of the swelling from isotropic into anisotropic. It is this latter characteristics which provides the means of tailoring the type and magnitude of the deformations of the integrated structure. The shape and deformation characteristics of the fibrous structure from untensioned to tensioned state is controlled by its fibre architecture. Even in its simplest form, such as braided cylinder, it is possible to implement actuation via axial contraction and radial expansion of the gel-braid assembly (tensile actuation) or, vice-versa, axial expansion and radial contraction (compressive actuation). An example of the former is shown in Figure 1. A commercial poly-acrylamide gel has been placed inside a cylindrical fibrous braid made of polyamide filaments. The fibres have a Young's modulus which is several orders of magnitude greater than that of the gel and can be considered virtually inextensible. The diameter of the cylinder is about 15 mm and its length about 100 mm (Figure 1a). The two ends are closed and to prevent the gel from extruding through the braid a very tightly knit but extremely deformable and very thin polyester fabric separates the gel from the braid. Once immersed in water, the gel expands and the expansion is converted into axial contraction and radial expansion of the structure (Figure 2b). Work is done in lifting the attached weights. Depending on the boundary conditions at the two end of the cylinder, one can design for free contraction (one end attached, no force applied at the other end), isotonic contraction (one end attached, constant force applied at the other end) or isometric conditions (imposed constant length). The first arrangement gives the maximum possible change of length and diameter, the last the maximum force which can be produced. For a given mass of gel, these quantities depend on the initial helical angle of the fibres in the braid. Different initial shapes of the confining fibre structure will produce different shape changes in the system upon swelling of the gel.

(a) (b)

Figure 1. Axial contraction of an active gel-fibrous braid structure

For cylindrical braided structures, which are easy to make in a wide range of diameters, fibre types and fibre contents using existing textile technology, the maximum internal volume of the cylinder corresponds to a helical fibre angle of about 55^0 with respect to the axis of the cylinder. Any increase or decrease in fibre angle implies a reduction of internal volume which is resisted by the near incompressibility of the swollen gel [21]. This means that to obtain substantial forces or deformations the initial helical fibre angle in the braid must differ significantly from the limit angle of 55^0. Experiments such as shown in Figure 1 confirm that the axial contraction of the gel stops when the braid fibre angle reaches that value, even if the gel is still capable of expanding further. When the system reaches this stage it acts as a very stiff element. If the deformation induced by the swelling gel is suppressed from the start (analogous to the isometric condition in muscles), large reacting forces are generated.

As well as axial extension or contraction, modified cylindrical braids can be used to generate bending deformations or torsion about the axis of the cylinder. The former is obtained by introducing axial fibres (helical angle zero) over a portion of the cylindrical braid. When the gel swells, the helical fibres rotate to allow increase in volume but the axial fibres introduce an asymmetry which results in the gel increasing or decreasing in length and bending at the same time. Similarly, if the left and right hand helical fibres in the braid have different extensibilities or different cross-sectional area, the swelling of the gel will occur with a twist about the axis.

3. MODELLING

Finite Element Modelling can be used to simulate the interaction between a swelling gel and a confining fibrous structure of virtually any shape and to assess the relative importance of the various parameters involved. The swelling of the gel can be simulated using thermal expansion on incompressible elements, using experimental results for the Young's modulus and the coefficient of hygro-expansion of the swollen polymer.

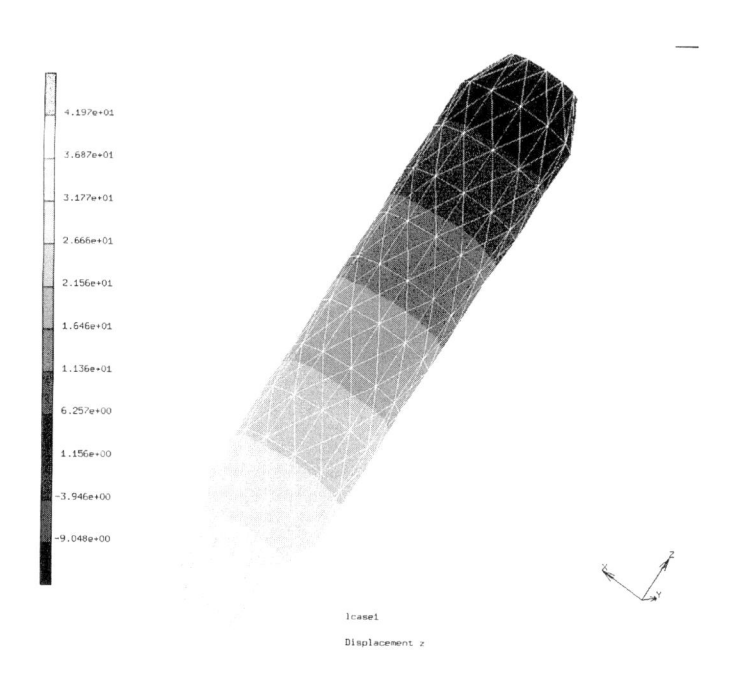

Figure 2. Axial contraction of an active gel-fibrous braid structure

The properties of the gel are described by a Mooney-type elastomeric material with three coefficients C_{10}, C_{01} and C_{11} obtained from experimental tests on the swollen gel. The braid fibres are modelled with 2-noded truss elements (axial loads only, no bending) with their nodes common to the end nodes of the outer tetrahedral elements used for the gel, aligned along the diagonals of the rectangular faces. The diffusion process can also be simulated using heat transfer which is governed by the same type of differential equation as diffusion, once the diffusion coefficient of the solvent has been measured. Owing to the large displacements and rotations involved a non-linear analysis is necessary.

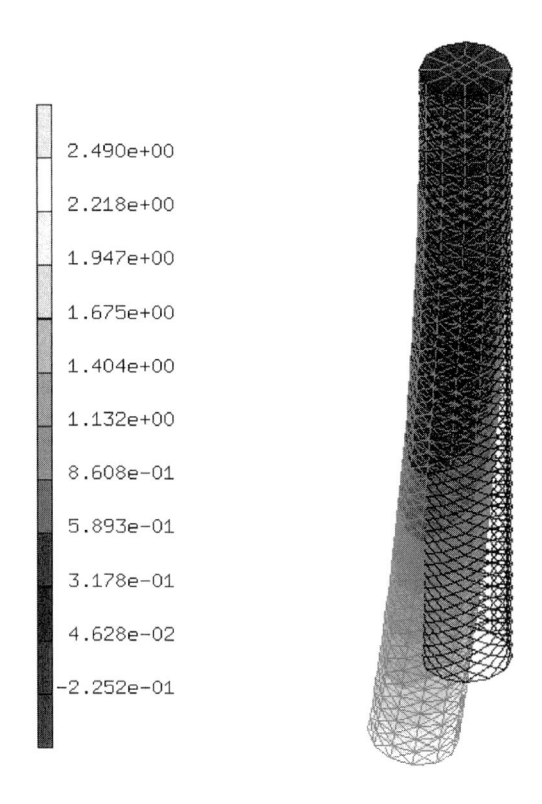

Figure 3. Axial expansion with bending deformations of an active gel-fibrous braid structure

Figure 2 shows a simulation of a gel-braid system similar to that of Figure 1 (free contraction condition). The initial braid fibre angle is about 20^0 and the maximum predicted axial contraction agrees reasonably well with the measured values. It is interesting to note that the FE simulation reproduces the maximum axial contraction when the braid fibre angle is 55^0. The modelling has been carried out using MSC/MARC software [22].

Figure 3 shows the bending deformations which can be induced by adding a subset of axial fibres to the helical fibres, on an arc sector only of the cylinder. The example shown illustrates bending coupled to axial extension where the initial braid fibre angle is about 70^0. Figure 4 illustrates the coupling of axial contraction and twisting deformation about the axis of the cylinder. This effect is induced when the two families of helical fibres, left and right, are not identical in terms of cross-sectional area or Young's modulus, at the same helical angle. A similar effect can be achieved by having different initial fibre angles in the left- and right-handed fibres with the same modulus and cross-sectional area.

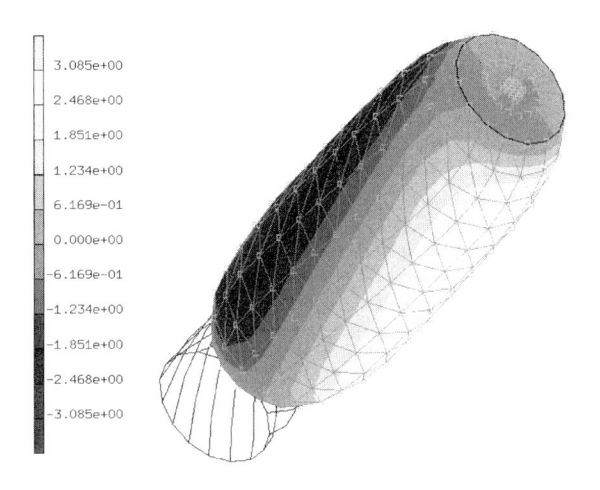

Figure 4. Induced twist in an active polymer gel-helical braid structure

In all the above examples one end of the structure is fixed in space (no translations and no rotations) and the other is free to move. This corresponds to the limit case of small force (vanishing small) and maximum deformation (free deformation). In order to assess the maximum force that these systems are capable of developing one can carry out a simulation of gel expansion within a braid under isometric conditions. Both ends are rigidly fixed in space and the gel is allowed to swell. The total reaction force is then calculated. Figure 5 shows that the force developed for a small structure of 4 mm diameter and 30 mm long is about 5 N, assuming maximum expansion of the gel which is of the same order as the force measured experimentally.

This kind of modelling is extremely valuable for assessing the effect of the many variables involved in controlling the type of actuation (shape change) and the magnitude of forces and displacements. It also allows to get some idea of the limitations imposed by size on the rate of swelling, and hence on the speed of response of the actuator, which is controlled by diffusion and hence by the ratio of volume to surface area of the gel unit.

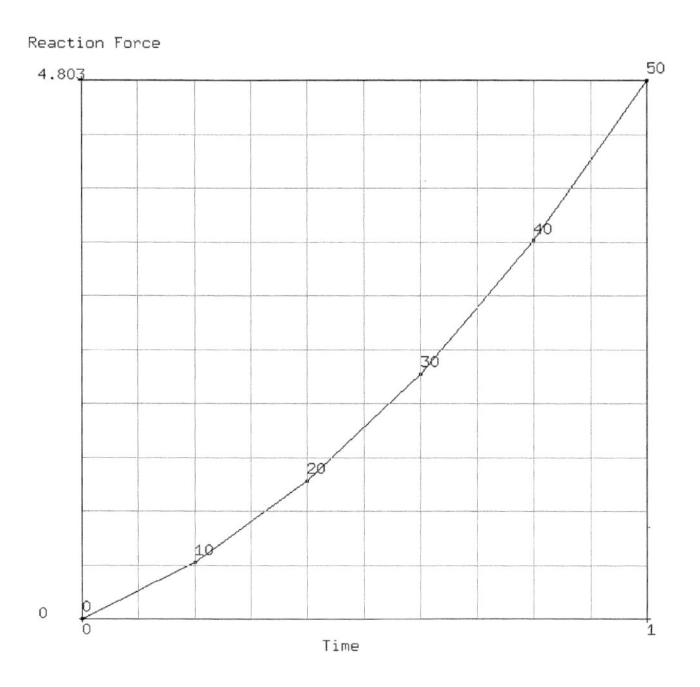

Figure 5. Induced twist in an active polymer gel-helical braid structure

4. CONCLUSIONS

When used in combination with flexible fibrous structures, active polymer gels can provide a very extensive range of actuator response, from high force-low deformation to low force-high deformation. The mechanical interaction between the expanding or contracting gel and the fibre architecture is the essential element for an effective and direct conversion of chemical into mechanical energy. Coupled deformation modes are also possible, depending on the fibre architectures used, allowing tailoring for specific requirements. Electro-active polymers can add greatly to the basic passive swelling behaviour of a gel by enhancing swelling rates, improving the reversibility of swelling/de-swelling, as well as offering the possibility of more controllable systems. The main limitation of active polymer gels for actuation is the diffusion controlled swelling rate. Although diffusion rates on the centimetre scale or above are slow, which requires typically times of the order of 10-20 minutes for structures with diameters of the order of 10 mm or so to react, the speed of response can be improved by decreasing the diameter of the electro-active polymer gel / fibrous structure combination down to a few millimetre, or below, and by coupling mini-actuators in parallel or series, according to needs. This is no different from the way in which the active units in muscle are arranged, providing a direct biomimetic analogue for a satisfactory solution. Making sub-millimetre systems is technologically possible although, perhaps, not sufficiently cost-effective at present [23]. The problem of system reversibility provides an additional challenge but there are indications that electro-active polymer gels in particular may be capable of increased levels of reversibility in their swelling behaviour. However, to realise the full potential of active polymer gels for actuation it will be necessary to make progress in the design aspects of systems as well in the physical-chemistry of the materials.

5. REFERENCES

1. DeRossi, D., Kajawara, K., Osada, Y. and Yamauchi, A. (1991) *Polymer Gels – Fundamentals and Biomedical Applications,* Plenum, New York.
2. Tanaka, T. (1981) Gels, *Scientific American* **244** 110-123.
3. Katayama, S., Hirokawa, Y. and Tanaka, T. (1984) Reentrant phase-transition in acrylamide-derivative copolymer gels, *Macromolecules* **17** 2641-643.
4. Hirokawa, Y. and Tanaka, T. (1984) Volume phase transition in a non-ionic gel, *J. Chem. Phys.* **81** 6379-6380.
5. Tanaka, T., Nishio, I., Sun, S.T., and Ueno-Nishio, S. (1982) Collapse of gels in an electric field, *Science* **218** 467-469.
6. Molloy, P.J., Smith, M.J. and Cowling, M.J. (2000) The effects of salinity and temperature on the behaviour of polyacryl-amide gels, *Mater. Des.* **21** 169-174.
7. Gong, J.P., Nitta, T. and Osada, Y. (1994) Electrokinetic Modeling of the Contractile Phenomena of Polyelectrolyte Gels. One-dimensional capillary Model. *J. Phys. Chem.* **98** 9583-9587.
8. T. Fernandez-Otero (2000) Biomimicking materials with smart polymer gels. In *Structural Biological Materials: Design and Structure-Property Relationships.* M. Elices, Ed., Pergamon, Oxford, pp. 189-220.
9. Osada, Y. and Hasebe, M. (1985) Electrically actuated mechanochemical devices using polyelectrolyte gels, *Chemistry Letters* 1285-1288.
10. Shiga, T., Hirose, Y., Okada, A., Karauchi, T. and Kanugai (1989) Design of biomimetic machinery systems with polymer gel, *Proc. 1st Int. SAMPE Symp., Nov 28-Dec 1,* pp. 659-663.
11. Suzuki, M., Tateishi, T., Ushida, T. and Fujishige (1986) An artificial muscle of polyvinyl alcohol hygrogel composites, *Biorheology* **23** 874-878.
12. Okuzaki, H. and Osada, Y. (1993) A chemomechanical polymer gel with electrically driven motility, *J. Intel. Matls. and Struct.* **4** 50-53.
13. Molloy, P.J. and Cowling, M.J. (2000) Volume and density changes in polymer gels in seawater environments, *Proc. Instn. Mech. Engrs.* **214L** 223-228.
14. Marra, S.P., Ramesh, K.T. and Douglas, A.S. (2001) The actuation of a biomimetic poly(vinyl alcohol)-poly(acrylic acid) gel, *Phil. Trans. Roy. Soc.* **360** 175-198.
15. Niklas, K.J. (1992) *Plant Biomechanics,* University of Chicago Press, Chicago.
16. McMahon, T.A. (1984) *Muscles, Reflexes, and Locomotion,* Princeton University Press.
17. Simons, P. (1992) *The Action Plant,* Blackwell, Oxford.

18. Huber, J.E., Fleck, N.A. and Ashby, M.F. (1997) The selection of mechanical actuators based on performance indices, *Proc. Roy. Soc. Lond.* A, **453** 2185-2205.
19. Jeronimidis, G. (1996) Integration of active polymer gels and fibre composite structures for "smart" actuation. In *5th Symposium on Intelligent Materials, The Society for Non-Traditional Technology, Intelligent Materials Forum,* Tokyo, pp.17-19.
20. Jeronimidis, G. (2001) Coupling between active polymer gels and fibre structures for tailored functionality. *Worshop on Electroactive Polymers and Biosystems: New Directions in Electroactive Polymer Materials for Biomimetic and Interactive Processes*, Il Ciocco, Italy, 30 Jul.- 3 Aug., pp.24-28.
21. Clark, R.B. and Cowey, J.B. (1958) Factors controlling the change of shape of some worms. *J. Exp. Biol.* **35** 731-748
22. MSC Software (2000) *MSC.Marc Mentat Manuals*, MSC. Software Corporation, Santa Ana, CA.
23. Suzuki, A. (1993) Phase transition in gels of sub-millimetre size induced by interaction with stimuli, Adv. Polym. Sci. **110** 199-240.

THERMALLY INDUCED VIBRATION OF AN INHOMOGENEOUS BEAM DUE TO A CYCLIC HEATING

RYUUSUKE KAWAMURA, YOSHINOBU TANIGAWA
Department of Mechanical Systems Engineering,
Graduate School of Engineering, Osaka Prefecture University,
Sakai, Osaka 599-8531 Japan
E-mail:kawamura@mecha.osakafu-u.ac.jp

RICHARD B. HETNARSKI
Department of Mechanical Engineering, Rochester Institute of Technology,
Rochester, NY 14623, U.S.A.
E-mail:632hetna@rit.edu

1. INTRODUCTION

With the improvement of the performance and the efficiency of machines, the use of thin-walled and lightweight structural members has been frequently required in machines operating in high temperatures and under high internal pressure. An application of inhomogeneous materials, such as functionally graded materials, in structures subjected to the extremely high temperatures has been anticipated and development of such materials is being actively continued. In general, it often happens that the thermally induced instability, such as vibration and buckling, creates problems in the thin-walled and lightweight structural members, leading to the reduction of stiffness.

Failures of machines and structural members subjected to both the repeated load and the temperature fluctuation do occur. Examples of such failures in the nuclear power plants are the rupture of the tube in a steam generator resulted from the resonance due to the flow induced vibration [1], and the crack initiation and propagation in the tee junction pipe resulted from the high-cycle thermal fatigue due to the thermal striping phenomenon [2]. In order to avoid failures resulted from the resonance or the high-cycle thermal fatigue it is important to evaluate the vibrations as well as the repeated thermal stress that are induced in the thin-walled structural members by the repeated load and the temperature fluctuation. The vibration problems in thermal environments for the structural members composed of homogeneous materials were studied in the past [3,4,5]. However, the studies related to the theoretical treatment of the thermally induced vibration problem for the inhomogeneous materials have not been seen in the literature.

This article is intended to be a theoretical analysis of the heat conduction and the thermally induced vibration problems for an inhomogeneous elastic beam subjected to a cyclic heat supply. The effects of the frequency of cyclic heat supply and the inhomogeneity in the material properties are examined by a numerical calculation.

2. ANALYTICAL DEVELOPMENT

2.1 Thermal Conduction Problem

We consider an inhomogeneous beam of length L whose cross-section is rectangle of width b, height h as shown in Figure 1. We assume that the beam is subjected on the upper surface $z = 0$ to the temperature change which is a sinusoidal function of time t, with the frequency ω and the amplitude T_0, and on the lower surface $z = h$ is thermally insulated.

Now, we assume that the thermal capacity $c(z)\rho(z)$ and the thermal conductivity $\lambda(z)$, possess the inhomogeneity given by the power product form of lateral coordinate z [6] as

$$c(z)\rho(z) = c_0\rho_0\left(1+\frac{z}{h}\right)^k , \quad \lambda(z) = \lambda_0\left(1+\frac{z}{h}\right)^l \tag{1}$$

where k and l are arbitrary parameters, and c_0, ρ_0 and λ_0 represent typical values of the specific heat $c(z)$, the density $\rho(z)$ and the thermal conductivity $\lambda(z)$, respectively.

The one-dimensional heat conduction equation in the lateral direction z, the initial condition and the thermal boundary conditions are given as

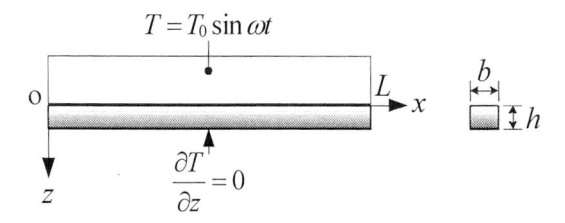

Figure 1. Coordinate system and thermal conditions

$$c(z)\rho(z)\frac{\partial T}{\partial t} = \frac{\partial}{\partial z}\left\{\lambda(z)\frac{\partial T}{\partial z}\right\}, \tag{2}$$

$$t = 0 \; : \; T = 0, \tag{3}$$

$$z = 0 \; : \; T = T_0 \sin\omega t, \quad z = h \; : \; \frac{\partial T}{\partial z} = 0. \tag{4}$$

Now, the dimensionless quantities are introduced:

$$\overline{T} = T/T_0, \; \zeta = 1 + z/h, \; \tau = \kappa_0 t/h^2, \; \kappa_0 = \lambda_0/c_0\rho_0, \; \overline{\omega} = \omega h^2/\kappa_0, \tag{5}$$

where T is the temperature change, and κ_0 is a typical value of the thermal diffusivity.

Solving the heat conduction equation (2) under the conditions (3) and (4) with the aid of the Laplace transform method, the temperature change is represented in the form of Bessel functions J_ν, Y_ν of the first and second kind of real order ν and complex arguments as

$$\begin{aligned}
\overline{T} = &-\sum_{j=1}^{\infty} \exp(-q_j^2 \tau) \frac{\overline{\omega}}{q_j^4 + \overline{\omega}^2} \frac{2q_j}{D_2'(q_j)} \\
&\times \zeta^{p_1} \Big[(p_1 + \nu p_2)\{Y_\nu(p_3 q_j 2^{p_2})J_\nu(p_3 q_j \zeta^{p_2}) - J_\nu(p_3 q_j 2^{p_2})Y_\nu(p_3 q_j \zeta^{p_2})\} \\
&- 2^{p_2} p_2 p_3 q_j \{Y_{\nu+1}(p_3 q_j 2^{p_2})J_\nu(p_3 q_j \zeta^{p_2}) - J_{\nu+1}(p_3 q_j 2^{p_2})Y_\nu(p_3 q_j \zeta^{p_2})\}\Big] \\
&+ \lim_{q \to \sqrt{\overline{\omega}/2}(1-i)} \Big\langle \frac{\cos\overline{\omega}\tau + i\sin\overline{\omega}\tau}{2i} \frac{1}{D_2(q)} \\
&\times \zeta^{p_1} \Big[(p_1 + \nu p_2)\{Y_\nu(p_3 q 2^{p_2})J_\nu(p_3 q \zeta^{p_2}) - J_\nu(p_3 q 2^{p_2})Y_\nu(p_3 q \zeta^{p_2})\} \\
&- 2^{p_2} p_2 p_3 q \{Y_{\nu+1}(p_3 q 2^{p_2})J_\nu(p_3 q \zeta^{p_2}) - J_{\nu+1}(p_3 q 2^{p_2})Y_\nu(p_3 q \zeta^{p_2})\}\Big] \Big\rangle \\
&- \lim_{q \to \sqrt{\overline{\omega}/2}(1+i)} \Big\langle \frac{\cos\overline{\omega}\tau - i\sin\overline{\omega}\tau}{2i} \frac{1}{D_2(q)} \\
&\times \zeta^{p_1} \Big[(p_1 + \nu p_2)\{Y_\nu(p_3 q 2^{p_2})J_\nu(p_3 q \zeta^{p_2}) - J_\nu(p_3 q 2^{p_2})Y_\nu(p_3 q \zeta^{p_2})\} \\
&- 2^{p_2} p_2 p_3 q \{Y_{\nu+1}(p_3 q 2^{p_2})J_\nu(p_3 q \zeta^{p_2}) - J_{\nu+1}(p_3 q 2^{p_2})Y_\nu(p_3 q \zeta^{p_2})\}\Big] \Big\rangle,
\end{aligned} \tag{6}$$

where i is the imaginary unit and q_j is the jth positive root which satisfies the transcendental equation $D_2(q) = 0$, with

$$D_2(q) = (p_1 + \nu p_2)\{Y_\nu(p_3 q 2^{p_2})J_\nu(p_3 q) - J_\nu(p_3 q 2^{p_2})Y_\nu(p_3 q)\}$$
$$- 2^{p_2} p_2 p_3 q \{Y_{\nu+1}(p_3 q 2^{p_2})J_\nu(p_3 q) - J_{\nu+1}(p_3 q 2^{p_2})Y_\nu(p_3 q)\} \tag{7}$$

$$p_1 = (1-l)/2, \; p_2 = (k-l+2)/2, \; p_3 = 1/|p_2|, \; \nu = |(l-1)/(k-l+2)| \tag{8}$$

If there is a relationship between k and l of the form

$$k = -l, \nu = 1/2 \ (-1 < k, l < 1), \tag{9}$$

the temperature change is represented by elementary functions as

$$\overline{T} = -\sum_{j=1}^{\infty} \exp(-q_j^2 \tau) \frac{2\overline{\omega} q_j}{q_j^4 + \overline{\omega}^2} (-1)^j \frac{\cos\{p_3(\zeta^{p_2} - 2^{p_2})q_j\}}{p_3(2^{p_2} - 1)}$$

$$+ \frac{1}{\cos^2 \beta \cosh^2 \beta + \sin^2 \beta \sinh^2 \beta}$$

$$\times \left[\{\sin \alpha \sinh \alpha \cos \beta \cosh \beta - \cos \alpha \cosh \alpha \sin \beta \sinh \beta\} \cos \overline{\omega} \tau \right.$$

$$\left. + \{\cos \alpha \cosh \alpha \cos \beta \cosh \beta + \sin \alpha \sinh \alpha \sin \beta \sinh \beta\} \sin \overline{\omega} \tau \right], \tag{10}$$

$$\alpha = p_3 \sqrt{\overline{\omega}/2} (\zeta^{p_2} - 2^{p_2}), \ \beta = p_3 \sqrt{\overline{\omega}/2} (1 - 2^{p_2}). \tag{11}$$

2.2 Thermally Induced Vibration Problem

We assume that, respectively, the longitudinal modulus of elasticity, the linear thermal expansion coefficient and the density are given by power products of coordinate z as

$$E(z) = E_0 \left(1 + \frac{z}{h}\right)^m, \ \alpha(z) = \alpha_0 \left(1 + \frac{z}{h}\right)^n, \ \rho(z) = \rho_0 \left(1 + \frac{z}{h}\right)^a \tag{12}$$

where m, n and a are arbitrary parameters representing the inhomogeneity, respectively, in the longitudinal modulus of elasticity, the linear thermal expansion coefficient, and the density, and E_0, α_0, ρ_0 represent typical values of appropriate material property.

Considering the equilibrium of forces and bending moments due to the distributed external loads p_x, q_x in a beam element, the equations of motion for the inhomogeneous beam are represented in the dimensionless forms as

$$\overline{C}_2 \overline{h}^4 \frac{\partial^4 \overline{w}}{\partial \overline{x}^4} + \overline{\mu} \frac{\partial^2 \overline{w}}{\partial \tau^2} = \overline{q}_x - \overline{h}^2 \frac{\partial^2 \overline{M}^T}{\partial \overline{x}^2}, \tag{13}$$

$$-\overline{C}_1 \overline{h}^2 \frac{\partial^2 \overline{u}_0}{\partial \overline{x}^2} + \overline{\mu} \frac{\partial^2 \overline{u}_0}{\partial \tau^2} = \overline{p}_x - \overline{h} \frac{\partial \overline{N}^T}{\partial \overline{x}} + \overline{\delta} \overline{h} \frac{\partial^2}{\partial \tau^2} \left(\frac{\partial \overline{w}}{\partial \overline{x}}\right), \tag{14}$$

where $\overline{u}_0$, $\overline{w}$ are the dimensionless forms of the in-plane displacement u_0 and the out-of-plane deflection w at the neutral axis; $\overline{N}^T$ and $\overline{M}^T$ are the dimensionless forms of the thermal resultant force and moment N^T, M^T, respectively,

$$\overline{N}^T = \int_1^2 \zeta^{m+n} \overline{T} d\zeta, \ \overline{M}^T = \int_1^2 \zeta^{m+n}(\zeta-1-\overline{\eta})\overline{T} d\zeta, \tag{15}$$

$$\left.\begin{array}{l} \overline{C}_1 = \int_1^2 \zeta^m d\zeta, \ \overline{C}_2 = \int_1^2 \zeta^m (\zeta-1-\overline{\eta})^2 d\zeta \\[6pt] \overline{\mu} = \dfrac{\kappa_0^2 \rho_0}{E_0 h^2} \int_1^2 \zeta^a d\zeta, \ \overline{\delta} = \dfrac{\kappa_0^2 \rho_0}{E_0 h^2} \int_1^2 \zeta^a (\zeta-1-\overline{\eta}) d\zeta \\[6pt] \overline{\eta} = \int_1^2 \zeta^m (\zeta-1) d\zeta \Big/ \int_1^2 \zeta^m d\zeta \end{array}\right\}. \tag{16}$$

Now, the dimensionless quantities are introduced:

$$\left.\begin{array}{l} (\overline{x}, \overline{h}) = (x, h)/L, \ (\overline{u}_0, \overline{w}) = (u_0, w)/\alpha_0 T_0 h, \ (\overline{p}_x, \overline{q}_x) = (p_x, q_x)/\alpha_0 T_0 E_0 b \\[4pt] \overline{N}^T = N^T/\alpha_0 T_0 E_0 bh, \ \overline{M}^T = M^T/\alpha_0 T_0 E_0 bh^2 \\[4pt] \overline{C}_1 = C_1/E_0 bh, \ \overline{C}_2 = C_2/E_0 bh^3 \\[4pt] \overline{\mu} = \mu/(E_0 bh^3/\kappa_0^2), \ \overline{\delta} = \delta/(E_0 bh^4/\kappa_0^2), \ \overline{\eta} = \eta/h, \ \overline{\sigma}_x = \sigma_x/\alpha_0 T_0 E_0 \end{array}\right\}, \tag{17}$$

where σ_x is the thermal stress in the axial direction of the beam.

Now, we assume the simply supported edges without axial restraint as the mechanical boundary conditions of the beam. Then, the mechanical boundary conditions are described in the dimensionless form are

$$\overline{x} = 0, 1 \ : \ \frac{\partial \overline{u}_0}{\partial \overline{x}} = \frac{\overline{N}^T}{\overline{C}_1 \overline{h}}, \ \overline{w} = 0, \ \frac{\partial^2 \overline{w}}{\partial \overline{x}^2} = -\frac{\overline{M}^T}{\overline{C}_2 \overline{h}^2}. \tag{18}$$

Disregarding the terms of both the distributed external loads and the inertia forces in the equations of motion, Eqs.(13) and (14), the quasi-static solutions u_0^q, w^q of the in-plane displacement and the out-of-plane deflection are obtained as

$$\overline{u}_0^q = \frac{\overline{N}^T}{\overline{C}_1 \overline{h}}\left(\overline{x}-\frac{1}{2}\right), \ \overline{w}^q = \frac{\overline{M}^T}{\overline{C}_2 \overline{h}^2}\frac{1}{2}\overline{x}(1-\overline{x}). \tag{19}$$

On the other hand, the dynamic solutions of the longitudinal and bending vibrations u_0^d, w^d are obtained by superimposing the free vibration solutions in terms of modal function of the natural vibration and the forced vibration solutions due to the temperature fluctuation under the initial conditions

$$\tau = 0 \ : \ \overline{u}_0(\overline{x}) = \frac{\partial}{\partial \tau}\overline{u}_0(\overline{x}) \equiv 0, \ \overline{w}(\overline{x}) = \frac{\partial}{\partial \tau}\overline{w}(\overline{x}) \equiv 0. \tag{20}$$

The dynamic solution w^d of the bending vibration is represented as

$$\overline{w}^d = \frac{1}{\overline{C}_2 \overline{h}^2} \sum_{i=1}^{\infty} \frac{1}{\alpha_i^2} \left\{ \sum_{j=1}^{\infty} \frac{C_{1ij}}{1+\left(q_j^2/\Omega_i\right)^2} \left(\cos\Omega_i\tau - \frac{q_j^2}{\Omega_i}\sin\Omega_i\tau \right) \right.$$
$$- \frac{1}{1-\left(\overline{\omega}/\Omega_i\right)^2}\left(C_{2i}\cos\Omega_i\tau - \frac{\overline{\omega}}{\Omega_i}C_{3i}\sin\Omega_i\tau \right)$$
$$\left. - \sum_{j=1}^{\infty} \frac{C_{1ij}\exp(-q_j^2\tau)}{1+\left(q_j^2/\Omega_i\right)^2} + \frac{1}{1-\left(\overline{\omega}/\Omega_i\right)^2}\left(C_{2i}\cos\overline{\omega}\tau + C_{3i}\sin\overline{\omega}\tau\right) \right\} \sin\alpha_i\overline{x}, \quad (21)$$

$$\Omega_i = \overline{h}^2\alpha_i^2\sqrt{\frac{\overline{C}_2}{\overline{\mu}}}, \quad \alpha_i = i\pi \ (i=1,2,\ldots) \quad (22)$$

where Ω_i is the dimensionless form of the natural frequency of bending vibration. The information on the coefficients C_{1ij}, C_{2i}, C_{3i} in Eq.(21) and the dynamic solution of the longitudinal vibration u_0^d are omitted here because of space limitations.

Then, the thermal stress σ_x is given as

$$\overline{\sigma}_x = \zeta^m \left\{ \overline{h}\frac{\partial \overline{u}_0}{\partial \overline{x}} - (\zeta - 1 - \overline{\eta})\overline{h}^2\frac{\partial^2 \overline{w}}{\partial \overline{x}^2} - \zeta^n \overline{T} \right\}, \quad (23)$$

which is the superposition of components of the in-plane displacement, the out-of-plane deflection and the thermal strain.

3. NUMERICAL RESULTS AND DISCUSSIONS

Numerical results of the thermally induced vibration are illustrated for a thin-walled flexural inhomogeneous beam subjected to the cyclic heat supply by the surrounding medium. The typical values of the material properties and the dimensions of the beam are shown in Table 1.

We suppose the parameters k and l are given as

Table 1. Material properties and dimensions of beam

Quantities			Unit
Longitudinal modulus of elasticity	E_0	70	[GPa]
Density	ρ_0	2.7×10^3	[kg/m^3]
Thermal diffusivity	κ_0	8.36×10^3	[m^2/s]
Height	h	5.0×10^{-3}	[m]
Length	L	14.82	[m]
Aspect ratio	h/L	3.4×10^{-4}	

$$k = l = 0. \tag{24}$$

The other parameters m, n and a are given as

$$\{m, n, a\} = -1, 0, 1. \tag{25}$$

The effects of the frequency of the cyclic heat supply and the inhomogeneity in the material properties on the thermoelastic dynamic response of the beam will now be examined. We assume the frequency of the temperature fluctuation ω is given by the product of the natural frequency of first-mode bending vibration Ω_1^h in the homogeneous beam and an arbitrary parameter ε

$$\bar{\omega} = \varepsilon \Omega_1^h, \tag{26}$$

The parameter ε is given by the following two values:

$$\varepsilon = \{0.3, 0.9\}. \tag{27}$$

3.1 Effect of Frequency of Cyclic Heat Supply

Figure 2 shows the time evolutions of the temperature change at the upper surface $\zeta = 1$ in the homogeneous beam. The larger the parameter ε, the shorter is the period. Figure 3 shows the lateral variation of the temperature change in the homogeneous beam for $\varepsilon = 0.9$. When the period of the cyclic heat supply is short, the temperature distribution in the lateral direction is almost constant.

Figure 4 shows the time evolutions of the out-of-plane deflection at the mid-span of the beam $x = L/2$. The amplitude of the dynamic solution w^d is larger than that of quasi-static one w^q. The ratio of amplitudes of the dynamic solution to the quasi-static one is approximately equal to 2.4 for $\varepsilon = 0.3$ [case (a)] and is approximately equal to 9 for $\varepsilon = 0.9$ [case (b)]. We thus observe the effect of inertia on the out-of-deflection from these results.

Figure 2. Time evolution of temperature change at upper surface

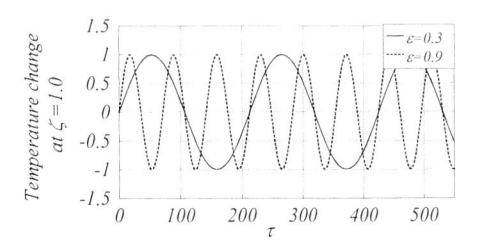

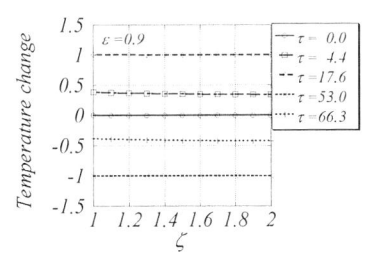

Figure 3. Lateral variation of temperature change

(a) ε = 0.3 *(a) ε = 0.3*

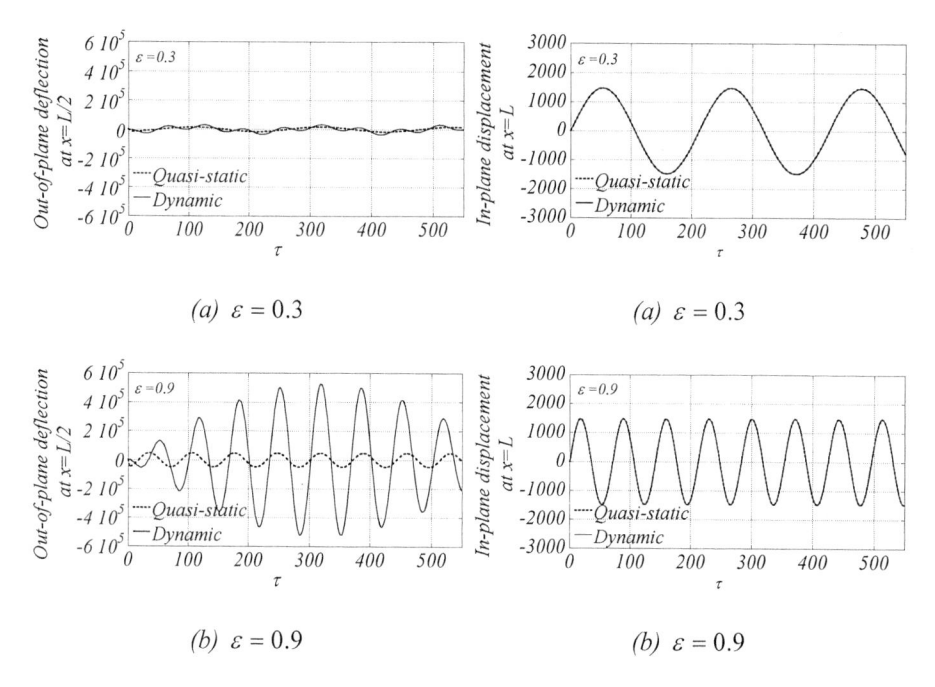

(b) ε = 0.9 *(b) ε = 0.9*

Figure 4. Time evolutions of out-of-plane deflection at mid-span

Figure 5. Time evolutions of in-plane displacement at edge

Figure 5 shows the time evolutions of the in-plane displacement at the right edge $x = L$. There is no difference in the values between the dynamic solution u_0^d and the quasi-static solution u_0^q. We do not observe the effect of inertia in the in-plane displacement within the given frequency domain of the cyclic heat supply. This is the reason why the first-order natural frequency of the longitudinal vibration is much larger than that of the bending vibration.

3.2 Effect of Inhomogeneity in Material Properties

The effects of inhomogeneity in the longitudinal modulus of elasticity, the linear thermal expansion coefficient, and the density on the dynamic response of the beam are now examined. While varying one of parameters in theses material properties, the other parameters are set to zero, in order to examine the effect of each material property independently.

Here, we give a short illustration by graphs of the numerical results concerned with the effect of the inhomogeneity in linear thermal expansion coefficient.
Figure 6 shows the lateral variation of the material properties in the inhomogeneous beam due to the parameter n. When the parameter is equal to -1, the value

Figure 6. Lateral variation of material property

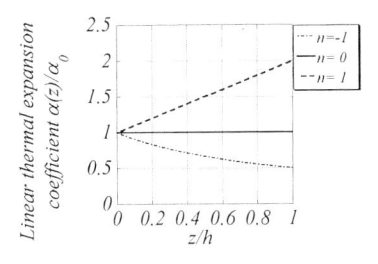

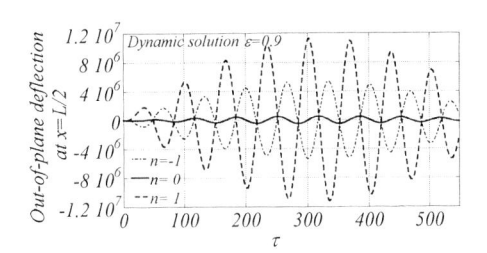

Figure 7. Effect of inhomogeneity in linear thermal expansion coefficient on out-of-plane deflection

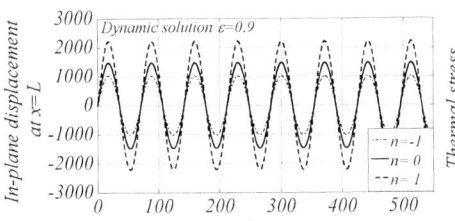

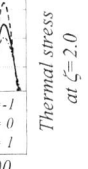

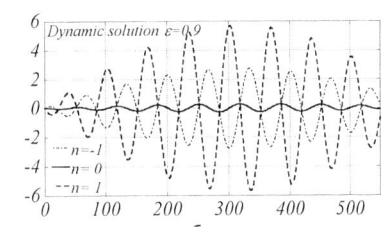

Figure 8. Effect of inhomogeneity in linear thermal expansion coefficient on in-plane displacement

Figure 9. Effect of inhomogeneity in linear thermal expansion coefficient on thermal stress

decreases with the increase of the lateral coordinate z. When the parameter is equal to 0, the value is constant with the change of the coordinate z. When the parameter is equal to 1, the value increases linearly with the increase of the coordinate z.

Figures 7, 8 and 9 show the effect of the inhomogeneity in the linear thermal expansion coefficient on the time evolutions of the out-of-plane deflection at the mid-span $x = L/2$, the in-plane displacement at the right edge $x = L$ and the thermal stress at the mid-span $x = L/2$ on the lower surface $\zeta = 2$. When the parameter n is equal to -1 or 1, the amplitudes of the out-of-plane deflection become large. This is the effect of the difference in the magnitude of the thermal expansion between the upper and lower surfaces in the beam. The larger the parameter n, the larger is the amplitude of the in-plane displacement. This is the reason why the magnitude of the thermal expansion in the beam becomes large with the increase of the parameter n. Furthermore, the larger the parameter n, the large is the amplitude of the thermal stress. This is the reason why the increase of the inhomogeneous parameter n increases the thermal strain component of the thermal stress as well as the components resulted from the in-plane displacement and the out-of-plane deflection.

4. CONCLUSIONS

In the present article we have dealt with problems of heat conduction and thermally induced vibration in an inhomogeneous beam due to a cyclic heat supply whose temperature change is a function of time of a sinusoidal form. The beam is composed of an inhomogeneous material with properties given by a power of a lateral coordinate z. The analytical solution of the temperature change and both the quasi-static and the dynamic thermoelastic solutions of the in-plane displacement, the out-of-plane deflection, and the thermal stress are derived under the mechanical boundary conditions of simply supported edges without axial restraint. The effects of the frequency of a cyclic heat supply and the inhomogeneity in the material properties on the thermoelastic dynamic responses are made revealed by the numerical results for the thin-walled flexural inhomogeneous beam subjected to a cyclic heat supply by the surrounding medium.

The present numerical results are summarized as follows:
1. The effect of inertia on the thermoelastic dynamic response of the beam is evident in particular in the bending vibration within the frequency domain of the cyclic heat supply.
2. The larger is the inhomogeneity parameter of the linear thermal expansion coefficient, the larger are the amplitudes of the out-of-plane deflection, the in-plane displacement, and the thermal stress.
3. The larger is the inhomogeneity parameter of the longitudinal modulus of elasticity, the smaller is the amplitude and the shorter is the period of the out-of-plane deflection. Thus, the amplitude of the thermal stress becomes smaller.
4. The larger is the inhomogeneity parameter of the density, the larger is the amplitude and the longer is the period of the out-of-plane deflection. Thus, the amplitude of the thermal stress becomes larger.
5. Among the material properties, the effect of the inhomogeneity parameter of the linear thermal expansion on the thermoelastic dynamic response of the beam is the most significant.

5. REFERENCES

1. Fukui Prefectural Government (2000). *Atomic energy in Fukui prefecture*. Fukui: Fukui Prefectural Government.(in Japanese)
2. Muramatsu, T. & Kasahara, N. (2000) Validation of numerical method for thermal striping phenomena with actual plant data. *Japan Nuclear Cycle Development Institute Technical Report, 6*, 81-92.(in Japanese)
3. Boley, B. A. & Weiner, J. H. (1997). *Theory of thermal stresses*. New York: Dover.
4. Timoshenko, S. P., Young, D. H. & Weaver, W. (1974). *Vibration problems in engineering Fourth Edition*. New York: John Wiley & Sons.
5. Noda, N., Hetnarski, R. B. & Tanigawa, Y. (2000). *Thermal stresses*. New York: Lastran.
6. Carslaw, H. S. & Jaeger, J. C. (1986). *Conduction of heat in solids Second Edition*. New York: Oxford University Press.

APPLICATION OF OPTICAL FIBER SENSORS TO SMART STRUCTURE

SANG-HOON KIM, DAE-CHEOL SEO and JUNG-JU LEE
*Department of Mechanical Engineering,
Korea Advanced Institute of Science and Technology,
Guseong-dong, Yuseong-gu, Daejeon 305-701, Korea (south)*
E-mail: jjlee@mail.kaist.ac.kr

1. INTRODUCTION

Optical fiber sensors have many advantages such as excellent sensitivity and resolution, immunity from electromagnetic interference, capability of multiplexing, dimensional compactness, etc. These optical fiber technologies are applicable to structural monitoring, acoustic emission sensing, impact and vibration detection and measurement of strain, temperature, internal deflection and various other physical quantities. Because optical fiber sensors are small, light and flexible, they can be easily inserted into or attached to engineering structures, resulting in their suitability to the sensing parts of smart structures [1] which were introduced in late 1980s.

Optical fiber sensors are classified into point-sensors and distributed sensors according to their measurement ranges. The point-sensors have small measurement range. The point-sensor such as an extrinsic Fabry-Perot interferometric (EFPI) sensor [2] can measure strain or temperature in point-like range. However, for the damage monitoring of large structures, the distributed sensors which have large measurement range are required. For the monitoring of large structures with point-sensors, large number of the sensors must be multiplexed. However, the possible number of the sensors is limited and the multiplexed sensor cannot measure distributed strain or temperature. On the other hand, distributed optical fiber sensors can measure at all positions of an optical fiber, and the continuous distribution of a measured value can be obtained. Moreover, they can be used in the monitoring of large structures such as bridges and buildings due to their measurement range reaching tens of kilometers. The distributed optical fiber sensors using the time-domain reflectometry have spatial resolution, and averaged strain or temperature within the spatial resolution range is measured. The resolution is generally tens of centimeters to several meters. Thus the sensor cannot measure localized strain or temperature. The point-sensors and distributed sensors have positive and negative attributes in the measurement range.

In this paper, the principles and applications of transmission-type EFPI (TEFPI) optical fiber sensors as point-sensors and Brillouin distributed optical fiber sensors as distributed sensors are presented. TEFPI optical fiber sensors show superior performance to other interferometric optical fiber sensors in distinguishing measurement directions. The TEFPI optical fiber sensors were applied to fatigue damage monitoring of aluminum plate patch with CFRP composites. The Brillouin distributed sensors were applied to deflection monitoring of a bending beam.

2. TRANSMISSION-TYPE EFPI OPTICAL FIBER SENSORS

Though most of the conventional interferometric optical fiber sensors (including the reflection-type EFPI optical fiber sensor) have good sensitivity and resolution due to the use of interferometric fringe counting, they have the disadvantage being unable to easily distinguish the direction of strain such as tension and compression. To overcome this drawback in conventional reflection-type EFPI optical fiber sensors, the quadrature phase-shifted EFPI system [3] and AEFPI (absolute EFPI) system [4], [5] using spectral analysis have been reported.
The recently developed transmission-type extrinsic Fabry-Perot interferometric (TEFPI) optical fiber sensor has both the advantages of reflection-type EFPI optical fiber sensors and a simpler and more effective function to distinguish strain direction than do reflection-type EFPI optical fiber sensors [6]. Therefore, this TEFPI optical fiber sensor could be easily applied to the fatigue test while the conventional reflection-type EFPI optical fiber sensor has difficulties in fatigue tests.

2.1. Sensing principle and sensor construction

The structure of the TEFPI optical fiber sensor is shown in *Figure 1*. Conventional reflection-type EFPI optical fiber sensors have only one single-mode fiber for transmitting incident and reflected light. On the other hand, the TEFPI optical fiber sensor consists of two single-mode fibers (Fiber 1 and 2) with an air gap between them. Most of the incident light of Fiber 1 is transmitted to Fiber 2 through the air gap (path 1). However, 3.5% of the incident light from Fiber 1 is reflected at the end of Fiber 2. Among this light, 3.5% of this reflected light is repeatedly reflected again at the end face of Fiber 1, and then is transmitted to Fiber 2 (path 2). The phase difference of these two light paths causes interference in the optical receiver. Thus, the change in gauge length caused by the applied physical quantity in the longitudinal direction of the glass capillary results in the interferometric fringes as the output signal when the sensors are applied to the measuring objects. Because this transmission-type EFPI structure differs from reflection-type EFPI optical fiber sensors, i.e. intensity loss from the spreading of light in the air gap occurs in both paths, the intensity of the transmitted light in the optical receiver clearly changes. Therefore, in this TEFPI optical fiber sensor system, the interferometric fringe counting indicates the measurement of the physical quantity and degree of intensity loss from the spreading of light through the air gap reveals the distinction of the direction of strain.

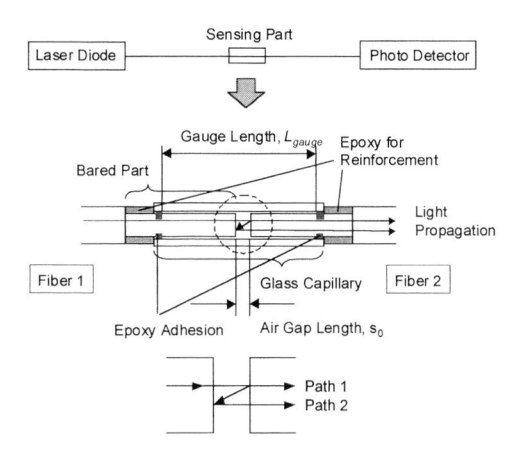

Figure 1. Structure of the TEFPI optical fiber sensor.

The optical fiber/air reflection factor r, the optical fiber/air transmission factor t_1 and the air/optical fiber transmission factor t_2 in Fresnel's equations are introduced into the wave equations of optical paths 1 and 2 of the TEFPI optical fiber sensor; the equations are then rearranged to the interference form by using the algebraic method. The transmittance T of the normalized intensity is described by

$$T(s) = t_1^2 t_2^2 \{L_1^2(s) + r^4 L_2^2(s) + 2r^2 L_1(s) L_2(s) \cos(2ks)\} \quad (1)$$

where s is the length of the air gap between the two optical fibers, and $L_1(s)$ and $L_2(s)$ are amplitude losses equivalent to intensity losses from the spreading of light in the two paths. k is the propagation constant.

The expression of $L_1(s)$ and $L_2(s)$ was based on the coupling loss caused by gap between two fibers with the application of the Gaussian approximation to the coupling efficiency equation derived by the overlap integral method formulated by Marcuse [7].

$$\eta(s_p) = L^2(s_p) = \frac{1}{1 + (0.5 s_p / x_R)^2} \quad (2)$$

In the above equation, s_p is the path length of light within the air gap. $L_1(s)$ and $L_2(s)$ are expressed with $L(s)$ and $L(3s)$.

2.2. Fatigue damage monitoring of aluminum plate patched with CFRP composites

2.2.1. Specimen and experimental setup

We used single-mode optical fibers which had a core diameter and a cladding diameter of 7.9 and 125μm, respectively. The glass capillary tube had approximately a 280μm outer diameter and a 128μm inner diameter. Leaving a 70μm air gap between two optical fibers, both ends of the capillary tube were bonded onto the optical fibers with epoxy adhesive. The interval between the bonded points defines the gage length of the sensor as shown in *Figure 1*. The gage length of the fabricated sensor was approximately 13 mm. The light source used in this experiment was a single-mode pigtailed laser diode of 1315 nm wavelength.

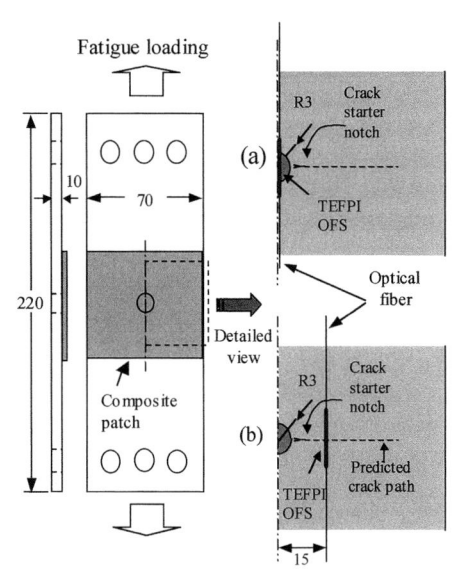

Figure 2. Dimensions of a center cracked tension specimen for the fatigue tests and attached locations of TEFPI sensors (a) at center point and (b) in front of the crack path.

Fatigue tests were performed to evaluate the monitoring technique of the fatigue crack growth using the TEFPI optical fiber sensors attached near the crack. The center cracked tension (CCT) aluminum specimen for the fatigue tests is shown in *Figure 2*. Its dimension is 220mm long, 70mm wide and 10mm thick. It has a 4mm-diameter hole with 1mm-long crack starter notches on both sides. Before the composite patch repair, fatigue loading for pre-cracking was applied until the crack propagated around 2 mm. Thus, the initial crack length, 2a, was 10 mm. The cracked specimen was repaired with a graphite/epoxy composite patch. The composite patch was made with 8 layers of prepreg and has 70x70x1mm dimension. The composite patch was cured using an autoclave and bonded to the cracked specimen using epoxy film adhesive.

A TEFPI optical fiber sensor was embedded to the composite patch during the curing process, and another TEFPI optical fiber sensor was bonded on the surface of patch at the same point of embedded sensor. The embedded TEFPI optical fiber sensor was inserted between the first and the second ply of the composite layer from the bonded surface.

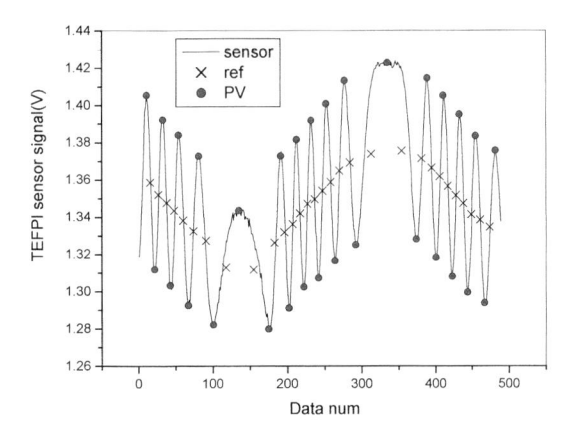

Figure 3. Typical output signal of the TEFPI optical fiber sensor.

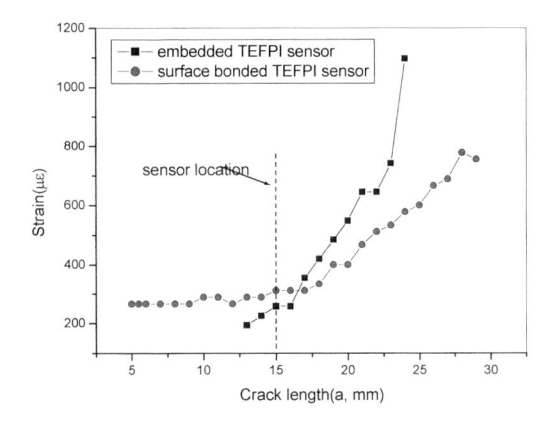

Figure 4. Crack length vs. strain data of the TEFPI sensor located at 15mm away from the center point.

The specimens were tested under a sinusoidal waveform loading with a frequency of 11Hz using a servo-hydraulic testing machine. The applied stress was 45MPa and the stress ratio was 0.1.

2.2.2. Experimental results

Figure 3 shows a typical TEFPI optical fiber sensor signal during 1.5 loading cycle. The output signal decreased or increased according to the tension and compression load. *Figure 4* shows the strain change of the embedded and surface bonded TEFPI optical fiber sensor located at 10mm ahead of initial crack (15mm away from center point) with respect to crack length. The rate of strain change was altered as the crack passed through the location of the sensor. Before the crack passed through the location of sensor, the rate of strain change was very small. However, as the crack passed through the location of sensor, the rate of strain change abruptly increased. Therefore, if the TEFPI optical fiber sensor is located where a fatigue crack may be initiated or ahead of an existing crack, the embedded or surface bonded sensor signal could indicate the crack initiation or the crack propagation.

3. BRILLOUIN DISTRIBUTED OPTICAL FIBER SENSORS

Most distributed optical fiber sensors use scattering occurring in optical fibers. The scattering in optical fibers is consisted of elastic one, where the wavelengths of the incident and scattering light are the same, and inelastic one, where the wavelengths of the incident and scattering light are different from each other. Brillouin scattering, a type of inelastic scattering, is suitable to the sensor principle due to the characteristic that the Brillouin frequency is shifted with strain and temperature.

Recently, a great deal of research on the application of Brillouin distributed optical fiber sensors to structural monitoring has been performed. Horiguchi, Kurashima et al. have reported the relationships between the Brillouin frequency shift and tensile strain and between the frequency shift and temperature [8]. DeMerchant [9] and Bao [10] have measured strain in a beam structure by using Brillouin distributed optical fiber sensors.

3.1. Strain measurement characteristics of Brillouin distributed optical fiber sensor

Brillouin scattering is induced by acoustic waves in an optical fiber. A thermally excited acoustic wave modulates the refractive index of an optical fiber core periodically, and Stokes and anti-Stoke Brillouin scattering occurs when the incident light is diffracted in this moving grating. Because the backward component is strongest in the scattering light, Brillouin backscattering is used in the sensor application. The Brillouin spectrum has a Lorenzian spectrum profile with a center

frequency of Brillouin frequency. The Brillouin frequency is shifted with strain or temperature, so that strain or temperature can be measured from finding the Brillouin frequency shift in the Brillouin spectrum.
The relationship between the Brillouin frequency shift and strain is shown in equation (3) [10].

$$\text{Strain coefficient,} \quad C \equiv \frac{1}{v_B(0)} \cdot \frac{\partial v_B(\varepsilon)}{\partial \varepsilon} \tag{3}$$

To obtain the Brillouin scattering signals, a pulsed light source is used. The position information is determined with the arriving time at an optical receiver after the pulse signal is launched. The length corresponding to the pulse duration in an optical fiber is the pulse length, and the measured value within half of the pulse length is observed at one detection time of the scattering light. This length is the spatial resolution, and strain within the spatial resolution range is averaged and shown at one position.

3.2. Beam deflection measurement using Brillouin distributed optical fiber sensors

The deflection of a structural bending beam is calculated with normal strain with respect to the length direction x as follows:

$$v(x) = \iint \left(-\frac{\varepsilon_x}{y} \right) dx dx \tag{4}$$

In the above equation, y is the distance from the beam neutral axis to the position, where strain ε_x is measured.
With distributed optical fiber sensors, the deflection can be calculated with integration by parts as follows:

$$f''(x) = \varepsilon(x) \tag{5}$$

$$v(x) = -\frac{1}{y} f(x)$$

$$= -\frac{1}{y} \{ f(0) + f'(0)x + \int_0^x \int_0^x f''(x) dx dx \} \tag{6}$$

The integration with discrete data can be processed as shown in equation (7). The interval Δx is related with sampling frequency in the sensor data acquisition.

$$\int_0^x f(x) dx = \sum_{i=1}^{\text{integer}(x/\Delta x_i)} f_i \Delta x_i \quad \text{when } \Delta x_i = x_i - x_{i-1} \tag{7}$$

$f(0)$ and $f'(0)$ can be obtained from the boundary conditions at $x=0$ and $x=L$, where L is the beam length.

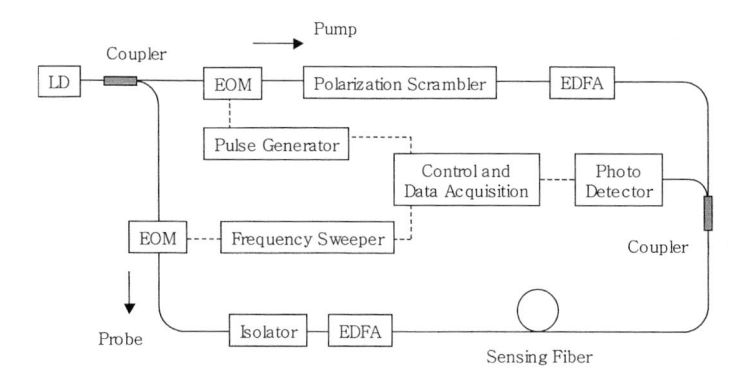

Figure 5. System configuration of the Brillouin distributed optical fiber sensor.

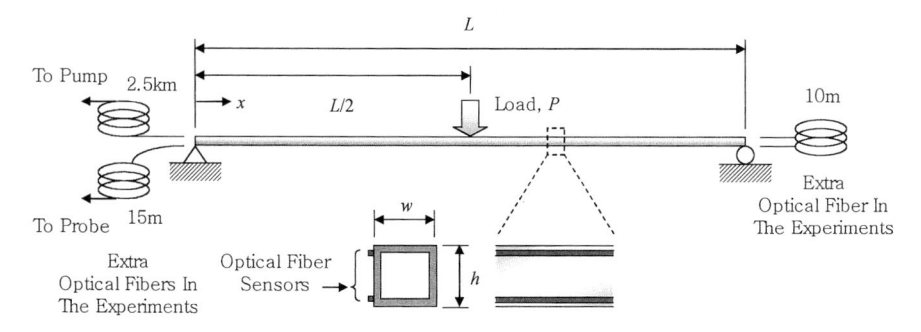

Figure 6. Three-point bending beam and attachment of optical fiber sensors (Extra optical fibers were connected in the experiments)

3.3. Experiment of deflection measurement

The system configuration of the Brillouin distributed optical fiber sensor used in the experiment is shown in *Figure 5*. The propagating light through an optical fiber from the laser diode is split into the pump and probe parts with a 3dB coupler. The pulsed pump light with an EOM (electro-optic modulator) is Brillouin-amplified by a frequency-modulated probe light, and the Brillouin spectrum is observed in the photo detector with the frequency difference between the two lights. Each time-domain signal is obtained at the modulated frequency.

To verify the beam deflection measurement of Brillouin distributed optical fiber sensors experimentally, a three-point bending load was applied to a square-tube beam with a width and height of 45mm and a thickness of 1.2mm, as shown in *Figure 6*. The beam length was 8.54m, and the span length between two lower supports was 8m. Optical fibers were attached to one side surface of the beam in two lines, which were positioned at ±15mm from the bending neutral axis. The beam in the experiment was of aluminum alloy 6063T5 with the modulus of elasticity of 68.9GPa and specific gravity of 2700kg/m^3. Also the concentrated load was applied at the beam center by using 1.22kg-weight.

The pulse duration of the Brillouin sensor system was 30nsec corresponding to 3m-spatial resolution, and the signal data were sampled at 100MHz. The modulation frequency in the probe light was swept with an interval of 1MHz.

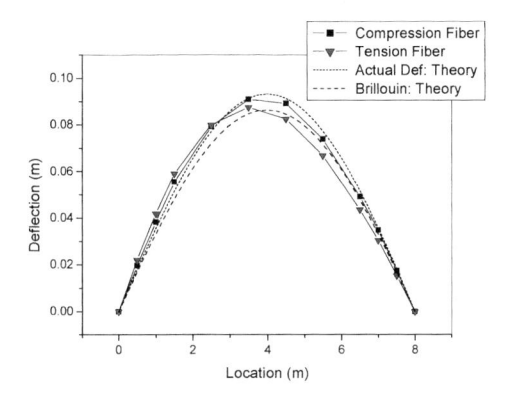

Figure 7. Beam deflection that was transformed from the measured strain in the case of additional center-concentrated loading.

The measured strain was transformed into the beam deflection as shown in *Figure 7*. The deflection in both the tension and compression sections approached the calculated deflection by beam theory. After the calculated strain by beam theory was applied to the strain measurement theory of Billouin distributed optical fiber sensors, it was transformed into the deflection. According to the results, it agreed with the measured deflection.

4. CONCLUSIONS

In this study of the fatigue crack growth monitoring of cracked thick aluminum plate repaired with composite patch using TEFPI optical fiber sensor, recently developed TEFPI optical fiber sensors were applied in the fatigue crack growth monitoring. The TEFPI optical fiber sensor was embedded and surface bonded to the composite patch that was bonded on the cracked aluminum plate. Both embedded and surface bonded sensors showed similar trend of strain change. But in careful comparison of the embedded and surface bonded sensors, the embedded sensor showed more sensitive strain change as the crack passed through the sensor location than that of the surface bonded sensor. The TEFPI optical fiber sensor located in front of the existing crack showed abrupt change in the rate of strain change as the crack passed through the location of sensor. This variation in the rate of strain change provides the detection of fatigue crack growth.

The strain measurement characteristics of Brillouin distributed optical fiber sensors were also studied. Because Brillouin distributed optical fiber sensors measure strain on the average within the spatial resolution range, they weren't much effective in measuring abruptly changing strain. This also induced measurement error of positions and strain, and thus the compensation method was required. The measured strain by Brillouin distributed optical fiber sensors was continuous with a position, so that the integral calculation in the beam deflection equation was possible, which was difficult with general sensors. Due to the strain measurement error at the beam center, the measured deflection was somewhat less than the actual deflection.

5. REFERENCES

1. Thompson, B.S. and Ghandi, M.V. (1990) *Smart Materials and Structures Technologies*, PA: Technomic Publication.
2. Sienkiewicz, F. and Shukla, A. (1994) Evaluation of a fiber-optic sensor for strain measurement and an application to contact mechanics, *Exper. Mechan.*, pp. 28-33.
3. Murphy, K.A., Gunther, M.F., Vengsarkar, A.M. and Claus, R.O. (1991) Quadrature phase-shifted extrinsic Fabry-Perot optical fiber sensors, *Optics Letters*, vol. 16, pp. 273-5.
4. Bhatia, V., Murphy, K.A., Claus, R.O., Tran, T.A. and Greene, J.A. (1994) Absolute strain and temperature measurements using high finesse EFPI cavities, *FEORC Fiber Opt.Rev. Conf. Blacksburg, VA*.
5. Liu, T., Brooks, D., Martin, A., Badcock, R. and Fernando, G.F. (1996) Design, fabrication and evaluation of an optical sensor for tensile & compressive strain measurements via the use of white light interferometry, in *Proc. SPIE*, vol. 2718, pp. 279-87.
6. Kim, S.H., Lee, J.J., Lee, D.C and Kwon, I.B. (1999) A study on the development of transmission-type extrinsic Fabry-Perot interferometric optical fiber sensor, *Journal of Lightwave Technology*, vol. 17, no. 10.
7. Marcuse, D. (1977) Loss analysis of single-mode fiber splices, *The Bell System Technical Journal*, vol. 56, pp. 703-18.
8. Horiguchi, T., Kurashima, T. and Tateda, M. (1989) Tensile strain dependence of Brillouin frequency shift in silica optical fiber, *IEEE Photonics Technology Letters*, vol. 1, no. 5, pp. 107-108.
9. DeMechant, M., Brown, A., Bao, X. and Bremner, T. (1999) Structural monitoring by use of a Brillouin distributed sensor, *Applied Optics*, vol. 38, no. 13, pp. 2755-2759.
10. Bao, X., DeMerchant, M., Brown, A. and Bremner, T. (2001) Tensile and compressive strain measurement in the lab and field with the distributed Brillouin scattering sensor, *Journal of Lightwave Technology*, vol. 19, no. 11, pp. 1698-1704.

APPLICATION OF STRESS AND STRAIN CONTROL TO LIVING TISSUES

VALENTIN KIRYUKHIN and YURIY NYASHIN
Department of Theoretical Mechanics, Perm State Technical University, Komsomolskii prospect, 29 a, Perm, 614600 Russia
E-mails: kvy@theormech.pstu.ac.ru, nyashin@theormech.pstu.ac.ru

1. INTRODUCTION

The problem of prescribed thermal and residual stresses after plastic deformations (for instance, hot rolling) is one of the important technical problems. Desired stress improves the practical properties of manufactured structure. In the outer space the structure can be deformed and the control procedure has to be applied to keep the structure characteristics. Other important application of the stress control deals with biomechanics problems. For example, in orthopedics the congenital maxillary anomaly is treated by means of apparatus which brings down the hard palate fragments from the nasal to the oral cavity. The problem is to find the technological parameters of optimal apparatus, which allow realizing the treatment stage by stress in the most desired way [1, 2]. One of the most actual problems is also the femur prosthesis design [3,4]. The remodelling process is interpreted as stress and strain control process by properties variation [5]. Many interesting aspects of biomechanics of growth can be found in the review by Taber [6].

In this study the problem of the optimal apparatus for the orthopedic treatment of the congenital maxillary cleft is discussed. Some partial solutions are received.

2. GROWTH STRAIN CONTROL PROBLEM IN ORTHOPEDIC TREATMENT OF THE CONGENITAL PALATE CLEFT IN CHILDREN

In spite of apparent proximity of the theory of stress and strain control in thermoelasticity to the technical problems, it turned out, that there are a lot of interesting and perspective problems of biomechanics that can be analyzed and described by this approach. Beforehand it can be said that from the point of result living tissue introduces more freedom. But the restrictions, nonlinearity, more difficult response of tissue, non-small strain arise in problems. Here we briefly describe actual medical problem of the orthopedic treatment of the congenital palate cleft in children.

It is well understood that children born with maxillary congenital anomaly need specialised emergency, which involves a very important stage of eliminating a disconnection of the nasal and the oral cavities. The success of measures aimed at recovering anatomic forms and functions of congenital deformed organs largely depends on such factors as the urgency of the first specialised aid and efficiency of methods used to correct pathomorphological, functional and anatomic disorders.

A mathematical substantiation of the optimal load on the bone tissue, which will not interfere with the growth of underdeveloped maxilla fragments is a key aspect to orthopedic reconstruction.

Nowadays the need of lending specialized aid to children with congenital maxillary defects is the problem of particular concern to physicians all over the world. With the establishment of rehabilitation and health centers and regional and municipal clinics at medical institutes for such children, an urgent aid to neonates with the congenital maxilla defects can be provided immediately at maternity home. This also extends the possibilities for designing effective methods of subsequent treatment, which is an effective combination of different therapeutic and surgical-cosmetic measures (early orthopedic reconstruction of abnormally developed maxilla, uranoplastic surgery) aimed at rehabilitation of preschool-age children.

The main stage in the treatment of the congenital maxillary defects is the employment of orthopedic apparatus which exerts a mechanical action on the separated sides of the cleft. The development of biomechanical grounds for control of the process is an important practical problem in the sense that they provide objective criteria for the choice of individual orthopedic apparatus for each patient and refinement of the existing treatment methods.

The simplest mathematical model for describing the behavior of the separated palate fragments taking into account the growth deformation has been first proposed by Masich *et al.* [7]. In that work, the palate fragment is viewed as a growing cantilever flexible beam.

In a more recent paper [8] the proposed mathematical model of growth deformation was refined to completely allow for the influence of the mechanical factor. Here we describe the two techniques used for rehabilitation of children and teenagers with congenital dentofacial pathology and serious speech disorders.

2.1. Methods of orthopedic treatment of the congenital palate cleft

The most essential stage of the methods is the application of orthopedic plate to separate the nasal and the oral cavities and exert mechanical action on the palate fragments. In such a way the fragments are subjected to pressure produced by the adhesion force of the applied plate and the work of a strong muscular masticatory organ – the tongue. Furthermore, the plate prevents the tongue from penetration to the cleft and, as clinical observations show, causes decreasing of the cleft.

First, consideration is given to an orthopedic remodeling technique used at the All-Russian scientific-practical centre of medical-social rehabilitation "Bonum"

(Ekaterinburg, Russia), Figure 1. In this method the plate closely copies the shape of the alveolar process connection with the palate fragment and is confined in the oral cavity due to adhesion in the contact zone. During this stage of treatment the ends of the palate fragments remain free, so that they can easily change their shape with the displacement from the nasal to the oral cavity.

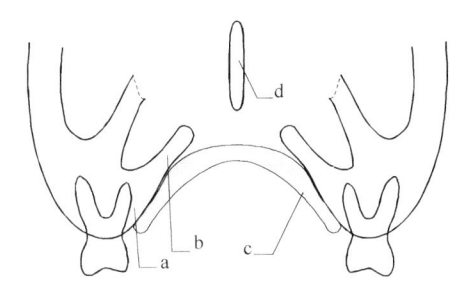

Figure 1. Schematic representation of the two-sided cleft and orthopedic plate: a – alveolar process, b – palate fragment, c – orthopedic plate, d – nasal plate.

The role of mechanical factor in this method is played by the orthopedic plate, which exerts pressure on the basis of underdeveloped palate fragments just at the place where the alveolar processes are transformed into palate fragments. The method is applied to breast-fed babies.

In all examined cases the early treatment leads to a visible growth of the palate fragments and a change in their position: there occurs an appreciable displacement of the fragments from the nasal to the oral cavity.

Another type of apparatus is considered further. Scharova and Simanovskaya [9] (Perm State Medical Academy) elaborated the method of stepwise preoperative orthopedic reconstruction of the defective maxilla. The key to this method is to bring down the hard palate fragments from the nasal to the oral cavity with the help of demountable orthopedic apparatus. The outline of its main elements is shown in Figure 2A. Figure 2B schematically shows the apparatus after applying to separated fragments in the case of the double-sided cleft. The apparatus consists of the nasal plate and the teeth-gum plate operated by a rubber ring with predetermined diameter. The mechanical force developed by the rubber ring mounted on six supporting loops is transferred to the nasal plate, which exerts a downward force on the palate. This results in a gradual bringing down the palate fragments from the nasal to the oral cavity and finally stimulates osteogenesis along their free edge. A period of use of the apparatus depends on the child's age, the kind of a cleft, the degree of deformation and underdevelopment of the palate fragments, as well as on the general state of a child's health and can last from 3 months to 1.5 year [9]. The efficiency of the apparatus is supported by the facts that the palate fragments change their position and the lateral dimensions of the cleft decrease. The basis for demountable apparatus is the principle of mechanical action. The main mechanical factor is the pulling force developed by the elastic rubber ring as soon as it has been mounted on the supporting loops.

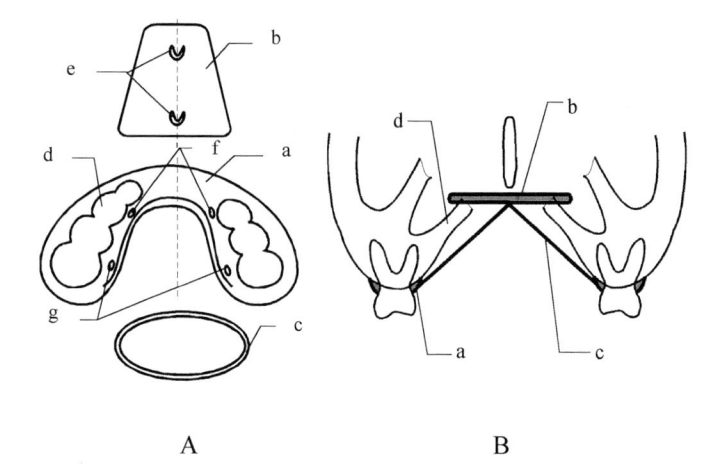

Figure 2. A. Schematic representation of the apparatus for correction of hard palate anomaly: a) teeth-gum plate; b) nasal plate; c) elastic ring; d) the teeth holes in the teeth-gum plate; e) supporting loops of nasal plate; f) medial supporting loops; g) distal supporting loops. B. Scheme for application of apparatus in the case of double-sided palate cleft: a) teeth-gum plate; b) nasal plate; c) elastic ring; d) palate fragment.

2.2. Mathematical formulation of the problem on biomechanical reconstruction of palate fragments

Here a mathematical model is proposed to describe the behavior of children's hard palate with the congenital cleft exposed to mechanical action of the orthopedic apparatus taking into account the growth strains. A general mathematical formulation of the problem on the deformation of growing elastic body has been described in detail in the earlier work [10]. This problem is defined by the following basic equations:

1. the equilibrium equation

$$\int_V \tilde{\sigma} \cdot \tilde{\varepsilon}(\vec{w}) dV - \int_V \vec{Q} \cdot \vec{w} dV - \int_{S_\sigma} \vec{P} \cdot \vec{w} dS = 0, \qquad (1)$$

where $\vec{w} \in \left(W_2^1(V)\right)^3$ and $\vec{w} = 0$ at $r \in S_u$;

2. the kinematic relations

$$\tilde{e} = \frac{1}{2}(\nabla \vec{v} + \nabla \vec{v}^T), \ \vec{r} \in \overline{\Omega}; \qquad (2)$$

3. the constitutive relations

$$\widetilde{e} = \widetilde{N} + \widetilde{M} : \widetilde{\sigma} + \frac{d}{dt}(\widetilde{C}^{-1} : \widetilde{\sigma}),$$

where
$$\widetilde{e}^g = \widetilde{N} + \widetilde{M} : \widetilde{\sigma}, \quad \widetilde{e}^e = \frac{d}{dt}\widetilde{\varepsilon}^e = \frac{d}{dt}(\widetilde{C}^{-1} : \widetilde{\sigma}); \quad (3)$$

4. the boundary conditions

$$\vec{v} = 0, \ \vec{r} \in S_v; \ \vec{n} \cdot \widetilde{\sigma} = \vec{P}, \ \forall \vec{r} \in S_\sigma, \quad (4)$$

5. the initial conditions

$$\widetilde{\sigma}(\vec{r}, 0) = 0. \quad (5)$$

In these equations $\widetilde{e}, \widetilde{e}^e, \widetilde{e}^g$ are the tensors of total, elastic and inelastic (growth) strain rates, respectively, $\vec{v}$ is the velocity vector of the medium, $\widetilde{C}^{-1}$ is the tensor of elastic compliance, $\widetilde{N}$ is the tensor characterizing the proper growth of the material (in the absence of stresses), $\widetilde{M}$ is the tensor responsible for the effect of stresses on the growth strain, $\Omega \in R^3$, $\overline{\Omega} = \Omega \cup S$, $S = S_v \cup S_\sigma$, part of the boundary S_v is subjected to kinematics boundary conditions and S_σ is subjected to forces.

2.3. Determination of growing cantilever beam configuration

As a biomechanical model capable of describing the clinical picture of the cleft palate response we have used a uniform growing beam with one end fixed and the other subjected to a load (Figure 3). This essentially simplifies the understanding the processes occurring in the growing medium and reduces the number of parameters for a mathematical model, many of which cannot be determined with sufficient accuracy.

To calculate deformation of the beam shown in Figure 3 we introduce local coordinates (ξ,η) related to an axial element of the beam (Figure 3). The global X-Y coordinates in the Figures 3, 5 and 6 indicate locations in the oral cavity. Neglecting the instantaneous elastic strain the constitutive relation (3) can be written as

$$e_\rho(\xi, \eta) = N + M\sigma(\xi, \eta), \quad (6)$$

where σ is the axial stress normal to the cross section, e_ρ is the strain rate at a point of the beam, N and M are the growth constants. For this relation we require the fulfilment of the hypothesis for the plane sections

$$e_\rho(\xi,\eta) = e_0(\xi) + \dot{\chi}(\xi)\eta \qquad (7)$$

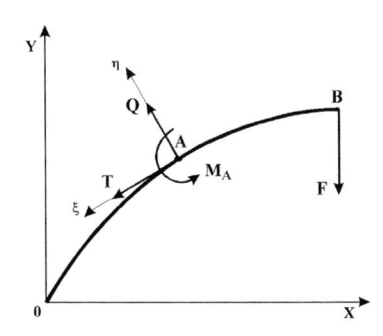

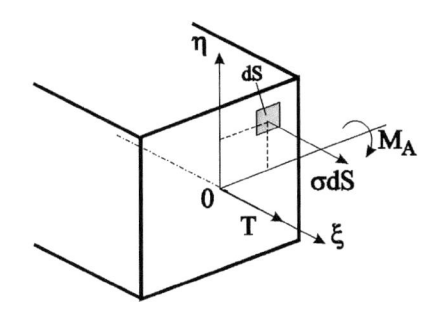

Figure 3. Local coordinates (ξ,η) related to an axial element of the axis; schematics of forces T, Q and moment M_A acting on imaginary cut section AB of the beam.

Figure 4. General equilibrium conditions under beam bending [12].

and neglect the shear strains. Here $e_0(\xi)$ defines the strain rate of the beam axial element $d\xi$ and $\dot{\chi}(\xi)$ is the bending strain rate of the cross-section.

The strain rate parameters $e_0(\xi)$ and $\dot{\chi}(\xi)$ are determined using general equilibrium conditions (Figure 4) [11]. The bending moment $M_A(\xi)$ and axial (longitudinal) force $T(\xi)$ are determined by equilibrium conditions on the separated section of the beam. Then we obtain:

$$e_0(\xi) = \frac{T(\xi) \cdot M}{S} + N, \quad \dot{\chi}(\xi) = \frac{M_A(\xi) \cdot M}{I_A}, \qquad (8)$$

where I_A and S are the moment of inertia and the area of the cross-section, respectively. Relation (8) determines configuration of the beam at each time instant.

Thus, at each current moment we can define the strain $\varepsilon_0(\xi)$ of the axial element $d\xi$ and its deflection $w(\xi)$ along local axis:

$$e_0(\xi) = \frac{d\varepsilon_0(\xi)}{dt}, \quad \chi(\xi) = \frac{\partial^2 w(\xi)}{\partial \xi^2}. \qquad (9)$$

The stated problem has been solved numerically for a beam of rectangular cross-section of the width $b = 15$mm and thickness $h = 3$mm. The load on the free end of the beam distributed over the width is 10g.

Then we have estimated the growth parameters for the case shown in Figure 5. The control points were obtained from the dental model of the patient D. made for a rear section of the one-sided cleft at the beginning of treatment ($t = 0$) and at the end ($t = 18$ months) and approximated by straight lines. The calculated values of the actual growth parameter N and growth viscosity M are, respectively, $N = 0.002543$ (1/month) and $M = 0.002096$ (mm^2/(g month)).

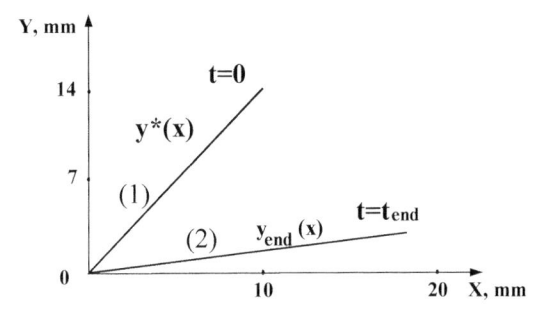

Figure 5. Experimental straight lines of the disconnected palate fragment obtained from the dental model of the patient D. having the one-sided palate cleft at the beginning of treatment and at the end.

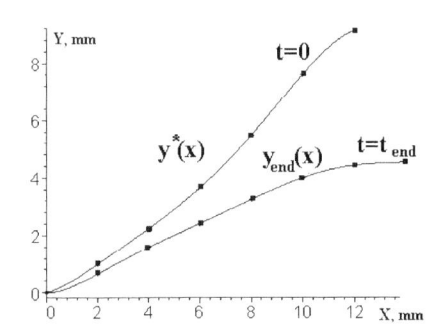

Figure 6. Experimental points for patient M. obtained from the rear section of double-sided cleft of the hard palate (symmetric one half).

The developed model can be also adapted to the case of real section geometry. Figure 6 illustrates the section configuration in the rear part of the double-sided cleft of the patient M. obtained with the help of the above procedure.

3. MATHEMATICAL FORMULATION OF OPTIMAL APPARATUS DESIGN PROBLEM. BIOMECHANICAL ANALYSIS

In the beginning we consider the application of the model to the method elaborated in Perm State Medical Academy. Once the growth parameters have been determined the question arises as to what magnitude of optimal force should be applied at the free end of the beam to form the palatal arch in the best possible way. Here of particular importance is the clinical fact that the maximum stress in the living structure produced by the action of applied external load must be not higher than 20 g/mm^2 [9], that would otherwise injure the mucous membrane and bone tissues of the hard palate. The maximum load F estimated under this physiological constraint was ~ 40 g. At present we have formulated and solved the problem of determining the optimal magnitude of the force F_{opt} with consideration of restrictions on the applied force F and the angle β, made by the force **F** with the horizontal axis x (Figure 7). From the biomechanical model in Figure 7 it follows that the vertical force **F** is decomposed into two components **T** and **N**. The compressive force **T** interferes with the growth. However, the actual tissue growth prevails over the action of the force **T** so that from clinical viewpoint we observe the osteogenesis processes. The force **N** turns the fragment and brings it to the position 2. The force **F** must be obviously applied in such a way as to help the actual physiological growth of the tissue and appropriately rotate the palate fragments, for example, to have the direction of the force **F'** as shown in Figure 7. From these considerations it follows that angle β has to be restricted to

$$0 \leq \beta \leq 90° - \alpha. \tag{10}$$

Hence, with the above restrictions the optimisation problem is reduced to finding F and β such that

$$\Psi(F,\beta) = \frac{D_1}{L_{end}^3} \int_0^L [\, y(x,F,\beta) - y_{end}(x) \,]^2 \, dx + D_2 [1 - \frac{L(F,\beta)}{L_{end}}]^2 \xrightarrow[F,\beta]{} \min, \tag{11}$$

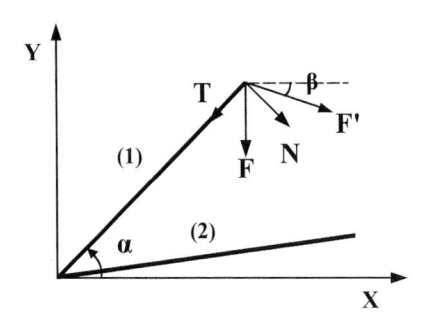

Figure 7. Graphical analysis of constraints on the force **F**.

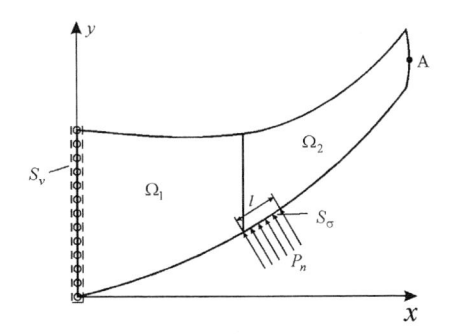

Figure 8. Mathematical model of the treatment elaborated in the centre "Bonum".

where D_1 and D_2 are the dimensionless weight coefficients. The solution has to be subjected to the constraints

1. $F \in [0, F_{max}]$, (12)
2. $\beta \in [0, 90° - \alpha]$, (13)
3. determination of the growing beam configuration according to (1)-(5). (14)

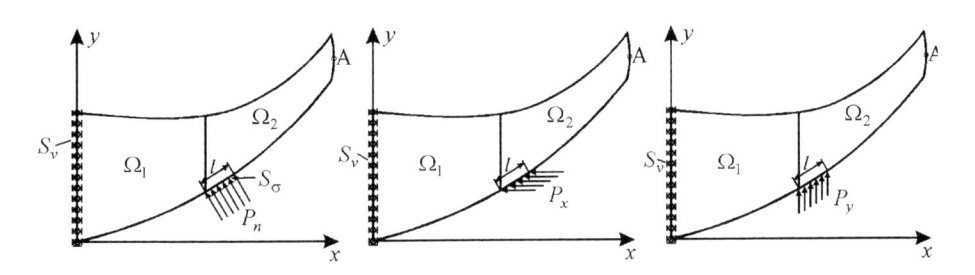

Figure 9. Three different pressure directions in the contact zone. Force P_n is normal to the contact zone. Force P_x is directed along the x-axis. Force P_y is directed along the y-axis.

The problem was solved by applying Hooke-Jeevs numerical optimisation method. The calculation gave the following values of $F_{opt} \approx 40$ g and $\beta_{opt} \approx 7.3°$.

Further we consider the application of the mathematical model for the method of the treatment elaborated in the centre "Bonum" (Figure 8). The developed mathematical model has been used to investigate the influence of the size of the contact zone on the variation of the fragments for the period of 6 months (t_{end}). The problem is solved as a plane-strain one. An approximate solution has been found using the finite element method [8].

Table 1. Values of displacement of point A

P	Value P (g)	$u_x (mm)$	$u_y (mm)$	$u(mm)$
P_y	8	1.75	1.20	2.12
P_n	8	2.10	0.79	2.24
P_x	8	2.56	-0.03	2.56

Furthermore the calculation shows the influence of the direction of the contact pressure (Figure 9) on the behavior of the fragments. Three variants of force directions are considered. The corresponding force components are assumed to be equal to 8g. The control parameter is the length of the contact zone l. For every case optimal problem is solved.

It can be noted that in the present problem the rotation of the fragment takes place not in the direction of the bending moment. It is connected with the fact that in the zone of action of force P the growth of the bone tissue is locally suppressed, whereas the remaining region the conditions of natural growth are preserved.

5. CONCLUSIONS

In the present work we have considered the two methods of the treatment of the children's palate cleft. We proposed an algorithm for determining individual optimal loads developed by the orthopedic apparatus in the process of bringing down the palate fragments.

The authors gratefully acknowledge Prof. Franz Ziegler for fruitful discussion. This research was supported by a grant of the Russian Fund of Basic Research (Ural), number 02-01-96416.

REFERENCES

1. Masich, A.G. and Nyashin, Y.I. (1999) Mathematical Modeling of Orthopedic Reconstruction of Children's Maxillary Anomaly, *Russian Journal of Biomechanics* **3** 101-109.
2. Masich, A.G., Simanovskaya, E.Yu., Chernopazov, S.A., Nyashin, Y.I., and Dolgopolova, G.V. (2000) The Role of Mechanical Factor in Orthopedic Treatment of Congenital Palate Cleft in Children, *Russian Journal of Biomechanics* **4** 33-42.
3. Allard, P., Trudeau, F., Prince, F., Dansereau, J., Labelle, H., Duhaime, M. (1995) Modelling and Gait Evaluation of Asymmetrical-Keel Foot Prosthesis, *Medical and Biological Engineering and Computing* **33** 2-7.
4. Dietrich, M. and Kedzior, K. (1999) Design and Manufacturing of the Human Bone Endoprostheses Using Computer-Aided Systems, *Journal of Theoretical and Applied Mechanics* **37** 481-503.
5. Lanyon, L.E. (1987) Functional Strain in Bone Tissue as an Objective, and Controlling Stimulus for Adaptive Bone Remodelling, *Journal of Biomechanics* **20** 1083-1093.
6. Taber, L.A. (1995) Biomechanics of Growth, Remodeling, and Morphogenesis, *Applied Mechanics Reviews* **48** 487-544.
7. Masich, A.G., Nyashin Y.I. (1999) Mathematical modelling of orthopedic reconstruction of children's congenital maxillary anomaly. *Russian Journal of Biomechanics* **3** 101-109.
8. Masich, A.G., Chernopazov, S.A., Nyashin, Y.I., Simanovskaya, E.Yu. (1999) Formulation of initial boundary-value problem and construction of computational algorithm in simulation of growing bone tissue. *Russian Journal of Biomechanics* **3** 32-38.
9. Sharova, T.V, Simanovskaya E.Yu. (1983) *Orthopedic reconstruction of children's congenital one- and double-sided palate cleft. Methodical recommendations*, Perm Medical Institute, Perm.
10. Stein, A.A. (1995) Axial compressive deformation of a rod made from growing biological material. *J Applied Mathematics and Mechanics* **59** 149-157.
11. Hsu, F.H. (1968) The influences of mechanical loads on the form of a growing elastic body. *J Biomech* **1** 303-312.
12. Birger, I.A., Mavljutov, R.R. (1986) *Material Strength: Educational supplies*, Nauka, Moscow.

UNFOLDING OF MORNING GLORY FLOWER AS A DEPLOYABLE STRUCTURE

HIDETOSHI KOBAYASHI, MASASHI DAIMARUYA and
HIROFUMI FUJITA
Department of Mechanical Systems Engineering,
Muroran Institute of Technology
Mizumoto, Muroran, Hokkaido, 050-8585 Japan
kobayasi@mmm.muroran-it.ac.jp

1. INTRODUCTION

Although there are some plant leaves which just grow from small size to normal size, most leaves or flowers not only grow but also deploy or unfold from their small buds. In the buds of leaves or flowers, therefore, various ideas to stow their bodies such as folding or rolling are often observed. For example, the leaves of hornbeam or common beech have a typical corrugated folding pattern [1,2] while the fan-type bellows pattern is observed in maple leaves [3]. As well-known, the petals or petal ribs of a morning glory flower mysteriously spiral in a bud. Beetles or grasshoppers have also folding patterns in thin hind wings under hard covering forewings which have been studied by many researchers [4-7]. The movements of beetle wings folding at rest were investigated and compared with those in flight [8]. In recent works[9,10], it was indicated that the key folding/unfolding system adopted in the hind wings of Coleopetra consists of four folding lines intersecting in a point. A number of deployable structures observed in nature are summarized and classified by Kresling [11]. The investigation of the folding/unfolding manner in plants [12] and insects, may be useful for in designing some artificial deployable structures [13-15], such as solar panels, antenna of satellites, or deployable roofs. However, there are little research concerning folding patterns in nature from mechanical point of view.

  In this research, the unfolding manner of morning glory flowers was studied as a good example of interesting folding patterns in nature. Using a video camera, the unfolding of morning glory flowers was observed. From the observation, the deploying radius of flowers was measured and the radius-time curves were obtained. The force and the energy required for deploying petals were estimated using a number of models with different deploying manners. In order to obtain the mechanical properties of flower materials, tensile tests were also performed.

2. MEASUREMENT OF MORNING GLORY FLOWER

2.1. Morning Glory Flower

The morning glory (*Ipompea nil*) is a lianoid therophyte in bellbind family whose original home is south China and south-east Asia. The bud of flowers has a spindle shape with spirally striped pattern as shown in Fig.1(a). In unfolding of a flower, the spiral stripes loosen and the petals are drawn from the chinks of the stripes. After 3 ~ 6 hours, a funnel shaped flower like Fig.1(b) appears. A sector drawn by white lines is one petal unit of a flower. The petal unit consists of two parts: one is a central part with pleats which is relatively hard and called "petal rib" here, and the other one is "petal" which is a thin membrane. Five fan-shaped petal units are jointed and make a morning glory flower. At the lower half of the bud body (Fig.1(b)), the petal ribs join together and construct a hollow tube. From the whole view of a flower, it appears that the petal ribs support the whole body of a flower.

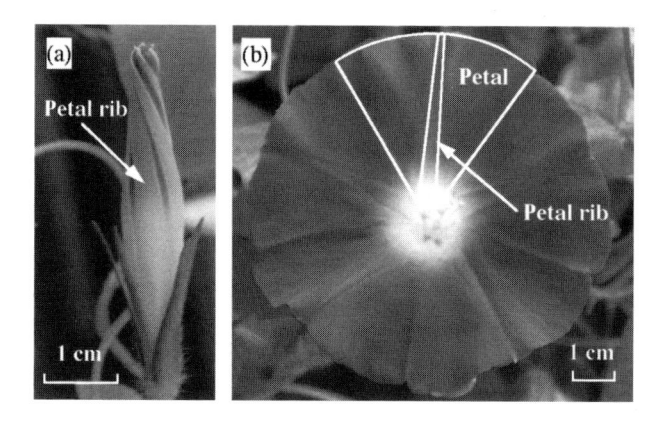

Figure 1. Morning glory flower: (a) side view of bud and (b) top view of flower opened.

2.2. Mechanical Properties of Petal Ribs and Petals

In order to obtain the mechanical properties of the petal ribs and petals, static tensile tests were carried out. For the tests, several rectangular tensile specimens whose gauge section is also a rectangle of 16 × 20 mm were prepared from these two parts, see Fig.2. The average measured thickness of these two parts are shown later in Table 1. After setting the specimen clamped by paper clips on the bed of a profile projector as shown in Fig.2(c), an initial load of 0.0049 N (0.5 gf) was applied by a dead weight. The longitudinal and transverse elongations of the specimen were measured using four painted markers every 0.0245 N (2.5 gf).

Fig.3 shows the nominal stress-strain curves of petals and petal ribs. Although these data were off-set due to an initial load, each data almost cluster on a line. The slope of these lines are 1.03 MPa and 3.0 MPa which correspond to Young's moduli of petals and petal ribs of morning glory flowers, respectively. Petal rib is about three times stiffer than petals. Other data obtained from the tests such as tensile strength and Poisson's ratio are listed in Table 1 with their density.

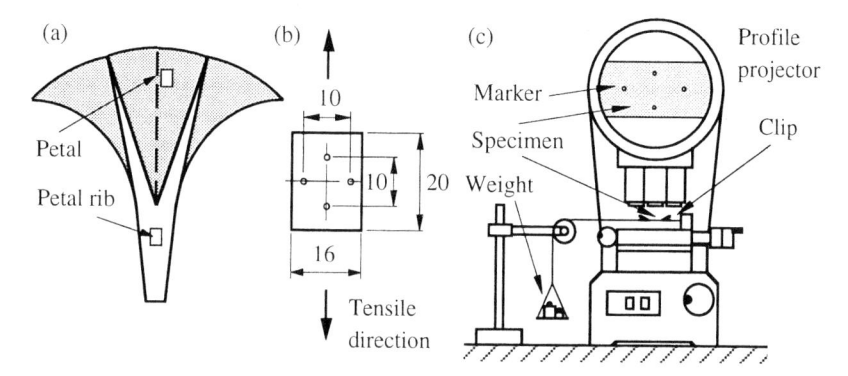

Figure 2. Setup for tension tests, (a) regions cut tingout specimens, (b) gauge section of a specimen and (c) experimental setup.

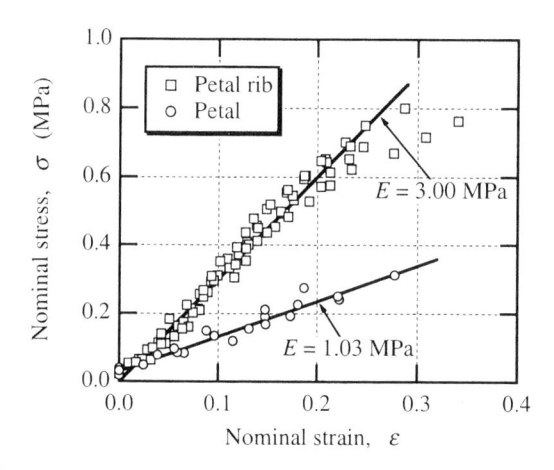

Figure 3. Nominal stress-strain curves of petal ribs and petals.

Table 1. Mechanical data of petals and petal ribs.

	Petal rib	Petal
Young's modulus (MPa)	3.00	1.03
Tensile strength (MPa)	0.78	0.27
Poisson's ratio	0.3	0.2
Density (kg/m^3)	890	730
Thickness (mm)	0.16	0.07

To understand the reason for the difference in the rigidity and the strength, the shape of cells in both area were examined by a microscope. Fig.4 shows the micrographs of petal surface and petal rib surface. The cells in the petal are almost round (Fig.4(a)) and have nearly same size, therefore, the petal seems to be homogeneous and isotropic. In the petal rib, however, the cells show long rectangles or hexagons along pleat lines observed by eyes and the cell walls make long lines, that is, the petal rib is probably reinforced in their longitudinal direction by cell walls.

From the results of tensile tests and the surface micrographs, it can be said that the petal rib takes a role to support wide thin petals and to give tensile stress to the petals for keeping the funnel shape of a morning glory flower.

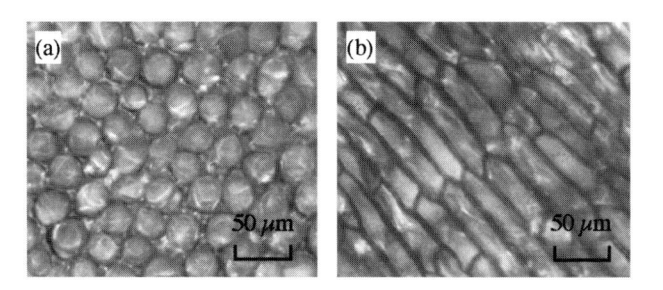

Figure 4. Micrographs of surfaces of (a) petal and (b) petal rib.

2.3. Deploying Radius of Flower

The deployment of morning glory flowers was observed by a video camera placed on the longitudinal axis of a flower bud to obtain the top view pictures. During the deployment of a flower, the tips of petal ribs move radially from the centre of the

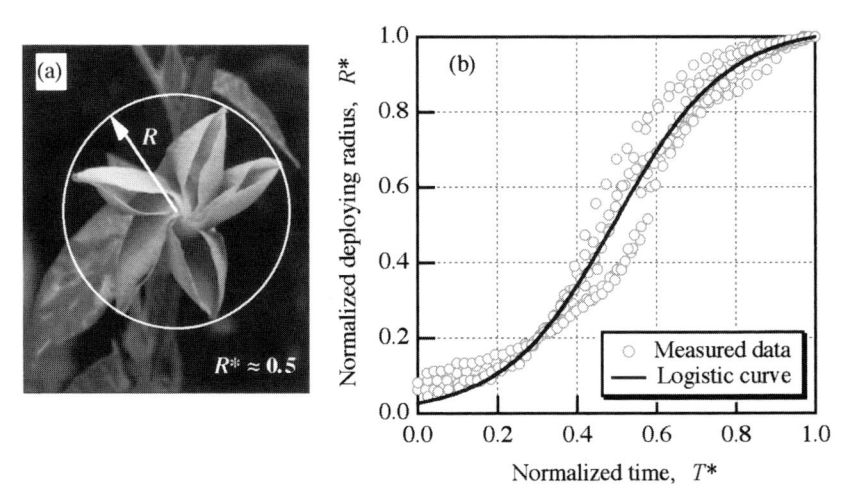

Figure 5. Deploying radius: (a) definition and (b) normalized radius-time curves.

bud to the outside as shown in Fig.5(a). We define the deploying radius R as the average distance between the tips of petal ribs and the centre of the deploying flower. To avoid the effect of flower size, it is convenient to use non-dimensional values. Using the deploying radius of the fully deployed flower R_0, therefore, we introduce a normalized deploying radius R^* ($= R/R_0$). The situation shown in Fig.5(a) is $R^* \approx$ 0.5 and five petals folded in two are still partially rolled up at the centre of the flower. In most flowers, five petals were released each other when $R^* \approx 0.6$.

The radius-time curves (R^*-T curves) were obtained from pictures taken every 5 or 10 minutes. The final time, T_0, i.e. the time when $R^* = 1.0$, may depends on the environment around the flower bud such as light and temperature. By introducing the normalized time, T^* ($= T/T_0$), therefore, we can get all data together as shown in Fig.5(b), if it is assumed that $T^* = 0.3$ when $R^* = 0.2$. It is clear that the deploying speed of a flower gradually increases in the early stage up to about $R^* = 0.6$ and then the unfolding decelerated in the latter stage until the flower fully deploys.

We can draw a curve passing through the centre of these data which is given by

$$R^* = \frac{c_1}{1 + c_2 \exp(-c_3 T^*)} \quad (1)$$

where c_1, c_2 and c_3 are constants which are $c_1 = 1.026$, $c_2 = 36.5$ and $c_3 = 7.23$. This curve is called a "logistic curve", which is usually used to present the growth of organisms. This means that the deployment of morning glory flower is probably related with the growth of cells in petal ribs due to the suction of water.

3. DISCUSSION DUE TO FLOWER MODELS

3.1 Flower Models

In order to understand the mechanical characteristics of the flower unfolding, a number of models with the different number of petal rib N were considered. Since only upper half of a bud deploys as shown in Fig.1, the upper half of a bud was replaced by a pyramid with a regular polygon as its base. The model shown in Fig.6 is the nearest model to a real bud. Fig.6(a) and (b) show a top view of the fully deployed model and a side view of the fully closed model, respectively. The numbers of the petal rib adopted here are 3, 4, 5, 6 and 8, i.e. the base polygons used are triangle, square, pentagon, hexagon and octagon. In all models, the area of a base polygon A_0, the maximum opening angle α and the height h of a side isosceles triangle, which corresponds to a petal rib, were fixed to be $A_0 = 100$ mm^2, $\alpha = 60°$ and $h = 60$ mm, respectively.

A petal part, shaded sector area in Fig.6(a), is folded in two and rolled to make a cone as shown in Fig.6(c). The rolled petals are overlapped each other in a bud. Therefore, the thickness of overlapped petals is an important geometrical factor of the models. Using the petal thickness, $t = 0.07$ mm, the maximum thickness of overlapped petals t_{max} was calculated. The rate of t_{max} to the radius R_1 of the circumscribed circle of the base polygon is shown in Fig.7. The overlapped thickness of the petals dramatically increases with the increase of N. Since the transverse size of the bud pyramid decreases toward its top, it may be difficult to put all rolled petals in the inside of the model with $N = 8$. This means that the small number of petal ribs is recommended from the shape of models fully closed.

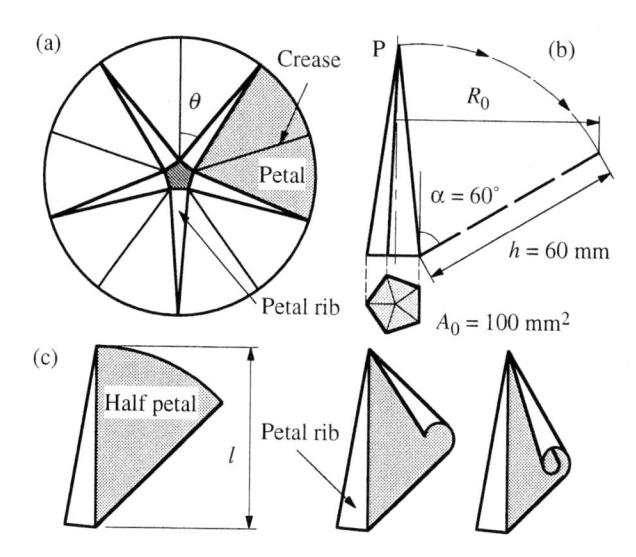

Figure 6. Model of upper half of bud: (a) top view of model fully deployed, (b) side view of model fully closed and (c) rolling manner of a half petal.

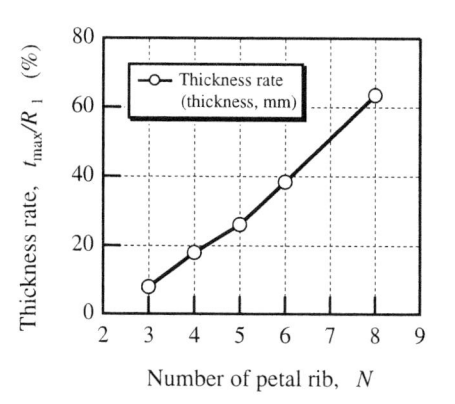

Figure 7. Maximum thickness of rolled petals in models.

3.2 Change of Contact Petal Area During Deployment

As mentioned before, the petals are rolled tightly and contact each other in a bud and pulled out from the inside due to the movement of the petal rib. Therefore, the contact area of petals in a bud may strongly relate to the force required for the deployment of flowers. Here, we try to estimate the contact petal area.

Since the base polygon of the pyramid bud models is regular, the N-th polygon can be divided into N congruent isosceles triangles (see Fig.6(b)). Thus, a half apex angle β, the height b and a base length a of the isosceles triangle are given by $\beta = \pi/N$, $b = \sqrt{A_0/(N\tan\beta)}$ and $a = 2b\tan\beta$, respectively. Because a petal rib is also an isosceles triangle which has the base length of a and the height of h. Thus, the side length l and a half apex angle γ of the petal rib are written by $l = \sqrt{(a/2)^2 + h^2}$ and $\gamma = \tan^{-1}(a/2h)$, respectively. When a half petal is a sector with the radius of l and the central angle of θ, the angle θ can be obtained from vector analysis, i.e. $\cos 2\theta = 1 - 2(R_0/l)^2\sin^2\beta$, where R_0 (= $b + h\cos\alpha$) is the maximum opening radius as shown in Fig.6(b). Since the number of petal rib N appears in the denominators of β and b, the θ decreases with the increase of N.

From the observation of the deployment of morning glory flowers, we take the following two assumptions about a boundary line between the contact part of petals and released parts: (1) when all petal ribs are fully closed, the boundary line consists with the symmetric line of a petal rib (an isosceles triangle) and (2) the boundary line is parallel to the initial line during unfolding.

Let us consider a half petal, a sector OAB, in the xy-coordinates as shown in Fig.8(a). A half apex angle of a petal rib is γ, thus the line AC represents the initial boundary line. During unfolding, the arc of the half petal AB always passes through the point P which is the apex of the initial pyramid (see Fig.6(b)). When the point D is at the apex, the boundary line DE parallel to AC can be drawn. Therefore, the shaded region BDE corresponds to the contact part of the half petal. The areas of the half petal OAB and the sector OAD are denoted by S_0 and S_1, respectively. If we take $\angle AOD = \phi$, S_0 and S_1 are obtained by $S_0 = (1/2)l^2\theta$ and $S_1 = (1/2)l^2\phi$. The x and y coordinates of the point D are $x_D = l\cos\phi$, $y_D = l\sin\phi$. From the line OB, $y = x\tan\theta$, and the line DE, $y = y_D - (x - x_D)\tan\gamma$, the point E ($x_E$, y_E) can be determined. When the lengths of OD and OE are L_1 and L_2, the area of $\triangle$ODE, S_2, is

$$S_2 = \frac{1}{2}L_1 L_2 K \sin(\theta - \phi) = \frac{1}{2}l^2 K \sin(\theta - \phi), \quad K = \frac{\sin\phi + \tan\gamma\cos\phi}{\sin\theta + \tan\gamma\cos\theta} \quad (2)$$

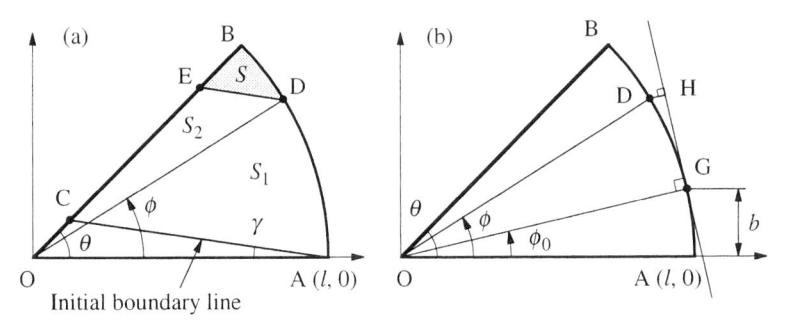

Figure 8. Contact area in a half petal and corresponding radius.

Therefore, the contact petal area (shaded portion) S is given by

$$S = S_0 - S_1 - S_2 = \frac{1}{2}l^2\{\theta - \phi - K\sin(\theta - \phi)\} \qquad (3)$$

Next, we obtain the deploying radius R corresponding to the point D. When the petal rib is perpendicular to the base polygon during unfolding, the distance between the central axis of the initial pyramid and the tip of the petal rib is equal to b. Then, the location of the apex of the pyramid P on the arc AB corresponds to the point G indicated in Fig.8(b) and the deploying radius of the point G is equal to b. Because the deploying radius is the distance on a tangent to the arc AB at G, the R of the point D is given by $b + GH$, that is,

$$R = l\sin(\phi - \phi_0) + b, \quad \phi_0 = \sin^{-1}\left(\frac{b}{l}\right) \qquad (4)$$

The relation between the contact petal area and the deploying radius of all models were obtained from Eqs.(3) and (4). Figure 9 shows the change of the non-dimensional contact petal area S^* ($= S/S_5$, S_5 : the initial contact petal area of the model with $N = 5$) for a single petal rib. The horizontal axis is the normalized deploying radius R^*. What is important in Fig.9 is that the R^* at $S^* = 0$ decreases with the increase of N, like 0.90, 0.73, 0.60, 0.51 and 0.39, although the total contact petal area of each model at the initial stage is almost same.

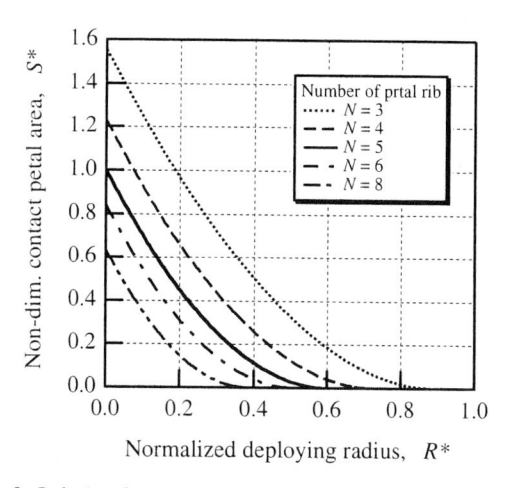

Figure 9. Relation between contact petal area and deploying radius.

3.3 Force and Energy During Deployment

Let us consider that the force F produced by a petal rib during the deployment of a morning glory flower mainly consists of the following two forces: the force F_1 needed for pulling and releasing petals and the force F_2 required for bending of the petal rib itself, because it can be presumed that the force for the motion of petal ribs and petals is extremely small comparing with F_1 and F_2.

To estimate the force F, we assume that the force F_1 decreases with the increase of R^* so as to be proportional to the contact petal area S. We also assume that the force F_2 increases linearly during unfolding and the magnitude of F_2 is proportional to the base size of petal rib a, because the bending of a petal rib results from the growth of cells in its bottom part. The third assumption is that in the model with $N = 5$, the ratio of the maximum F_1 and F_2 is five, i.e. $F_1/F_2 = 5$, taking account of tore petals during unfolding and so on.

The relation of $F = F_1 + F_2$ offers us the estimation of the force produced by a petal rib during unfolding a flower, as shown in Fig.10(a). Since the maximum F_2 of the model with $N = 5$ is equal to 1, these forces are non-dimensional. It can be found that the force of the models with large N, e.g. $N = 6$ or 8, is smaller than that of small number models, although it must be multiplied by N to obtain the total force required for the deployment of a flower.

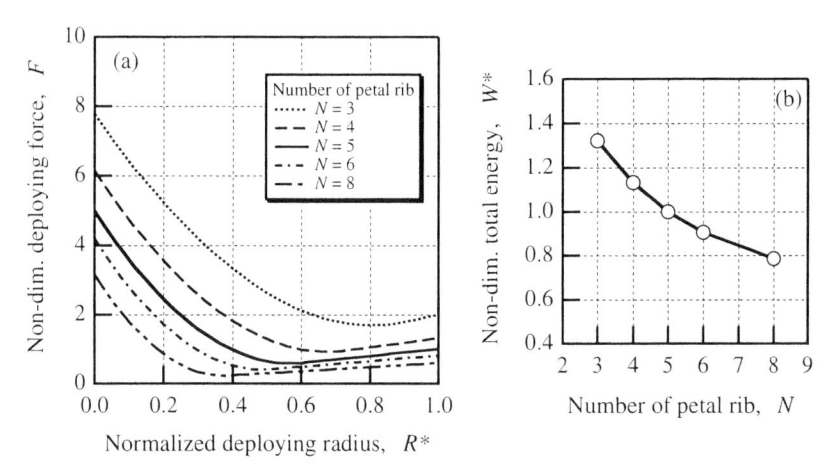

Figure 10. Force and energy during deployment, (a) forces and (b) total energies.

The curves in Fig.10(a) represent the force-moving distance curves. Therefore, the integration of the lower parts of these curves may give us the energy required for the deployment of a single petal rib. The total energy W needed for the unfolding of a flower can be obtained from multiplying by the petal rib number N. Fig.10(b) shows the non-dimensional total energy W^* taking W^* of the model with $N = 5$ equal to one. The total energy W^* decreases with the increase of the number of petal rib. This means that the model with a large petal rib number is more effective from the view point of total energy required for the deployment of a flower.

Concerning the overlapped thickness, however, the model with a large petal rib number is not recommended because of so much overlapping of rolled petals in a bud. Therefore, it can be said that the fact of five petal ribs being in morning glory flowers results from the compromise between geometrical and mechanical factors and appears to be a quite reasonable choice.

4. CONCLUSION

We conclude as follows: (1) the rigidity and the strength of petal ribs are about three times greater than those of petals because of the difference of the cell shape, i.e. the cells in petal ribs are slender and reinforced by the cell walls, while the cells in petals is round and isotropic, (2) the deploying radius of a flower increase with time along a logistic curve, (3) when the number of petal ribs in a morning glory flower increases, the overlapped petal thickness increases dramatically and consequently it becomes difficult to put the rolled petals in a small bud, (4) the energy required for the deployment of a morning glory flower decreases with the increase of the number of petal ribs and (5) the fact of five petal ribs adopted in morning glory flowers results from the compromise between geometrical and mechanical factors and appears to be a quite reasonable choice.

Finally, we deeply appreciate financial supports due to Grant-in-Aid for Scientific Research (c), No.11650074 and No.13650070 from Japanese government.

REFERENCES

1. Kobayashi H., Kresling B. and Vincent J.F.V. (1998) The geometry of unfolding tree leaves, *Proc. Roy. Soc. Lond. Ser. B*, **265** 147-154.
2. Kobayashi H., Daimaruya M. and Vincent J.F.V. (1999) Effect of crease interval on unfolding manner of corrugated tree leaves, *JSME Int. J. Ser.C*, **42** 759-767.
3. Kobayashi H., Daimaruya M. and Vincent J.F.V. (2000) Folding/unfolding manner of tree leaves as deployable structures, in *IUTAM-IASS Sympo. on Deployable Structure : Theory and Applications*, edited by Pellegrino S. and Guest S.D., Kluwer Academic Pub., London, pp.211-220.
4. Forbes W.T.M. (1926) The wing folding patterns of the Coleoptera, *J. New York Entomological Soc.*, **34** 42-115.
5. Wootton R.J. (1981) Support and deformability in insect wings, *J. Zool. Lond.*, **193** 447-468.
6. Wootton R.J. (1992) Functional morphology of insect wings, *Annu. Rev. Entomol.*, **37** 113-140.
7. Nachtigall W. and Kesel A.B. (1994) The insect wing - A multifunctional mechanical system, *Proc. 3rd Int. Sympo. SFB 230*, Stüttgart, pp.181-184.
8. Brackenbury J. H. (1994) Wing folding and free-flight kinematics in Coleoptera (Insecta): A comparative study, *J. Zool. Lond.*, **232** 253-283.
9. Haas F. and Wootton R.J. (1996) Two basic mechanisums in insect wing folding, *Proc. Roy. Soc. Lond. Ser. B*, **263** 1651-1658.
10. Haas F. (2000) Wing folding in insects : A natural deployable structure, in *IUTAM-IASS Sympo. on Deployable Structure : Theory and Applications*, edited by Pellegrino S. and Guest S.D., Kluwer Academic Pub., London, pp.137-142.
11. Kresling B. (2000) Coupled mechanisums in biological deployable structure, in *IUTAM-IASS Sympo. on Deployable Structure : Theory and Applications*, edited by Pellegrino S. and Guest S.D., Kluwer Academic Pub., London, pp.229-238.
12. Delarue J. M. (1994) Generation of Geodesic Folding Patterns, *Proc. 3nd Int. Sympo. SFB 230*, Stüttgart, pp.25-30.
13. You Z. and Pellegrino S.(1996) Cable-Stiffened Pantographic Deployable Structures part 1: Triangular Mast, *AIAA Journal*, **34** 813-820.
14. Miura K. and Natori M.(1985) 2-D Array Experiment on Board a Space Flyer Unit, *Space Solar Power Rev.*, **5** 345-356.
15. Guest S. D. and Pellegrino S. (1992) Inextensional Wrapping of Flat Membranes, *Proc. 1st Int. Semi. Struct. Morphol.*, edited by Motro R. and Wester T., Montpellier, pp.203-215.

MECHANICS OF PLASMA MEMBRANE VESICLES IN CELLS

TADASHI KOSAWADA
Department of Mechanical Engineering
Yamagata University, Yonezawa 992-8510, Japan.
E-mail: kosawada@yz.yamagata-u.ac.jp

1. INTRODUCTION

The mechanical characteristics of biological membranes are of fundamental importance in cellular biology. Many cellular processes, such as endocytosis, exocytosis and cell fusion, are strongly involved with large mechanical deformations of the membrane accompanied by changes in curvature [1]. One of the remarkable features in this respect are local membrane regions with high curvature, such as the clathrin-coated pits, vesicles, chained vesicles and channels which facilitate the cell to transport specific macromolecules. In this study, mechanics of plasma membrane vesicles in cells have theoretically been investigated based on minimization of bending and in-plane shear strain energy of the membrane. Effects of outer surrounding cytoplasmic flat membrane upon mechanically stable shapes of the vesicles were revealed as well as the effects of the in-plane shear elasticity.

Basic assumptions, which are characteristic to this problem[2]-[5], are: 1) A lipid bilayer membrane vesicle has an axisymmetric shell configuration and homogenous isotropic properties. 2) The cytoplasmic side of lipid bilayer membrane of the vesicle has a molecular layer to induce a spontaneous curvature in the resting state. 3) The membrane bending elasticity and the in-plane shear elasticity are involved to form elastic strain energy of the vesicle. 4) The membrane surface area remains constant during deformation due to fixed lipid bilayer thickness.

2. BASIC EQUATIONS

Figure 1 shows initial (dotted line) and deformed cross section of an axisymmetric chained vesicles connected with outer membrane. The radius r_0, the surface area

A_0 are defined on an initial spherical vesicle, and its arbitrary point Q_0 corresponds to point Q in the deformed cross section. T is the radial tension (in x direction), p the pressure difference between external and internal compartment, A the surface area, V the volume, μ the shear modulus, D the bending stiffness, c_1, c_2 the principal curvatures after deformation, c_0 the spontaneous curvature in the resting state. A general set of boundary conditions for which different vesicle shapes can be computed are:

$$\theta(0) = \theta_i \ , \ \theta(A_t) = \theta_t \ , \ r(0) = r_i \ , \ r(A_t) = r_t \tag{1}$$

where θ_i, r_i and θ_t, r_t are the constants to be determined according to the starting and ending boundary conditions, respectively. The strain energy functional for the membrane vesicle with respect to independent variable A is defined as follows [5].

$$E = \int_0^{A_t} \frac{\mu}{2}\left(\frac{\Gamma}{r^2} + \frac{r^2}{\Gamma} - 2\right) dA + \int_0^{A_t} \frac{D}{2}(c_1 + c_2 - c_0)^2 dA$$

$$+ p\left(\int_0^{A_t} \frac{1}{2} r \cos\theta dA - V_t\right) + T_t\left(\int_0^{A_t} \sin\theta dA - \pi r_t^2\right) \ ,$$

$$\Gamma = R^2 = \left(\frac{2}{c_0}\right)^2 - \left(\frac{A}{2\pi r} - \frac{2}{c_0}\right)^2 \tag{2}$$

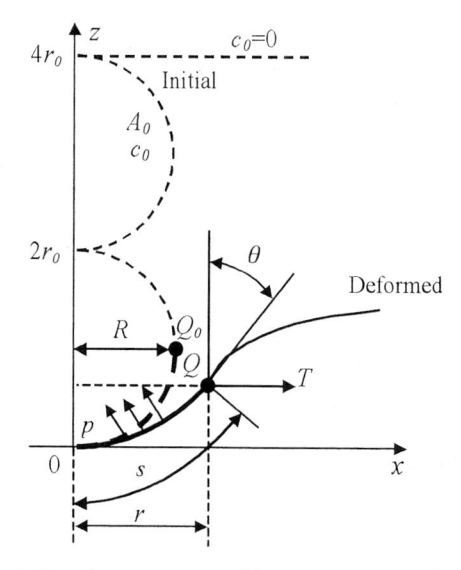

Figure 1. Coordinate system and basic parameters taken on the cross section of an axisymmetric membrane connected system.

where, the first term is the in-plane shear strain energy, the second term is the bending strain energy, and the third and the fourth terms are the constraints with respect to volume and radius, respectively. Also, p and T_t serve as Lagrange multipliers. The constraint of constant membrane surface area is automatically satisfied by using A as an independent variable. The variational principle $\delta E = 0$ yields the following nonlinear simultaneous Euler equations:

$$\frac{dT}{dA} = \frac{1}{2\pi r}\left\{\mu\left(\frac{\Gamma}{r^3}-\frac{r}{\Gamma}\right)+D\left(\frac{2B\cos\theta}{r^2}-\frac{B^2}{r}-\frac{Bc_0}{r}\right)-\frac{1}{2}p\cos\theta\right\}$$

$$\frac{dB}{dA}=\frac{1}{2\pi r D}\left(\frac{1}{2}pr\sin\theta - T\cos\theta\right), \quad B = c_1 + c_2 - c_0 \qquad (3)$$

The geometrical variables are related to each other by the following equations:

$$\frac{d\theta}{dA} = \frac{1}{2\pi r}\left(\frac{\cos\theta}{r}-c_0-B\right) \quad , \quad \frac{dr}{dA}=\frac{\sin\theta}{2\pi r}$$

$$\frac{d\Gamma}{dA}=\frac{8\pi - c_0^2 A}{8\pi^2} \qquad (A \leq A_0)$$

$$\frac{d\Gamma}{dA}=\frac{8\pi - c_0^2(A-(n-1)A_0)}{8\pi^2} \qquad (A \geq A_0)$$

(4)

where, n is the number of chained vesicles. The following identities are used to compute membrane vesicle shapes, surface area, volume and strain energy:

$$\frac{dz}{dA}=\frac{\cos\theta}{2\pi} \,,\, \frac{ds}{dA}=\frac{1}{2\pi r} \,,\, \frac{dV}{dA}=\frac{1}{2}r\cos\theta \,,\, \frac{dE}{dA}=\frac{\mu}{2}\left(\frac{\Gamma}{r^2}+\frac{r^2}{\Gamma}-2\right)+\frac{D}{2}B^2$$

(5)

3. METHOD OF NUMERICAL COMPUTATION

The simultaneous nonlinear differential Eqs. (3) and (4) are solved by the fourth-order Runge-Kutta method. The initial values of θ and r are geometrically determined from the boundary conditions (1). On the other hand, T_i and B_i, the initial values of T and B, can not be determined directly from the boundary conditions. They should be iteratively specified to satisfy the boundary conditions. This type of problem is called a two point boundary value problem and the shooting method [7] is used here to solve the problem numerically.

We first select values of T_i and B_i randomly, then Eqs. (3) and (4) are integrated with the aid of the Runge-Kutta technique so as to determine the values of that boundary condition. The technique utilizes an adaptive step-size control through which an overall tolerance level of $\pm 10^{-6}$ can be achieved. The discrepancies between the calculated boundary conditions and the desired boundary conditions are defined as $F(T_i, B_i)$. Then the Newton-Raphson method is used in order to seek values of T_i and B_i which ultimately satisfy the equation $F(T_i, B_i) = 0$. By the nature of the method, small changes in the values of T_i and B_i produce relatively large changes in the results. By using computed values of T_i and B_i, Eqs. (3), (4) and (5) are solved again by the Runge-Kutta technique to yield the deformed shapes and their parameters. The computations presented here were carried out with double precision arithmetic where the number of significant figures were kept at 15 or 16.

As a numerical example, we assume a connected system of which two axisymmetric spherical membrane chained vesicles with radius r_0 and spontaneous curvature c_0 are connected with an outer flat circular membrane without spontaneous curvature. Initially the spherical vesicles are unstressed, and their spontaneous curvature c_0 is given by $2/r_0$ [3]-[5]. They are connected with each other through holes at the mutual contact point. These holes are assumed to be infinitely small and connected smoothly with each other. The surface areas of the connected total system A_t were set as $6A_0$, $7A_0$ and $8A_0$. The spontaneous curvature of the vesicle section is set as $c_0 = 6$, while that of the outer flat circular membrane section is set as $c_0 = 0$. The radius of the outer perimeter is denoted by r_t. We set geometrical boundary conditions of this connected system as follows:

$$r(0) = 0, \quad \theta(0) = \frac{\pi}{2}, \quad r(A_t) = r_t, \quad \theta(A_t) = \theta_t \tag{6}$$

As for initial state, we have,

$$r_t = 2r_0\sqrt{A_1/A_0} \tag{7}$$

The total surface area of the system is held constant at A_t during computation. Then the membrane tension is applied to open the outer perimeter of the circular membrane until the connected system become almost a flat membrane disk. We looked for the minimum energy equilibrium configuration under the condition that the slope and the curvature are smoothly continuing at the connected point. At the end of the computation, by considering experimental evidence, the radius r_0, the shear modulus μ, bending rigidity D, pressure difference p (external minus internal) are scaled with 33.3 nm (corresponding to $c_0 = 6.0$), (0, 0.066, 0.132)$\times 10^{-2}$ dyn/cm(10^{-5}N/m), 10^{-12}dyncm(10^{-19}Nm) [6], 1.0cmH$_2$0(98Pa) [2], respectively, so as to obtain the order of magnitude of possible tensions necessary to unfold the connected system.

4. RESULTS AND DISCUSSIONS

In Fig. 2(a), two-dimensional membrane contours of equilibrium shape of the connected system are shown with various opening radii in the case where $\mu = 0$ and

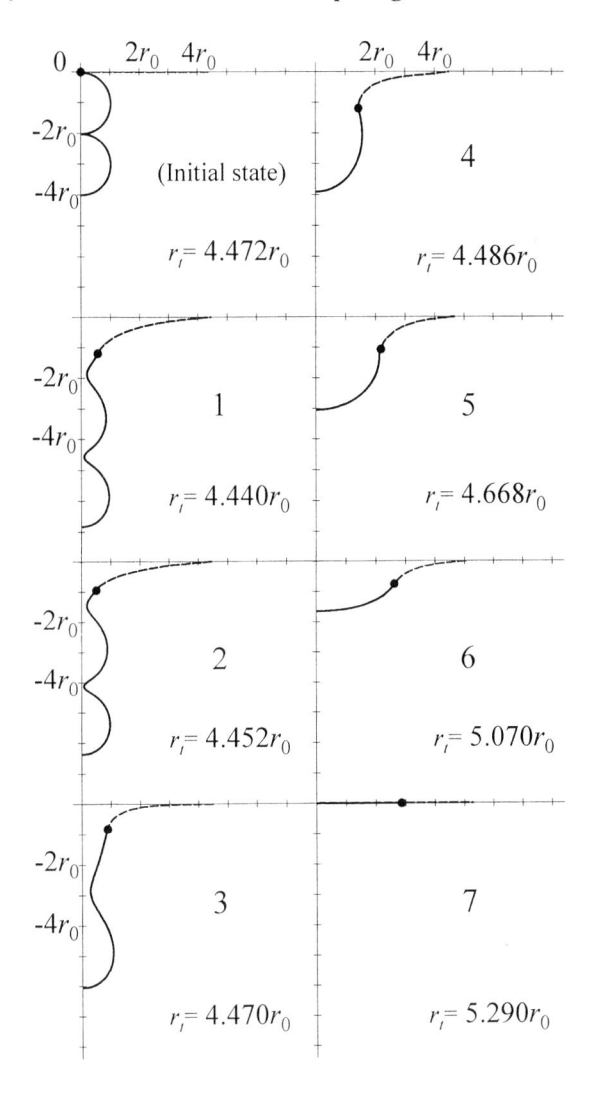

Figure 2(a). Two-dimensional membrane contours of equilibrium shape of vesicles.

$A_t = 7A_0$. The position of the upper vesicle perimeter is intentionally aligned along the x-axis ($z = 0$) so that the membrane contour shapes can be compared. Only the right side of each membrane cross section is shown. The shear modulus μ was set as zero in order to obtain results for wide range of r_t. The initial state is shown along with the computed shapes by configuration numbers 1 to 7. The black circle on the line denotes the boundary between the upper outer membrane(broken line) and the chained vesicles(solid line). Smooth but dramatic shape changes from the chained

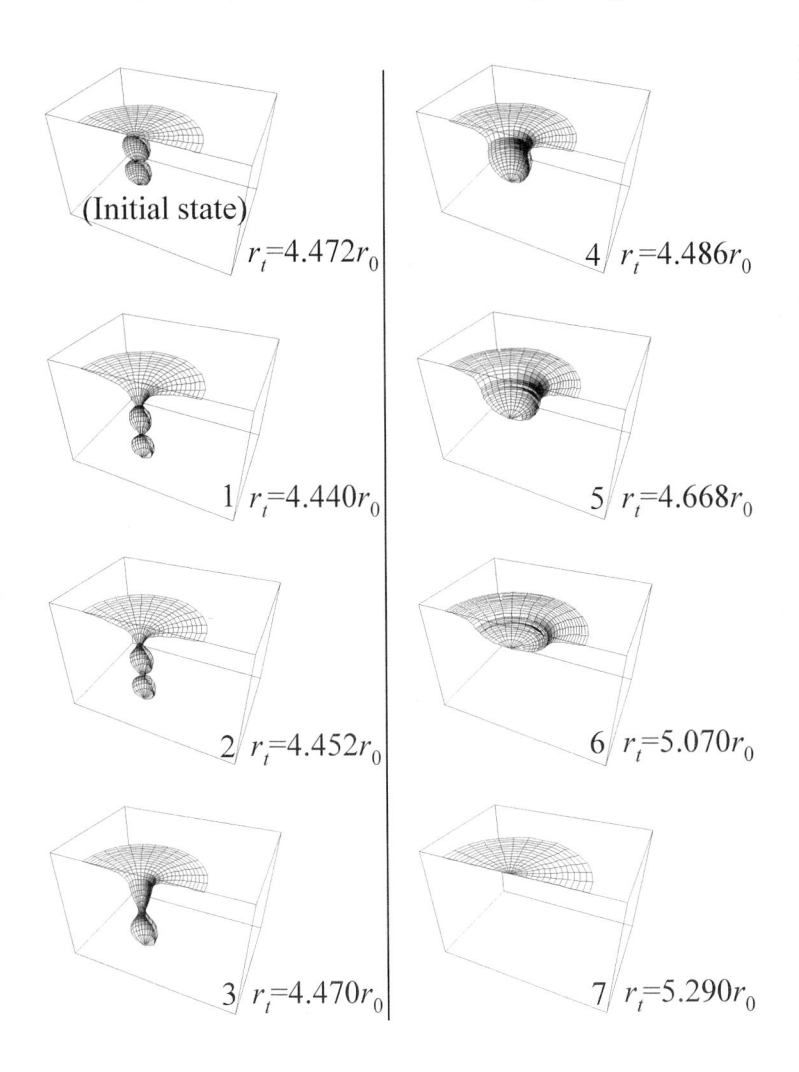

Figure 2(b). Three-dimensional membrane contours of equilibrium shape of vesicles.

vesicles to the flat circular membrane disk are clearly observed. At number 1, the outer circular membrane dips downward while the chained vesicles section are kept unchanged. Then the lowest energy equilibrium configuration is given at number 2. Its configuration is mechanically stable and may most frequently be formed in plasma membrane. The chained vesicles starts to rise, while their shapes remain spherical. At number 3, the upper vesicle is opened up halfway while the lower vesicle remains spherical with a slight increase in its radius. At number 4, suddenly, a constricted part of the vesicle disappears. The vesicle is opened up halfway and becomes somewhat larger compared to the vesicles of numbers 1 to 3. From number 4 to 7, the vesicle unfolds further and merges into the outer membrane, and finally becomes almost a flat membrane disk. The computed opening radius at number 7 became $r_t = 5.290 r_0$. Theoretically, the maximum opening radius is $r_t = 5.292 r_0$, while the boundary between the outer membrane and the chained vesicles is $r_t = 2.828 r_0$, both of these are deduced directly from Eq. (7). Figure 2(b) shows three-dimensional membrane contours of the results. They were drawn by using Mathematica. The meshes in the figure were automatically generated by Mathematica so as to show three-dimensional images effectively.

Figures 3, 4 and 5 show the relationships between the non-dimensional opening radius and in-plane tension, strain energy, total surface area and volume in case for three different total surface areas, $6A_0$, $7A_0$ and $8A_0$ ($\mu = 0$). In Fig. 3, in-plane tension shows similar tendency in each case and does not change substantially until very last stages of the opening radius. However, it shows a dramatic as the connected system is unfolded into a flat membrane disk. In Fig. 4, there exists a step-like increase, for each case almost at the same level in strain energy, accompanying the transition from a spherical vesicle into a somewhat large and hemispherical vesicle as mentioned in Fig. 2. Therefore, it seems that the above transition phenomena does not depend on the size of the outer flat circular membrane but on the sizes of the

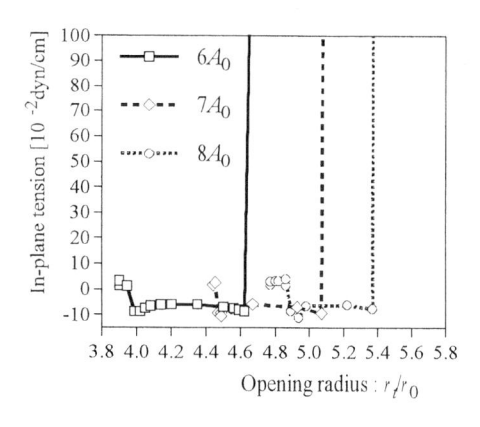

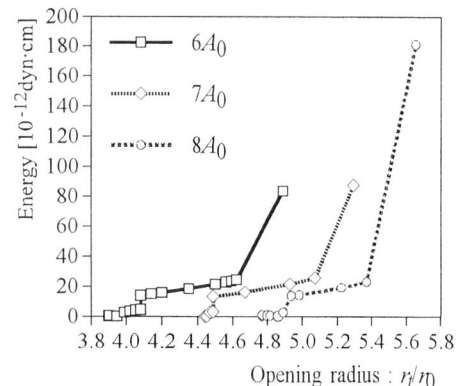

Figure 3. In-plane tension vs opening radius.   *Figure 4. Strain energy vs opening radius.*

two spherical membrane chained vesicles. In Fig. 5, the total surface area remains constant as assumed. As the opening radius increases, the volume in each case has local maximum and minimum and then converges to zero.

Figure 6 shows that the computed shape of the connected system (the lowest energy equilibrium configuration, $\mu = 0$) is relatively in good agreement with the schematic drawing traced directly from an electron micrograph published by Palade and Bruns [8] in endothelium of a rat tongue blood capillary. Several of the vesicle shapes encountered in the numerical computations are similar to those seen on electron micrographs. Although we saw a good agreement as a whole between the computed shape and the experimentally observed one, few discrepancies were still seen in the constricted part and in the neck part of the chained vesicles. The cause of these discrepancies is unknown.

Few details are known about the value of in-plane shear modulus μ in the plasma membrane vesicles. In order to estimate the effects of in-plane shear elasticity in equilibrium configuration, Fig. 7 shows two-dimensional membrane contours of the connected system with various shear moduli at $A_t = 7A_0$. In this study, the values of

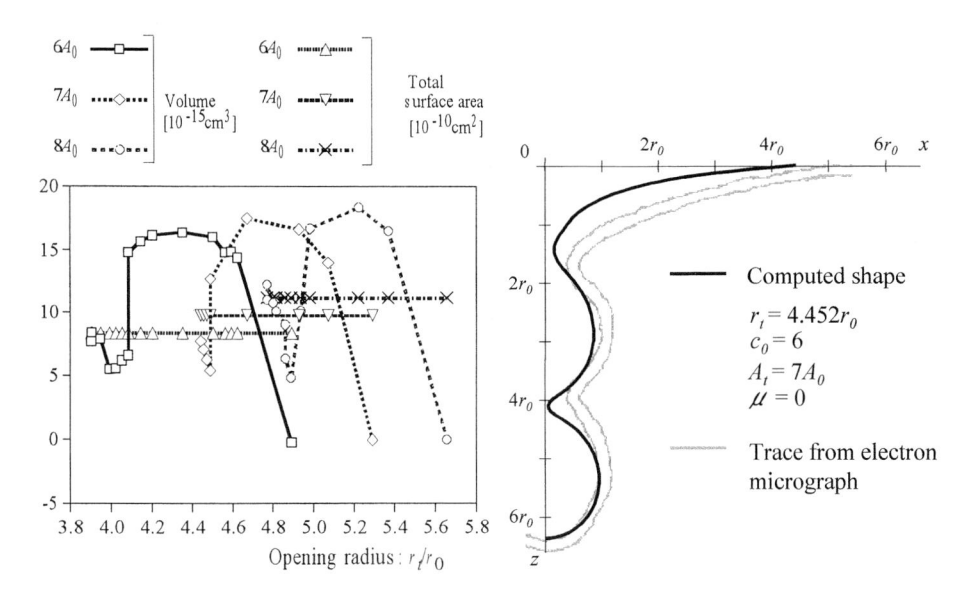

Figure 5. Volume, total surface area vs opening radius.

Figure 6. Computed shape and experimental observation.

μ were tentatively set up to 20% of μ for the red blood cell membrane. The figures show cross section at each opening radius. As seen from the figures, effects of the shear modulus μ upon the equilibrium shape of the connected system are quite significant particularly in the vicinity of the neck part and the constricted part of the chained vesicles where curvature changes dramatically. In case of (a) for example, as the value of μ increases, the upper vesicle is unfolded while the lower vesicle tends to be kept in its spherical configuration. This tendency is also seen in case of (b).

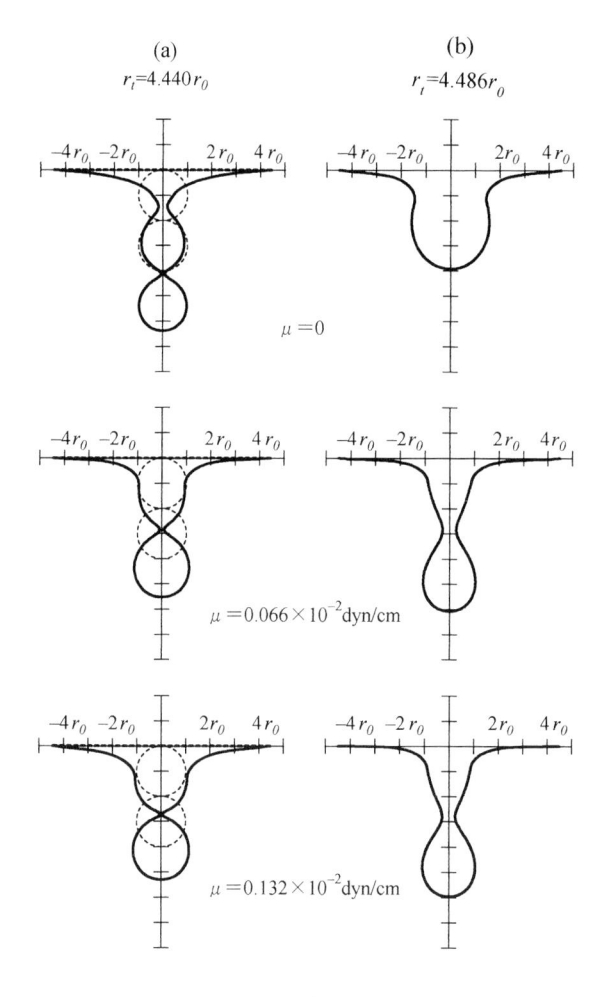

Figure 7. Effects of in-plane shear modulus μ in equilibrium shape of vesicles.

5. CONCLUSIONS

In this study, mechanics of plasma membrane vesicles in cells have been theoretically investigated based on minimization of bending and in-plane shear strain energy of the membrane. The results are: i) The vesicles change their shapes dramatically as the opening radius increases. During the large deformation process, the sudden shape changes due to spontaneous curvature were observed accompanying with a step-like increase in strain energy. ii) Effects of the in-plane shear modulus μ upon the equilibrium shape of the connected system were observed particularly in the vicinity of the neck part and the constricted part of the vesicles where curvature changes dramatically. iii) The computed shape given by the present method is relatively in good agreement with the results observed on electron micrographs.

6. ACKNOWLEDGMENTS

This work was supported in part by research grants from the Ministry of Education, Science, Sports and Culture, Japan, Grants-in-Aids for Scientific Research (B), Nos. 09558111 and 12450093, and also by Suzuken Memorial Foundation.

7. REFERENCES

1. Alberts, B., Bray, D., Lewis, J., Raff, M., Roberts, K. and Watson, J.D. (1994) *Molecular Biology of the Cell*, 3rd ed., Garland Pub., pp. 599-651.
2. Schmid-Schonbein, G.W., Kosawada, T., Skalak, R. and Chien, S. (1995) "Membrane Model of Endothelial Cells and Leukocytes. A Proposal for the Origin of a Cortical Stress," *ASME Journal of Biomechanical Engineering*, 117-2, pp.171-178.
3. Kosawada, T., Yoshida, O., Skalak, R. and Schmid-Schonbein, G.W. (1999) "Generation Mechanism of Vascular Endothelial Chained Vesicles and Transendothelial Channel," *JSME International Journal*, Series C, 42-3, pp.796-803.
4. Kosawada, T., Skalak, R. and Schmid-Schonbein, G.W. (1999) "Chained Vesicles in Vascular Endothelial Cells," *ASME Journal of Biomechanical Engineering*, 121-5, pp.472-479.
5. Kosawada, T., Sanada, K. and Takano, T. (2001) "Large Deformation Mechanics of Plasma Membrane Chained Vesicles in Cells," *JSME International Journal*, Series C, 44-4, pp.928-936.
6. Evans, E. A. and Skalak, R. (1980) *Mechanics and Thermodynamics of Biomembranes*, CRC Press, Inc.,pp.141-180.
7. Press, W.H., Teukolsky, S.A., Vetterling, W.T. and Flannery, B.P. (1992) *Numerical Recipes in Fortran*, 2nd ed., Cambridge Univ. Press, pp.745-778.
8. Palade, G.E. and Bruns, R.R. (1968) "Structural Modulations of Plasmalemmal Vesicles," *The Journal of Cell Biology*, Vol.37, pp.633-646.

CONTROL OF STRUCTURES BY MEANS OF HIGH-FREQUENCY VIBRATION

AGNESSA KOVALEVA
Russian Academy of Sciences,
Mechanical Engineering Research Institute,
Moscow 101990 Russia
E-mail: a.kovaleva@ru.net

1. INTRODUCTION

In recent years, there has been increasing interest in applications of vibrational control theory [4], [6], [10] to problems of stabilization in mechanical structures. It has been found that high-frequency periodic or quasi-periodic excitation ("vibrational control") can stabilize an unstable equilibrium position ("vibrational stabilization"). The effect of high-frequency vibration is similar to modification of the "effective potential" of the system and can be achieved either by high-frequency programme control, or by nonlinear feedback control. However, feedback control requires measurement of the system state and costly signal processing, whereas vibrational control can be applied as a pregiven programme. This makes it a powerful tool for stabilization of complicated mechanical systems.

It follows from general results [1], [4] that a system subjected to high-frequency perturbations of parameters not only can be stabilized, but also can acquire additional stable positions not found in the unperturbed system. This phenomenon has been studied in detail for specific mechanical systems [1], [2], [3], [7]. Investigation has been based on transformation of the equations of motion to the form allowing averaging, with further analysis of fixed points of the averaged system [1], [4]. The direct application of this approach to the Lagrange equations requires preliminary transformations of the equations and does not reveal an explicit connection between the conditions of stability and the configuration of the system. This paper uses concepts presented in [8] for the direct analysis of the Lagrange equations of motion. It is proved that the effect of fast perturbations is similar to a modification of the potential in the averaged system. The modified potential can give rise to new fixed points different from the equilibrium positions of the unperturbed system. This phenomenon remains valid for both potential and non-potential perturbations.

The approach developed is illustrated by examples. Appearance of new equilibrium positions is demonstrated for a pendulum suspended in the inclined position. It is shown that fast vibration of the pivot entails the change of the stable angle of inclination of the suspended pendulum. Vibrational control as a tool of stabilization of a flexible beam is proposed. It is shown that fast quasi-periodic modulation of the flexural stiffness results in an increase of the critical value of the bending force and can prevent buckling.

2. MAIN ASSUMPTIONS AND EQUATIONS OF MOTION

The equations of motion have the form of the Lagrange equations

$$\frac{d}{dt}\frac{\partial T}{\partial \dot{x}} - \frac{\partial T}{\partial x} + \varepsilon^2 Q(x) = \varepsilon S(t,x), \qquad x \in D \subset R^n \tag{1}$$

In (1) we denote $\dot{x} = dx/dt$, the scalar product $T(x, \dot{x}) = \frac{1}{2}(\dot{x}, A(x)\dot{x})$ is kinetic energy of the system, the kinetic energy matrix $A(x)$ is symmetric and positive-definite in D. An admissible domain D is an open set in R^n. Potential energy of the system is denoted as $\Phi(x)$, $Q(x) = d\Phi(x)/dx$ is the vector of potential forces. Perturbation $S(t, x)$ is a quasi-periodic zero mean process of t. The functions $A(x)$, $Q(x)$ and $S(t, x)$ are assumed to be sufficiently smooth in all variables to allow for necessary transformations. Introduction of the small parameter $\varepsilon > 0$ indicates that perturbation $S(t, x)$ is considered as a fast process compared to motion of the system.

To simplify necessary transformations, we introduce the change of variables

$$x = q, \quad p = A(q)\dot{q} \tag{2}$$

or

$$\dot{q}(p,q) = v(p,q) = a(p,q)p, \quad a(q) = A^{-1}(q) \tag{3}$$

In this case the total energy of the system can be written as

$$E(p,q,\varepsilon) = G(p,q) + \varepsilon^2 \Phi(q), \quad G(p,q) = T(q, v(p,q)) \tag{4}$$

where, by (1) – (3), the scalar product $G(p,q) = \frac{1}{2}(p, a(q)p)$. The equations of motion can thus be written in the form

$$\dot{q} = \frac{\partial E}{\partial p}, \quad \dot{p} = -\frac{\partial E}{\partial q} + S(t,q) \tag{5}$$

In order to make the time scales transparent, we introduce the slow variable y by the formula

$$p = \varepsilon y + \varepsilon V(t,q) \tag{6}$$

where $V(t,q)$ is defined by the relations

$$S(t,q) = \frac{\partial V(t,q)}{\partial t}$$

$$<V(t,q)> = \lim_{T \to \infty} \frac{1}{T} \int_0^T V(t,q)dt = 0 \tag{7}$$

The latter condition in (7) defines a unique function $V(t,q)$ corresponding to a given function $S(t,q)$.

From (5), (6), (7) we obtain the equations in the standard form for the variables q, y

$$\dot{q} = \varepsilon a(q)y + \varepsilon F_1(t,q)$$
$$\dot{y} = -\varepsilon \frac{\partial}{\partial q}[\Phi(q) + F(q) + G(y,q)] + \varepsilon F_2(t,q,y) \tag{8}$$

where $F(q) = <F(t,q)>$, $F(t,q) = \frac{1}{2}(V(t,q), a(q)V(t,q))$ and $G(y,q) = \frac{1}{2}(y, a(q)y)$.

The coefficient $F(t,q)$ can be simplified, if $S(t,q) = \dot{v}(t)Y(q)$, where $\dot{v}(t)$ is a scalar. In this case we obtain

$$F(t,q) = \frac{1}{2}v^2(t)f(q), \quad F(t,q) = \frac{1}{2}\sigma^2 f(q)$$
$$f(q) = \frac{1}{2}(Y(q), a(q)Y(q)), \quad \sigma^2 = <v^2(t)> \tag{9}$$

Transformations (5), (6), (7) define the coefficients F_1 and F_2 as quasi-periodic zero mean functions of t, that is $<F_{1,2}(t,q,y)> = 0$. In addition, we assume that the right hand sides of (8) are sufficiently smooth. Under these assumptions, the averaging principle [5] can be used for the asymptotic analysis of Eq. (8).

Together with Eq. (8), we consider the truncated averaged system

$$\dot{q}_0 = \varepsilon a(q_0)y_0$$
$$\dot{y}_0 = -\varepsilon \frac{\partial}{\partial q_0}[\Phi(q_0) + F(q_0) + G(y_0, q_0)] \tag{10}$$

Eq. (10) describe the dynamics of a conservative system with Hamiltonian $H(y_0, q_0) = G(y_0, q_0) + U(q_0)$, where q_0 and y_0 are the vectors of the generalized coordinates and momenta, respectively, $U(q_0) = \Phi(q_0) + F(q_0)$ is potential energy of this system, the scalar product $G(y_0, q_0) = \frac{1}{2}(y_0, a(q_0)y_0)$ is kinetic energy. The function $U(q_0)$ is said to be "the effective potential". The term $F(q_0)$ corresponds to the contribution of fast perturbations in the effective potential, the derivative $K(q_0) = \partial F(q_0)/\partial q_0$ defines additional potential forces. The effect of fast perturbations is thus similar to the modification of the potential forces compared with the unperturbed system.

Steady-state solutions of Eq. (10) can be found and examined by making use of the standard machinery of averaging [5]. If Eq. (10) does not include cyclic coordinates [12], fixed points of system (10) can be found as extrema of the potential $U(q_0)$, i.e. from the equation

$$\frac{\partial U}{\partial q_0} = R(q_0) = Q(q_0) + K(q_0) = 0 \qquad (11)$$

Let Eq. (11) have a solution q^*. Stationary quasi-periodic motion of the original system (1) can be presented as small quasi-periodic oscillations in a neighbourhood of the fixed point q^* [5]. If the solution q^* corresponds to a strict minimum of $U(q_0)$, then q^* is a stable fixed point of the system (10), and quasi-periodic motion of system (1) remains in a neighbourhood of the equilibrium position q^* at least over the time interval $t \sim 1/\varepsilon$. A detailed analysis of convergence requires consideration of higher order asymptotic approximations. If the solution q^* corresponds to a strict maximum of $U(q_0)$, then q^* is an unstable fixed point of the system (10), and the steady motion of system (1) is unstable [5].

Let the coefficients $a(q)$, $\Phi(q)$, $F(q)$ depend only on coordinates q_i, where $i = 1,\ldots, m < n$. In this case system (10) allows separation of the cyclic coordinates. The truncated system for the positional coordinates retains the form (10), but the function $\Phi(q)$ is interpreted as the Routh potential [12]. Fixed points q^* can be obtained from Eq. (11) as extrema of a relevant "effective Routh potential".

2.1. Analysis of the Equilibrium Positions

In the general case of nonlinear systems, fixed points found from Eq. (11) do not coincide with the equilibrium positions of the unperturbed system. Consider some special cases.

Suppose the kinetic energy matrix A is independent of x, and $S(t,x) = S(t)$. Then $F(t,q) \equiv 0$, $K(q) \equiv 0$. From (11) it follows that an additive fast perturbation does not change the equilibrium positions of a nonlinear system and does not affect the conditions of stability.

Suppose the unperturbed system does not include the cyclic coordinates. From Eq. (11) it follows that the unperturbed equilibrium positions persist if the roots of the equation $K(q) = 0$ are equal to the roots of the equation $Q(q) = 0$. However, the stability conditions for perturbed and unperturbed equilibrium positions can be different.

As an example, we consider a linear system in the form

$$\ddot{x} + \varepsilon^2 Cx + \varepsilon S(t)x = 0, \qquad S(t) = \dot{V}(t) \qquad (12)$$

If the matrix C is positive-definite, the unperturbed system ($S(t) = 0$) has the stable equilibrium $x = 0$.

Truncated system (10) takes the form

$$\dot{q}_0 = \varepsilon y_0, \quad \dot{y}_0 = -(C + D)q_0 \qquad (13)$$

where, by (8), D is the matrix with the components $D_{ij} = <V_i(t)V_j(t)>$. If the matrix $C + D$ is non-degenerate, system (13) has the equilibrium position $q_0 = 0$. The stability conditions is determined by properties of the characteristic equation

$$\det |Ip^2 + (C + D)| = 0 \qquad (14)$$

From (14) it follows that, for a correct choice of the perturbation intensity, the matrix $(C + D)$ becomes positive-definite, even if the matrix C does not possesses this property. An unstable equilibrium position can thus be stabilized by a fast parametric perturbation (see [6] for details and examples).

3. STABILITY OF A SUSPENDED PENDULUM

As an example, we will examine the equilibrium states of a pendulum OC with an elastic suspension DC (Figure 1).

The pendulum consists of the weightless rod with the mass m on its edge C. The effect of gravity is ignored. Axis of rotation O is located under the fixing point D of the suspension; without loss of generality we let $OD = OC = l$. Kinetic energy of the pendulum is $T = \frac{1}{2} m(l\theta')^2$, that is $A = ml^2$, $a = 1/ml^2$, θ is the angle between the pendulum and the axis OD. The pivot O oscillates with acceleration $w(t)$ directed along the vertical axis. Following D'Alembert's principle, we change kinematic excitation to the forces of inertia, and consider relative motion as oscillations of the

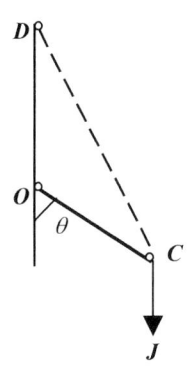

Figure 1: Suspended pendulum

pendulum with the fixed pivot O but acted upon by the force of inertia $J = -mw(t)$ applied at the center of mass (Figure 1). This implies that vibrational control corresponds to the force of inetrtia.

Potential energy of the elastic suspension is $\Pi(\theta) = 2m(lk)^2[\cos(\theta/2) - \lambda]^2$, where $\lambda = l_0/2l$, l_0 is the length of the non-deformed suspension, k is the frequency of small linear oscillations of the pendulum, $k^2 = c/m$. The moment of the force of inertia is $L = -mw(t)l\sin\theta$.

Let the base oscillations be harmonic, with acceleration $w(t) = -a\omega^2\sin\omega t$, and this process be fast compared to small linear oscillations of the pendulum. In this case a dimensionless small parameter can be chosen as $\varepsilon = k/\omega \ll 1$. Then, since the amplitude of the base vibration is supposed to be small compared to the length of the pendulum, we write $a/l = \varepsilon\rho$.

Potential energy of the system and the generalized force applied to the system can be written in the forms

$$\Pi(\theta) = \varepsilon^2 \Phi(\theta) = 2\varepsilon m(l\omega)^2[\cos(\theta/2) - \lambda]^2$$

$$L(t, \theta) = \varepsilon S(t, \theta) = -\varepsilon m(l\omega\rho \sin\theta)^2 \sin\omega t$$

Following (7), (8), (9), we obtain $F(\theta) = \frac{1}{2}m(l\omega\rho)^2\sin^2\theta$. Then we have

$$U(\theta) = \Phi(\theta) + F(\theta) = m(l\omega)^2\{2[\cos(\theta/2) - \lambda]^2 + \frac{1}{2}(\rho\sin\theta)^2\} \tag{15}$$

Extrema of the effective potential (15) can be found as roots of the equation

$$R(\theta) = -2\sin\frac{\theta}{2}(\cos\frac{\theta}{2} - \lambda) + \rho^2\sin\theta\cos\theta = 0 \tag{16}$$

(a positive constant coefficient is omitted). In the absence of perturbations ($\rho = 0$) the equilibrium positions of the inclined pendulum can be defined as

$$\theta_0 = 0, \quad \theta_1 = 2\arccos\lambda \tag{17}$$

An analysis shows that the lower equilibrium $\theta_0 = 0$ is unstable, whereas the upper position θ_1 is stable. To calculate the perturbed equilibrium positions, we transform Eq. (16) to the form

$$R(\theta) = 2\sin\frac{\theta}{2}\{-[z(\theta) - \lambda] + \rho^2 z(\theta)[2z^2(\theta) - 1]\} = 0, \quad z(\theta) = \cos\frac{\theta}{2} \tag{18}$$

Equation (18) retains the root $\theta_0 = 0$. It can be verified that this position is stable for $\rho^2 > 1$. Additional equilibrium positions can be found as the roots of the equation

$$f(z) = 2\rho^2 z^3 - (1 + \rho^2)z = -\lambda \tag{19}$$

If the inequality

$$\frac{2}{3\sqrt{6}}(1 + \mu^2)^{3/2} > \mu^2 \lambda, \quad \mu = \rho^{-1} \tag{20}$$

holds, Eq. (19) has three real roots Z_i, two of which are positive. In Figure 2 these roots are defined as the points of intersection of the straight line *1*, corresponding to $\lambda = \lambda_1$, with the graph of function (19). If condition (20) is not satisfied, Eq. (19) has a unique real negative root Z_3^*, defined as the point of intersection of the straight line *2*, corresponding to $\lambda = \lambda_2$, with the graph of function (19). This root is not taken into account in the analysis of stability.

Condition (20) is valid if $\mu = \rho^{-1}$ is small enough and $\lambda < 1$. In this case the roots of Eq. (20) and the corresponding equilibrium positions have the form

$$z_1 \approx \frac{\sqrt{2}}{2} + \mu^2(\frac{\sqrt{2}}{4} - \lambda), \; z_2 \approx \mu^2 \lambda,$$

$$\theta_1 \approx \frac{\pi}{2} - \mu^2(1 - 2\sqrt{2}\lambda), \; \theta_2 \approx \pi - 2\mu^2 \lambda \tag{21}$$

It can be proved that the positions $\theta_0 = 0$ and θ_2 are stable, and the position θ_1 is unstable for $\mu < 1, \rho > 1$.

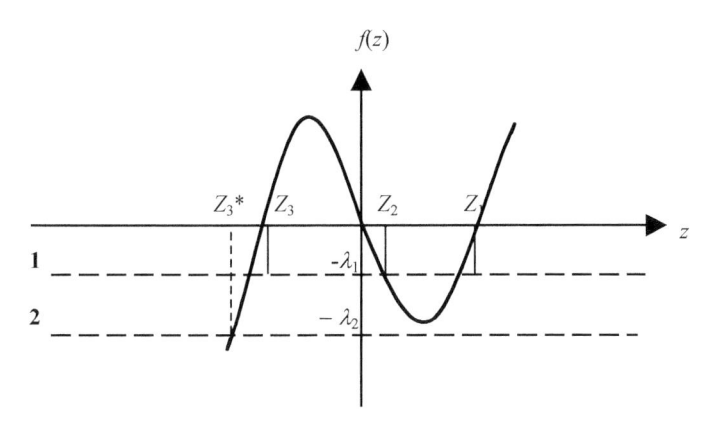

Figure 2. Roots of Eq (19)

If $\rho \ll 1$, condition (20) holds but the maximum root $z_1 > 1$. In this case two equilibrium positions exist: $\theta_0 = 0$, $\theta_2 \approx 2(\arccos\lambda + \rho\sqrt{8})$. It can be proved that the lower position $\theta_0 = 0$ is unstable and the upper position θ_2 is stable. This implies that weak perturbations ($\rho \ll 1$) result in a small displacement of the equilibrium positions, whereas intensive perturbations ($\rho \gg 1$) can result in occurrence of new stable equilibria.

5. VIBRATIONAL CONTROL OF A FLEXIBLE BEAM

Consider a flexible beam loaded by a bending force (Figure 3). Stability of flexible systems subjected to a constant bending force has been studied thoroughly, see, i.e., [11], [13]. It is known that a load, exceeding a critical Euler force, entails the loss of elastic stability, i.e. buckling. In this paper we consider a concept of parametric control of vibration. Control is interpreted as periodic modulation of the flexural stiffness. We show that fast periodic modulation of stiffness increases a critical value of the bending force and can prevent buckling in the case of instability of non-controlled system. Implementation of stiffness modulations is beyond the scope of this paper. It can be effected, e.g., by embedding of microscopic piezoelectric pins or micro-inclusions from shape memory alloys into the core of the beam. These inclusions would modify the bending stiffness under effect of the periodic electromagnetic or heating field, respectively (see discussion in [9]).

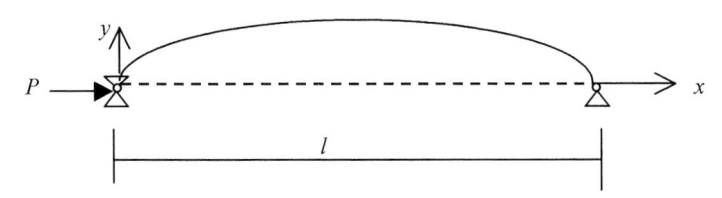

Figure 3. Bending oscillations of a flexible beam

The linearized equation and the boundary conditions of a hinged beam undergoing bending vibrations can be written as [11]

$$\mu \frac{\partial^2 y}{\partial t^2} + \alpha(t) \frac{\partial^4 y}{\partial x^4} + P \frac{\partial^2 y}{\partial x^2} = 0, \quad 0 < x < l$$

$$y = 0, \quad \frac{\partial^2 y}{\partial x^2} = 0, \quad x = 0, l \tag{22}$$

where $\alpha(t) = \alpha_0 + \alpha_1 \cos\omega t$ is the reduced flexural stiffness, μ is the mass per unit length, P is the bending load applied at the edge $x = 0$.

Eqs (22) can be converted into a set of ordinary second-order differential equations by means of Galerkin's method [11]. To this end, the function $y(x,t)$ can be expressed as

$$w(t) = \sum_{m=1}^{n} \varphi_m(t)\sin\lambda_m x, \quad \lambda_m = \pi m/l \tag{23}$$

where $\varphi_m(t)$ are generalized coordinates. In theory, n can be infinite, but in practice only a finite number of terms is used. Inserting Eq. (23) into Eq. (22), we obtain

$$\frac{d^2\varphi_m}{dt^2} + \varepsilon^2 k_m^2(\alpha_0 - Pl^2)\varphi_m + \varepsilon k_m^2 \alpha_1 \sin\omega t = 0$$

$$k_m^2 = \mu\lambda_m^2/l^2 \tag{24}$$

Compare Eq. (24) with Eq. (12). By making use of notations of Section 1, we write the parameters

$$C_m = k_m^2(\alpha_0 - Pl^2)$$

$$S_m(t) = k_m^2 \alpha_1 \sin\omega t \quad V_m(t) = -k_m^2 \alpha_1 \cos\omega t/\omega \tag{25}$$

This defines

$$D_m = k_m^4 \alpha_1^2/2\omega^2 \tag{32}$$

Thus, for the m-th mode of oscillations we have

$$C_m + D_m = k_m^2(\alpha_0 - Pl^2) + 2(k_m^2 \alpha_1/2\omega)^2 \tag{26}$$

From Eqs. (14), (26) we obtain the estimate of the critical load

$$P < P^* = \frac{\alpha_0}{l^2}(1 + \frac{\alpha_1^2 k_m^2}{2\alpha_0 \omega^2}) \tag{27}$$

Condition (26) implies that any bending force exceeding P^* entails the loss of stability. The critical value $P^0 = \alpha_0/l^2$ corresponds to the system with non-modulated stiffness. It is easy to see that $P^0 < P^*$. This implies that fast periodic modulation of the flexural stiffness can prevent buckling even if the non-modulated system is unstable.

5. CONCLUSIONS

The paper extends theory of vibrational stabilization to nonlinear systems with quasi-periodic parametric perturbations. A special change of variable is proposed to reduce the Lagrangian equations of motion to the form allowing averaging. The averaged system has the canonical form even if the perturbations are non-potential. A detailed analysis of the averaged system indicates that the effect of fast perturbations on Lagrangian systems is similar to the change of the potential forces acting upon the system. Extrema of the averaged potential differ from the equilibrium positions of the unperturbed system. This leads to the change in the stability conditions. The method developed is illustrated by examples.

6. ACKNOWLEDGEMENT

This research was partially supported by RFBR, grant 02-01-00011, and INTAS, grant 97-1140.

7. REFERENCES

1. Akulenko, L. D. (1994) Asymptotic Analysis of Dynamical Systems Subjected to High- Frequency Excitations. *J. Appl. Math. Mech.*, **58**, 23 –31.
2. Akulenko, L. D. (2000) Control of Relative Motion for the Pendulum on the Rotating Base. *J. Appl. Math. Mech.*, **64**, 204 –216.
3. Anderson, G. L. and Tadgjbakhsh, I. G. (1989) Stabilization of Ziegler's Pendulum by Means of the Method of Vibrational Control. *J. Math. Anal. Appl.*, **143**, 198-223.
4. Bellman, R., Bentsman, J. and Meerkov, S. (1985) Vibrational Control of Nonlinear Systems: Vibrational Stabilizability. *IEEE Trans. Automat. Control.*, **30**, 289-291.
5. Guckenheimer, J and Holmes, P. (1986) *Nonlinear Oscillations, Dynamical Systems, and Bifurcations of Vector Fields*. Springer-Verlag. New York:.
6. Kabamda, P. T., Meerkov, S., and Poh, E. (1998) Poles Placement Capabilities of Vibrational Control. *IEEE Trans. Automat. Control.*, **43**, 1256-1262.
7. Kholostova, O. V. (1999) Dynamics of the Lagrange Top with the Vibrating Pivot. *J. Appl. Math. Mech.*, , **63**, 5, 785-796 .
8. Kovaleva, A.S. (2001) Stabilization of a Quasiconservative System subjected to High-Frequency Excitation. *J. Appl. Math. Mech.*, **65**, 925-936.
9. Krylov, V. and Sorokin, S. (1997) Dynamics of Elastic Beams with controlled distributed stiffness parameters. *Smart Materials and Structures*, **6**, 573 – 582.
10. Meerkov, S. (1980) Principle of Vibrational Control, Theory and Applications. *IEEE Trans. Automat. Control.*, , **25**, 4, 755-762.
11. Meirovitch, L. (1980) *Computational Methods in Structural Dynamics*. Sijthoff-Nordhoff, The Netherlands.
12. Whittaker, E. T. (1964) *Treatise on Analytical Dynamics of Particles and Rigid Bodies*. Cambridge University Press, Cambridge.
13. Ziegler, H. (1968) Principles *of Structural Stability*. Blaisdell, London.

NUMERICAL MODELING OF SMART DEVICES [*]

R. LERCH and H. LANDES and M. KALTENBACHER
Friedrich–Alexander–University, Erlangen–Nuremberg,
Department of Sensor Technology, Paul–Gordan–Str. 3/5,
D–91052 Erlangen, Germany
E–mail: reinhard.lerch@lse.e–technik.uni–erlangen.de

1. INTRODUCTION

Technical components and systems often suffer from parasitic vibrations which are basically induced by mechanical or electromagnetic energy necessarily present within the structure. These vibrations partly result in damage of the structures or at least reduce their life time cycles due to permanent heavy loads. For many other cases, the vibrations of technical equipment result in emission of sound which is often disturbing the environment and has to be reduced [4, 5].

For the reduction or even elimination of these parasitic vibrations we utilize adaptive or smart structures exhibiting some intelligence in form of sensors, controlers and actuators. The general problem is demonstrated in fig. 1. The vibrations to be reduced are either generated by mechanical energy present within the components or stem from an electromechanical actuating mechanism. The presence of these primary energies is on the other hand necessary for the operation of such technical equipment. Typical examples for such systems are electrical transformers or Magnetic Resonance Imaging (MRI) devices. In both, the emitted parasitic noise results from the partial and unintentional conversion of primarily electrical to vibrational energy of the structure.

It is the purpose of adaptive or smart systems to reduce these unintentionally generated vibrations. This is performed by detecting the vibrations first by utilising an appropriate electromechanical sensor. This may be an acceleration sensor, a velocity sensor, a displacement sensor, or a microphone. Such sensors provide electrical signals which should be proportional to their measurable quantities (acceleration, velocity, displacement or, sound pressure). In many cases these sensors are directly integrated into the electromechanical structure or are even an inherent part of it.

The electrical signals of the sensors are forwarded to a controller which is implemented in hardware or in software on dedicated realtime processor (fig. 1). This controller delivers a signal which feeds an electromechanical actuator, mostly via a power amplifier. The purpose of the actuator is to apply mechanical loads to the structure in order to reduce its parasitic vibrations. The transducing mechanisms of these actuators are often based on piezoelectric or magnetomechanical conversion.

---

[*] This work was supported by 'Deutsche Forschungsgemeinschaft', Sonderforschungsbereich SFB 603

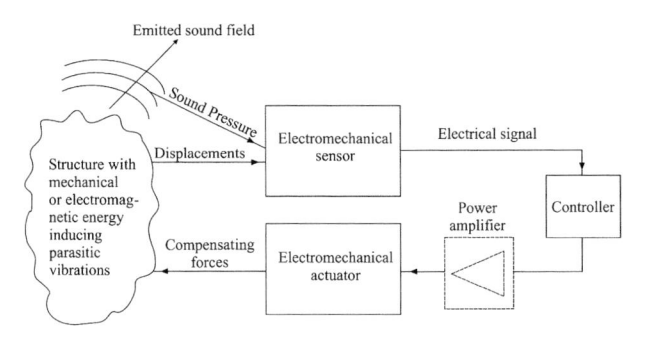

Figure 1. Adaptive System for the reduction of parasitic mechanical vibrations and accompanied sound fields

The main goal of our research work is to model adaptive systems like the one shown in fig. 1. Besides beeing generally applicable to all major sensor and actuator principles this modeling should be precise without restrictive assumptions or approximations. The computer modeling presented here is based on numerical simulation schemes which we have developed during the last years. These schemes and their according software implementations are especially dedicated to the numerical computation of coupled field problems as arising within the controlled loop shown in figure 1.

This loop contains at least two and sometimes three or even four coupling mechanisms:

1. Coupling mechanisms which are **anytime** present:

   Sensor: Coupling of mechanical and electromagnetic field
   Actuator: Coupling of electromagnetic and mechanical field

2. Coupling mechanisms which are **partly** present:

   Structure: Coupling of solid vibrations and sound field
   Structure: Coupling of electromagnetic field and mechanical field

The main focus of our recent work was directed to the simultaneous numerical computation of the complete chain "structure-sensor-controller-actuator-structure".

In the following chapters the basics and major requirements for such computations are reported first. Then we consider some practical examples which consist of a servo-acceleration sensor, plate vibrations and, an epoxy tube which is electromagnetically excited and piezoelectrically damped.

It should be noted that the numerical modeling of smart structures is an emerging field considered by several other scientific groups around the world [1, 5, 7, 8, 15, 19, 21, 24].

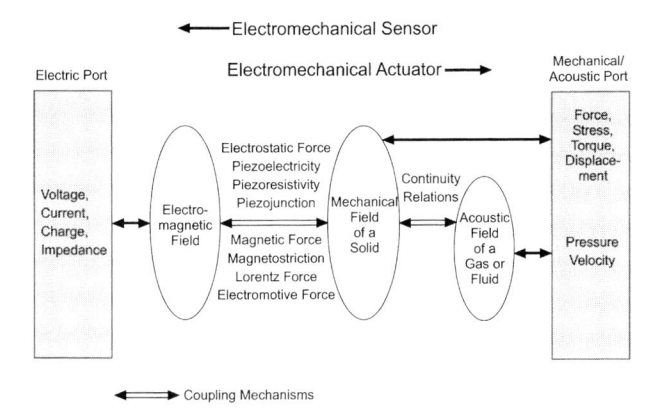

Figure 2. Coupling of fields within electromechanical transducers

2. BASICS OF COMPUTER SIMULATIONS FOR SMART DEVICES

The numerical computation of electromechanical transducers (sensors and actuators) should be precise in accuracy, efficient considering required computer resources as well as flexible in respect to non-regular geometrical shapes. In order to meet these requirements numerical discretization techniques such as appropriate finite element (FEM) and boundary element (BEM) methods have to be applied. Furthermore, for several problems arising in transducer technology and smart structures a combination of both, a FEM-BEM coupling, is the best choice. These finite elements and boundary elements cover the following transducing mechanisms and couplings (see also figure 2)

- electrostatic (capacitive) transducers

- piezoelectric transducers

- nearly all types of magnetomechanical transducers

- fluid-solid coupling around electromechanical transducers

- sound emission and reception of electroacoustic transducers (sonic as well as ultrasonic)

- sound wave propagation (including weak shock waves) in fluid or gaseous media (media at rest as well as flowing media).

The mathematical and algorithmic basis of all these finite and boundary element procedures has already been reported in our earlier publications [3, 4, 6, 9–14, 18, 22]. A variety of application examples, ranging from simple verification tests to the final optimization of commercial transducers, is also found in these papers.

In this paper we report on a new feature which has been added to the existing simulation environment. This add–on consists of an interface connecting and synchronizing two or even more simultaneous finite element simulations, explicitely that of the detecting sensors and that of the actuating transducer. Therefore, a supervising module (fig. 3) is needed to steer all submodules, which are FEM, BEM or FEM-BEM codes. This is a great step forward reflecting the fact that we are now able to compute the complete chain of a smart device.

The applied controller [2, 17, 20] has to be also simulated by an appropriate software module, the input of which stemming from the FE run simulating the sensor. The output signal of the controller module is send as input (excitation) to the finite element run simulating the actuator considering the correct real time sequence (fig. 3). The device itself is simulated in a further finite element procedure running simultaneously with the other FE runs.

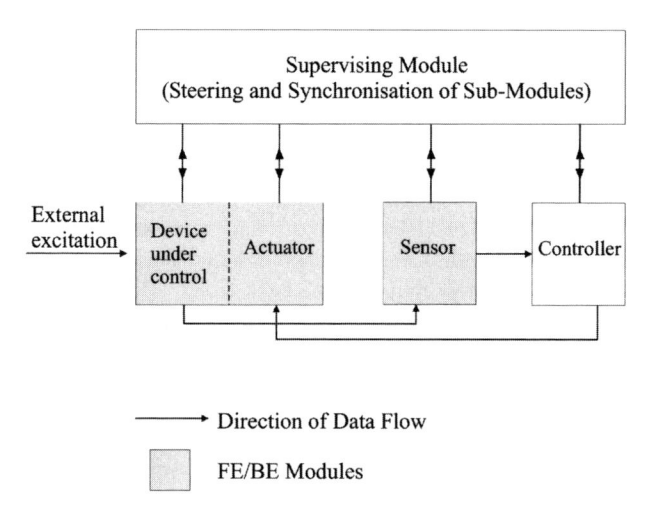

Figure 3. Numerical Computation of a smart device

One of the most important tasks during steering and controlling all these finite element runs is their proper precise time synchronisation which is required for correct results. Otherwise, the errors may accumulate from time step to time step leading to eventually unstabilities. Therefore, we apply transient analysis by utilizing appropriate time step algorithms. The various modules of fig. 3 may work with different time step quantities. In order to save further computer resources the algorithms partly rely on adaptive steering of the time step lengths as well.

3. APPLICATIONS

3.1. *Servo-Acceleration Sensor*

Figure 4 shows the principle of a controlled acceleration sensor. Such a sensor is a smart device in the sense that a sensor, a controller and an actuator operate in a sequential chain. During acceleration the capacitance meter detects a change in the capacitance defined by the two electrodes. In a non-controlled operation mode of the sensor this capacitance is proportional to the measured quantity (acceleration).

However, the non-controlled version exhibits the drawback of nonlinearity errors when high acceleration levels are present. The linearity as well as the dynamic behavior can be tremendously improved by the controlling mechanism shown in fig. 4. Here, the controller puts an electrical voltage to the electrodes which produces a mechanical force just compensating the acceleration force. Therewith, the seismic mass does not deflect in steady state. Only during the transient phase some displacement will be observed, but still limited to a range, where nonlinearity errors do not occur. Now the controller output voltage resembles a direct measure of the applied acceleration.

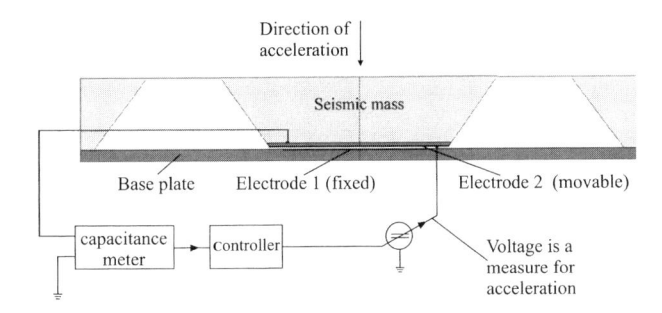

Figure 4. Silicon micromechanical capacitive acceleration sensor

The purpose of numerical computations is to come to an optimum design of the silicon structure in combination with appropriate parameters of the nonlinear controller. Here, we use a combined FEM-BEM model (fig. 5). The finite element model is used to compute the mechanical behavior whereas the boundary elements model the electrical field in the capacitor and its surrounding. Finite and boundary elements are coupled by the electrostatic force. This FEM-BEM approach has the advantage that deformations of the movable electrode will not cause a deformation of surrounding finite elements which would be necessary to describe the electric field in the case of a pure finite element modeling.

Figures 6 and 7 show results of numerical simulations, in which the seismic mass was excited by an acceleration step. The displacement response for this mechanical pulse is shown in fig. 6. Due to symmetry the finite element mesh can be restricted to the part right (or left) of the axis of symmetry. The final result is demonstrated in fig. 7 comparing the controlled and the non-controlled operation mode. The simulation results which have been confirmed by experiments reveal that the dynamic behavior

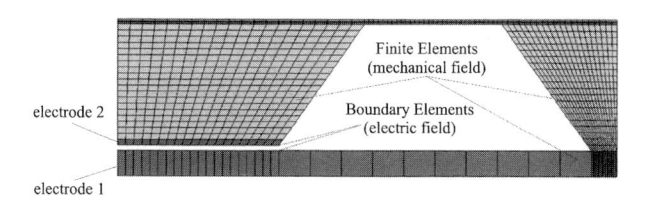

Figure 5. FEM/BEM discretization of a capacitive acceleration sensor

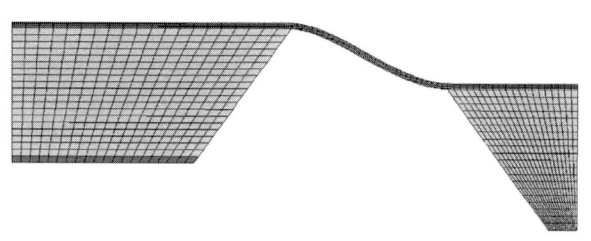

Figure 6. Deformations of the moving part of the acceleration sensor due to an acceleration step

is tremendously enhanced by applying the control mechanism. While the seismic mass exhibits slowly damped vibrations in its eigenmodes for the non-controlled operation mode ending up in a constant deformation, the controlled version of the sensor comes back to its initial state within a very short time period (smaller than 100 μs).

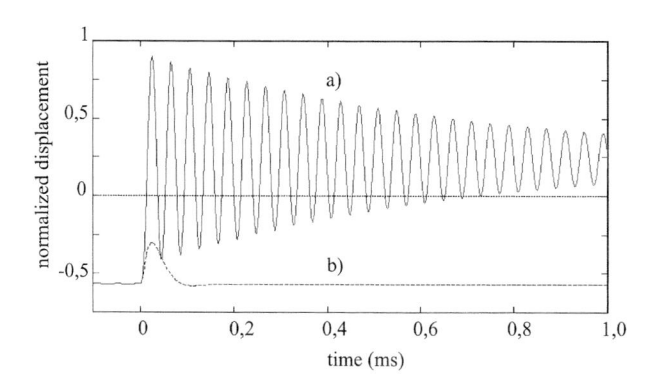

Figure 7. a) Non-controlled and b) controlled dynamic behavior due to an acceleration step

3.2. Active Damping of Plate Vibrations

In a second application a thin circular aluminum disc which is clamped at its outer diameter was excited by an electrostatic force resulting from the electric field between the vibrating disc and a non-movable metallic base plate (fig. 8). The electric voltage applied to the electrodes is a short high voltage pulse which results in an equivalent mechanical pressure load. Therewith, the fundamental eigenmode (bending mode) is excited within the aluminum disc.

In order to damp out this vibration we detect the average mechanical displacement of the disc by measuring the capacitance between the movable aluminum disc and the metallic base plate (fig. 8). This capacitance is a measure for the amplitude of the disc vibrations. Its actual values are fed to a controller which delivers an appropriate output voltage via an electronic amplifier. Its output voltage is applied to the electrodes of a piezoceramic actuator which is directly attached to the top side of the aluminum disc (fig. 9). The purpose of this piezoceramic ring actuator is to produce a mechanical load representing a counterpart to the vibrational energy of the eigenmode of the aluminum disc.

In order to prove the principle and to find an optimum arrangement of the controller and the actuator the complete device was numerically computed by utilizing our software environment. Table I shows, how the transducing mechanisms are numerically modeled. The results of this numerical modeling are demonstrated in the following figures. Figure 10 shows the mechanical displacement of the aluminum disc at its center point from its beginning up to 10 ms, whereas a later time period is presented in fig. 11. These figures clearly demonstrate the proper function of the controlled device. Compared to the non-controlled structure, the vibration amplitude is reduced to a few percent after about four milliseconds.

Table I. Numerical Modeling of Transducing Mechanisms of an adaptive Disc (fig. 8)

Sensing/Actuating	Coupled field problem to be solved	Numerical Procedure
Primary excitation of aluminum disc	electrostatic force resulting from electrostatic field	coupled FEM-BEM FEM: mechanical field BEM: electric field
Detection of disc deflection	capacitance between vibrating aluminum disc and metallic base plate	BEM (as above)
Actuation of compensating mechanical loads	piezoelectric vibrator	FEM module for piezoelectricity

Figure 12 and 13 compare controlled with non-controlled operation. Figure 12 shows the displacement in an early phase where the piezoelectric actuator is generating a bending moment which works against the basic eigenmode vibration of the plate. In later time phases (fig. 13) the compensation has reached nearly steady state whereas the non-controlled plate continues vibrating with high amplitudes.

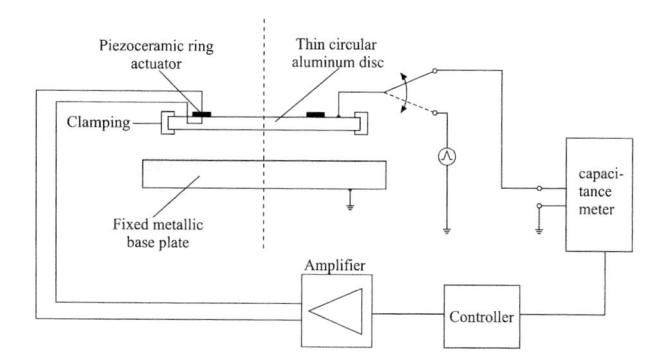

Figure 8. Adaptive structure - Test case

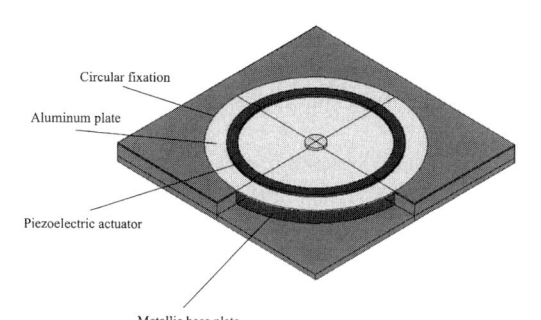

Figure 9. Geometrical model of adaptive structure (fig. 8)

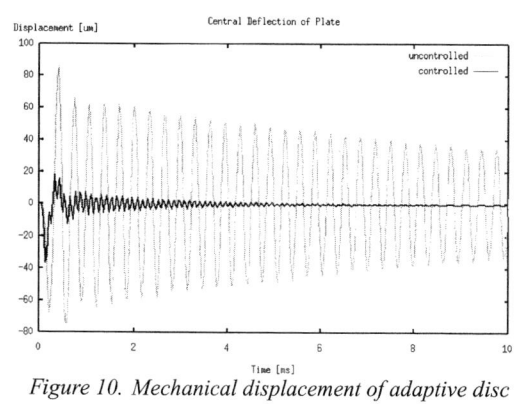

Figure 10. Mechanical displacement of adaptive disc

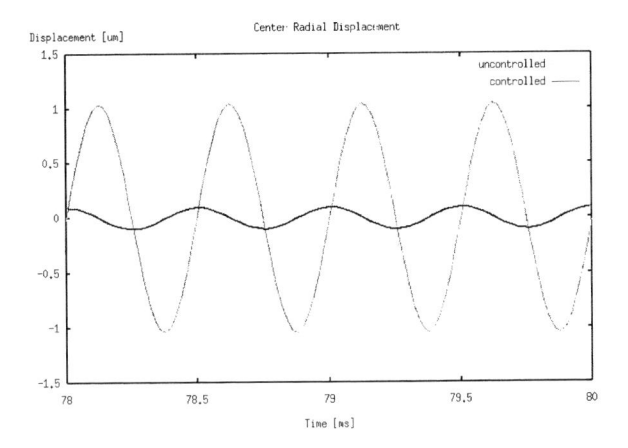

Figure 11. Mechanical displacement of adaptive disc for later time interval

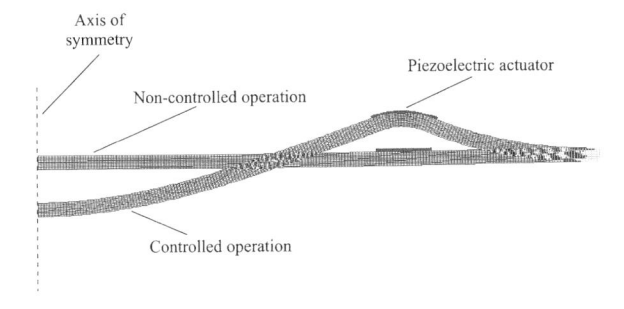

Figure 12. Vibrations of adaptive disc - early time phase

3.3. Active Damping of Tube Vibrations

In many electric devices electromagnetic energy is coupled to parasitic vibrations via various electromechanical transducing mechanisms. In case that these vibrations are in the audible frequency range, the radiated noise may be very disturbing. Therefore, we have investigated the passive and active damping of such devices. First, we have examined power transformers as used in electrical energy generation and distribution. The results have been published elsewhere [16].

Here, we report another effort aiming at the noise reduction of Magnetic Resonance Imaging (MRI) equipment. In a standard MRI device we find an epoxy tube where the patient is placed in during the imaging process. Within this tube a horizontally oriented magnetic field is generated by superconducting magnets. This static base field exhibits inductions of 1 Tesla or even more. For the imaging process several other magnetic coils loaded by transient electrical currents are required. These coils are located around the tube (fig. 14) at various positions. The electrical currents of these coils in combination with the static base field generate Lorentz forces which

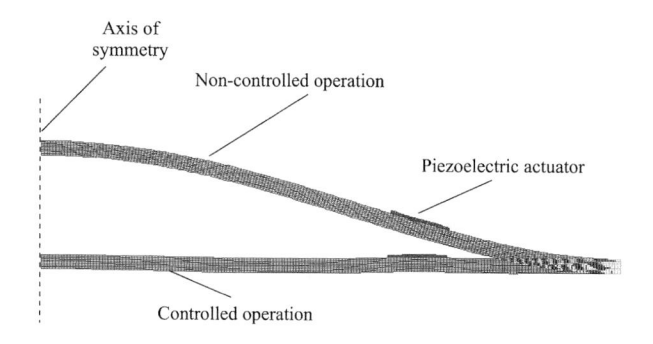

Figure 13. Vibrations of adaptive disc in "steady" state

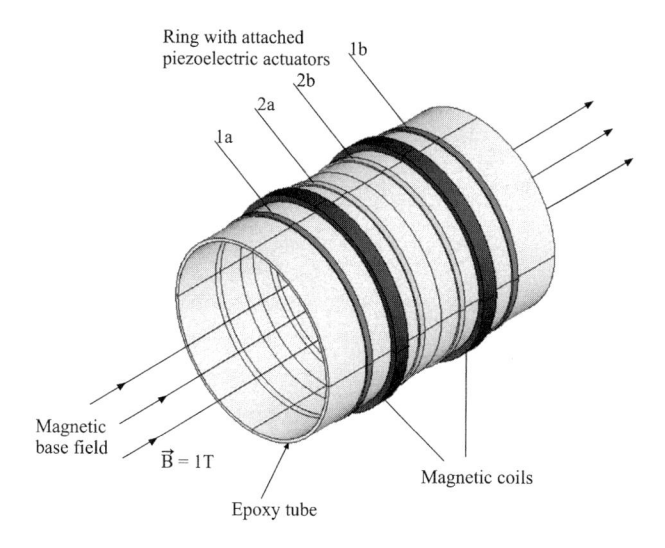

Figure 14. Principle of smart Magnetic Resonance Imaging (MRI) device

mechanically excite the tube and attached mechanical structures resulting in strong vibrations. These vibrations in turn emit sound into the tube interior as well as to its surrounding. Sound pressure levels of more than 100 dB(A) may result, which is very disturbing both for the patient as well as the operating person.

The idea we followed here is to damp these vibrations by active means. Therefore, we put piezoelectric actuators around the tube. In order to produce a stress distribution, which is homogeneous over the polar angle, one first wraps a ring around the tube. The piezoelectric actuators are tightly attached to this ring. We used four of these ring actuators and located them in a symmetric manner at various heights of the tube (fig. 14). The two inner rings and the two outer rings are excited in phase.

The mechanical displacement of the tube was detected by using optical sensors. The optical transducers offer the great advantage that they are insensitive to the

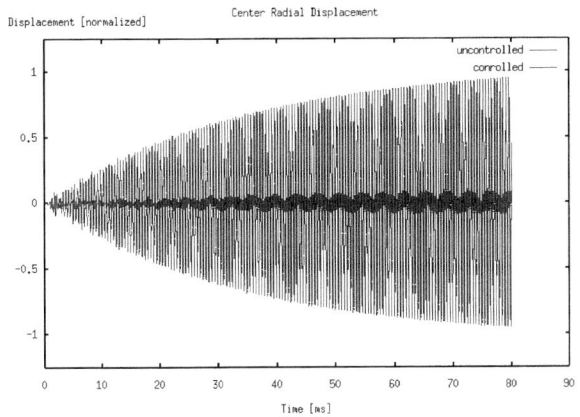

Figure 15. Radial mechanical displacement of tube for controlled and non-controlled operation

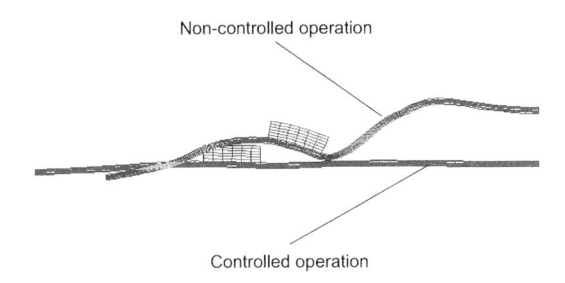

Figure 16. Maximum Deformation for controlled and non-controlled tube

strong magnetic fields around them during the operation in MRI devices. Such optical sensors exhibit a nearly ideal behavior in respect to the proportionality between output signal and measured quantity. Therefore, we handle them as devices with ideal characteristics within the modeling process.

Again, we simultaneously modeled the complete chain of the smart device, beginning with the basic mechanical excitation due to magnetomechanical coupling and ending with the piezoceramic actuators. The coupled field problems to be solved within the modeling process of such a device are presented in Table II. In case that the detecting sensors (displacement or sound pressure) are not ideal, they have to be modeled, too.

The numerical computations have been performed on a PC with a PENTIUM III processor (1100 MHz, 1.5 GB memory). The results are presented in figures 15 and 16. Figure 15 shows the radial mechanical displacement of the tube in its center height for the controlled and non-controlled case. One recognizes a proper operation of the controller after a short transient period (about 2 ms) at the beginning. In steady state the vibrations are damped down to less than 10 % of their amplitudes occur-

Table II. Coupled field problems arising in smart MR devices

Sensing/Actuating	Coupled field problem to be solved	Numerical Procedure
Excitation of parasitic vibrations	Generation of Lorentz forces due to current loaded coils embedded in a static magnetic field	FEM [16]
Emission of sound	Fluid-Solid Coupling (Air-Tube)	FEM or FEM-BEM [11]
Piezoelectric actuators	Piezoelectricity including non-linearities (geometry and material)	FEM [23]

ing during non-controlled operation. Figure 16 shows these maximum deformations comparing the controlled and non-controlled device.

4. CONCLUSION

The numerical computation of the complete dynamic behavior of smart devices asks for a very proper modeling approach. This means that all transducing mechanisms present within the device have to be precisely covered by appropriate modules. Each transducing mechanism is mathematically represented by a coupled field problem, e. g. the coupling of an electrostatic field and a mechanical field. Due to basic requirements (flexibility in geometry and efficiency in respect to computer resources) we implemented FEM, BEM as well as FEM-BEM algorithms, which allow the precise numerical computation of these coupled fields, even in nonlinear cases.

Furthermore, the controller module has to be integrated into the computations. Thereby, the correct synchronisation in time of all software modules asking for input and delivering output asks for special care.

5. REFERENCES

1. Berger, H., H. Koeppe, U. Gabbert, and F. Seeger (2000), 'On Finite Element Analysis of Piezoelectric Controlled Smart Structures'. In: U. Gabbert and H. S. Tzou (eds.): *IUTAM-Symposium on Smart Structures and Structronic Systems*. pp. 189–196. ISBN 0-7923-6968-8.
2. Eccardt, P.-C., M. Knoth, G. Ebest, H. Landes, C. Clau, and S. Wnsche (1996) a, 'Coupled Finite Element and Network Simulation for Microsystem Components'. In: *Proceedings of Micro Systems Technologies '96*. pp. 145–150. Potsdam, 17.-19.09.1996.
3. Eccardt, P.-C., H. Landes, and R. Lerch (1996) b, 'Finite Element Simulation of Acoustic Wave Propagation within Flowing Media'. In: *Proceedings of the IEEE Ultrasonics Symposium*. pp. 991–994. San Antonio, USA.
4. Exler, R. (2002), 'Laermminderung in Lueftungssystemen durch optimierte Anwendung passiver und aktiver Schalldaempfer'. Ph.D. thesis, Johannes Kepler University of Linz. submitted.

5. Gopinathan, S. V., V. V. Varadan, and V. K. Varadan (2000), 'Active Noise Control Studies using the Rayleith-Ritz Method'. In: U. Gabbert and H. S. Tzou (eds.): *IUTAM-Symposium on Smart Structures and Structronic Systems*. pp. 169–177. ISBN 0-7923-6968-8.
6. Hoffelner, J., H. Landes, and R. Lerch (2000), 'Calculation of Acoustic Streaming Velocity and Radiation Force based on Finite Element Simulation of Nonlinear Wave Propagation'. In: *Proceedings of the IEEE Ultrasonics Symposium*, Vol. 1. pp. 585–588.
7. Irschik, H., M. Krommer, and P. U. (2000), 'Collocative Control of Beam Vibrations with Piezoelectric Self-Sensing Layers'. In: U. Gabbert and H. S. Tzou (eds.): *IUTAM-Symposium on Smart Structures and Structronic Systems*. pp. 315–322. ISBN 0-7923-6968-8.
8. Irschik, H. and U. Pichler (2002), 'Maysel's Formula for Small Vibrations Superimposed upon large Static Deformations of Piezoelastic Structures'. In: *IUTAM-Symposium on Dynamics of Advanced Materials and Smart Structures*. Yonezawa, Japan, 20.-24.05.2002, submitted.
9. Kaltenbacher, M., H. Landes, and R. Lerch (1997), 'An Efficient Calculation Scheme for the Numerical Simulation of Coupled Magnetomechanical Systems'. *IEEE Transactions on Magnetics* **33**(2), 1646–1649.
10. Lerch, R. (1988), 'Finite element analysis of piezoelectric transducers'. In: *Proceedings of the IEEE Ultrasonics Symposium*. pp. 643–654.
11. Lerch, R., M. Kaltenbacher, H. Landes, and P.-C. Eccardt (2000) a, 'Combination of Finite Element and Boundary Element Methods in Computational Acoustics and Coupled Field Problems of Electro-Acoustic Transducers'. In: O. von Estorff (ed.): *Boundary Elements in Acoustics (Advances and Applications)*. WIT-Press, pp. 337–375. ISBN 1-85312-556-3.
12. Lerch, R., H. Landes, W. Friedrich, R.-M. Hebel, and H. Kaarmann (1992), 'Modelling of Acoustic Antennas with a Combined Finite– Element–Boundary–Element-Method'. In: *Proceedings of the IEEE Ultrasonics Symposium*. pp. 581–584.
13. Lerch, R., H. Landes, and H. Kaarmann (1994), 'Finite Element Modeling of the Pulse-Echo Behavior of Ultrasound Transducers'. In: *Proceedings of the IEEE Ultrasonics Symposium*. pp. 1021–1025. Cannes, Nov. 1994, ISBN 0-780-32012-3.
14. Lerch, R., H. Landes, R. Simkovics, and M. Kaltenbacher (2000) b, 'Numerical Analysis of Nonlinear and Controlled Electromechanical Transducers'. In: U. Gabbert and H. S. Tzou (eds.): *IUTAM-Symposium on Smart Structures and Structronic Systems*. pp. 25–32. ISBN 0-7923-6968-8.
15. Liu, G. R., C. Cai, K. Y. Lam, and V. V. K. (2002), 'A Review of Simulation Methods for Smart Structures with piezoelectric Material'. In: *IUTAM-Symposium on Dynamics of Advanced Materials and Smart Structures*. Yonezawa, Japan, 20.-24.05.2002, submitted.
16. Rausch, M., M. Kaltenbacher, H. Landes, and R. Lerch (2001), 'Numerical Computation of the Emitted Noise of Power Transformers'. *COMPEL - The International Journal for Computation and Mathematics in Electrical and Electronic Engineering* **20**(2), 636–648.
17. Reitz, S., J. Bastian, J. Haase, P. Schneider, and P. Schwarz (2002), 'System level modeling of mecrosystems using order reductions methods '. In: *Symposium of Design, Test, Integration and Packaging of MEMS/MOEMS*. Cannes, Frankreich, 05.-08.05.2002, submitted.
18. Schinnerl, M., J. Schberl, and M. Kaltenbacher (2000), 'Nested Multigrid Methods for the Fast Numerical Computation of 3D Magnetic Fields'. *IEEE Transactions on Magnetics* **36**(4), 1557–1560.
19. Schlacher, K. and A. Kugi (2002), 'Active Control of Smart Structures using Port Controlled Hamiltonian Systems'. In: *IUTAM-Symposium on Dynamics of Advanced Materials and Smart Structures*. Yonezawa, Japan, 20.-24.05.2002, submitted.
20. Schwarz, P. and P. Schneider (2001), 'Model Library and Tool Support for MEMS Simulation'. In: *Proceedings of the SPIE's Microelectronic and MEMS Technology*, Vol. 4407. Edinburgh, Scotland, 30.05.-01.06.2001.
21. Shibuya, Y. and S. Watanabe (2002), 'Numerical Simulation for Control of Progressive Plastic Buckling with Defects on Axisymmetric Shell Structure'. In: *IUTAM-Symposium on Dynamics of Advanced Materials and Smart Structures*. Yonezawa, Japan, 20.-24.05.2002, submitted.
22. Simkovics, R., H. Landes, M. Kaltenbacher, and R. Lerch (2000), 'Finite Element Analysis of Ferroelectric Hysteresis Effects in Piezoelectric Transducers'. In: *Proceedings of the IEEE Ultrasonics Symposium*, Vol. 2. pp. 1081–1084. San Juan, Puerto Rico, 22.-25.10.2000.
23. Simkovics, R., H. Landes, M. Kaltenbacher, and R. Lerch (2005), 'Analysis of the Nonlinear Behavior of Piezoceramic Multilayer Stack Actuators'. *IEEE Transactions on Ultrasonics, Ferroelectics, and Frequency Control* **999**.

24. Sunar, M. (2002), 'Modeling of Piezoelectric/Magnetostrictive Materials for SmartStructures'. In: *IUTAM-Symposium on Dynamics of Advanced Materials and Smart Structures*. Yonezawa, Japan, 20.-24.05.2002, submitted.

A REVIEW OF SIMULATION METHODS FOR SMART STRUCTURES WITH PIEZOELECTRIC MATERIALS

G. R. LIU[1,2], C. CAI[3], K. Y. LAM[2,3] and V. K. VARADAN[4]

[1] SMA Fellow, Singapore-MIT Alliance, National University of Singapore
[2] Centre for Advanced Computations in Engineering Science (ACES), Department of Mechanical Engineering, National University of Singapore, 9 Engineering Drive 1, Singapore 117576
[3] Institute of High Performance Computing, 1 Science Park Road #01-01 The Capricorn, Singapore Science Park II, Singapore 117528
[4] Centre for the Engineering of Electronic and Acoustic Materials, Pennsylvania State University, State College, PA 16801, USA

1. INTRODUCTION

In order to improve the performance of the existing piezoelectric devices and develop novel piezoelectric materials, it is necessary to have accurate and reliable models for their analysis and design. Most investigators extensively use classic, numerical approaches to predict and analyse characteristics of smart structures. Continuous demand for more precise systems, however, has pushed worldwide investigators to search for more accurate and efficient computational models. It is generally recognized that theoretical models are effective only for well-defined geometries and boundary conditions. In practical applications, finite element method usually provides an alternative in modelling, simulation and analysis of smart structures.

In the viewpoint of applications, there are two major means to apply the smart structures. One uses distributed piezoelectric devices that cover or embed the entire structure (laminated-type smart structures). The other uses discrete piezoelectric devices that occupy a relatively small area of structures (discrete-type smart structures). The modelling approaches and analysis techniques differ considerably between the laminated-type and discrete type smart structures.

The paper is to review current simulation methods for smart structures with piezoelectric materials. It is divided into the following sections: (1) linear theory of electrodynamics, (2) methods for laminated-type smart structures, (3) methods for discrete type smart structures and (4) potential new research areas for smart structures where modelling and simulation are of importance.

2. BASICS OF LINEAR THEORY OF ELECTRODYNAMICS

The linear theory of electrodynamics treats all the elastic, piezoelectric, and dielectric coefficients as constants independent of the magnitude and frequency of applied mechanical stresses and electric fields. Governing equation in the absence of the body forces and body charges consists of stress equations of motion [1, 2]:

$$\rho \ddot{\mathbf{U}} - \mathbf{L}_d^T \boldsymbol{\sigma} = 0 \qquad (1)$$

The charge equation of electrostatics and the electric field-electric potential relations are

$$\nabla \cdot \mathbf{D} = 0, \quad \mathbf{E} = -\nabla \phi \qquad (2, 3)$$

where the symbolic dot denotes differentiation with respect to time t and the superscript "T" the transposition. ρ is the density of the material. U is the particle displacement vector. $\boldsymbol{\sigma}$ is the stress vector. D is the electric displacement vector. E is the electric field vector. ϕ is the scalar electric potential. In addition,

$$\mathbf{L}_d^T = \begin{bmatrix} \dfrac{\partial}{\partial x} & 0 & 0 & 0 & \dfrac{\partial}{\partial z} & \dfrac{\partial}{\partial y} \\ 0 & \dfrac{\partial}{\partial y} & 0 & \dfrac{\partial}{\partial z} & 0 & \dfrac{\partial}{\partial x} \\ 0 & 0 & \dfrac{\partial}{\partial z} & \dfrac{\partial}{\partial y} & \dfrac{\partial}{\partial x} & 0 \end{bmatrix}$$

Equation (1) is coupled with Eq. (2) by means of the piezoelectric constants. In piezoelectric theory, the full electromagnetic equations are not usually needed. The quasielectrostatic approximation is adequate because the phase velocities of elastic waves are approximately five orders of magnitude less than the velocities of electromagnetic waves.

The constitutive relations for a piezoelectric layer direction are

$$\boldsymbol{\sigma} = \mathbf{c}\boldsymbol{\varepsilon} - \mathbf{e}^T \mathbf{E}, \quad \mathbf{D} = \mathbf{e}\boldsymbol{\varepsilon} + \mathbf{g}\mathbf{E} \qquad (4, 5)$$

where c is the stiffness matrix at constant electric field strength. $\boldsymbol{\varepsilon}$ is the strain vector. e is the piezoelectric matrix. g is the dielectric matrix at constant mechanical strain. In this paper, the piezoelectric material with hexagonal 6mm is applied.

Combination of the equations above yields the three-dimensional differential equations of the linear piezoelectric continuum as

$$\mathbf{L}_d^T \mathbf{c} \mathbf{L}_d \mathbf{U} + \mathbf{L}_d^T \mathbf{e}^T \nabla \phi = \rho \ddot{\mathbf{U}}, \quad \nabla \cdot (\mathbf{e} \mathbf{L}_d \mathbf{U} - \mathbf{g} \nabla \phi) = 0 \qquad (6, 7)$$

In the presence of boundaries, the appropriate boundary conditions must be adjoined to Eqs. (6) and (7). If there is a material surface of discontinuity, across the surface there are the continuity conditions of stresses, displacements, electric displacements and electric potentials.

3. METHODS FOR LAMINATED-TYPE SMART STRUCTURES

The modelling of piezoelectric laminates is carried out in two aspects: more accurate mechanic models to account for the characteristics of composite laminates and more accurate electroelastic models to account for the coupling effects between the electrical and mechanical fields inside the piezoelectric devices. The fundamental work of [1, 3] provides much of the necessary theoretical development for the static and dynamic behaviour for laminated-type smart structures.

3.1 Elastic theory

3.1.1 Transfer matrix method

This method is a classical approach for dealing with laminated structures. It starts with building system equations for each layer, and then the continuity of displacements, stresses and electric parameters at the interfaces of different layers is imposed. On the top and bottom surfaces of the laminate, boundary conditions for normal displacements, stresses and electric variables are used. There are different ways of formulating the global system equations for the entire laminate, and the key is to separate the wave-models (or eigen-modes) according to the direction of wave propagation to avoid the so-called numerical truncation problems (see, section 4.4 in Ref. [24]). This ensures the efficiency of the method for laminates with large number of layers. The method has been successfully used in dealing with wave propagation in layered composite materials with smart materials [4, 5].

3.1.2 Surface impedance tensor approach

Honein [6] proposed a systematic method of formulation utilizing the surface impedance tensor. It is proposed to overcome a numerical difficulty that exists when getting the solution for many layers as the product of the solutions of each layer. When the plane harmonic motions in 2-D domains are assumed, the governing equations can be represented as a set of 1st order ordinary differential equations

$$\frac{d}{dz}\bar{\mathbf{s}}(z) = -i\mathbf{N}\bar{\mathbf{s}}(z) \tag{8}$$

where $\bar{\mathbf{s}}$ is the state vector defined as $\bar{\mathbf{s}}(z) = \{\bar{u}(z) \quad i\bar{\sigma}_z \quad i\bar{D}_z(z) \quad i\bar{\tau}_{xz} \quad \bar{w}(z) \quad \bar{\phi}(z)\}^T$. z is the coordinate in the direction of the layer's thickness. Matrix $\mathbf{N}$ can be found from reference [6].

The solution of Eq. (8) can be written immediately as

$$\bar{\mathbf{s}}(z) = \mathbf{M}(z)\bar{\mathbf{s}}(0) \tag{9}$$

where $\mathbf{M}(z) = e^{-i\mathbf{N}z}$ is called the propagator matrix.

Similar to the transfer matrix method, one needs to separate the field variables into two parts, the upward and downward to ensure the numerical stability [7]. Similarly, the concept of the surface impedance matrices relating the "generalized displacement" vector $\mathbf{U}(z) = \{\bar{u}(z) \quad \bar{w}(z) \quad i\bar{D}_z(z)\}^T$ and the "generalized traction" vector $\mathbf{V}(z) = \{i\bar{\sigma}_z \quad i\bar{\tau}_{xz} \quad \bar{\phi}(z)\}^T$ at the normal plane of the layer is introduced. After a surface impedance matrix for a single layer is evaluated, a simple recursive algorithm can be written down to evaluate the surface impedance matrix for many layers.

3.2 Classic laminated plate theory (CLPT)

For problems concerns with lower frequency whose wavelength is sufficiently larger than the thickness of the laminates, one can treat the laminate as a single layer with equivalent properties. These equivalent properties can be obtained using the CLPT that is an extension of the classical plate theory to composite laminates [8]. Lee and Moon [9], and Lee [10] used the assumptions of CLPT to derive a simple theory for piezoelectric laminate. Reddy [11] presented the theoretical formulation of laminated plates with piezoelectric layers as sensors or actuators. Many investigators [12, 13] used CLPT model and its variations to design piezoelectric laminates for different applications. These models use simplifying approximations in characterizing the induced strain field and electric fields generated due to an applied voltage and/or external load. Since the CLPT is based on the Kirchhoff hypothesis holds for the laminate, it applies only to thin plates. Key equations of the CLPT are listed below.

3.2.1 Strain displacement relationship
The assumption leads to the displacement field as:
$$u(x,y,z,t) = u^0(x,y,t) - z\,\partial w^0(x,y,t)/\partial x \tag{10a}$$
$$v(x,y,z,t) = v^0(x,y,t) - z\,\partial w^0(x,y,t)/\partial y \tag{10b}$$
$$w(x,y,z,t) = w^0(x,y,t) \tag{10c}$$
where u, v and w are the displacement components in the x, y and z directions, respectively, and u^0, v^0 and w^0 are the displacement components on the mid-plane.

3.2.2 Lamina constitutive relationship
Assuming $E_x = E_y = 0$, the linear constitutive relations with piezoelectric effect for the kth layer are

$$\begin{Bmatrix} \sigma_x \\ \sigma_y \\ \tau_{xy} \end{Bmatrix}_k = \begin{bmatrix} Q_{11} & Q_{12} & 0 \\ Q_{12} & Q_{22} & 0 \\ 0 & 0 & Q_{66} \end{bmatrix}_k \begin{Bmatrix} \varepsilon_x \\ \varepsilon_y \\ \gamma_{xy} \end{Bmatrix}_k + \begin{Bmatrix} c_{13}e_{33}c_{33}^{-1} - e_{31} \\ c_{23}e_{33}c_{33}^{-1} - e_{32} \\ 0 \end{Bmatrix}_k (E_z)_k \tag{11a}$$

$$(D_z)_k = \bar{e}_{31}\varepsilon_x + \bar{e}_{32}\varepsilon_y + \left(e_{33}^2 c_{33}^{-1} + g_{33}\right)(E_z)_k \tag{11b}$$

where $(Q_{ij})_k$ are the plane stress-reduced stiffness and $\bar{e}_{3i}$ are the plane stress-reduced piezo constants. $Q_{ij} = c_{ij} - c_{i3}c_{j3}c_{33}^{-1}$ and $\bar{e}_{3i} = e_{3i} - c_{3i}c_{33}^{-1}e_{33}$ (i, j=1,2). $Q_{66} = c_{66}$.

3.3 First-order shear deformation theory (FSDT)

FSDT [11] takes into account the effects of shear deformation. It means that the transverse normals do not remain perpendicular to the mid-surface after deformation.
$$\begin{aligned} u(x,y,z,t) &= u^0(x,y,t) - z\varphi_x(x,y,t) \\ v(x,y,z,t) &= v^0(x,y,t) - z\varphi_y(x,y,t) \\ w(x,y,z,t) &= w^0(x,y,t) \end{aligned} \tag{12}$$
where φ_x and φ_y are the rotations of a transverse normal about the y and x axes.

With the strain displacement relationship, the lamina constitutive relationship and laminate constitute equations can be derived in a similar way for CLPT. It is noted that shear correction factors are normally adopted in computing the transverse shear force resultants. Due to the consideration of the shear deformation, FSDT can be used for thicker laminates [51].

3.4 Third-order shear deformation theory (TSDT)

The shear correction factors used in the FSDT are not easy to determine, and the shear stresses produced by FSDT will not satisfy the free stress condition on the laminate surface. Therefore, higher-order expansions of the displacement field such as the TSDT are developed. The TSDT ensures the vanishing of transverse shear strain on the top and bottom surfaces of a laminate [11, 14], and no shear correction factor is needed. It assumes that the displacement field has the form of

$$u(x,y,z,t) = u^0(x,y,t) - z\varphi_x(x,y,t) - z^3\theta_x(x,y,t)$$
$$v(x,y,z,t) = v^0(x,y,t) - z\varphi_y(x,y,t) - z^3\theta_y(x,y,t) \qquad (13)$$
$$w(x,y,z,t) = w^0(x,y,t)$$

where $\theta_x = 4(\partial w/\partial x - \varphi_x)/3h^2$, $\theta_y = 4(\partial w/\partial y - \varphi_y)/3h^2$ and h is the laminate thickness. Compared to the FSDT, TSDT is much more effective for thick laminates. A comparison study on all these plate theories for smart laminates is performed by Liu and Zhou [14].

3.5 Coupled layerwise theory

Discrete layer theories [15] and layerwise theories [16] are developed for the static and dynamic analysis of piezoelectric laminates. The mechanical displacements and the electric potential are assumed to be piecewise continuous across the thickness of the laminate in the layerwise theory. The theory provides a much more kinematically accurate representation of cross sectional warping. Also it can capture non-linear variation of electric potential through the thickness associated with thick laminates. The developments of layerwise laminate theory for a laminate with embedded piezoelectric devices were presented by [16, 17]. Comparisons of the predicted free vibration results from the layerwise theory with the exact solutions for a simply supported piezoelectric laminate reveals the accuracy and robustness of the layerwise theory over CLPT and FSDT [18].

3.6 Layered Element Method for piezoelectrics

Tani [21] and Liu [22] proposed a layered element method for investigating the surface waves in functionally gradient piezoelectric plates based on works for layered elastic materials by Dong and Nelson [19], Waas [52], Kausel [20] and many others. This method was recently extended for analyzing surface waves in multi-layered piezoelectric circular cylinders [25]. The method works as follows.

The layered piezoelectric is divided into N infinite layered elements. An element has lower, middle and upper nodal lines, and two degrees of freedom per nodal line, u and v. Hence the vector of the unknown nodal displacement amplitudes of the element is expressed as $\mathbf{U}^e = \{u_L \quad v_L \quad u_M \quad v_M \quad u_U \quad v_U\}^T$.

The displacement $\mathbf{U}$ and electrostatic potential ϕ in an element are assumed in the form of $\mathbf{U}(x,y,t) = \mathbf{N}_d(y)\mathbf{U}^e(x)\exp(-i\omega t)$ and $\phi(x,y,t) = \mathbf{N}_\varphi(y)\phi^e(x)\exp(-i\omega t)$. The vector of the unknown electrostatic potential of an element is expressed as

$\phi^e = \{\phi_L \ \phi_M \ \phi_U\}^T$, where ϕ_L, ϕ_M and ϕ_U are nodal electrostatic potentials on the lower, middle and upper nodal lines of the element, respectively. $\mathbf{N}_d$ and $\mathbf{N}_\phi$ are given by

$$\mathbf{N}_d(y) = [(1 - 3\bar{y} + 2\bar{y}^2)\mathbf{I} \quad 4(\bar{y} - \bar{y}^2)\mathbf{I} \quad (2\bar{y}^2 - \bar{y})\mathbf{I}] \quad (14)$$

$$\mathbf{N}_\phi(y) = [(1 - 3\bar{y} + 2\bar{y}^2) \quad 4(\bar{y} - \bar{y}^2) \quad (2\bar{y}^2 - \bar{y})] \quad (15)$$

where $\bar{y} = (y - y_L)/h_n$ and $\mathbf{I}$ is a 2 by 2 identity matrix. y_L and h_n are the coordinate of the lower nodal line in the y direction and the nth element thickness, respectively.

The governing equations of the nth element could be developed by means of Hamilton's principle. Finally, a set of differential equations for the nth element is obtained

$$\mathbf{T} = \mathbf{M}\ddot{\psi} + \mathbf{K}_D \psi \quad (16)$$

where

$$\mathbf{K}_D = \begin{bmatrix} \mathbf{A}_D & \mathbf{C}_D \\ \mathbf{C}_D^T & -\mathbf{G}_D \end{bmatrix}, \qquad \mathbf{M} = \begin{bmatrix} \mathbf{M}_s & 0 \\ 0 & 0 \end{bmatrix} \quad (17)$$

$$\mathbf{T}^T = \{\mathbf{F}^T \ \mathbf{D}_y^T\}, \ \psi^T = \{\mathbf{d}^T \ \phi^T\}, \quad (18)$$

Matrices $\mathbf{A}_D$, $\mathbf{M}_s$, $\mathbf{C}_D$, $\mathbf{G}_D$ can be found from reference [22, 24]. Assembling the matrices of N layer elements, the global system differential equations for the entire domain $\mathbf{D}_p$ are obtained:

$$\mathbf{T}_t = \mathbf{M}_t \ddot{\psi}_t + \mathbf{K}_{Dt} \psi_t \quad (19)$$

where the subscript t indicates the matrices correspond to the entire domain and

$$\mathbf{K}_{Dt} = \begin{bmatrix} \mathbf{A}_{Dt} & \mathbf{C}_{Dt} \\ \mathbf{C}_{Dt}^T & -\mathbf{G}_{Dt} \end{bmatrix}, \mathbf{M}_t = \begin{bmatrix} \mathbf{M}_{st} & 0 \\ 0 & 0 \end{bmatrix}, \ \mathbf{T}_t^T = \{\mathbf{F}_t^T \ \mathbf{D}_{yt}^T\}, \psi_t^T = \{\mathbf{d}_t^T \ \phi_t^T\} \quad (20)$$

Equation (19) can then solved using the Fourier transform method, model analysis technique for the time integration, and numerical integration for the inverse Fourier transform [22, 24].

4. METHODS FOR DISCRETE TYPE SMART STRUCTURES

4.1 Equivalent line moment approach

It is demonstrated that the actuator symmetrically and perfectly bonded on a structure is equivalent to external line moments acting along its boundaries. The representative papers are from Crawley [26] and Dimitriadis [27]. In their models the usual assumption is that unless an electric field is applied the presence of the piezoelectric material on or in the substrate does not alter the overall structural properties significantly.

Crawley [26] presented a rigorous study of the stress-strain-voltage behaviour of piezoelectric elements bonded to and imbedded in one-dimensional beams. An important observation of them is that the effectiveness moments resulting from the piezoactuators can be seen as concentrated on the two ends of the actuators when the bonding layer is assumed infinitely thin. Dimitriadis [27] developed the finite two-dimensional piezoelectric elements perfectly bonded to the upper and lower surfaces of elastic plate through static and dynamic analysis. The loads induced by the piezoelectric actuator to the supporting thin elastic structure are estimated in their research. Then, the equivalent magnitude of the edge moments is applied to the plate to replace the actuator patch such that the bending stress at the surface of the plate is equal to the plate's interface stress when the patch is activated.

The approach uses the induced strain by the piezoelectric actuators as an applied strain that contributes to the total strain of the non-active structures, similar to a thermal strain contribution. Strictly speaking, it is not a fully coupled analysis between mechanical and piezoelectric structures, and only considers the converse piezoelectric effect. Many investigators have used this analytical model for various applications where piezoelectric patches are used for controlling beams, plates and shells. In many device geometries, all elements of the piezoelectric tensors are coupled with the strain.

4.2 Hamilton's principle with a Rayleigh-Ritz formulation

Hagood [28] studied the damping of structural vibrations with piezoelectric materials and passive electric networks. They derived an analytical model for an electroelastic system with piezoelectric materials using the Hamilton's principle. The Hamilton variational principle or energy methods will effectively include all coupling between piezoelectric devices and substrate. The physics of the entire structure has been fully accounted for in the energy integrals and there is no need to derive equations based on forces and moments. The resultant equation of motion is solved using the Rayleigh-Ritz method.

Hagood [28] found that the piezoelectric energy transformation properties highly couple to the dynamics of the electric circuit and elastic system. They studied the coupling how to affect the damping effects on the structural modes of a cantilevered beam. Gibbs [29] developed an analytical static model to describe the response of an infinite beam subjected to an asymmetric actuation induced by a perfectly bonded piezoelectric element. Plantier [30] derived a dynamical model for a beam driven by a single asymmetric piezoelectric actuator. Their model includes the effect of the adhesive bonding layer, i.e., the actuator is not assumed to be perfectly bonded to the base structure.

4.3 Finite element method

For discrete piezoelectric patches, investigators have mostly used numerical methods such as the FEM because obtaining exact solutions is difficult.

FEMs were put into application for piezoelectric structures since Allik [31]. Rao [32] developed FEMs to study the dynamic as well as the static response of plates containing distributed piezoelectric devices based on the variational principles. Hwang [13] used a four-node quadrilateral element based on classic laminate theory with the induced strain actuation and Hamilton's principle. They assumed that no stress field is applied to the actuator layer and accordingly the equivalent actuator moments per unit length are found as external excitation loads. Lam [33] developed a finite element model based on the classic laminated plate theory for the active vibration control of a composite plate containing distributed piezoelectric sensors and actuators. Liu and Zhou [14] used the same concept to establish finite element models of piezoelectric composite laminates based on Hamilton's principle as well as CLPT, FSDT and TSDT respectively. Chandrashekhara [34] developed a nine-node shear flexible finite element to study the dynamics of the laminated plate with actuators and sensors. Shen [35] developed a one-dimensional finite element formulation for the flexural motion of a beam containing distributed piezoelectric devices. In his work, the generalized variational principle is used to include the virtual work done by the inertial and electric forces to get the functional. Chen [36] presented the general finite element formulations for piezoelectric sensors and actuators by using the virtual work principle. Kim [37] developed a transition element to connect the three-dimensional solid elements in the piezoelectric region to the flat-shell elements used for the plate. They adopted some special techniques to overcome the disadvantages and inaccuracy of modelling a plate with three-dimensional elements.

4.4 Mesh free method

Mesh free methods requires no mesh for simulations, and hence have achieved remarkable progress in recent years. A number of methods have been developed by many researchers as detailed in the monograph on meshfree methods by Liu [39].

Liu et al. [38, 39] originated the point interpolation method (PIM) with the so-called matrix triangulation algorithm (MTA) that uses the nodal values in the local support domain to interpolate the shape functions. Recently, Liu et al. [40] dealt with piezoelectric device analysis with mechanical and electrical coupled problems using PIM. The problem domain is represented by a set of arbitrarily distributed nodes. A polynomial basis is used to construct the shape functions, which possess delta function properties. Piezoelectric structure with arbitrary shape is formulated using polynomial PIM in combination with variational principle and linear constitutive piezoelectric equations.

PIM obtains its approximation by letting the problem function pass through each scattered nodes within the support domain. Consider a function $u(\mathbf{x})$ defined in the problem domain Ω that is represented by a set of properly scattered nodes. The PIM interpolates $u(\mathbf{x})$ with the representation of

$$u^h(\mathbf{x},\mathbf{x}_Q) = \sum_{i=1}^{n} p_i(\mathbf{x})a_i(\mathbf{x}_Q) = \mathbf{p}^T(\mathbf{x})\mathbf{a}(\mathbf{x}_Q) \tag{21}$$

where $u^h(\mathbf{x},\mathbf{x}_Q)$ represents the approximation of function $u(\mathbf{x})$ within the support domain of point $\mathbf{x}_Q$. $p_i(\mathbf{x})$ is the basis function of monomials in space coordinates. $\mathbf{x}^T = \{x,y\}$ for two-dimensional problems. n is the number of nodes in the support domain, and $a_i(\mathbf{x}_Q)$ is the coefficient for the monomial $p_i(\mathbf{x})$ corresponding to $\mathbf{x}_Q$. $\mathbf{a}$ is defined as $\mathbf{a}^T(\mathbf{x}_Q) = \{a_1, a_2, \cdots\cdots, a_n\}$.

The basis function $p_i(\mathbf{x})$ in Eq. (21) is created using the Pascal's triangle [39] for two-dimensional problems. A basis in two-dimension has the form of

$$\mathbf{p}^T = \{1, x, y, xy, x^2, y^2, x^2y, xy^2, x^3, y^3, \cdots\} \quad (22)$$

In general, polynomials selected should satisfy the completeness or quasi-completeness. At node *i*, we can get

$$u_i = \mathbf{p}^T(\mathbf{x}_i)\mathbf{a} \quad (i=1,2,\cdots,n) \text{ or } \mathbf{u}^e = \mathbf{P}_Q\mathbf{a} \quad (23)$$

where u_i is the nodal value of function $u(\mathbf{x})$ at $\mathbf{x} = \mathbf{x}_i$. Solving for **a** leads to the formulations of the PIM shape functions, with ways to ensure non-singular $\mathbf{P}_Q$ [39]. Once the PIM shape function is obtained, a procedure similar that used in the conventional FEM is used to obtain the system equations:

$$\mathbf{m}\ddot{\hat{\mathbf{u}}} + \mathbf{k}_{uu}\hat{\mathbf{u}} + \mathbf{k}_{u\phi}\hat{\phi} = \hat{\mathbf{F}} \quad (24)$$

$$\mathbf{k}_{u\phi}^T\hat{\mathbf{u}} + \mathbf{k}_{\phi\phi}\hat{\phi} = \hat{\mathbf{Q}} \quad (25)$$

where **u** and **u̇** are the vectors of mechanical displacement and its time derivatives. ϕ is the electric potential vector.

5. POTENTIAL RESEARCH AREAS IN SMART STRUCTURES

5.1 Non-linear characteristics of the piezoelectric materials

Among the currently available sensors and actuators, the smallest ones are of the order of few millimetres. However, progress towards smart structures requires us to develop the systems that are of the order of a few microns. The reduction in size has tremendous technological benefit, however, clear understanding of reliability and system integrity is vital to the efficient and optimum use of smart material systems. As dimensions get smaller, induced electro-mechanical fields get larger. Therefore, the convenience of linearity in modelling should be abandoned, and material and geometric nonlinearities should be accounted for. In addition, piezoelectric materials exhibit a linear relationship between the electric field and strain for low field values (up to *100V/mm*). The relationship behaves non-linear for large fields, the material exhibits hysteresis, furthermore, piezoelectric materials show dielectric aging and hence lack reproducibility of strains.

By incorporating the effect of large deflection and in-plane loads due to the piezo-actuators, Kim [41] derived a non-linear governing motion equation of a composite beam and therefore proposed a semi-active control strategy using embedded piezoelectric devices to modify the vibration response of the beam. Royston [42] presented the non-linear vibratory behaviour of a 1-3 piezoceramic composite theoretically and experimentally.

5.2 Accurate description of electric potentials in the piezoelectric layers

Generally, a constant electric potential distribution of the whole piezoelectric actuator in its length dimension and a linear distribution in its thickness dimension is assumed. However, to satisfy the Maxwell equation, a constant electric potential distribution in the longitudinal direction may not be assumed, but rather obtained by solving the coupled governing equations [43].

The field inside layers of a piezoelectric laminate is previously examined by Roh [44] and Heyliger [45] using full elasticity theories, without any approximations on the mechanical and electrical fields. The exact solution obtained using the exact elasticity theory indicated that the electric and elastic field distributions are often poorly modelled using the simplified theories.

To increase the order of variation of electrical and mechanical fields inside the layers, higher order theories are employed in the laminate models. Yang [46] included higher order (quadratic) electric potential variation through the thickness of the actuators and obtained two-dimensional equations for the bending motion of elastic plates with partially electroded piezoelectric actuators attached to the top and bottom surfaces of a thick plate. Although negligible for thin actuators, the effect of higher order electrical behaviour through the thickness is considered for thick actuators. Using a variational formulation, Krommer [47] observed that for a Timoshenko type smart piezoelectric beam, the potential inside the smart beam could be expressed by a quadratic function in the thickness coordinate. Wang [48, 49] assumed the distribution in the transverse direction to be sinusoidal for the short-circuited electrodes case when considering a long and thin beam embedded piezoelectric actuators. They concluded that the longitudinal distribution of the electric potential in the piezoelectric layer is not constant.

5.3 Nanostructured ceramic

Nanostructured ceramic [50] has been put into the research stage because the nanoscale microstructures may result in changes in properties when the feature size is less than a particular level. The potential changes in properties (improving strength and toughness etc.) have driven researches in nanoceramics over the last decade.

6. CLOSING REMARKS

Several simulation methods on laminated-type smart structures are reviewed in this paper. These are (1) elastic theory, (2) classic laminated plate theory, (3) first-order shear deformation theory, (4) third-order shear deformation theory, (5) coupled layerwise theory, and (6) layered element method. Three simulation methods on discrete-type smart structures are (1) equivalent line moment approach, (2) Hamilton's principle with a Rayleigh-Ritz formulation, (3) finite element approach, and (4) Meshfree method. There are several areas of simulation methods that need attention in the years to come. These include (1) non-linearity of the piezoelectric materials, (2) accurate description of electric fields in the piezoelectric materials and (3) new piezoelectric materials using nano technologies.

7. REFERENCES

1. Tiersten H. F. (1969) Linear piezoelectric plate vibrations, Plenum, New York
2. Nayfeh A H. (1995) Wave propagation in layered anisotropic media with applications to composites. North-Holland series in Applied Mathematics and Mechanics, Elsevier

3. Kraut E. A. (1969) New Mathematical formulations for Piezoelectric Wave Propagation. *Physical Review* **188**(3) pp. 1450-1455
4. Cai C., Liu G. R. and Lam K. Y. (2001) A transfer matrix method to analyse the acoustic properties of active tiles. *Journal of Sound and Vibration* **248** (1) pp.71-89
5. Cai C., Liu G. R. and Lam K. Y. (2001) A technique for modelling multiple piezoelectric layers. *Smart Material and Structures* **10** pp. 689-694
6. Honein B., Braga A. M. B., Barbone P. and Herrmann G. (1991) Wave Propagation in Piezoelectric Layered Media with Some Applications. *Journal of Intelligent Material Systems and Structures* **2** pp. 542-557,
7. Liu G. R., Cai C. and Lam K. Y. (2000) Sound Reflection and Transmission of Compliant Plate-like Structures by a Plane sound Wave Excitation. *Journal of Sound and Vibration* **230** (4) pp. 809-824
8. Reddy, J. N. (1997) Mechanics of laminated composite plates: theory and analysis. Boca Raton : CRC Press
9. Lee C. K. and Moon F. C (1989) Laminated piezopolymer plates for torsion and bending sensors and actuators *Journal of the Acoustical Society of America* **85** pp. 2432-2439
10. Lee C. K. (1990) Theory of laminated piezoelectric plates for the design of distributed sensors and actuators. Part I: governing equations and reciprocal relations. *Journal of the Acoustical Society of America* **87** pp. 1144-1158
11. Reddy, J. N. (1999) On laminated composite plates with integrated sensors and actuators. *Engineering Structures* **21** (7) pp.568-593
12. Lam K. Y. and Ng. T. Y. (1999) Active control of composite plates with integrated piezoelectric sensors and actuators under various dynamic loading conditions. *Smart Materials and Structures* **8**(2) pp. 223-237
13. Hwang Woo-Seok Park, Hyun Chul (1993) Finite element modelling of piezoelectric sensors and actuators. *AIAA JOURNAL* **31** (5) pp. 930-937
14. Zhou Y. L. (1999) Finite element models of piezoelectric composite laminates. *A Thesis for the degree of master of engineering, National University of Singapore*
15. Mitchell J. A. and Reddy J. N. (1995) A refined hybrid plate theory for composite laminates with piezoelectric laminate. *International Journal of Solids and Structures* **32** pp. 2345-2367
16. Saravanos, Dimitris A., Heyliger, Paul R., Hopkins, Dale A. (1997) Layerwise mechanics and finite element for the dynamic analysis of piezoelectric composite plates. *International Journal of Solids and Structures* **34**(3) pp. 359-378
17. Saravanos, Dimitris A. and Heyliger, Paul R. (1995) Coupled layerwise analysis of composite beams with embedded piezoelectric sensors and actuators *Journal of Intelligent Material Systems and Structures* **6** pp. 350-363
18. Gopinathan S. V., Varadan V. V., Varadan V. K. (2000) A review and critique of theories for piezoelectric laminates. *Smart Material and Structures* **9** (1) pp 24-48
19. Dong, S. B. and Nelson, R. B. (1972) On natural vibrations and waves in laminated orthotropic plates, *ASME Journal of Applied Mechanics*, 39 pp 739
20. Kausel, E. (1986) Wave propagation in anisotropic layered media, *Int. J. Numer. Methods Eng.*, 23, 1567, 1986 Kausel, E., Wave propagation in anisotropic layered media, *Int. J. Numer. Methods Eng.*, 23, pp 1567
21. Tani J. and Liu G. R. (1993) SH surface waves in Functionally gradient piezoelectric plates. JSME International Journal **36** pp. 152-155
22. Liu G. R. and Tani J. (1994) Surface waves in Functionally gradient piezoelectric plates. *Transactions of the ASME Journal of Vibration and Acoustics* **116** pp. 440-448
23. Liu G. R. and Achenbach J. D. (1994) A Strip Element Method for Stress Analysis of Anisotropic Linearly Elastic Solids *ASME J. Appl. Mech.* **61** pp. 270-277
24. Liu G. R and Xi Z. C. (2001) Elastic Waves in Anisotropic Laminates, CRC Press. Pp. 235-253
25. Liu G. R. and Han X. (2002) Dispersion and characteristic surfaces of waves in hybrid multilayered piezoelectric circular cylinder. *To be submitted for publication*
26. Crawley E. F. and de Luis J. (1987) Use of piezoelectric actuators as elements of intelligent structures. *AIAA JOURNAL* **25** pp. 1373-1385
27. Dimitriadis E. K., Fuller C. R. and Rogers C. A. (1991) Piezoelectric actuators for distributed vibration excitation of thin plates. *Transactions of the ASME* **113** pp. 100-107
28. Hagood N. W., Chung W. H. and Von Flotow A. (1990) Modelling of piezoelectric actuator dynamics for active structural control. AIAA Pap, AIAA-90-1087-CP pp. 2242-2256

29. Gibbs G. P. and Fuller C. R. (1992) Excitation of thin beams using asymmetric piezoelectric actuators. *Journal of the Acoustical Society of America* **92** pp. 3221-3227
30. Plantier G., Guigou C., Nicolas J., Piaud J. B. and Charette F. (1995) Variational analysis of a thin finite beam excitation with a single asymmetric piezoelectric actuator including bonding layer and dynamical effects. *Acta Acustica* **3** pp. 135-151
31. Allik H. and Hughes T. J. R. (1970) Finite element method for piezoelectric vibration. *International Journal for Numerical Methods in Engineering* **2** pp. 151-157
32. Rao S. S. and Sunar M. (1993) Analysis of Distributed thermopiezoelectric sensors and actuators in advanced intelligent structures. *AIAA Journal* **31**(7) pp. 1280-1286
33. Lam K. Y., Peng X. Q., Liu G. R. and Reddy J. N. (1997) A finite element model for piezoelectric composite laminates. *Smart Materials and Structures* **6** pp. 583-591
34. Chandrashekhara K. and Agarwal A. (1993) Active vibration control of laminated composite plates using piezoelectric device: a finite element approach. *Journal of Intelligent Material Systems and Structures* **4** pp. 496-508
35. Shen M-H H. (1994) Analysis of beams containing piezoelectric sensors and actuators. *Smart Materials and Structures* **3** pp. 439-447
36. Chen C. Q., Wang X. M and Shen Y. P. (1996) Finite element approach of vibration control using self-sensing piezoelectric actuators. *Computers & Structures* **60**(3) pp. 505-512
37. Kim J., Varadan V. V. and Varadan V. K. (1997) Finite element modelling of structures including piezoelectric active devices. *International Journal for Numerical Methods in Engineering* **40**(5) pp. 817-832
38. Liu G. R. and Gu Y. T. (2001) A point interpolation method for two-dimensional solids. *Int. J. Muner. Meth. Engrg*, **50** pp. 937-951
39. Liu G. R. (2002) Mesh Free Methods: moving beyond the finite element method, *CRC Press*.
40. Liu G. R., Dai K. Y., Lim K. M. and Gu Y. T. (2002) A point interpolation mesh free method for static and frequency analysis of two-dimensional piezoelectric structures. *submitted to Computational Mechanics*
41. Kim S. J. and Jones J. D. (1991) Semi-active control of a composite beam using embedded piezoelectric actuators. *Smart Struct. Mater., Winter Annu. Meeting ASME (Atlanta, GA) AD* vol 24/AMD **123** pp. 131-138
42. Royston T. J and Houston B. H. (1998) Modelling and measurement of nonlinear dynamic behavior in piezoelectric ceramics with application to 1-3 composites. *Journal of the Acoustical Society of America* **104**(5) pp. 2814-2827
43. Gopinathan S. V. (2001) Modeling of piezoelectric smart structures for active vibration and noise control applications *PhD Thesis* The Pennsylvania State University
44. Roh Y. R., Varadan V. V. and Varadan V. K. (1996) Optimal design of immersed SAW devices for high- and low-frequency applications. *Smart Materials and Structures* **5** pp. 369-378
45. Heyliger P. (1997) Exact solutions for simply supported laminated piezoelectric plates. *J. Appl. Mech. Trans. ASME* **64** pp. 299-306
46. Yang J. S. (1999) Equations for thick elastic plates with partially electroded piezoelectric actuators and higher order electric fields. *Smart Materials and Structures* **8** pp. 73-82
47. Krommer M. and Irschik H. (1998) On the influence of the electric field on free transverse vibrations of smart beams. *Proc. of SPIE* **3323** pp. 180-189
48. Wang Q. and Quek S. T. (2000) Flexural vibration analysis of sandwich beam coupled with piezoelectric actuator. *Smart Materials and Structures* **9** pp. 103-109
49. Quek S. T and Wang Q. (2000) On dispersion relations in piezoelectric coupled-plate structures. *Smart Materials and Structures* **9** pp. 859-867
50. Cain M. and Morrell R. (2001) Nanostructured ceramics: a review of their potential Applied Organometallic. *Chemistry* **15** pp. 321-330
51. Liu, G R, Zhou Y L, Lam, K Y, Peng X Q and Tani J (1997) Finite element modelling of piezoelectric sensors and actuators bonded in thick composite laminates. *Proc. of the 8th Inter. Conf. on Adaptive Structures and Technologies*, Technomic Publishing, pp. 113-122
52. Waas, G., Linear two-dimensional analysis of soil dynamics problems in semi-infinite layer media, *Ph.D. Thesis*, University of California, Berkeley, California, 1972.

APPLICATION OF TRANSFER MATRIX METHOD IN ANALYZING THE INHOMOGENEOUS INITIAL STRESS PROBLEM IN PRESTRESSED LAYERED PIEZOELECTRIC MEDIA

HUA LIU, ZHEN-BANG KUANG and ZHENG MIN CAI
*Department of Engineering Mechanics, Shanghai Jiaotong University,
Shanghai, 200240, P.R. China
E-mail:zbkuang@mail.sjtu.edu.cn*

1. BASIC EQUATIONS AND TRANSFER MATRIX

Researches into the propagation of surface acoustic waves in layered piezoelectric media have long been of great interest [1]. As we known, there exit initial stresses in the layer during the manufacture process of acoustic surface wave devices. The present study involves the application of the matrix method to analyze the effect of the inhomogenous initial stresses on the Love wave mode and dispersion behavior in a layered piezoelectric ceramic. The initial stresses in the thin layer are assumed to have depth dependent properties.

On the basis of the nonlinear continuum mechanics [2], the equations of motion and corresponding boundary conditions for a piezoelectric medium are

$$[(\sigma_{ij}^0 + \sigma_{ij}) + (u_{j,k}^0 + u_{j,k})(\sigma_{ik}^0 + \sigma_{ik})]_{,i} = \rho_0 \ddot{u}_j, (D_i + D_i^0)_{,i} = 0 \quad (1)$$

where $i, j, k = 1,2,3$, σ_{ij}, D_i and u_j are the unknown second Piola-Kirchhoff stress tensor, electric displacement and mechanical displacement respectively. They are the physical variables produced by the electroelastic wave. All the physical variables in the biasing state are designated by a superscript label "0". $\sigma_{ij}^0 + \sigma_{ij}$, $u_j^0 + u_j$, $D_i + D_i^0$ etc. are total physical variables. ρ_0 is the mass density in the natural undeformed configuration.

Due to the biasing state is a static equilibrium state, from Eq. (1) we can get the governing equations for the unknown variables as

$$(\sigma_{ij} + u_{j,k}\sigma_{ik}^0)_{,i} = \rho_0 \ddot{u}_j, \quad D_{i,i} = 0 \qquad (2a)$$

Analogously the boundary conditions for the unknowns are

$$(\sigma_{ij} + u_{j,k}\sigma_{ik}^0)N_i = T_j, \quad D_i N_i = \sigma^* \qquad (2b)$$

where N_i is the outward unit in the natural undeformed configuration. T_j and σ^* are the applied surface traction and electric charge density acting on per unit natural surface area respectively. In Eq. (2), the constitutive equations are read as

$$\sigma_{ij} = c^*_{ijpl} u_{p,l}, \quad D_i = e^*_{ipl} u_{p,l}, \quad (i,j,l = 1,2,3, \; p = 1,2,3,4) \qquad (3)$$

where $u_4 = \Phi$, Φ is electric potential. As $p = 4$, let $c^*_{ij4l} = e^*_{lij}$, $e^*_{i4l} = -\epsilon^*_{il}$, c^*_{ijkl}, e^*_{mij} and ϵ^*_{mn} are effective elastic, piezoelectric and dielectric constants respectively, and they are related to the initial values u_m^0 and Φ^0 by

$$c^*_{ijkl} = c_{ijkl} + (c_{ijnl}\delta_{km} + c_{inkl}\delta_{jm} + c_{ijklmn})u^0_{m,n} + e_{mijkl}\Phi^0_{,m}$$
$$e^*_{mij} = e_{mij} + (e_{mil}\delta_{jk} + e_{mijkl})u^0_{k,l} - l_{mnij}\Phi^0_{,n}, \epsilon^*_{mn} = \epsilon_{mn} + l_{mnij}u^0_{i,j} - \epsilon_{mnk}\Phi^0_{,k}$$
(4)

where $i,j,k,l,m,n = 1,2,3$, c_{ijkl} and c_{ijklmn} are the second-and third-order elastic constants, e_{mij} and e_{mijkl} are the second-and third-order piezoelectric constants, ϵ_{mn} and ϵ_{mnk} are the second-and third-order dielectric constants, l_{mnij} is the electrostrictive constants.

The most commonly used layered piezoelectric structures comprise a layer with a thickness of h and a substrate rigidly bonded at their interface and stacked normal to the x axis. The symbols $x = x_1$, $y = x_2$, $z = x_3$ are also employed. The layer is in the region $-h < x < 0$ and the substrate $x > 0$. The surface of the layer is free of stresses and in it the inhomogeneously distributed initial stresses vary along the depth x and get their peak values near the surface. It is assumed that only σ^0_{22} exists in the layer. The thickness of the substrate is considerably larger than h and can be treated as a half space, the initial stresses in the substrate are negligible. The layer can be further divided into N sub-layers. In each sub-layer m, the initial stresses keep constant. The substrate is noted by 0 and $N+1$ is the air.

The basis for the transfer matrix method is to develop a transfer matrix for each sub-layer m which maps displacements, stress tractions, electric potential and electric charge from the lower surface of the sub-layer m to it's upper surface [3]. Successive application of the transfer matrix through media 0 to $N+1$ and invoking corresponding interface continuity conditions at their interfaces yields a set of equations relating the boundary conditions at the first interface to the last interface. After introducing the external boundary conditions at the last interface, the equations of motion can be handled.

Because the initial stresses are functions of depth x_1 and neglecting the components of initial stresses between laminas, Eq. (2a) is reduced to

$$\sigma_{ij,i} + u_{j,22}\sigma^0_{22} = \rho_0 \ddot{u}_j, \quad D_{i,i} = 0 \qquad (5)$$

The solutions for the displacement u_p are in the form

$$u_p = A_p(x)\exp[i(\kappa_y y + \kappa_z z - \omega t)] = A_p(x)\exp[i(\kappa_\alpha x_\alpha - \omega t)] \qquad (6)$$

where $p = 1,2,3,4, \alpha = 2,3, i = \sqrt{-1}$, A_p is the amplitude of each partial wave, κ_y and κ_z are wave numbers in the y and z directions respectively, ω is the circular frequency. If we let

$$\left.\begin{array}{l}\sigma_{ij} = \tilde{\sigma}_{ij}(x)\exp[i(\kappa_\alpha x_\alpha - \omega t)] \\ D_i = T_{i+6}(x)\exp[i(\kappa_\alpha x_\alpha - \omega t)]\end{array}\right\}, i,j = 1,2,3, \alpha = 2,3 \qquad (7)$$

and substitute Eqs. (6) and (3) into the first equation of (5), this will result in

$$\tilde{\sigma}_{1j,1} + i\kappa_\beta c^*_{\beta j p1} A_{p,1} = (-\rho_0 \omega^2 + \kappa_2^2 \sigma^0_{22})A_j - i\kappa_\beta i\kappa_\gamma c^*_{\beta j p\gamma} A_p \qquad (8a)$$

In a similar way, the second equation of (5) and Eq. (3) may also be expressed as

$$T_{7,1} + i\kappa_\beta e^*_{\beta p1} A_{p,1} = -i\kappa_\beta i\kappa_\gamma e^*_{\beta p\gamma} A_p, \quad c^*_{1jp1} A_{p,1} = \tilde{\sigma}_{1j} - i\kappa_\beta c^*_{1jp\beta} A_p,$$

$$e^*_{1p1} A_{p,1} = T_7 - i\kappa_\beta e^*_{1p\beta} A_p, \quad j = 1,2,3, \alpha,\beta,\gamma = 2,3, p = 1,2,3,4 \qquad (8b)$$

In order to write Eq. (8) in a matrix form, we let $\sigma_n = T_n(x)\exp[i(\kappa_\alpha x_\alpha - \omega t)]$ ($n = 1,2,\cdots,6$) correspond to $\sigma_x, \sigma_y, \sigma_z, \sigma_{yz}, \sigma_{zx}$ and σ_{xy}, respectively. Eq. (8) is the equation of state in a state space [4] and can be written as

$$\left(B_m(x)\frac{d}{dx} - F_m(x)\right)v_m(x) = 0 \quad \text{or} \quad \left(\frac{d}{dx} - B_m^{-1}(x)F_m(x)\right)v_m(x) = 0 \qquad (9)$$

where v_m is the state vector with eight components in the sub-layer m, i.e.,

$$v_m(x) = (A_{1m}, A_{2m}, A_{3m}, A_{4m}, T_{1m}, T_{6m}, T_{5m}, T_{7m})^T \qquad (10)$$

and $B_m^{-1}(x)F_m(x)$ is the state matrix of the sub-layer m. Eq.(9) can be solved quite easily and it has the solution

$$v_m(x) = Q_m R_m(x) a_m \qquad (11)$$

where $Q_m = (h_{1m}, h_{2m}, h_{3m}, h_{4m}, h_{5m}, h_{6m}, h_{7m}, h_{8m})$ \qquad (12)

$$R_m(x) = diag[\exp(b_{1m}x), \exp(b_{2m}x), \cdots, \exp(b_{8m}x)], a_m = (a_{1m}, a_{2m}, \cdots, a_{8m})^T \quad (13)$$

h_{sm} and b_{sm} ($s=1$-8) are the eigenvectors and eigenvalues of the state matrix respectively. a_{sm} ($s=1$-8) are undetermined coefficients, respectively. Let I denote the identify matrix, then the eigenvalue equation can be written as

$$[b_m I - B_m^{-1} F_m] \{Q_m\} = 0 \qquad (14)$$

The transfer matrix P_m (x_m–d_m, x_m) can be used to relate the state vector at the bottom of the sub-layer m to those at it's surface, i.e.,

$$v_m(x_m - d_m) = P_m(x_m - d_m, x_m) v_m(x_m) \qquad (15)$$

$$v_N(-h) = \lim_{\max\{d_m\} \to 0} \prod_{m=1}^{N} P_m(x_m - d_m, x_m) v_1(0) \qquad (16)$$

where x_m is located in the bottom plane of the sub-layer m. d_m is the thickness of the sub-layer m. $v_N(-h)$ and $v_1(0)$ respectively are the state vectors at the upper and lower surfaces of the layer.

From Eqs. (9) to (15), we are capable to write the transfer matrix in the form

$$P_m(x_m - d_m, x_m) = Q_m R_m(-d_m) Q_m^{-1} \qquad (17)$$

Now we discuss the Love wave propagation in a transversely isotropic piezoelectric medium with initial stresses. Here the piezoelectric medium is polarized along the z-axis. The wave propagates in the positive direction of the y-axis, and only one constant prestress component $\sigma_y^0(x)$ exists in the layer. Thus for Love waves, we get $u = v = 0$, $w = w(x,y,t)$, $\Phi = \Phi(x,y,t)$, $\kappa_y = k$, $\kappa_z = 0$, u, v and w are the components of displacement along the x, y and z axis, respectively. Then for Love waves, the state vectors of Eqs. (9) and (10) are

$$v_m(x) = (A_{3m}, A_{4m}, T_{5m}, T_{7m})^T \qquad (18)$$

$$B_m(x) = \begin{bmatrix} 0 & 0 & 1 & 0 \\ 0 & 0 & 0 & 1 \\ c_{55}^* & e_{15}^* & 0 & 0 \\ e_{15}^* & -\epsilon_{11}^* & 0 & 0 \end{bmatrix}, \quad F_m(x) = \begin{bmatrix} -\rho_0 \omega^2 + [c_{44}^* + \sigma_y^0(x_m)]\kappa^2 & e_{24}^* \kappa^2 & 0 & 0 \\ e_{24}^* \kappa^2 & -\epsilon_{22}^* \kappa^2 & 0 & 0 \\ 0 & 0 & 1 & 0 \\ 0 & 0 & 0 & 1 \end{bmatrix} \qquad (19)$$

The initial stress in each sub-layer is uniform and assumed to equal to $\sigma_y^0(x_m)$. For ceramics from 6mm class in the absence of stresses, we have $c_{44}=c_{55}$, $e_{24}=e_{15}$, $\epsilon_{22}=\epsilon_{11}$. Hear the differences between c_{44}^* and c_{55}^*, e_{24}^* and e_{15}^*, ϵ_{22}^* and ϵ_{11}^* are respectively neglected. The eigenvalues and corresponding eigenvectors of the state matrix are

$$b_{1m,2m} = \pm \kappa, \quad b_{3m,4m} = \pm \kappa q_m, \quad q_m = \sqrt{1 - \frac{\rho c^2 - \sigma_y^0(x_m)}{\bar{c}_{55}}} \qquad (20)$$

$$Q_m = \begin{bmatrix} 0 & 0 & 1 & 1 \\ 1 & 1 & e_{15}^*/\epsilon_{11}^* & e_{15}^*/\epsilon_{11}^* \\ -e_{15}^*\kappa & e_{15}^*\kappa & -\bar{c}_{55}q_m\kappa & \bar{c}_{55}q_m\kappa \\ \epsilon_{11}^*\kappa & -\epsilon_{11}^*\kappa & 0 & 0 \end{bmatrix} \quad (21)$$

where $\bar{c}_{55} = c_{55}^* + (e_{15}^*)^2/\epsilon_{11}^*$, c is the phase velocity and given by $c = \omega/\kappa$. And then the transfer matrix of the sub-layer m is

$$P(x_m - d_m, x_m) = \begin{bmatrix} \cosh(\kappa q_m d_m) & 0 & -\dfrac{\sinh(\kappa q_m d_m)}{\bar{c}_{55}\kappa q_m} & 0 \\ \dfrac{e_{15}^*}{\epsilon_{11}^*}\cosh(\kappa q_m d_m) & \cosh(\kappa d_m) & -\dfrac{e_{15}^*\sinh(\kappa q_m d_m)}{\bar{c}_{55}\epsilon_{11}^*\kappa q_m} & \dfrac{\sinh(\kappa d_m)}{\epsilon_{11}^*\kappa} \\ \bar{c}_{55}\kappa q_m \sinh(\kappa q_m d_m) & -e_{15}^*\kappa \sinh(\kappa d_m) & \cosh(\kappa q_m d_m) & -\dfrac{e_{15}^*}{\epsilon_{11}^*}\cosh(\kappa d_m) \\ 0 & \epsilon_{11}^*\kappa \sinh(\kappa d_m) & 0 & \cosh(\kappa d_m) \end{bmatrix} \quad (22)$$

It is well known that the major disturbance of Love wave motion is confined to the region near the surface. Thus the terms containing b_{10} and b_{30} in $R_0(x)$ disappear. Q_0 can be obtained by substitution of the material constants of the substrate into Eq.(21) and the state vector at $x = 0$ becomes

$$v_0(0) = Q_0(0)(0, a_{20}, 0, a_{40})^T \quad (23)$$

The electric potential Φ_{N+1} and electric displacement $D_x^{(N+1)}$ in the air ($x < -h$) can be expressed as

$$\Phi_{N+1}(x,t) = a_{N+1}\exp(\kappa x)\exp[i(\kappa_\alpha x_\alpha - \omega t)] \ (\alpha = 2,3), \ D_x^{(N+1)} = -\epsilon_0 \Phi_{N+1,x} \quad (24)$$

where a_{N+1} is the undetermined coefficient, $\in_0$ is the dielectric constant of air.

The boundary and continuity conditions are

$$\begin{aligned}&\sigma_{zx} = 0, (x = -h)\\ &D_x = D_x^{(N+1)}, \quad \Phi = \Phi_{N+1}, \quad (x = -h, \text{free})\\ &\Phi = 0, \quad (x = -h, \text{shorted})\\ &[\sigma_{zx}] = 0, \; [w] = 0, \; [D_x] = 0, \; [\Phi] = 0, \quad (x = 0, \text{continuity condition})\end{aligned} \quad (25)$$

From Eqs. (25) and (16), one obtains the state vectors at $x = -h$

$$v_N(-h) = \boldsymbol{H}(0, a_{20}, 0, a_{40})^T, \boldsymbol{H} = \lim_{\max\{d_m\}\to 0} \prod_{m=1}^{N} \boldsymbol{P}_m(x_m - d_m, x_m)\boldsymbol{Q}_0 \quad (26)$$

Substituting Eqs. (24) and (26) into the boundary conditions at $x = -h$ for the electrically free case, we obtain a set of three algebraic equations in the unknowns a_{20}, a_{40} and a_{N+1}. To obtain the nontrivial solutions the determinant of the boundary condition matrix should vanish. Then one obtains the phase velocity equation for the electrically free case,

$$(H_{72} + \kappa \in_0 H_{42})H_{54} - (H_{74} + \kappa \in_0 H_{44})H_{52} = 0 \quad (27)$$

The phase velocity for the shorted case can be obtained in a similar way

$$H_{42}H_{54} - H_{52}H_{44} = 0 \quad (28)$$

2. DISCUSSIONS AND NUMERICAL ILLUSTRATIONS

A numerical example is discussed below. The system is a PZT-1 ceramic substrate with a totally depolarized surface layer (e_{15}=0). The material constants of PZT-1 ceramics are from [5].

It is found from Eq. (4) that for the plain strain problems, the effective material constants c_{55}^*, e_{15}^* and $\in_{11}^*$ equal to c_{55}, e_{15} and $\in_{11}$ respectively when the third-order material constants are not considered. We assume that the initial stress $\sigma_y^0(x)$ in the layer varies exponentially with x. It reaches the maximum at the surface of the layer and approaches zero at the bottom of the layer and whole substrate. The layer is equally divided into N ($N \geq 20$) sub-layers.

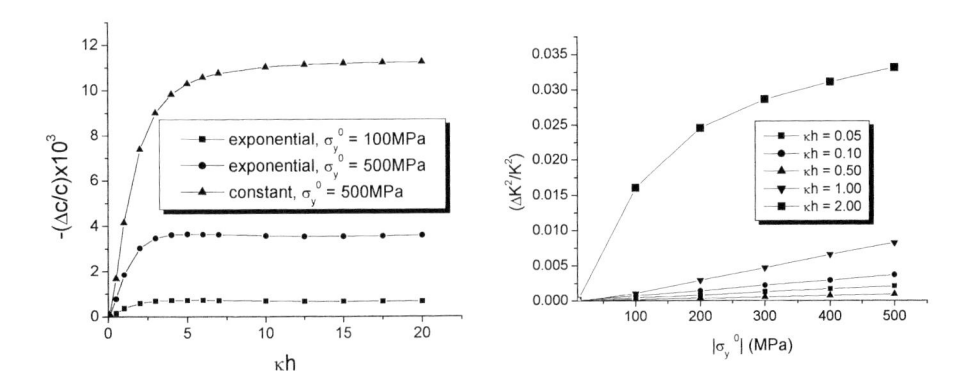

Figure 1. Variations of the fractional phase velocity change $\Delta c / c$ with κh for different magnitudes of the initial stresses.

Figure 2. The factional change in the electromechanical coupling coefficient K^2 with $|\sigma_y^0|$ for different value of κh.

Existence of the initial stress in the layer leads to a reduction of the phase velocity of Love waves in the layered PZT-1 ceramics, as shown in Figure 1. We note that for a given value of κh, the fractional phase velocity change $\Delta c / c$ increases with increasing the absolute value of the initial stress. The maximal fractional velocity change $\Delta c / c$ that we obtained is 3.65×10^{-3} corresponding to $|\sigma_y^0(-h)| = 500$ MPa at $\kappa h = 5$. For a fixed magnitude of the initial stress, $\Delta c / c$ enhances very fast as κh changes from 0 to 5, and reaches its maximum at approximately $\kappa h = 5$, then decreases a little and the curve tends asymptotically to a horizontal line as $\kappa h \to \infty$. Figure 1 also shows the fractional velocity changes $\Delta c / c$ corresponding to a constant initial stress with $\sigma_y^0 = 500$ MPa in the layer. It can be seen that $\Delta c / c$ is dependent on the characteristics of the inhomogeneity of the initial stress. The fractional velocity change due to the exponentially gradient initial stress in less than that of the constant initial stress in the layer.

Effects of the initial stress on the electromechanical coupling coefficient K^2 are shown in Figure 2, which is defined as $K^2 = 2(c_f - c_s)/c_f$. It is found that the maximally improved fractional change in K^2 is 3.309% corresponding to $|\sigma_y^0(-h)| = 500$ Mpa (at $\kappa h = 2$). Thus one may state that the application of proper prestress can enhance the electromechanical coupling behavior to a certain extent.

ACKNOWLEDGEMENT

This work is supported by the National Science Foundation through Grants No. 10132010 and 10072033.

REFERENCES

1. Nayfeh, A. H. (1991) The general problem of elastic wave propagation in multilayered anisotropic media, *J. Acoust. Soc. A.,* **89** 1521-1531.
2. Kuang, Z. B. (2002) *Nonlinear continuum mechanics*, Shanghai Jiaotong University Press, Shanghai.
3. Stewart, J. T. and Yong, Y. K. (1994) Exact analysis of the propagation of acoustic waves in multilayered anisotropic piezoelectric plates, *IEEE Trans. Ultrason. Ferroelect. Freq. Contr.,* **41** 375-390.
4. Liang, W. (2000) Investigation of the crack problems in soft ferro-magnetic materials and acoustic propagation in SAW sensors, Ph.D thesis of Xian Jiaotong University, pp.72-84.
5. Jin, F., Wang, Z. K. and Wang, T. J. (2000) The B-G wave in the piezoelectric layered half-space for the layer and the substrate in opposite polarization, *Acta. Mechanica Solida Sinica.* **21** 95-101.

INFINITESIMAL MECHANISM MODES OF TENSEGRITY MODULES

HIDENORI MURAKAMI and YOSHITAKA NISHIMURA
Department of Mechanical and Aerospace Engineering
University of California at San Diego
9500 Gilman Drive, La Jolla, CA 92093-0411, U.S.A.
E-mail: hmurakami@ucsd.edu

1. INTRODUCTION

Tensegrity structures, invented by Snelson in 1948, are a class of truss structures consisting of a continuous set of tension members and a discrete set of compression members [1, 2]. Ingber [3] stated that the assembly pattern of organic structures such as cells, tissues, organs, and animals are described by a universal building rule of self-assembly at different scales based upon geodesic tensegrity. Furthermore, animal mechanisms, including that of humans, do not include any pins or gears that are representative components of machine. Therefore, in order to develop smart deployable structures as well as understand the role of cytoskeleton in cell mechanotransduction, it is necessary to understand infinitesimal mechanism modes of tensegrity modules [4, 5, 6]. The objectives of the paper are to analytically find initial configurations and pre-stress modes for regular truncated octahedral and cubic tensegrity modules and to classify a large number of infinitesimal mechanism modes by conducting modal analyses. In what follows, the static and dynamic characterization procedure proposed by the authors is employed using the same notation as that of [7, 8].

2. REGULAR TRUNCATED OCTAHEDRL TENSEGRITY MODULES

2.1 Maxwell Number and the Number of Infinitesimal Mechanisms

The regular octahedron, shown in Fig. 1a, consists of six vertices ($v = 6$) and eight equilateral triangular faces ($f = 8$). Four triangles meet at each vertex. There are twelve edges according to Euler's relation: $e = f + v - 2 = 12$. If a regular octahedron is truncated by cutting off each vertex along a plane perpendicular to the radius through the vertex, the faces that are created become a set of squares, as shown in Fig. 1b. The truncated octahedral tensegrity module in Fig. 1c has the same number of nodes as the truncated regular octahedron. Therefore, the number of nodes in the tensegrity module becomes $n_N = 24$. Since bars connect 24 nodes, there are 12 bars. Further, the connection of cables is the same as the edges of the truncated octahedron in Fig. 1b. As a result, there are 12 vertical cables and 4 x 6 cables along the edges of the truncating triangles. Thus the number of truss elements become $n_E = 48$. The Maxwell number $Mx = n_E - (3 n_N - 6)$ [4] of a truncated octahedral tensegrity module becomes -18. The tensegrity module exists under a pre-stressed condition. Calladine's relation [5] shows that the tensegrity modules possess 19 infinitesimal mechanisms, *i.e.*, $n_M = 1 - Mx = 19$. In the tensegrity module, each truncating square only exists at a twisted state with respect to the radius through the truncated vertex.

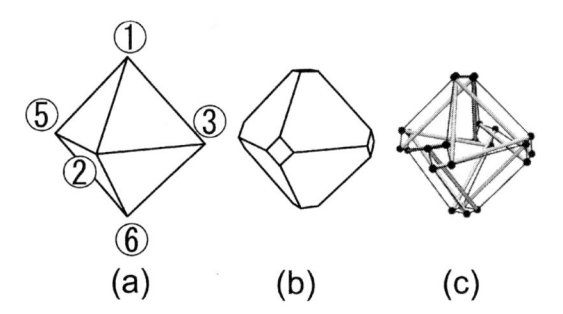

Figure 1. (a) A regular octahedron, (b) a regular truncated octahedron, and (c) a regular truncated octahedral tensegrity module.

2.2 Initial Shape Finding

For a truss structure with n_E elements and n_N nodes with 6 linearly-independent displacement constraints, there are $n_V = 3 n_N - 6$ unknown displacement components. Let the internal element-force vector be denoted by **s**, an n_E x 1 column matrix, and the external nodal-force vector by **f**, an n_V x 1 column matrix. The equilibrium equation for quasi-static loading **f** is expressed as:

$$\mathbf{A\,s} = \mathbf{f}, \tag{1a}$$

where **A** is an n_V x n_E matrix consisting of direction cosines of truss elements. A nontrivial internal element-force **s** without external forces, *i.e.*, **f** = **0**, is referred to as a pre-stress mode. The pre-stress mode exists if and only if

$$\det(\mathbf{A}^T \mathbf{A}) = 0. \tag{1b}$$

Fuller's spherical tensegrity modules have three cables and one bar at each node. In order to find the twist angle, it is necessary to impose the existence condition (1b) of a pre-stress mode on the initial equilibrium matrix **A**. To form **A**, nodal coordinates must be analytically computed for an unknown twist angle α.

In order to define nodal coordinates, the congruent motions of a regular octahedron become extremely helpful. There are 24 congruent configurations of a regular octahedron. They form the octahedral group S_4 [9]. The congruent motions of a regular octahedron in Fig. 1a are described by the product of two generators: **R**, the rotation by $\pi/2$ with respect to a diagonal joining two opposite vertices, and **F**, the flip by π with respect to a median defined by connecting the centers of two facing edges. The octahedral group also describes the congruent motions of both a regular truncated octahedron and a regular truncated octahedral tensegrity module. Three diagonals form an orthogonal triad. Therefore, they are selected as the body-fixed Cartesian x'-, y'-, and z'-axes, as well as the inertial Cartesian x-, y-, and z-axes, as shown in Fig. 2.

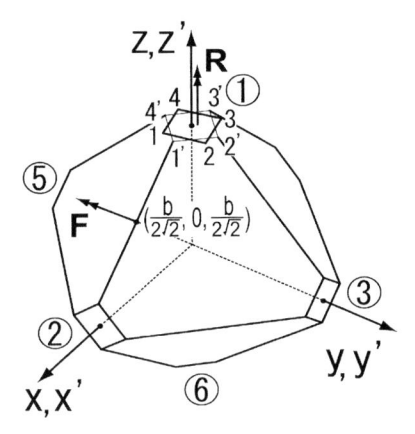

Figure 2. A body-fixed Cartesian coordinate system $\{x', y', z'\}$ and the inertial Cartesian coordinate system $\{x, y, z\}$ erected for a regular truncated octahedron.

In the figure, both coordinate frames coincide. However, as **R** and **F** are applied, the body-fixed frame $\{x', y', z'\}$ rotates with the octahedron while the inertial frame $\{x, y, z\}$ remains stationary. Let the length of the edges of a regular octahedron be denoted by b. The edge length of the truncating square is denoted by a. The edges of the regular octahedron intercept the x'-, y'-, and z'- axes at $\pm b/\sqrt{2}$. Figure 2 also shows the positive direction of the twist angle with respect to square 1'-2'-3'-4' twisted in the "clockwise direction" by α to become square 1-2-3-4. The rotation **R** about a fixed vector $\mathbf{u} = [u_1, u_2, u_3]^T$ by angle ϕ is described by using the Euler parameters $\{e_0, e_1, e_2, e_3\}$ as follows:

$$e_0 = \cos\frac{\phi}{2}, \quad e_i = u_i \sin\frac{\phi}{2} \quad i = 1, 2, 3, \tag{2a}$$

$$\mathbf{R} = 2 \begin{bmatrix} e_0^2 + e_1^2 - \frac{1}{2} & e_1 e_2 - e_0 e_3 & e_1 e_3 + e_0 e_2 \\ e_1 e_2 + e_0 e_3 & e_0^2 + e_2^2 - \frac{1}{2} & e_2 e_3 - e_0 e_1 \\ e_1 e_3 - e_0 e_2 & e_2 e_3 + e_0 e_1 & e_0^2 + e_3^2 - \frac{1}{2} \end{bmatrix}. \tag{2b}$$

The rotation matrix $\mathbf{R}$ with respect to the z'-axis through vertex 1 is described by the axis $[0, 0, 1]^T$ and the rotation angle $\phi = \pi/2$ with respect to the body-fixed frame {x', y', z'}. The flip $\mathbf{F}$ is the rotation with respect to $\mathbf{w}' = [1/\sqrt{2} \ \ 0 \ \ 1/\sqrt{2}]^T$ by π. The matrices $\mathbf{R}$ and $\mathbf{F}$ with respect to the body-fixed frame become

$$\mathbf{R} = \begin{bmatrix} 0 & -1 & 0 \\ 1 & 0 & 0 \\ 0 & 0 & 1 \end{bmatrix}, \quad \mathbf{F} = \begin{bmatrix} 0 & 0 & 1 \\ 0 & -1 & 0 \\ 1 & 0 & 0 \end{bmatrix}. \tag{3}$$

Figure 3a shows a graph of the octahedral group S_4, which represents all congruent motions of a regular octahedron, shown in Fig. 1a [9]. In the graph, nodes represent different configurations of a regular octahedron and edges represent congruent motions. Solid lines with arrows indicate $\mathbf{R}$ with respect to the z'-axis by $\pi/2$ and dashed lines denote the flip $\mathbf{F}$ with respect to the w'-axis.

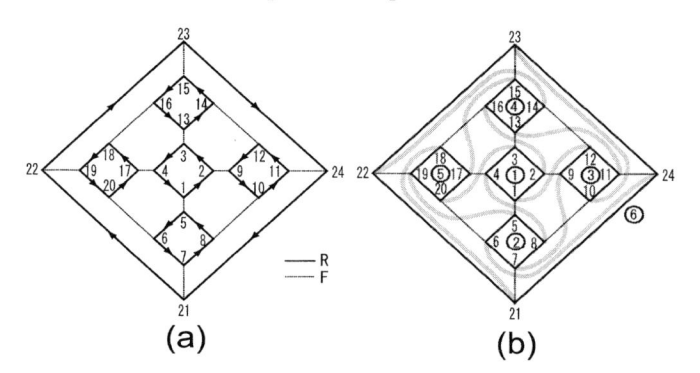

Figure 3 (a) A graph of the octahedral group, (b) a connectivity diagram of a truncated octahedral tensegrity module with a positive twist angle.

Let the identity of the group be denoted by $\mathbf{I}$. The generators, $\mathbf{R}$ and $\mathbf{F}$, satisfy the following relations:

$$\mathbf{RRRR} = \mathbf{I}, \quad \mathbf{FF} = \mathbf{I}, \quad \mathbf{FRFRFR} = \mathbf{I}, \quad \mathbf{R}^{-1} = \mathbf{R}^T, \quad \mathbf{F}^{-1} = \mathbf{F}^T. \tag{4}$$

The above group operations are expressed by matrix multiplication, *i.e.*, the operators are applied from the right to the left.

The graph of the octahedral group will be used: (i) to generate all nodal coordinates from the coordinate of node 1 of the twisted square 1-2-3-4 in Fig. 2 and (ii) to define the connections of cable elements. Let the coordinates be defined at node 1 in Fig. 2. By following the lines in Fig. 3a and performing the **R** or **F** operations, one can move node 1 to any desired node identified in the original configuration. For example, node 1 can be moved to the node 3 position by **RR** or $\mathbf{R}^T\mathbf{R}^T$ as shown in Fig. 3a. The motion to node 15 is achieved by multiplying **RRFRR**. It is noted that the number of congruent configurations is the same as the number of nodes of both the truncated octahedron in Fig. 1b and the truncated octahedral tensegrity module in Fig. 1c. In order to take advantage of this coincidence between the nodes of the tensegrity module and the congruent configurations of the octahedral group, node numbers are assigned to the nodes of the graph in Fig. 3a. An examination of cable connections of the tensegrity module reveals that there is a one-to-one correspondence between cable elements and the edges of the truncated octahedron in Fig. 1b. Furthermore, the edges of the truncated octahedron in Fig. 1b have a one-to-one correspondence with the paths in Fig. 3a. Therefore, the paths of the graph show the connection of cables of truncated octahedral tensegrity modules. A connectivity diagram of tensegrity elements is shown for a positive twist angle α in Fig. 3b. In the figure, each truncating square is identified by the corresponding vertex of the regular octahedron. Thick free-hand curves show the connection of bars.

In order to compute nodal coordinates with respect to the inertial frame $\{x, y, z\}$, one must keep track of the orientation of the body-fixed frame $\{x', y', z'\}$ with respect to the inertial frame $\{x, y, z\}$ at each **R** or **F** operation. Therefore, a shortcut is made to generate nodal coordinates by using rotational operations defined with respect to the inertial frame $\{x, y, z\}$. Let the position vector of node i be denoted by $\mathbf{r}_i$. Let the truncation ratio be defined as $h = a/b$. The coordinates of node 1' of square 1'-2'-3'-4' are defined with respect to the inertial frame $\{x, y, z\}$ as:

$$\mathbf{r}_{1'} = \frac{b}{\sqrt{2}} \begin{bmatrix} h & 0 & 1-h \end{bmatrix}^T. \tag{5a}$$

By rotating $\mathbf{r}_{1'}$ with the z-axis in the clockwise direction α according to (2a, b) with $\phi = -\alpha$, the coordinates of node 1 are obtained:

$$\mathbf{r}_1 = \frac{b}{\sqrt{2}} \begin{bmatrix} h\cos\alpha & -h\sin\alpha & 1-h \end{bmatrix}^T. \tag{5b}$$

Let the rotation with respect to the z-axis by $\pi/2$ described in the inertial frame $\{x, y, z\}$ be denoted by $\mathbf{R}_G$ and the flip with respect to $\mathbf{w} = [0]^T$ be denoted by $\mathbf{F}_G$. The coordinates of nodes 2, 3, and 4 can easily be computed by

$$\mathbf{r}_2 = \mathbf{R}_G \mathbf{r}_1, \quad \mathbf{r}_3 = \mathbf{R}_G \mathbf{r}_2, \quad \mathbf{r}_4 = \mathbf{R}_G \mathbf{r}_3. \tag{5c-f}$$

It is noted that the paths in Fig. 3a are neatly identified by **R** and **F** with respect to the body-fixed frame $\{x', y', z'\}$. With respect to the inertial frame $\{x, y, z\}$, however, the paths are described by more complicated combinations of $\mathbf{R}_G$ and $\mathbf{F}_G$. For example, node 1 is moved to node 7 by $\mathbf{F}_G\mathbf{R}_G\mathbf{R}_G$ instead of **RRF**.

At this stage, the nodal coordinates of the truncated octahedral tensegrity module are all defined with respect to the inertial frame $\{x, y, z\}$ and the connection of elements are defined by Fig. 3b. The unconstrained initial equilibrium matrix **A** becomes 72 x 48, and a straightforward application of (1b) yields only numerical solutions for α. To obtain an analytical solution for the twist angle α, the spherical symmetry of the tensegrity module defined by congruent motions is utilized by assuming that the tensegrity elements in the same family take the same pre-stress. Since there are only three families of tensegrity elements including bars, vertical cables, and truncating edge cables, the unknown element forces reduce to $\mathbf{s}' = \{s_b\ s_v\ s_t\}$. Let the equilibrium equation at node 1 be expressed as follows:

$$\mathbf{A}_1 \mathbf{s}' = \mathbf{0}, \tag{6a}$$

where $\mathbf{A}_1$ is a 3 x 3 matrix. The columns of $\mathbf{A}_1$ consist of the direction cosines of the bar and cables emanating from node 1. It is noted that if local frames are erected at all nodes by rotating the local frame at node 1 according to the congruent motions, the same equilibrium equation (6a) holds at all nodes [8]. As a result, under the assumption of the same element forces for the same family of elements, the existence condition of a pre-stress mode (1b) reduces to

$$\det \mathbf{A}_1 = 0. \tag{6b}$$

Let the length of bars, vertical cables, and truncating edge cables be denoted by l_b, l_v, and l_t, respectively. They are defined as follows:

$$l_b = \|\mathbf{r}_{12} - \mathbf{r}_1\|, \quad l_v = \|\mathbf{r}_5 - \mathbf{r}_1\|, \quad l_t = \|\mathbf{r}_2 - \mathbf{r}_1\|. \tag{7}$$

From Fig. 3b the equilibrium equation (6a) at node 1 becomes

$$(\mathbf{r}_{12} - \mathbf{r}_1)\frac{s_b}{l_b} + (\mathbf{r}_5 - \mathbf{r}_1)\frac{s_v}{l_v} + (\mathbf{r}_2 - \mathbf{r}_1 + \mathbf{r}_4 - \mathbf{r}_1)\frac{s_t}{l_t} = \mathbf{0}. \tag{8}$$

The characteristic equation (1b) becomes

$$h^2(-\cos^3\alpha - \sin\alpha\cos^2\alpha - 2\sin\alpha\cos\alpha + \sin\alpha - 1)$$
$$+ h\{2(1+\sin\alpha)\cos\alpha - 2\sin\alpha + 1\} - \cos\alpha + \sin\alpha = 0, \quad (9)$$

with the pre-stress mode:

$$\left\{\frac{s_v}{l_v}, \frac{s_t}{l_t}\right\} = -\frac{s_b}{l_b}\left\{\frac{1+h(-1+\sin\alpha)}{1-h(1+\cos\alpha)}, \frac{1+h(-1+2\cos\alpha+\sin\alpha)}{2h\cos\alpha}\right\}. \quad (10)$$

Figure 4a shows the variation of the twist angle α as a function of the truncation ratio h (= a/b). The corresponding pre-stress mode [$-s_t/s_b$, $-s_v/s_b$] is plotted in Fig. 4b.

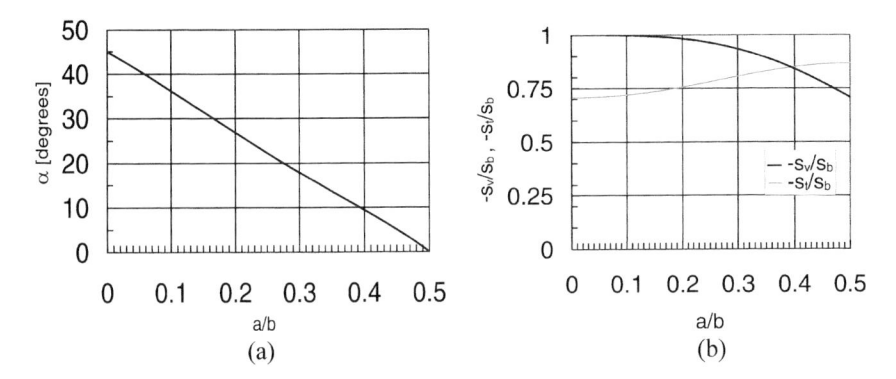

Figure 4. (a) The twist angle α versus the truncation ratio a/b and (b) the pre-stress mode versus the truncation ratio a/b for regular truncated octahedral tensegrity modules.

The pre-stress modes in equation (10) take tension in cables and compression in bars. Figure 4b shows that cable tension is always less than the absolute value of bar compression. Furthermore, at the optimum truncation ratio h = 0.39 and α = 9.9°, the tension in vertical cables and truncating edge cables become equal and the maximum cable tension becomes minimum.

2.3 Modal Analyses of Regular Truncated Octahedral Tensegrity Modules

For a given bar force s_b and twist angle α = 30°, the element forces were determined by using (9) and (10). Modal analyses were conducted for pre-stressed configurations of increasing pre-stress amplitudes |s_b| [7, 8]. For the subsequent numerical examples, steel bars and cables with Young's modulus Y_0 = 200 GPa and mass density $\rho = 7860$ kg/m^3 are considered. The diameter of cables is 0.003m. Bars are hollow circular cylinders with outer and inner radii, 0.011m and 0.009m, respectively. The length of bars is $l_b = 0.5$m.

Table 1. *Natural Frequencies and Algebraic Multiplicity of Regular Truncated Octahedral Tensegrity Modules for Increasing Pre-stress Amplitudes*

$|s_b|$	10[N]	50[N]	100[N]	500[N]	1000[N]
1st	2.46(2)	5.50(2)	7.77(2)	17.38(2)	24.57(2)
2nd	3.79(3)	8.46(3)	11.97(3)	26.76(3)	37.84(3)
3rd	3.82(1)	8.54(1)	12.07(1)	27.00(1)	38.18(1)
4th	4.37(3)	9.76(3)	13.80(3)	30.85(3)	43.60(3)
5th	4.41(1)	9.86(1)	13.94(1)	31.17(1)	44.09(1)
6th	4.45(3)	9.95(3)	14.07(3)	31.46(3)	44.49(3)
7th	4.64(3)	10.38(3)	14.68(3)	32.82(3)	46.41(3)
8th	5.42(2)	12.12(2)	17.14(2)	38.31(2)	54.18(2)
9th	7.13(1)	15.95(1)	22.55(1)	50.42(1)	71.30(1)
10th	596.33(3)	596.38(3)	596.45(3)	597.00(3)	597.69(3)
11th	615.21(3)	615.28(3)	615.38(3)	616.15(3)	617.12(3)
12th	740.98(3)	741.07(3)	741.19(3)	742.10(3)	743.23(3)

Results of the modal analyses are summarized in Table 1. Included are the twelve lowest natural frequencies in Hz. The nine lowest natural frequencies are associated with 19 infinitesimal mechanism modes. Their algebraic multiplicities, shown in parentheses, are also included in Table 1. The natural frequencies of infinitesimal mechanism modes, whose stiffness is due to pre-stress, increase proportionally to the square root of the amplitude of pre-stress. The natural frequencies higher than the ninth correspond to deformation modes which exhibit nonzero element elongation. Therefore, the tenth and higher natural frequencies remain almost constant with increasing pre-stress amplitudes.

Figures 5a, b, and c show one of two modes associated with the first natural frequency. The mode exhibits an ellipsoidal deformation. The orthogonal axes of the ellipsoidal deformation are the diagonals of a regular octahedron. The length of one of the diagonals remains constant, while the remaining diagonals elongate and contract to form an ellipsoid. Figures 5d, e, and f illustrate one of three modes associated with the second natural frequency. In these mode, the length of diagonals do not change. However, they are twisted. The mode associated with the third natural frequency, shown in Figs. 5g, h, and i becomes the spherically symmetric mode without any multiplicity. The mode shapes and number of modes remain essentially the same even if the twist angle was considerably changed, except for minor switching of the orders of higher frequency modes.

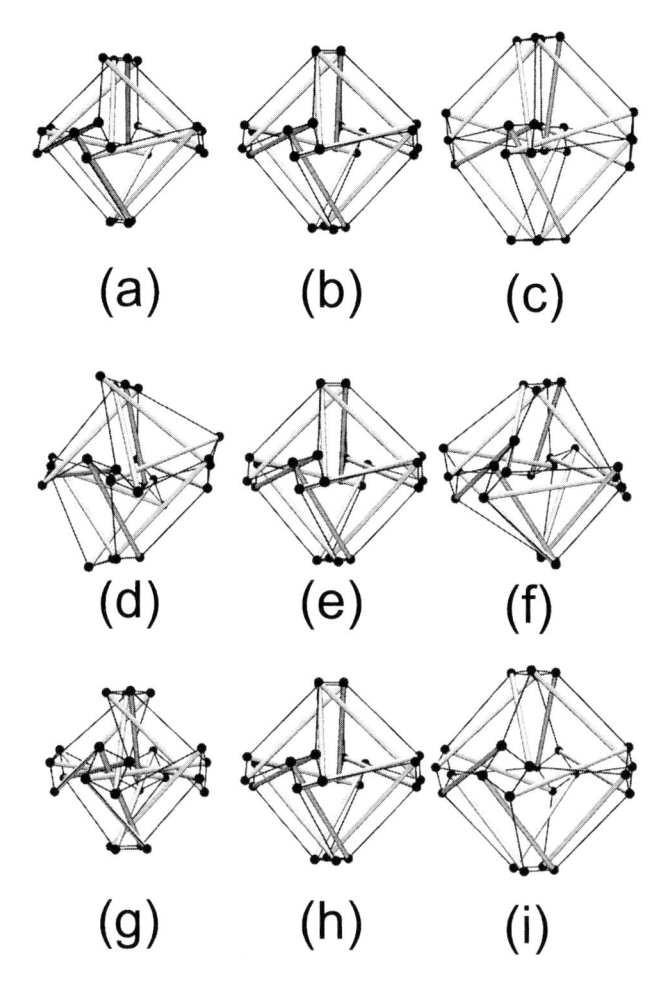

Figure 5 Mode shapes of a regular truncated octahedral tensegrity module associated with the first natural frequency (a, b, c), the second natural frequency (d, e, f), and the third natural frequency (g, h, i).

3. REGULAR TRUNCATED CUBIC TENSEGRITY MODULES

3.1 Maxwell Numbers and the Number of Infinitesimal Mechanisms

If the cube in Fig. 6a is truncated by cutting off each vertex along a plane perpendicular to the radius at the vertex, equilateral triangles are created as shown in Fig. 6b. The truncating equilateral triangles in the tensegrity module in Fig. 6c are twisted by a specific angle α compared to the equilateral triangles in Fig. 6b. It can be easily shown that the numbers of nodes and elements of the cubic tensegrity modules are the same as those of the octahedral tensegrity modules. Consequently, the Maxwell number $Mx = -18$ and the number of infinitesimal mechanisms $n_M = 19$ also remain the same.

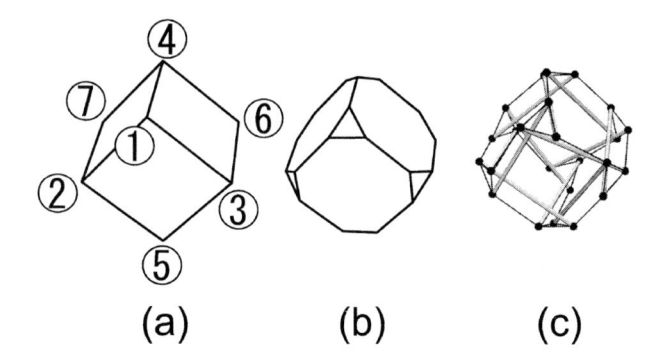

Figure 6. (a) A cube, (b) a regular truncated cube, and (c) a regular truncated cubic tensegrity module.

3.2 Initial Shape Finding

The center of each face of the octahedron corresponds to a vertex of the cube. This indicates that a cube and a regular octahedron are dual. There are 24 congruent configurations of a cube. They form the octahedral group S_4 [9]. The octahedral group also describes the congruent motions of both a regular truncated cube and a regular truncated cubic tensegrity module.

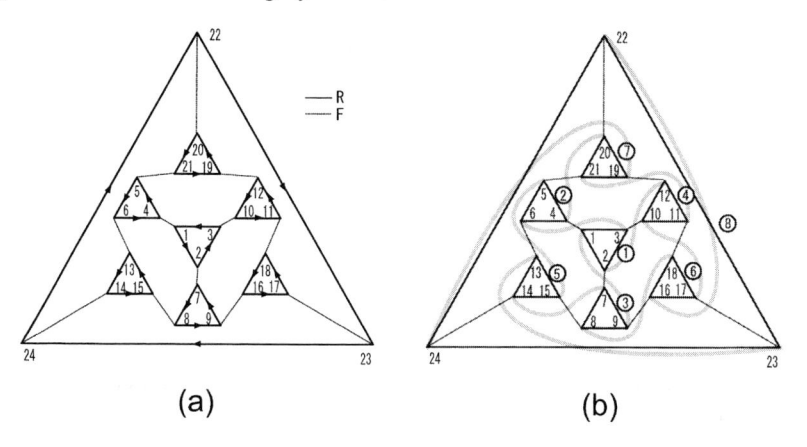

Figure 7. (a) A graph of the octahedral group and (b) a connectivity diagram of a truncated cubic tensegrity module with a positive twist angle.

There are two generators **R** and **F**, which satisfy relation (4). In the cube, the operator **R** becomes the rotation with respect to the axis joining the centers of two facing squares. However, for a truncated cube and a regular truncated cubic tensegrity module, it is convenient to use rotation **R'** with respect the radius through a vertex by angle $2\pi/3$, i.e., **R'R'R' = I**. The flip **F** is defined again as the π rotation with respect to a median connecting a pair of facing edges. Figure 7a shows a graph of the octahedral group, redrawn for the generators **R'** and **F**. The solid line **R'** in Fig. 7a corresponds to a combination of a solid line and a dashed line: **FR** in Fig. 3a. In the graph, nodes represent different configurations of a cube, and edges represent congruent motions.

Due to page limitations, the derivation of nodal coordinates is not shown here. For the nodal coordinates readers are referred to the thesis by Nishimura [10]. Let the length of bars, vertical cables, and truncating edge cables be denoted by:

$$l_b = \|\mathbf{r}_9 - \mathbf{r}_1\|, \quad l_v = \|\mathbf{r}_4 - \mathbf{r}_1\|, \quad l_t = \|\mathbf{r}_2 - \mathbf{r}_1\|. \tag{11}$$

From Fig. 7b, the equilibrium equation (6a) at node 1 becomes

$$(\mathbf{r}_9 - \mathbf{r}_1)\frac{s_b}{l_b} + (\mathbf{r}_4 - \mathbf{r}_1)\frac{s_v}{l_v} + (\mathbf{r}_2 - \mathbf{r}_1 + \mathbf{r}_3 - \mathbf{r}_1)\frac{s_t}{l_t} = \mathbf{0}. \tag{12}$$

Let the truncation ratio be defined as: $h = a/b$ where $\sqrt{2}a$ is the length of the truncating equilateral triangle in Fig. 6b and b is the length of the edges of the cube in Fig. 6a. The characteristic equation obtained from (12) reveals the relationship between the twist angle α and the truncation ratio h:

$$2h^2\{-8\cos^3\alpha - 2\cos^2\alpha + 2(2 - \sqrt{3}\sin\alpha)\cos\alpha + 2\sqrt{3}\sin\alpha - 3\}$$
$$+ 3h\{2\cos^2\alpha + 2(2 + \sqrt{3}\sin\alpha)\cos\alpha - 4\sqrt{3}\sin\alpha + 3\} \tag{13}$$
$$- 9(\cos\alpha - \sqrt{3}\sin\alpha) = 0.$$

The solution of (13) reveals that the twist angle α tends to $3\pi/10$ as $h \to 0$ while α tends to zero as $h \to 1/2$.

The corresponding pre-stress mode satisfies the tensegrity condition with tension in cables and compression in bars. The mode shape becomes

$$\frac{s_v}{l_v} = -\frac{s_b}{l_b}\frac{h(6\cos^2\alpha + 2\sqrt{3}\sin\alpha\cos\alpha - 3)}{3(\cos\alpha - \sqrt{3}\sin\alpha) - 2h(2\cos^2\alpha + \cos\alpha - \sqrt{3}\sin\alpha)}, \tag{14a}$$

$$\frac{s_t}{l_t} = -\frac{s_b}{l_b} \frac{1}{9(\cos\alpha - \sqrt{3}\sin\alpha) - 6h(2\cos^2\alpha + \cos\alpha - \sqrt{3}\sin\alpha)}$$
$$\times [3(3\cos\alpha - \sqrt{3}\sin\alpha) \qquad (14b)$$
$$+ 2h\{-9\cos^2\alpha - (3+\sqrt{3}\sin\alpha)\cos\alpha + \sqrt{3}\sin\alpha + 3)\}].$$

The numerical results for natural frequencies and mode shapes, similar to those in Section 2.3, may be found in [10].

4. CONCLUSIONS

In order to define the nodal coordinates and connectivity of regular octahedral and cubic tensegrity modules, the congruent motions of the octahedral group were utilized. By introducing a reduced equilibrium matrix, the initial configurations and pre-stress modes were analytically obtained for regular truncated octahedral and cubic tensegrity modules. Furthermore, modal analyses were performed to classify nineteen infinitesimal mechanism modes into subspaces based upon natural frequencies. Spherically symmetric modes appear without algebraic multiplicity.

5. REFERENCES

1. Marks, R. and Fuller, B. R. (1973) *The Dymaxion World of Buckminster Fuller*, Anchor Books, Garden City, NY.
2. Pugh, A. (1976) *An Introduction to Tensegrity*, University of California Press, Berkeley, CA.
3. Ingber, D. E. (1998) The Architecture of Life, *Scientific American*, January 48-57.
4. Maxwell, J. C. (1864) On the Calculation of the Equilibrium and Stiffness of Frames, *Philosophical Magazine* **27**, 250-256.
5. Calladine, C. R. (1978) Buckminster Fuller's "Tensegrity" Structures and Clerk Maxwell's Rules for the Construction of Stiff Frames, *International Journal of Solids and Structures* **14**, 161-172.
6. Pellegrino, S. and Calladine, C. R. (1986) Matrix Analysis of Statically and Kinematically Indeterminate Frameworks, *International Journal of Solids and Structures* **22** 409-428.
7. Murakami, H. and Nishimura, Y. (2001) Static and dynamic characterization of some tensegrity modules, *Journal of Applied Mechanics* **68** 19-27.
8. Murakami, H. and Nishimura, Y. (2001) Static and dynamic characterization of regular truncated icosahedral and dodecahedral tensegrity modules, *International Journal of Solids and Structures* **38** 9359-9381.
9. Grossman, I., and Magnus, W. (1964) *Groups and Their Graphs*, The Mathematical Association of America, Washington.
10. Nishimura, Y. (2000) *Static and Dynamic Analyses of Tensegrity Structures*, Ph.D. Thesis, University of California at San Diego, La Jolla, CA.

SHAPE AND STRESS CONTROL IN ELASTIC AND INELASTIC STRUCTURES

YURIY NYASHIN and VALENTIN KIRYUKHIN

Department of Theoretical Mechanics, Perm State Technical University, Komsomolskii prospect, 29 a, 614600, Perm, Russia
E-mail: nyashin@theormech.pstu.ac.ru

1. INTRODUCTION

Problems of stress and strain control arise in many important technical problems. Such severe problems take place in space constructions (platforms, antennas, telescopes, etc.) which are inevitably exposed to large temperature gradients. The lightweight and flexible structures respond with considerable shape distortion. Currently there is much interest in the possibility of designing large-diameter telescopes and antennas with one of the more stringent requirements of surface accuracy. One of the major problems to be solved in this context involves the proper design, fabrication, and maintenance of the telescope's primary mirror with a diameter of one or more meter, assuming the surface geometry to be controlled to a root mean square tolerance of about 10^{-8} m. Similarly, the parabolic surface of a microwave antenna of the order of 100 m in diameter must be maintained at a tolerance of less than 1 mm [1, 2, 3, 4]. To minimize those unwanted effects twofold strategies are developed: in the stage of manufacturing an optimal design possibly using graded material and specifically designed residual stress distribution, in the sense of a "feedforward control" should be available, accompanied by a sensor-actuator based optimal feedback control during the life-time of the structure.

Technological processes of material production and, e.g., welding of structure, are commonly accompanied by (unwanted) residual stresses. They influence the characteristics of the finished workpieces in different ways. In many cases, the residual stress deteriorate the strength properties of structures, reduce their stability limit, distort their form, etc. Such examples are well known. On the other hand, specifically created residual stresses can improve the workpiece characteristics. Among them are such processes as autofrettage, shot peening, preliminary plastic

---

The paper is dedicated to Professor Franz Ziegler on the occasion of 65th birthday.

tension of rods and tubes, creation of compressive residual stresses on coatings and others [5, 6].

It is interesting to note that problems of shape and stress control arise in living tissues also. Many orthopedic appliances are used in medical treatment, e.g. for treatment of congenital maxillary defects. Here, an optimal scheme of the design is very important for the choice of the individual orthopedic apparatus of every patient and refinement of existing treatment methods [7].

This investigation considers the general statement and way of solution of problem of strain and stress control in anisotropic inhomogeneous elastic and inelastic structures. Elaborated approach can be used equally to living and nonliving tissues.

The work is separated logically in two parts.

First, the quasi-static mathematical model is proposed that is based on the generalized solution of the thermoelastic initial boundary-value problem. Thermoelastic properties of the structure can be anisotropic and inhomogeneous, e.g. for a graded material. For theses problems the existence and uniqueness of generalized solutions are proven [8].

The following problems based on the action of thermal strain, or piezostrain, or other types of eigenstrain and/or variation of material properties are considered:
1. To obtain prescribed stress (in particular residual stress) distribution.
2. To obtain prescribed strain distribution.
3. To obtain prescribed displacements in a subregion of the structure.

All prescribed fields must be admissible in linear elasticity. Some combinations of the above control problems can be accepted.

Second, the control of quasi-static strain and stress in structures with inelastic effects is considered. In this case the loading history has to be taken into consideration, i.e. change of structural configuration and strain accumulation have to be taken into account.

To solve the problem the process of plastic deformation (growth in living tissues) is separated by steps of time. In each time-step the control problem is solved under the condition of small increment of the total, elastic, thermal and plastic (growth) strain. The configuration of a solid after one time-step defines the initial configuration for the following small time interval. Stepwise solution represents the consequent solutions of the problem of controlled plastic deformation.

Many scientists of different countries (S. Timoshenko, W.Nowacki, A.Kovalenko, B. Boley, J. Weiner, J. Podstrigach, J. Burak, etc.) investigated theoretical aspects of thermal stress and strain modeling. In particular, the question on thermal heating, which does not produce thermal stress was studied (recently such thermal strain was named as impotent). Scientists of Vienna Technical University (E. Melan, H. Parkus, F. Ziegler, H. Irschik, F. Rammerstorfer, etc.) made the great contribution to the development of the theory of mathematical modeling and optimal control of thermal stress, strain and displacement. Since the 30th of the past century some monographs devoted to modeling of thermal stress in thermoelasticity have been published [9, 10, 11]. Later the new approach for determination of thermal stress, strain and displacement was elaborated. This theory was founded on Maysel's method, 1941 [12]. Along with the concept of impotent thermal strain the new concept of nilpotent thermal strain was introduced. In this case the total strain vanishes. The important results in thermal stress and strain control by thermal and piezoactuation were received by H. Irschik and F. Ziegler [13, 14, 15].

Some authors investigated the problem of residual stress optimization. This problem for metal plates and shells during their thermoelastic heating has been posed and solved in [16, 17]. The elastic strain energy has been used as an objective function, force and temperature loads have been chosen as control parameters. The power of the heat sources in a body's volume is introduced as an objective function in [18], thereby the solution results in the identification of residual stresses which minimize the energy expenditure. Analogous solution have been found for viscoelastic bodies [19]. Melan's shakedown theorem, which demonstrates the elastic shakedown of a system [20], is another example of the usage of a favorable residual stress distribution.

Mechanists of Perm (Russia) elaborated a novel approach of residual stress control in problems of thermoelastoplasticity (A. Pozdeev, Y. Nyashin, P. Trusov) which is founded on the concepts of Hilbert spaces. Using this method the problems of residual stress control in metal treatment were solved (hot shape rolling, wire drawing, etc.).

Since 1995 common research of scientists of Vienna, Perm and Linz Technical Universities in the field of modeling and control of thermal stress and strain has been carried out initially according to the agreement on collaboration between the Universities and later using support of INTAS-ESA grant (project coordinator Professor Franz Ziegler). Some results of this collaboration are included in this article.

2. POSING THE PROBLEM

Let the solid being investigated occupy a region V in three-dimensional Euclidian space E^3, the region boundary being S ($\overline{V} = V \cup S$). If inelastic strains $\tilde{\varepsilon}^*$ are prescribed, the symmetric stress tensor $\tilde{\sigma}$ and the strain tensor $\tilde{\varepsilon}$ are found by solving the quasi-static boundary-value problem (time can be considered as a parameter):

$$Div\tilde{\sigma} + \mathbf{Q} = 0, \quad \mathbf{x} \in V, \tag{1}$$

$$\tilde{\sigma} = \tilde{C} \cdot \cdot (\tilde{\varepsilon} - \tilde{\varepsilon}^*), \quad \mathbf{x} \in \overline{V}, \tag{2}$$

$$\tilde{\varepsilon}(\mathbf{u}) = \frac{1}{2}(\nabla \mathbf{u} + \mathbf{u}\nabla), \quad \mathbf{x} \in \overline{V}, \tag{3}$$

$$\mathbf{u} = 0, \quad \mathbf{x} \in S_u, \tag{4}$$

$$\mathbf{n} \cdot \tilde{\sigma} = \mathbf{P}, \quad \mathbf{x} \in S_\sigma. \tag{5}$$

Here $\tilde{\varepsilon}^* \in (C^1(\overline{V}))^6$ are the components of the tensor of elastic coefficients $C_{ijkl} \in C^1(\overline{V})$, $\tilde{\sigma} \in (C^1(\overline{V}))^6$, $\mathbf{u} \in (C^2(\overline{V}))^3$. Equations (1) and (5) reflect the static equilibrium conditions, Eq. (2) is Hooke's law, relating the stress and elastic strain, Eq. (3) is linearized strain-displacement relation, Eq. (4) – boundary condition on immovable supports. Here the notations $(C^2(\overline{V}))^3$, etc. designate the space $C^2(\overline{V})$ for every component of a vector.

The tensor $\tilde{\sigma}$ defined as a result of the solution of the system Eqns. (1)-(5) is called the classical solution of the problem of stress (in particular residual stress if $\mathbf{Q} = \mathbf{P} = 0$) determination.

It should be noted that it is very difficult (sometimes even impossible) to find the classical solution of most practical problems. Therefore a generalized solution of the problem is introduced.

The stress field $\tilde{\sigma}$ is the generalized solution of the thermoelasticity problem if it satisfies the relation

$$\tilde{\sigma} = \tilde{C} \cdot \cdot (\tilde{\varepsilon}(\mathbf{u}) - \tilde{\varepsilon}^*) \tag{6}$$

where $\mathbf{u} \in (W_2^1(V))^3$, $\mathbf{u} = 0$, $\mathbf{x} \in S_u$ and

$$\int_V \tilde{\sigma} \cdot \cdot \tilde{\varepsilon}(\mathbf{w}) \, dV - \int_{S_\sigma} \mathbf{P} \cdot \mathbf{w} \, dS - \int_V \mathbf{Q} \cdot \mathbf{w} \, dV = 0,$$

$$\forall \mathbf{w} \in (W_2^1(V))^3, \quad \mathbf{w} = 0, \quad \mathbf{x} \in S_u. \tag{7}$$

Here $W_2^1(V)$ is Sobolev functional space of square-summable functions with generalized square-summable derivatives; the norm in $W_2^1(V)$ is defined by

$$\|u\|_{W_2^1(V)} = \left[\int_V u^2(\mathbf{r}) \, dV + \int_V \sum_{k=1}^3 \left(\frac{\partial u}{\partial x_k} \right) dV \right]^{\frac{1}{2}}.$$

Here $\tilde{\varepsilon}(\mathbf{u})$ and $\tilde{\varepsilon}(\mathbf{w})$ are both defined through the linearized geometric relations (3) where the derivatives are considered to be generalized. It can be shown that the classical solution is the generalized solution and, in the case of adequate smoothness, the generalized solution becomes the classical solution. The uniqueness and existence of generalized solution is proven by means of theorem 3.5 in [8].

3. BASIC THEOREM OF THERMAL STRESS CONTROL

We suppose that the optimal (in some sense) distribution of thermal stress is known for the system under investigation. This optimal problem can be solved if we know loads acting upon the system. The aim of control is to generate this prescribed distribution of thermal stress.

It is important to stress that any other inelastic strain can be considered instead thermal strain. We say thermal strain for the sake of definiteness only.

This problem is formulated in terms of an optimization problem of solid mechanics. The prescribed thermal stress distribution results from a stationary temperature field. It is interesting to note that the proposed approach allows one to determine optimal control parameters without solving a problem of thermoelasticity. Solely, the appropriate problem of heat transfer has to be solved.

We introduce the space H_u of symmetric and compatible tensors of the second rank that are defined in the region $\overline{V}$; that is if $\tilde{f} \in H_u$ then $\tilde{f}$ can be represented in a form

$$\tilde{f} = \frac{1}{2}(\nabla \mathbf{u} + \mathbf{u}\nabla) \tag{8}$$

where the derivatives are considered to be generalized and $\mathbf{u} = 0$, $\mathbf{x} \in S_u$. Then we formulate the following theorem.

Theorem 1. Let $\tilde{\sigma}_0$ be a symmetric tensor of the second order that satisfies Eq. (7) (that stress tensor can be named as statically admissible). Further we introduce the tensor $\tilde{f} = \tilde{\varepsilon}^T + \tilde{C}^{-1} \cdot \cdot \, \tilde{\sigma}_0$. The condition $\tilde{f} \in H_u$ is the necessary and sufficient one such that the stress $\tilde{\sigma} = \tilde{\sigma}_0$ has been created in a solid occupying a region $\overline{V}$ in space E^3.

Once more we note that any inelastic strain tensor $\tilde{\varepsilon}^*$ can be considered instead of the thermal strain tensor $\tilde{\varepsilon}^T$.

The proof of theorem 1 is performed in two steps.

Proof of necessity. Let $\tilde{\sigma} = \tilde{\sigma}_0$; then considering the positive definiteness of matrix $\tilde{C}$, it can be concluded from Eq. (6):

$$\tilde{f} = def\,\mathbf{u} = \tilde{\varepsilon}^T + \tilde{C}^{-1} \cdot \cdot \, \tilde{\sigma}_0,$$

and $\tilde{f} \in H_u$.

Proof of sufficiency. Let $\tilde{f} = \tilde{\varepsilon}^T + \tilde{C}^{-1} \cdot \cdot \, \tilde{\sigma}_0 \in H_u$, where tensor $\tilde{\sigma}_0$ is statically admissible, i.e. it satisfies Eq. (7). Then considering the uniqueness of solution of problem (6, 7) it is concluded that $\tilde{\sigma} = \tilde{\sigma}_0$ almost everywhere.

The sense of this theorem is close to the Castigliano's principle. The solution of the boundary-value problem is reduced to the search of compatible strain. The difference from the known principle is the consideration of generalized solutions taking into account inelastic strain.

Thus it can be shown that in order to obtain a thermal stress distribution the thermoelasticity process should be controlled in such a manner that a compatible strain $\tilde{\varepsilon}^T + \tilde{C}^{-1} \cdot \cdot \, \tilde{\sigma}_0$ in the solid under prescribed thermal stress results.

4. COROLLARIES OF THEOREM 1

Five corollaries can be derived from theorem 1. Corollary 1 is connected with the problem of impotent temperature field. Which temperature does not influence the stress in a solid? Many authors discuss this problem in the case of a body which is free to expand.

<u>Corollary 1.</u> The condition $\tilde{\varepsilon}^T \in H_u$ is the necessary and sufficient one that the stress vanishes in a solid, i.e. $\tilde{\sigma}_0(\mathbf{r}) = 0$, $\forall \mathbf{r} \in \overline{V}$.

Proof. The corollary is derived immediately from theorem 1. Let the stress tensor be prescribed over the volume as $\tilde{\sigma}_0(\mathbf{r}) = 0$, $\forall \mathbf{r} \in \overline{V}$. The tensor $\tilde{f}$ is simplified to the expression $\tilde{f} = \tilde{\varepsilon}^T$. So the sufficient and necessary condition to obtain the desired zero stress is the condition $\tilde{f} = \tilde{\varepsilon}^T \in H_u$.

<u>Corollary 2.</u> Let the components of the strain tensor have the second derivatives with respect to the spatial Cartesian coordinates, i.e. $\tilde{\varepsilon}^T \in (C^2)^6$. The only linear temperature dependence on space Cartesian coordinates

$$T(x, y, z, t) = a_1(t)x + a_2(t)y + a_3(t)z + a_4(t) \tag{9}$$

does not give rise to the thermal stress in thermally homogeneous and isotropic unsupported solid. It follows from the compatibility equations and corollary 1. For the supported body it is the necessary condition only.

This result was received earlier by different authors.

<u>Corollary 3.</u> The compatibility equations of $\tilde{\varepsilon}^T$ are the necessary and sufficient conditions that the thermal stress vanishes in a statically determined body. It is only a necessary condition if a statically undetermined body is analyzed.

Proof. The thermal stress vanishes if the thermal strain is compatible and the displacement vanishes at the supports, or at the points of surface S_u. The displacement determined by strain is not unique. Permissible displacements can only differ by rigid body motion. It means that the displacement expression contains the constants describing a rigid body motion. Their number equals the number of degrees of freedom of the body as rigid body. (3-in 2D motion, 6-in 3D motion) and equals the number of constraints of supports in statically determined system. Therefore, such constants can be determined that displacements at supports vanish. For redundant (statically undetermined) systems there are displacements which can not vanish with any constants.

Next corollaries do not to relate to the problem of impotent temperature but they are very useful in algorithm of solving the problems of the control and the optimal design.

<u>Corollary 4.</u> If the stress $\tilde{\sigma}(\mathbf{r}) = \tilde{\sigma}_0(\mathbf{r})$ arises in region $\overline{V}$ then $\tilde{f} \in H_u$ and for the total strain $\tilde{\varepsilon}$ the relation takes place: $\tilde{\varepsilon} = \tilde{f}$, $\forall \mathbf{r} \in \overline{V}$.

<u>Corollary 5.</u> Any strain $\varepsilon_{add} \in H_u$ imposing to the body does not give rise to the change of the stress tensor in a body.

5. THEOREM 2

The next theorem is formulated to express complimentary properties of strain entailed no stress. The proof of theorem 2 is omitted here.

Theorem 2. The strain tensor $\tilde{\varepsilon}$ does not change the stress (or the strain tensor $\tilde{\varepsilon}$ belongs to space H_u) if and only if there exist such volume force **Q** and surface traction **P** which produce the strain $\tilde{\varepsilon}^F$ equaled $\tilde{\varepsilon}$:

$$\tilde{\varepsilon}^F(\mathbf{r}) = \tilde{\varepsilon}(\mathbf{r}), \quad \forall \mathbf{r} \in \overline{V}. \tag{10}$$

Considering corollaries 1 and 5 (theorem 1) we can say that the strain does not result in the stress if and only if the tensor of the strain belongs to space H_u. Therefore we can say now that $\tilde{\varepsilon} \in H_u$ if and only if there exist such volume force **Q** and surface traction **P** which produce the strain $\tilde{\varepsilon}^F$ equal to $\tilde{\varepsilon}$, $\forall \mathbf{r} \in \overline{V}$.

This theorem allows to solve very important problem of construction of the space H_u in practical problems. The space H_u is an infinite-dimensional functional space. The finite-dimensional subspace has to be constructed when practical problem is analyzed. The subspace is characterized by a set of basic elements. The basic elements determination is not an easy problem in general. But theorem 2 allows to elaborate a common approach to determine the basis with finite dimension. Every basic element is determined by imposing an arbitrary force field. The chosen actual force distributions produce basic elements. Then the procedure of orthogonalization is realized. Created basis is utilized for solving the control problems.

6. OPTIMIZATION OF UNIAXIAL THERMAL STRESS DISTRIBUTION

We introduce the objective function of the optimization problem of thermoelasticity that allows us to determine optimal control parameters in the case of an uniaxial stress state for systems of bars [21]. The statically determinate structure is considered. This situation is met quite often in applications. In particular, as experiments demonstrate, longitudinal thermal stress is considerably larger than the other components of the stress tensor in the case of shaping by hot rolling (e.g. I-beams and rails). We introduce again the tensor $\tilde{f}$ as

$$\tilde{f} = \tilde{\varepsilon}^T + \tilde{C}^{-1} \cdot \cdot \tilde{\sigma}_0. \tag{11}$$

Below we shall prove that this tensor can be used to evaluate the deviation of actual thermal stress from prescribed one. In the uniaxial stress state only components σ_{xx} and σ_{0xx} differ from zero. With respect to strain, $\varepsilon_{ij} = 0$ if $i \neq j$.

It is natural to consider that the deviation between σ_{xx} and σ_{0xx} is determined by the incompatibility of the strain components with indices "xx".

Let f_{xx}^m denote the volumetric mean value of f_{xx} then the centered strain $\Delta f = f_{xx} - f_{xx}^m$ becomes

$$\Delta f = f_{xx} - f_{xx}^m = \varepsilon_{xx}^T + \frac{\sigma_{0xx}}{E} - \frac{1}{\Omega}\int_V (\varepsilon_{xx}^T + \frac{\sigma_{0xx}}{E}) dV . \qquad (12)$$

Here Ω is the volume of the region $\overline{V}$. The following conditions should be taken into account:
1. Naturally $\int_V \Delta f_{xx} dV = 0$.
2. The condition of the equilibrium : $\int_V \sigma_{0xx} dV = P_0$, $\int_V \sigma_{xx} dV = P$, $P = P_0$.
3. The compatibility of strain of the form $f_{xx} = const$.

Hence, using the generalized Hooke's law,

$$\sigma_{xx} = E(\varepsilon_{xx} - \varepsilon_{xx}^T) = E(\varepsilon_{xx} - f_{xx} + \frac{\sigma_{0xx}}{E}),$$

the norm becomes

$$\|(\sigma_{xx} - \sigma_{0xx})\|^2 = \int_V (\sigma_{xx} - \sigma_{0xx})^2 dV = (\varepsilon_{xx} - f_{xx}^m)(P - P_0) -$$

$$- E^2 \int_V \Delta f_{xx}(\varepsilon_{xx} - f_{xx}^m - \Delta f_{xx}) dV = E^2 \int_V (\Delta f_{xx})^2 dV. \qquad (13)$$

Using the definition

$$\|A\| = \sqrt{\int_V A^2 dV} \qquad (14)$$

the result can be written in the form

$$\|(\sigma_{xx} - \sigma_{0xx})\| = E \|(\Delta f_{xx})\|. \qquad (15)$$

Thus the objective function of the problem of obtaining the prescribed thermal stress distribution (in the uniaxial stress state for system of bars) is rewritten as

$$\Phi(\varepsilon_{xx}^T) = E \|\Delta f_{xx}\|. \qquad (16)$$

The values of the control parameters for the thermoelastic deformation should be chosen such that

$$\Phi(\varepsilon_{xx}^T) \to \inf . \qquad (17)$$

If $\Phi(\varepsilon_{xx}^T) = 0$, then from Eqs. (16) and (17) it follows $\sigma_{xx} = \sigma_{0xx}$ almost everywhere in $\overline{V}$ and the problem is solved. Due to existing restrictions such an ideal solution usually can not be achieved. However, any reduction of $\Phi(\varepsilon_{xx}^T)$ means that σ_{xx} approaches $\sigma_{\alpha xx}$.

The theory of shape and stress control suggested in this article can be used for different inhomogeneous and anisotropic, living and nonliving systems.

7. FAVOURABLE RESIDUAL STRESSES IN HOT ROLLED I-BEAMS

Forced water-air cooling of the profile surfaces is used as a controlling influence in the problem.

This cooling is made after the beam rolling has been completed. The control function to be found is the coefficient of heat emission that is sought as a function of the coordinates and time.

While solving the problem the assumption is made that stresses and strains are absent after rolling is complete, but that there is a considerable temperature gradient through the beam's cross-section that approaches 150-200 K in large beams. As thermal strains that correspond to the cooling-down from the rolling completion temperature to the environmental temperature are known, the prescribed residual stress distribution can only be achieved with the help of a favorable plastic strain distribution.

Residual stresses developing in beams during their natural cooling-down in a cooler are usually of an unfavorable character. Longitudinal residual stresses can be up to 75% of the yield limit and can considerably reduce the buckling resistance of the profiles (by 1/3) if they are used as columns. However, residual stress distributions can be found for which the profile buckling resistance is even increased. This work is not aimed at finding optimal residual stress distributions.

The aim of this part is to find thermoelastoplastic deformation control resulting in residual stresses that approach the prescribed ones.

In the existing rolling technology the longitudinal residual stresses at characteristic points of the beam's cross-section are equal to: $\rho_1 = -135$ MPa (the web center), $\rho_2 = 155$ MPa (the region center shift from the web to the flanges), $\rho_3 = -130$ MPa (the web edge). The prescribed residual stress distribution is: $\rho_1 = -65$ MPa, $\rho_2 = -23$ MPa, $\rho_3 = 125$ MPa and that differs qualitatively from the existing one. A residual stress distribution of such a character increases the beam stability.

Some results from the solution of the problem of controlling the residual stresses are shown in Figure 1 and Figure 2. Figure 1 illustrates the dimensions of cross-section of the I-Beam. The following notation is used in Figure 2: solid lines (———) designate the residual stresses during natural cooling-down; dot-and-dash lines (—·—) designate the prescribed residual stress distribution; dashed lines (------) designate the residual stress distribution obtained by solving the control problem.

The proposed theory allows us to find the variation with time of the heat emission coefficient (control parameter) in the process of forced cooling-down. The solution was made with the help of the objective function Φ (17) where plastic strain was included instead thermal strain. The plastic strain was found in each time-step from the boundary-value problem of thermoelasoplasticity. Details of the solution were described in [5, 21].

As a result, the following residual stresses have been obtained: $\rho_1 = -10$ MPa, $\rho_2 = -30$ MPa, $\rho_3 = 90$ MPa. They coincide qualitatively with the prescribed residual stresses. The relay character o the variation change of the heat emission coefficient in the process of control was found. This regime is especially favorable for a rapid creation of necessary plastic strains.

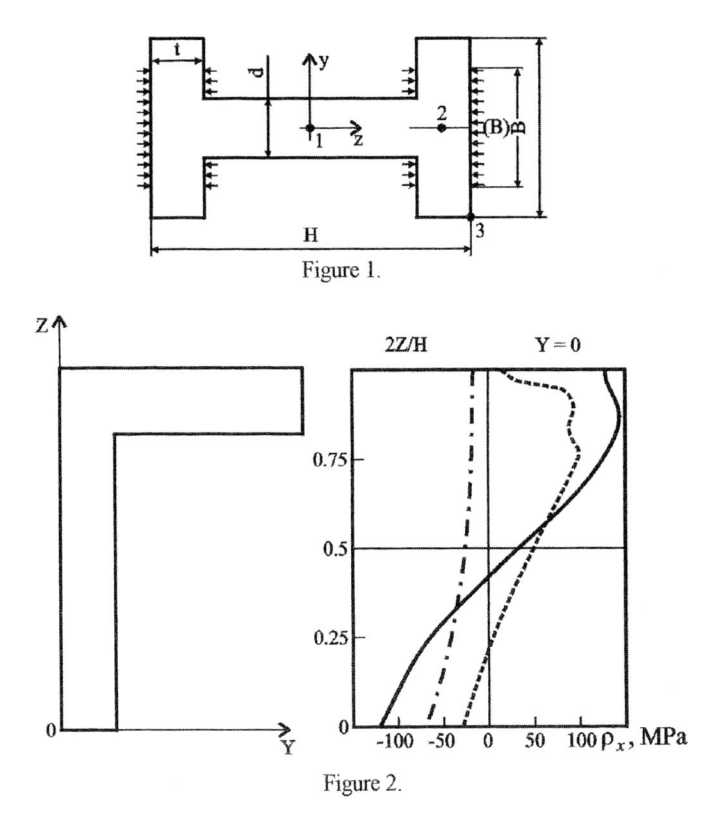

Figure 1.

Figure 2.

The analysis of the results obtained indicates that introducing control immediately after the rolling has been completed is more preferable when the beam temperature is high and the yield limit is still low. Therefore it can be shown that the prescribed residual stress distribution is achieved with adequate accuracy. The using regulated forced cool-down is one of the real technological methods which is applied in practice. Detailed description of other examples can be found in the papers [5, 6, 7, 21]. In [21] some examples of thermal stress and strain control in thermoelasticity are presented. In [6] the residual stress control in technological processes of thermoplastic deformation are discussed. In [7] the application of the theory to living tissues is presented.

In conclusion the authors want to stress that the problem of investigation and control of residual stresses in different organs and tissues of a human organism is very interesting and perspective. We hope that the proposed theory can be useful for analysis of residual stresses and search for their optimal distribution.

8. CONCLUSIONS

The theorems describing the necessary and sufficient conditions for obtaining the prescribed thermal stress and strain distribution in a solid during the thermoelastoplastic deformation have been proven. The problem of the control of thermal stress distribution is posed as an optimization problem in the mechanics of deformable solids. In order to solve the problem an objective function describing the thermal stress level in terms of current parameters of the process has been formulated and the algorithm for searching for an optimal solution has been developed.

The authors gratefully acknowledge a grant from INTAS-ESA, number 99-10 00185 on «Optimal design of space structures: stress and strain control», project coordinator Professor Franz Ziegler. We further acknowledge fruitful discussion with the coordinator in cause of development of the paper. This research was supported also by a grant of Russian Foundation for Basic Research (Ural), number 02-01-96416.

9. REFERENCES

1. Bushell, D. (1979) Control of Surface Configuration by Application of Concentrated Loads, *AIAA Journal* **17** 71-77.
2. Haftka, P.T. and Adelman, H.M. (1985) An Analytical Investigation of Static Shape Control of Large Space Structures by Applied Temperature, *AIAA Journal* **23** 450-457.
3. Austin, F., Rossi, M.J., Van Nostrad, W., Knowles, G., Jameson, A. (1994) Static Shape Control of Adaptive Wings, *AIAA Journal* **32** 1895-1901.
4. Sunar, M. And Rao, S.S. (1997) Thermopiezoelectric Control Design and Actuator Placement, *AIAA Journal* **35** 534-539.
5. Pozdeev, A.A., Nyashin, Y.I. and Trusov, P.V. (1982) *Residual Stresses: Theory and Applications,* Nauka, Moscow.
6. Nyashin, Y.I. and Ilialov, O.R. (1995) Optimization Problem for Obtaining a Prescribed Residual Stress Distribution: Formulation and Solution, *International Journal of Mechanical Sciences* **37** 485-493.
7. Kiryukhin,V.Y. and Nyashin, Y.I. (2002) Application of Stress and Strain Control Theory to Living Tissues (in this issue).
8. Duvant, G. and Lions, T.-L. (1972) *Les Inequations en Mecanique et en Physique*, D. Dunod, Paris.

9. Boley, B.A. and Weiner, J.H. (1960) *Theory of Thermal Stresses*, Wiley, New York.
10. Parkus, H. (1976) *Thermoelasticity,* Springer, Vienna, New York.
11. Ziegler, F. and Irschik, H. (1987) Thermal Stress Analysis Based on Maysel's Formula, in *Thermal Stress* II, edited by Hetnarcki R.B., Elsevier, Amsterdam, p. 120-128.
12. Maysel, B.M. (1941) Generalization of Betti's-Maxwell Theorem to the Thermal State and Some Applications, *Dokladi of Academy of Science of the USSR* **30** 115.
13. Irschik, H and Ziegler, F. (2001) Eigenstrain without stress and static shape control of structures, *AIAA Journal* **39** 1985-1999
14. Irschik, H. and Pichler, U. (2001) Dynamic Shape Control of Solids and Structures by Thermal Expansion Strains, *Journal of Thermal Stresses* **24** 565-576.
15. Irschik, H. (2002) A Review on Static and Dynamic Shape Control of Structures by Piezoelectric Actuation, *Engineering Structures* **24** 5-11.
16. Shablii, O.N. and Zaretsky, V.I. (1981) Optimal Control of the Stress-Strain State of a Disk, *Soviet Applied Mechanics* **17** 755-759.
17. Shablii, O.N. and Garapyuk, T.R. (1987) Optimizing the Process of Residual Stress Removal in Solids, *Soviet Applied Mechanics* **23** 678-682.
18. Grigoluk, E.I., Podstrigach, .Y.S. and Burak Y.I. (1979) *Optimization of Shells and Plates Heating,* Naukova Dumka, Kiev.
19. Weistman, Y. Residual Thermal Stresses Due to Cool-Down of Epoxy-Resin Composites, *Journal of Applied Mechanics* **46** 563-567.
20. Koiter, W. *General Theorems for Elastic-Plastic Solids* (1961) North-Holland, Amsterdam.
21. Nyashin, Y.I., Kiryukhin, V.Y. and Ziegler, F. (2000) Control of Thermal Stress and Strain, *Journal of Thermal Stresses* **23** 309-326.

TRANSIENT PIEZOTHERMOELASTICITY FOR A CYLINDRICAL COMPOSITE PANEL

YOSHIHIRO OOTAO and YOSHINOBU TANIGAWA
Department of Mechanical Systems Engineering,
Graduate School of Engineering, Osaka Prefecture University,
1-1 Gakuen-cho, Sakai, 599-8531 Japan
E-mail: ootao@mecha.osakafu-u.ac.jp

1. INTRODUCTION

Recently smart composite materials composed piezoelectric materials have received attention. A basic element of these smart composite materials is a laminated piezoelectric structure. One of cause of damage in this laminated piezoelectric structure includes delamination. In order to evaluate this phenomenon, it is necessary to analyze the piezothermoelastic problems taking into account the transverse stress components. Since cylindrical panels and cylindrical shells are used in various industrial fields as structure elements, piezothermoelastic analysis of these structural models becomes important as well as those of plate models. Then, there are a few exact analyses concerned with the piezothermoelastic problems of laminated cylindrical panels and shells taking into account transverse stress components [1-4]. These papers, however, treated only the piezothermoelastic problems under the steady temperature distribution. To the author's knowledge, the exact analysis for a transient piezothermoelastic problem of laminated cylindrical panel has not been reported.

In the present article, we have treated exactly the transient piezothermoelastic problem of a simply supported cylindrical composite panel due to a nonuniform heat supply in the circumferential direction.

2. ANALYSIS

We consider an infinitely long, angle-ply laminated cylindrical panel to which a piezoelectric layer of crystal class mm2 is perfectly bonded, the length of the side in the circumferential direction of which is denoted by θ_0. The combined panel's inner and outer radii are designated a and c, respectively. Moreover, b is the coordinate of interface between the angle-ply laminate and the piezoelectric layer. Throughout this article, the quantities with subscripts $i=1,2,\cdots,N$ and $i=N+1$ denote those for ith layer of the angle-ply laminate and piezoelectric layer, respectively. It is assumed that each layer of the angle-ply laminate maintains the orthotropic material properties and the fiber direction in the ith layer is alternated with ply angle ϕ_i to the z axis. It is assumed that the principal axes of the piezoelectric layer are parallel to the axes of the cylindrical coordinate, and the piezoelectric layer is poled in the radial direction.

2.1. Heat Conduction Problem

It is assumed that the combined panel is initially at zero temperature and is suddenly heated partially from the free surface of angle-ply laminate by surrounding media, the temperature of which is denoted by the function $T_a f_a(\theta)$. The relative heat transfer coefficients on the inner and outer surfaces of the combined panel are designated h_a and h_c, respectively. We assume that the edges of the combined cylindrical panel are held at zero temperature. The transient heat conduction equation for the ith layer and the initial and thermal boundary conditions in dimensionless form are taken in the following forms:

$$\frac{\partial \bar{T}_i}{\partial \tau} = \bar{\kappa}_{ri}\left(\frac{\partial^2 \bar{T}_i}{\partial \rho^2} + \frac{1}{\rho}\frac{\partial \bar{T}_i}{\partial \rho}\right) + \frac{\bar{\kappa}_{\theta i}}{\rho^2}\frac{\partial^2 \bar{T}_i}{\partial \theta^2}; \quad i = 1 \sim (N+1) \tag{1}$$

$$\tau = 0; \bar{T}_i = 0 \; ; i = 1 \sim (N+1) \tag{2}$$

$$\rho = \bar{a}; \frac{\partial \bar{T}_1}{\partial \rho} - H_a \bar{T}_1 = -H_a \bar{T}_a f_a(\theta) \tag{3}$$

$$\rho = R_i; \bar{T}_i = \bar{T}_{i+1}, \bar{\lambda}_{ri}\frac{\partial \bar{T}_i}{\partial \rho} = \bar{\lambda}_{r,i+1}\frac{\partial \bar{T}_{i+1}}{\partial \rho}; \quad i = 1 \sim N \tag{4}$$

$$\rho = \bar{c}; \frac{\partial \bar{T}_{N+1}}{\partial \rho} + H_c \bar{T}_{N+1} = 0 \tag{5}$$

$$\theta = 0, \theta_0; \bar{T}_i = 0 \; ; i = 1 \sim (N+1) \tag{6}$$

where

$$\bar{\kappa}_{ri} = \bar{\kappa}_{Ti}, \bar{\kappa}_{\theta i} = \bar{\kappa}_{Li}\sin^2\phi_i + \bar{\kappa}_{Ti}\cos^2\phi_i, \bar{\lambda}_{ri} = \bar{\lambda}_{Ti}\;;\; i = 2 \sim (N+1) \tag{7}$$

In expressions (1)-(7), we have introduced the following dimensionless values:

$$(\bar{T}_i, \bar{T}_a) = (T_i, T_a)/T_0, \; (\rho, R_i, \bar{a}, \bar{c}) = (r, r_i, a, c)/b, \; (H_a, H_c) = (h_a, h_c)b,$$
$$\bar{\kappa}_{ki} = \kappa_{ki}/\kappa_0, \; (\bar{\lambda}_{ri}, \bar{\lambda}_{Ti}) = (\lambda_{ri}, \lambda_{Ti})/\lambda_0, \; \tau = \kappa_0 t/b^2 \tag{8}$$

where T_i is the temperature change of the ith layer; κ_{ri} and $\kappa_{\theta i}$ are thermal diffusivities in the r and θ directions, respectively; λ_{ri} is thermal conductivity in the r direction ; t is time; and T_0, κ_0, and λ_0 are typical values of temperature, thermal diffusivity and thermal conductivity, respectively. In Eq.(7), the subscripts L and T denote the fiber and transverse directions, respectively. Moreover, r_i $(i = 1, 2, \cdots, N)$ are the coordinates of interface of the laminated cylindrical panel.

Introducing the finite sine transformation with respect to the variable θ and Laplace transformation with respect to the variable τ, the solution of equation (1) can be obtained so as to satisfy the conditions (2)-(6). This solution is shown as follows:

$$\overline{T}_i = \sum_{k=1}^{\infty} \overline{T}_{ik}(\rho, \tau) \sin q_k \theta \ ; \ i = 1 \sim (N+1) \tag{9}$$

where

$$\overline{T}_{ik}(\rho, \tau) = \frac{2}{\theta_0} [\frac{1}{F}(\overline{A}'_i \rho^{\gamma_i} + \overline{B}'_i \rho^{-\gamma_i})$$

$$+ \sum_{j=1}^{\infty} \frac{2\exp(-\mu_j^2 \tau)}{\mu_j \Delta'(\mu_j)} \{\overline{A}_i J_{\gamma_i}(\beta_i \mu_j \rho) + \overline{B}_i Y_{\gamma_i}(\beta_i \mu_j \rho)\}] \tag{10}$$

where $J_\gamma(\ )$ and $Y_\gamma(\ )$ are the Bessel function of the first and second kind of order γ, respectively; Δ and F are the determinants of $2(N+1) \times 2(N+1)$ matrix $[a_{kl}]$ and $[e_{kl}]$, respectively; the coefficients $\overline{A}_i$ and $\overline{B}_i$ are defined as the determinants of the matrix similar to the coefficient matrix $[a_{kl}]$, in which the $(2i-1)$th column or $2i$th column is replaced by the constant vector $\{c_k\}$, respectively; similarly, the coefficients $\overline{A}'_i$ and $\overline{B}'_i$ are defined as the determinants of the matrix similar to the coefficient matrix $[e_{kl}]$, in which the $(2i-1)$th column or $2i$th column is replaced by the constant vector $\{c_k\}$, respectively. The nonzero elements a_{kl}, e_{kl} and c_k are given by Eqs. (3)-(5). Furthermore, In Eqs.(9) and (10), $\Delta'(\mu_j)$, q_k, β_i and γ_i are

$$\Delta'(\mu_j) = d\Delta/d\mu\big|_{\mu=\mu_j}, \quad q_k = k\pi/\theta_0, \quad \beta_i = 1/\sqrt{\overline{\kappa}_{ri}}, \quad \gamma_i = \sqrt{\overline{\kappa}_{\theta i}/\overline{\kappa}_{ri}} \, q_k \tag{11}$$

and μ_j represent the jth positive roots of the following transcendental equation

$$\Delta(\mu) = 0 \tag{12}$$

2.2. Piezothermoelastic Problem

We develop the analysis for transient piezothermoelasticity of a simply supported cylindrical composite panel composed of angle-ply laminae and piezoelectric material as a generalized plane deformation problem. We assume that the displacement components for the global coordinate system (r,θ,z) are independent z. In the case of the piezoelectric layer of crystal class mm2, the stress-strain relations are expressed in dimensionless form as follows:

$$\begin{aligned}
\bar{\sigma}_{rri} &= \bar{C}_{11i}\bar{\varepsilon}_{rri} + \bar{C}_{12i}\bar{\varepsilon}_{\theta\theta i} - \bar{\beta}_{ri}\bar{T}_i - \bar{e}_1\bar{E}_r, \\
\bar{\sigma}_{\theta\theta i} &= \bar{C}_{12i}\bar{\varepsilon}_{rri} + \bar{C}_{22i}\bar{\varepsilon}_{\theta\theta i} - \bar{\beta}_{\theta i}\bar{T}_i - \bar{e}_2\bar{E}_r, \\
\bar{\sigma}_{zzi} &= \bar{C}_{13i}\bar{\varepsilon}_{rri} + \bar{C}_{23i}\bar{\varepsilon}_{\theta\theta i} - \bar{\beta}_{zi}\bar{T}_i - \bar{e}_3\bar{E}_r, \\
\bar{\sigma}_{\theta z i} &= \bar{C}_{44i}\bar{\gamma}_{\theta z i}, \quad \bar{\sigma}_{rzi} = \bar{C}_{55 6i}\bar{\gamma}_{rzi} - \bar{e}_5\bar{E}_z, \\
\bar{\sigma}_{r\theta i} &= \bar{C}_{66i}\bar{\gamma}_{r\theta i} - \bar{e}_6\bar{E}_\theta \quad ; i = N+1
\end{aligned} \qquad (13)$$

where

$$\begin{aligned}
\bar{\beta}_{ri} &= \bar{C}_{11i}\bar{\alpha}_{ri} + \bar{C}_{12i}\bar{\alpha}_{\theta i} + \bar{C}_{13i}\bar{\alpha}_{zi}, \\
\bar{\beta}_{\theta i} &= \bar{C}_{12i}\bar{\alpha}_{ri} + \bar{C}_{22i}\bar{\alpha}_{\theta i} + \bar{C}_{23i}\bar{\alpha}_{zi}, \\
\bar{\beta}_{zi} &= \bar{C}_{13i}\bar{\alpha}_{ri} + \bar{C}_{23i}\bar{\alpha}_{\theta i} + \bar{C}_{33i}\bar{\alpha}_{zi}
\end{aligned} \qquad (14)$$

In the case of the angle-ply laminate $(i = 1 \sim N)$, the stress-strain relations for the global coordinate system are expressed in dimensionless form as follows:

$$\begin{aligned}
\bar{\sigma}_{zzi} &= \bar{Q}^*_{12i}\bar{\varepsilon}_{\theta\theta i} + \bar{Q}^*_{13i}\bar{\varepsilon}_{rri} + \bar{Q}^*_{16i}\bar{\gamma}_{\theta z i} - \bar{\beta}_{zi}\bar{T}_i, \\
\bar{\sigma}_{\theta\theta i} &= \bar{Q}^*_{22i}\bar{\varepsilon}_{\theta\theta i} + \bar{Q}^*_{23i}\bar{\varepsilon}_{rri} + \bar{Q}^*_{26i}\bar{\gamma}_{\theta z i} - \bar{\beta}_{\theta i}\bar{T}_i, \\
\bar{\sigma}_{rri} &= \bar{Q}^*_{23i}\bar{\varepsilon}_{\theta\theta i} + \bar{Q}^*_{33i}\bar{\varepsilon}_{rri} + \bar{Q}^*_{36i}\bar{\gamma}_{\theta z i} - \bar{\beta}_{ri}\bar{T}_i, \\
\bar{\sigma}_{r\theta i} &= \bar{Q}^*_{44i}\bar{\gamma}_{r\theta i} + \bar{Q}^*_{45i}\bar{\gamma}_{rzi}, \bar{\sigma}_{rzi} = \bar{Q}^*_{45i}\bar{\gamma}_{r\theta i} + \bar{Q}^*_{44i}\bar{\gamma}_{rzi}, \\
\bar{\sigma}_{\theta z i} &= \bar{Q}^*_{26i}\bar{\varepsilon}_{\theta\theta i} + \bar{Q}^*_{36i}\bar{\varepsilon}_{rri} + \bar{Q}^*_{66i}\bar{\gamma}_{\theta z i} - \bar{\beta}_{\theta z i}\bar{T}_i
\end{aligned} \qquad (15)$$

where

$$\begin{aligned}
\bar{\beta}_{zi} &= \bar{Q}^*_{11i}\bar{\alpha}_{zi} + \bar{Q}^*_{12i}\bar{\alpha}_{\theta i} + \bar{Q}^*_{13i}\bar{\alpha}_{ri} + \bar{Q}^*_{16i}\bar{\alpha}_{\theta z i}, \\
\bar{\beta}_{\theta i} &= \bar{Q}^*_{12i}\bar{\alpha}_{zi} + \bar{Q}^*_{22i}\bar{\alpha}_{\theta i} + \bar{Q}^*_{23i}\bar{\alpha}_{ri} + \bar{Q}^*_{26i}\bar{\alpha}_{\theta z i}, \\
\bar{\beta}_{ri} &= \bar{Q}^*_{13i}\bar{\alpha}_{zi} + \bar{Q}^*_{23i}\bar{\alpha}_{\theta i} + \bar{Q}^*_{33i}\bar{\alpha}_{ri} + \bar{Q}^*_{36i}\bar{\alpha}_{\theta z i}, \\
\bar{\beta}_{\theta z i} &= \bar{Q}^*_{16i}\bar{\alpha}_{zi} + \bar{Q}^*_{26i}\bar{\alpha}_{\theta i} + \bar{Q}^*_{36i}\bar{\alpha}_{ri} + \bar{Q}^*_{66i}\bar{\alpha}_{\theta z i}
\end{aligned} \qquad (16)$$

The constitutive equations for the electric field in dimensionless form are given as

$$\bar{D}_r = \bar{e}_1 \bar{\varepsilon}_{rr,N+1} + \bar{e}_2 \bar{\varepsilon}_{\theta\theta,N+1} + \bar{\eta}_1 \bar{E}_r + \bar{p}_1 \bar{T}_{N+1},$$
$$\bar{D}_\theta = \bar{e}_6 \bar{\gamma}_{r\theta,N+1} + \bar{\eta}_2 \bar{E}_\theta, \quad \bar{D}_z = \bar{e}_5 \bar{\gamma}_{rz,N+1} + \bar{\eta}_3 \bar{E}_z \quad (17)$$

The relations between the electric field intensities and the electric potential ϕ in dimensionless form are defined by

$$\bar{E}_r = -\bar{\phi}_{,\rho}, \quad \bar{E}_\theta = -\rho^{-1}\bar{\phi}_{,\theta}, \quad \bar{E}_z = -\bar{\phi}_{,\bar{z}} = 0 \quad (18)$$

where a comma denotes partial differentiation with respect to the variable that follows. If the free charge is absent, the equation of electrostatics is expressed in dimensionless form as follows:

$$\bar{D}_{r,\rho} + \rho^{-1}(\bar{D}_r + \bar{D}_{\theta,\theta}) = 0 \quad (19)$$

In expressions (13)-(19), the following dimensionless values have been introduced:

$$\bar{\sigma}_{kli} = \frac{\sigma_{kli}}{\alpha_0 Y_0 T_0}, \quad (\bar{\varepsilon}_{kli}, \bar{\gamma}_{kli}) = \frac{(\varepsilon_{kli}, \gamma_{kli})}{\alpha_0 T_0}, \quad \bar{u}_{ki} = \frac{u_{ki}}{\alpha_0 T_0 b}, \quad (\bar{\alpha}_{ki}, \bar{\alpha}_{\theta zi}) = \frac{(\alpha_{ki}, \alpha_{\theta zi})}{\alpha_0},$$

$$(\bar{C}_{kli}, \bar{Q}_{kli}^*) = \frac{(C_{kli}, Q_{kli}^*)}{Y_0}, \quad \bar{D}_k = \frac{D_k}{\alpha_0 Y_0 T_0 |d_1|}, \quad \bar{E}_k = \frac{E_k |d_1|}{\alpha_0 T_0}, \quad \bar{\phi} = \frac{\phi |d_1|}{\alpha_0 T_0 b},$$

$$\bar{e}_k = \frac{e_k}{Y_0 |d_1|}, \quad \bar{\eta}_k = \frac{\eta_k}{Y_0 |d_1|^2}, \quad \bar{p}_1 = \frac{p_1}{\alpha_0 Y_0 |d_1|} \quad (20)$$

where σ_{kli} are the stress components, ε_{kli} are the normal strain components, γ_{kli} are the shearing strain, u_{ki} are the displacement components, α_{ki} and $\alpha_{\theta zi}$ are the coefficients of linear thermal expansion, C_{kli} are the elastic stiffness constants, Q_{kli}^* are the transformed elastic stiffness constants, D_k are the electric displacement components, e_k are the piezoelectric coefficients, η_k are the dielectric constants, p_1 is the pyroelectric constant, d_1 is the piezoelectric modulus and α_0 and Y_0 are the typical values of the coefficient of linear thermal expansion and Young's modulus of elasticity, respectively.

In the case of the piezoelectric layer ($i = N+1$), substituting the displacement-strain relations and (18) into Eqs.(13) and (17), and later into Eqs.(19) and the equilibrium equations, the governing equations of the displacement components and the electric potential in dimensionless form are written as

$$\bar{C}_{11i}(\bar{u}_{ri,\rho\rho} + \rho^{-1}\bar{u}_{ri,\rho}) - \rho^{-2}(\bar{C}_{22i}\bar{u}_{ri} - \bar{C}_{66i}\bar{u}_{ri,\theta\theta}) + (\bar{C}_{12i} + \bar{C}_{66i})\rho^{-1}\bar{u}_{\theta i,\rho\theta}$$

$$-(\overline{C}_{22i}+\overline{C}_{66i})\rho^{-2}\overline{u}_{\theta i,\theta}+\overline{e}_1\overline{\phi}_{,\rho\rho}+\overline{e}_6\rho^{-2}\overline{\phi}_{,\theta\theta}+(\overline{e}_1-\overline{e}_2)\rho^{-1}\overline{\phi}_{,\rho}$$
$$=\overline{\beta}_{ri}\overline{T}_{i,\rho}+\rho^{-1}(\overline{\beta}_{ri}-\overline{\beta}_{\theta i})\overline{T}_i \tag{21}$$

$$(\overline{C}_{66i}+\overline{C}_{12i})\rho^{-1}\overline{u}_{ri,\rho\theta}+(\overline{C}_{66i}+\overline{C}_{22i})\rho^{-2}\overline{u}_{ri,\theta}+\overline{C}_{66i}(\rho^{-1}\overline{u}_{\theta i,\rho}-\rho^{-2}\overline{u}_{\theta i}+\overline{u}_{\theta i,\rho\rho})$$
$$+\overline{C}_{22i}\rho^{-2}\overline{u}_{\theta i,\theta\theta}+\rho^{-2}\overline{e}_6\overline{\phi}_{,\theta}+(\overline{e}_6+\overline{e}_2)\rho^{-1}\overline{\phi}_{,\rho\theta}=\rho^{-1}\overline{\beta}_{\theta i}\overline{T}_{i,\theta} \tag{22}$$

$$\overline{C}_{55i}(\overline{u}_{zi,\rho\rho}+\rho^{-1}\overline{u}_{zi,\rho})+\overline{C}_{44i}\rho^{-2}\overline{u}_{zi,\theta\theta}=0 \tag{23}$$

$$\overline{e}_1\overline{u}_{ri,\rho\rho}+(\overline{e}_1+\overline{e}_2)\rho^{-1}\overline{u}_{ri,\rho}+\overline{e}_6\rho^{-2}\overline{u}_{ri,\theta\theta}+(\overline{e}_2+\overline{e}_6)\rho^{-1}\overline{u}_{\theta i,\rho\theta}-\overline{e}_6\rho^{-2}\overline{u}_{\theta i,\theta}$$
$$-\overline{\eta}_1(\overline{\phi}_{,\rho\rho}+\rho^{-1}\overline{\phi}_{,\rho})-\overline{\eta}_2\rho^{-2}\overline{\phi}_{,\theta\theta}=-\overline{p}_1(\overline{T}_{i,\rho}+\rho^{-1}\overline{T}_i) \tag{24}$$

In the case of the angle-ply laminate ($i = 1 \sim N$), substituting the displacement-strain relations into Eq.(15), and later into the equilibrium equations, the governing equations of the displacement components in dimensionless form can be obtained. If the inner and outer surfaces of the combined panel are traction free, and the interfaces of the each layer are perfectly bonded, then the boundary conditions of inner and outer surfaces and the conditions of continuity at the interfaces can be represented as follows:

$$\rho=\overline{a};\ \overline{\sigma}_{rr1}=0,\ \overline{\sigma}_{r\theta 1}=0,\ \overline{\sigma}_{rz1}=0,$$
$$\rho=\overline{c};\ \overline{\sigma}_{rr,N+1}=0,\ \overline{\sigma}_{r\theta,N+1}=0,\ \overline{\sigma}_{rz,N+1}=0,$$
$$\rho=R_i;\ \overline{\sigma}_{rri}=\overline{\sigma}_{rr,i+1},\ \overline{\sigma}_{r\theta i}=\overline{\sigma}_{r\theta,i+1},\ \overline{\sigma}_{rzi}=\overline{\sigma}_{rz,i+1},$$
$$\overline{u}_{ri}=\overline{u}_{r,i+1},\ \overline{u}_{\theta i}=\overline{u}_{\theta,i+1},\ \overline{u}_{zi}=\overline{u}_{z,i+1}\ ;\ i=1\sim N \tag{25}$$

The boundary conditions in the radial direction for the electric field are expresses by

$$\rho=\overline{c};\ \overline{D}_r=0,\quad \rho=1;\ \overline{\phi}=0 \tag{26}$$

We now consider the case of a simply supported panel and assume that the edges of the piezoelectric layer are electrically grounded. The boundary conditions are given as follows:

$$\theta=0,\ \theta_0;\ \overline{\sigma}_{\theta\theta i}=0,\ \overline{\sigma}_{\theta z i}=0,\ \overline{u}_{ri}=0,\ \overline{\phi}=0 \tag{27}$$

We assume the solutions of the displacement components and electric potential in order to satisfy Eq.(27) in the following form.

$$\overline{u}_{ri}=\sum_{k=1}^{\infty}\{U_{rcik}(\rho)+U_{rpik}(\rho)\}\sin q_k\theta,\ \overline{u}_{\theta i}=\sum_{k=1}^{\infty}\{U_{\theta cik}(\rho)+U_{\theta pik}(\rho)\}\cos q_k\theta,$$
$$\overline{u}_{zi}=\sum_{k=1}^{\infty}\{U_{zcik}(\rho)+U_{zpik}(\rho)\}\cos q_k\theta,\ \overline{\phi}=\sum_{k=1}^{\infty}\{\Phi_{ck}(\rho)+\Phi_{pk}(\rho)\}\sin q_k\theta \tag{28}$$

In expressions (28), the first term on the right side gives the homogeneous solution and the second term of right side gives the particular solution. However, since Eq.(23) has not the term of the temperature, the particular solution $U_{zpik}(\rho)$ of the piezoelectric layer doesn't exist. We now consider the homogeneous solution and introduce the following equation.

$$\rho = \exp(s) \qquad (29)$$

Substituting the first term on the right side of Eq.(28) into the homogeneous expression of the governing equations of Eqs.(21)-(24), and later changing a variable with the use of Eq.(29), we obtain the simultaneous ordinary different equations of U_{rcik}, $U_{\theta cik}$, U_{zcik} and Φ_{ck}. We show U_{rcik}, $U_{\theta cik}$, U_{zcik} and Φ_{ck} as follows:

$$(U_{rcik}, U_{\theta cik}, U_{zcik}, \Phi_{ck}) = (U_{rcik}^0, U_{\theta cik}^0, U_{zcik}^0, \Phi_{ck}^0)\exp(\lambda_i s); \; i = N+1 \qquad (30)$$

Substituting Eq.(33) into the simultaneous ordinary different equations, the condition that a non-trivial solution of $(U_{rcik}^0, U_{\theta cik}^0, \Phi_{ck}^0)$ exists leads to the following equation.

$$p_i^3 + d^{(i)} p_i + f^{(i)} = 0 \qquad (31)$$

where

$$p_i = \lambda_i^2 - B^{(i)}/3A^{(i)}, \quad A^{(i)} = \overline{C}_{66i}(\overline{e}_1^2 + \overline{\eta}_1 \overline{C}_{11i}),$$
$$B^{(i)} = q_k^2[\overline{\eta}_1(\overline{C}_{11i}\overline{C}_{22i} - \overline{C}_{12i}^2 - 2\overline{C}_{12i}\overline{C}_{66i}) + \overline{\eta}_2 \overline{C}_{11i}\overline{C}_{66i} + \overline{C}_{11i}(\overline{e}_2 + \overline{e}_6)^2$$
$$-2\overline{e}_1 \overline{e}_2 (\overline{C}_{12i} + \overline{C}_{66i}) - 2\overline{C}_{12i}\overline{e}_1 \overline{e}_6 + \overline{e}_1^2 \overline{C}_{22i}]$$
$$+\overline{C}_{66i}[(\overline{C}_{11i} + \overline{C}_{22i})\overline{\eta}_1 + \overline{e}_1^2 + \overline{e}_2^2] \qquad (32)$$

In Eq.(31), $d^{(i)}$ and $f^{(i)}$ are the functions of the material constants and q_k, the details are omitted here. From Eq.(31), there might be three distinct real roots, three real roots with at least two of them being equal or a real root in conjunction with one pair of conjugate complex roots depending $H_i \equiv (f^{(i)})^2/4 + (d^{(i)})^3/27$ is negative, zero or positive, respectively. For instance, $U_{rcik}(\rho)$, $U_{\theta cik}(\rho)$ and $\Phi_{ck}(\rho)$ can be expressed as follows when $H_i < 0$ and $p_{iJ} + B^{(i)}/3A^{(i)} > 0$:

$$U_{rcik}(\rho) = \sum_{J=1}^{3} U_{rcik}^J(\rho), \; U_{\theta cik}(\rho) = \sum_{J=1}^{3} U_{\theta cik}^J(\rho), \; \Phi_{ck}(\rho) = \sum_{J=1}^{3} \Phi_{ck}^J(\rho) \qquad (33)$$

where

$$U_{rcik}^{J}(\rho) = F_{rJ}^{(i)}\rho^{m_{Ji}} + S_{rJ}^{(i)}\rho^{-m_{Ji}},$$

$$U_{\theta cik}^{J}(\rho) = L_{kiJ}(m_{Ji})F_{rJ}^{(i)}\rho^{m_{Ji}} + L_{kiJ}(-m_{Ji})S_{rJ}^{(i)}\rho^{-m_{Ji}},$$

$$\Phi_{ck}^{J}(\rho) = R_{kiJ}(m_{Ji})F_{rJ}^{(i)}\rho^{m_{Ji}} + R_{kiJ}(-m_{Ji})S_{rJ}^{(i)}\rho^{-m_{Ji}}, \quad m_{Ji} = \sqrt{p_{iJ} + \frac{B^{(i)}}{3A^{(i)}}} \quad (34)$$

From Eq.(23), $U_{zcik}(\rho)$ can be expressed as follows

$$U_{zcik}(\rho) = F_z\rho^{m_{4i}} + S_z\rho^{-m_{4i}} \quad ; \quad i = N+1 \quad (35)$$

where

$$m_{4i} = \pm\sqrt{\overline{C}_{44i}/\overline{C}_{55i}}\, q_k \quad (36)$$

In Eqs.(35) and (36), $F_{rJ}^{(i)}$, $S_{rJ}^{(i)}$, F_z and S_z are unknown constants. In the case of angle-ply laminate, $U_{rcik}(\rho)$, $U_{\theta cik}(\rho)$ and $U_{zcik}(\rho)$ are obtained in the same way as the case of the piezoelectric layer [5].

Next, in order to obtain the particular solution, we use the series expansions of the Bessel functions. Since the order γ_i of the Bessel function in Eq.(10) is not integer in general except ply angle $\phi_i = 0°$, Eq.(10) can be written as the following expression.

$$\overline{T}_{ik}(\rho,\tau) = \sum_{n=0}^{\infty}\{a_{ni}(\tau)\rho^{2n+\gamma_i} + b_{ni}(\tau)\rho^{2n-\gamma_i}\} \quad (37)$$

$U_{rpik}(\rho)$, $U_{\theta pik}(\rho)$, $U_{zpik}(\rho)$ and $\Phi_{pk}(\rho)$ of the particular solutions are obtained as the function systems like Eq.(37). Then, the stress components and the electric displacements can be evaluated from the displacement components and the electric potential. The unknown constants in the homogeneous solutions such as Eqs. (34) and (35) are determined so as to satisfy the boundary conditions (25) and (26).

3. NUMERICAL RESULTS

We consider the piezoelectric layer composed of a cadmium selenide solid and the angle-ply laminate composed of alumina fiber reinforced aluminum composite. We assume that each layer of angle-ply laminated panel consists of the same orthotropic material, and consider a 2-layered anti-symmetric angle-ply laminated panel with the fiber-orientation ($60°/-60°$) and the same thickness. We assume that the combined cylindrical panel is heated by surrounding media, the temperature of which is denoted by the symmetric function with respect to the center of the panel ($\theta = \theta_0 / 2$). Then, numerical calculative parameters of heat condition and shape are presented as follows:

$$H_a = H_c = 1.0, \ \bar{T}_a = 1, \ \theta_0 = 90°, \ \bar{a} = 0.7, \ \bar{c} = 1.05,$$
$$f_a(\theta) = \left(1 - \theta'^2 / \theta_a^2\right) H(\theta_a - |\theta'|), \ \theta_a = 15°, \ \theta' = \theta - \theta_0 / 2 \qquad (38)$$

where $H(x)$ is Heaviside's function. The material constants for cadmium selenide are taken as

$$\alpha_\theta = \alpha_z = 4.396 \times 10^{-6} \ 1/K, \ \alpha_r = 2.458 \times 10^{-6} \ 1/K, \ C_{11} = 83.6 \ GPa,$$
$$C_{22} = C_{33} = 74.1 \ GPa, \ C_{23} = 45.2 \ GPa, \ C_{12} = C_{13} = 39.3 \ GPa,$$
$$C_{66} = 13.17 \ GPa, \ e_1 = 0.347 \ C/m^2, \ e_2 = e_3 = -0.16 \ C/m^2,$$
$$e_6 = -0.138 \ C/m^2, \ \eta_1 = 9.03 \times 10^{-11} \ C^2/Nm^2, \ \eta_2 = 8.25 \times 10^{-11} \ C^2/Nm^2,$$
$$p_1 = -2.94 \times 10^{-6} \ C/m^2K, \ d_1 = -3.92 \times 10^{-12} \ C/N, \ \lambda_\theta = 8.6 \ W/mK,$$
$$\lambda_r = 1.5\lambda_\theta, \ \kappa_\theta = 3.28 \times 10^{-6} \ m^2/s, \ \kappa_r = 1.5\kappa_\theta \qquad (39)$$

and those for alumina fiber reinforced aluminum composite are taken in [5]. The typical values of material properties such as κ_0, λ_0, α_0 and Y_0, used to normalize the numerical data, are based on those of cadmium selenide as follows:

$$\kappa_0 = \kappa_\theta, \ \lambda_0 = \lambda_\theta, \ \alpha_0 = \alpha_\theta, \ Y_0 = 42.8 \ GPa \qquad (40)$$

Figure 1 shows the variation of the temperature change at the midpoint of the panel. Figure 2 shows the variation of the electric potential on the outer surface. Figure 3 shows the the variation of the transverse stress $\bar{\sigma}_{r\theta}$ on the interface between

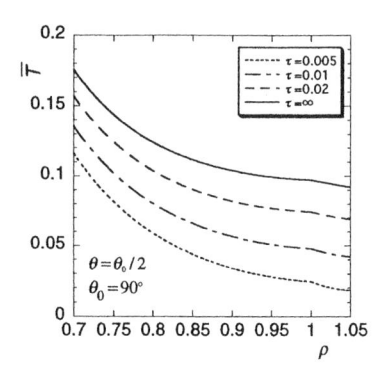

Figure 1. Temperature change.

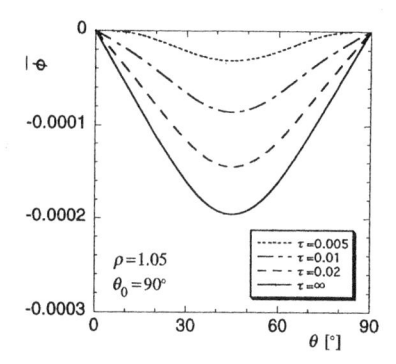

Figure 2. Electric potential

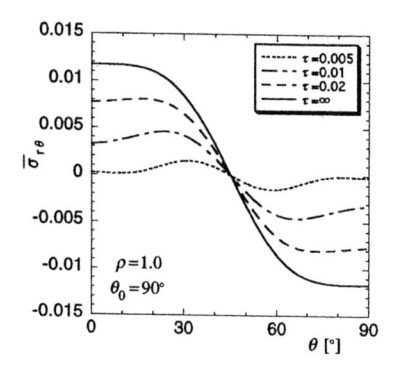

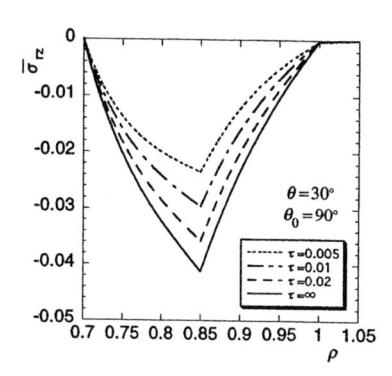

Figure 3. Thermal stress $\bar{\sigma}_{r\theta}$

Figure 4. Thermal stress $\bar{\sigma}_{rz}$

the piezoelectric layer and the angle-ply laminae. Figure 4 shows the variation of the transverse stress $\bar{\sigma}_{r\theta}$ on the cross section ($\theta = 30°$). From Figs.3 and 4, it can be seen that the maximum values of $\bar{\sigma}_{r\theta}$ and $\bar{\sigma}_{rz}$ occur in the steady state.

4.CONCLUSIONS

In the present article, we obtained the exact solution for the transient temperature and transient piezothermoelastic response of a simply supported cylindrical composite panel composed of angle-ply laminate and piezoelectric layer of crystal class mm2 due to a nonuniform heat supply in the circumferential direction. We conclude that we can evaluate not only the all transverse stress components of the combined cylindrical panel, but also the electric field of the piezoelectric layer quantitatively in a transient state.

5. REFERENCES

1 Chen C.-Q. and Shen, Y.-P. (1996) Piezothermoelasticity analysis for a circular cylindrical shell under the state of axisymmetric deformation. *Int. J. Engng. Sci.* **34** 1585-1600.
2. Xu, K. and Noor, A. K. (1996) Three-dimensional analytical solutions for coupled thermoelectroelastic esponse of multilayered cylindrical shells. *AIAA J.* **34** 802-812.
3. Kapuria, S., Sengupta, S. and Dumir, P.C (1997) Three-dimensional solution for a hybrid cylindrical shell under axisymmetric thermoelectric load. *Arch. Appl. Mech.* **67** 320-330.
4. Kapuria, S., Sengupta, S. and Dumir, P.C. (1997) Three-dimensional piezothermoelastic solution for shape control of cylindrical panel. *J.Thermal Stresses* **20** 67-85.
5. Ootao, Y. and Tanigawa, Y. (2002) Transient thermal stresses of angle-ply laminated cylindrical panel due to nonuniform heat supply in the circumferential direction. *Compos. Struct.* **55** 95-103.

ACTIVE DAMPING OF TORSIONAL VIBRATION IN A PIEZOELECTRIC FIBER COMPOSITE SHAFT

PIOTR M. PRZYBYŁOWICZ
Warsaw University of Technology
Institute of Machine Design Fundamentals
Narbutta 84, 02-524 Warsaw, Poland
E-mail: piotrp@ipbm.simr.pw.edu.pl

1. INTRODUCTION

Problem of active damping of torsional vibration with the help of piezoelectric materials has been taken up since late 80s. The most popular piezoelectrics applied to control systems were piezoceramics based on lead zirconate titanate (PZT) and piezopolymers made of polyvinylidene fluoride (PVDF). Admittedly, they enjoy a special favour up to present.

In majority of papers discussing the efficiency of piezoelectric elements in torsional systems it is assumed that they are elements of negligible mass and are perfectly attached to the given structure. On that assumption Meng-Kao Yeh and Chih-Yuan Chin [1] studied the sensing capability of piezoelements, and Chia-Chi Sung et al. [2] investigated a torsional system with a closed control loop. The shear piezoelectric effect was considered in their work. An interesting approach was proposed by Spearritt and Asokanthan [3] who took advantage of the longitudinal piezoelectric effect for controlling torsional vibration. They wrapped PVDF actuators around a tube at 45^0.

The feasibility of inducing twist in a plate using piezoelectric elements was also demonstrated by Lee and Moon [4]. The desired effect was achieved by bonding layers of PVDF films with skew angles 45^0 on both sides of the plate.

The effort to develop an effective method of torsional vibration control is especially important in the case of high-responsibility structures, e.g. aircrafts, helicopters, etc. Büter and Breitbach [5] showed that a piezoelectrically controlled tension-torsional coupling results in higher aerodynamic efficiency of helicopter blades. They obtained a 1.5^0 piezo-induced twist of the blades near the resonance frequency.

Another concept of generating torsional, bending and longitudinal control loads in a thin-walled structural member was described by Kawiecki et al. [6]. The torque was produced by extension/compression action of a set of piezoelectric actuators bonded to the given host structure. The capability of damping the torsional modes was well-proved in closed cross-section members.

An effort to create a theoretical model for a closed-loop control of torsional modes in a fixed-free tube was taken up by Przybyłowicz [7], where the author rejected the assumption of an undisturbed mechanical connection between the active PZT ring and the host structure. The bonding layer of glue was no longer rigid and could not secure perfect attachment between these elements. Some viscoelastic properties of the interlayer as well as the inertia of the actuator itself were taken into account.

Recently, the problem of vibration control in torsional systems has been extended to systems incorporating laminates reinforced with active piezoelectric fibers, however the idea of piezoelectric fiber composites (PFCs) emerged in early 80s, see Newnham [8]. Zhu et al. [9] presented a dynamic analytical model of an anisotropic piezoelectric laminate controlled by the extension-twisting coupling effect. The authors gave analytical solutions to the model with harmonic excitations. They reduced a three-dimensional problem into a one-dimensional formulation without loosing the complexity of the undertaken task. A one-dimensional model derived from three-dimensional equations was also presented by Vidoli and Batra [10] in the case of torsional deformations coupled with extensional ones in a piezoelectric cylinder.

2. CONSTITUTIVE EQUATIONS

In this section analysis of pure torsion in a PFC laminated shaft is considered. The shaft is modelled as a closed cylindrical shell with a regular structure (symmetric or anti-symmetric, cross- or angle-plied). Consider now the static equilibrium of an infinitesimal segment of a thin-walled shell in the $x-y$ plane, see Fig. 1.

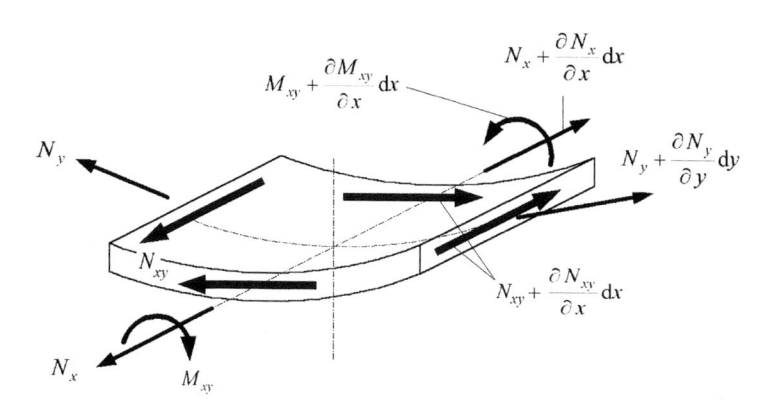

Figure 1. Internal forces and moments in a thin-walled cylindrical shaft (pure bending)

The equilibrium equations are:

$$\frac{\partial N_x}{\partial x} + \frac{\partial N_{xy}}{\partial y} = 0 \Rightarrow \frac{dN_x}{dx} = 0 \qquad (1)$$

$$\frac{\partial N_y}{\partial y} + \frac{\partial N_{xy}}{\partial x} = 0 \Rightarrow \frac{dN_{xy}}{dx} = 0 \qquad (2)$$

$$\frac{\partial^2 M_x}{\partial x^2} + 2\frac{\partial^2 M_{xy}}{\partial x \partial y} + \frac{\partial^2 M_y}{\partial y^2} - \frac{N_y}{r} = 0 \Rightarrow \frac{d^2 M_x}{dx^2} - \frac{N_y}{r} = 0 \qquad (3)$$

on the assumption that the internal forces and moments are independent of the co-ordinate following the circumference of the shaft cross-section, denoted by y. Taking into account that for torsional modes the longitudinal strain in the x-direction is $\varepsilon_{x0} = du_0/dx$, the transverse strain (radial direction) $\varepsilon_{y0} = w_0/r$, the shear $\gamma_{xy0} = dv_0/dx$, the curvatures: $\kappa_x = d^2w_0/dx^2$, $\kappa_y = 0$ and the torsion is $\kappa_{xy} = \gamma_{xy0}/r$ (where u_0 stands for the axial displacement of the mid-surface in the x-direction, v_0 - circumferential displacement, and w_0 - radial one), and additionally assuming that the transverse displacement is negligible with respect to the radial size of the given shaft ($w_0 \ll r$), and finally considering pure torsion only, i.e. $dw_0/dx = 0$, $d^2w_0/dx^2 = 0$, $d^3w_0/dx^3 = 0$, one obtains the following form of the constitutive relationship for an active symmetric laminate, see Zhu et al. [9]:

$$\begin{bmatrix} N_x \\ N_y \\ N_{xy} \\ M_x \\ M_y \\ M_{xy} \end{bmatrix} = \begin{bmatrix} A_{11} & A_{12} & A_{16} & 0 & 0 & 0 \\ A_{12} & A_{22} & A_{26} & 0 & 0 & 0 \\ A_{16} & A_{26} & A_{66} & 0 & 0 & 0 \\ 0 & 0 & 0 & D_{11} & D_{12} & D_{16} \\ 0 & 0 & 0 & D_{12} & D_{22} & D_{26} \\ 0 & 0 & 0 & D_{16} & D_{26} & D_{66} \end{bmatrix} \begin{bmatrix} \dfrac{du_0}{dx} \\ 0 \\ \dfrac{1}{2}\dfrac{dv_0}{dx} \\ 0 \\ 0 \\ \dfrac{1}{2r}\dfrac{dv_0}{dx} \end{bmatrix} - \begin{bmatrix} N_x^A \\ N_y^A \\ N_{xy}^A \\ 0 \\ 0 \\ 0 \end{bmatrix} \qquad (4)$$

as symmetric laminates lack the elasticity coupling matrix **B** and the "active" moments **M**^{A}. Symbols with the superscript "A" are relative to piezoelectric fibers. Substitution of (4) into equilibrium equations yields:

$$\begin{bmatrix} A_{11} & \dfrac{1}{2}A_{16} \\ A_{16} & \dfrac{1}{2}A_{66} \end{bmatrix} \begin{bmatrix} \dfrac{d^2 u_0}{dx^2} \\ \dfrac{d^2 v_0}{dx^2} \end{bmatrix} = \begin{bmatrix} \dfrac{dN_x^A}{dx} \\ \dfrac{dN_{xy}^A}{dx} \end{bmatrix} \qquad (5)$$

Integrating the solution to (5) over the length of the shaft, one gets:

$$v_0(x) = \frac{A_{11} N_{xy}^A - A_{16} N_x^A}{A_{11} A_{66} - A_{16}^2} \left[xH(x) - (x-l)H(x-l) \right] + C_1 x + C_2 \tag{6}$$

provided that the stiffness coefficients satisfy $A_{11} A_{66} \neq A_{16}^2$. The function $H(.)$ is Heaviside's step one – its presence ensues from the fact that the actuating electrodes can cover only a part of the shaft structure, however in the analysed case they do the entire length l. The integration constants C_1, C_2 can be found from boundary conditions which are the following: no angular displacement v_0 and the longitudinal one u_0 at the beginning (fixed) point $x = 0$ of the shaft, no axial force there, and application of the external torque M_s balancing the internal torsional moments:

$$v_0(0) = 0, \quad u_0(0) = 0, \quad N_x(0) = 0, \quad M_{xy}(0) + r N_{xy}(0) = \frac{M_s}{2\pi r} \tag{7}$$

Having them substituted into (6) and then dividing by the radius r, the static torsional displacement can be determined:

$$\varphi(x) = \frac{v_0}{r} = \frac{x}{J_0 G_{TS}} \left[M_s + 2\pi r^2 \left(N_{xy}^A - \frac{\overline{Q}_{16}}{N \overline{Q}_{11}} N_x^A \right) \right] \tag{8}$$

where N denotes the number of active layers the shaft is made of, $J_0 = 2\pi r^3 h$ - second moment of the cross-section, $\overline{Q}_{ij}$ - stiffness coefficients of a single lamina in a transformed (rotated with respect to principal anisotropy axes) co-ordinate system, G_{TS} - equivalent shear modulus: $G_{TS} = \overline{Q}_{66} - \overline{Q}_{16}^2 / (N^2 \overline{Q}_{11})$. In (8) it is noted that $xH(x) - (x-l)H(x-l) \equiv x$ for $x \in (0, l)$. Proceeding in an analogous way, in the case of anti-symmetric composites one obtains:

$$\varphi(x) = \frac{x}{J_0 G_{TA}} \left(M_s + \pi r^2 \frac{\overline{Q}_{66}}{\overline{Q}_{11}} N_x^A \right) \quad \text{where} \quad G_{TA} = \overline{Q}_{66} - \frac{h^2 \overline{Q}_{16}^2}{4 N^2 r^2 \overline{Q}_{11}} \tag{9}$$

3. APPLICATION OF THE ELECTRIC FIELD

According to Nye [11] the constitutive equations of piezoelectricity, the stress-strain-electric field relation is:

$$\begin{bmatrix} \varepsilon_1 \\ \varepsilon_2 \\ \varepsilon_3 \\ \varepsilon_4 \\ \varepsilon_5 \\ \varepsilon_6 \end{bmatrix} = \begin{bmatrix} s_{11} & s_{12} & s_{13} & 0 & 0 & 0 \\ s_{12} & s_{22} & s_{13} & 0 & 0 & 0 \\ s_{13} & s_{13} & s_{33} & 0 & 0 & 0 \\ 0 & 0 & 0 & s_{44} & 0 & 0 \\ 0 & 0 & 0 & 0 & s_{44} & 0 \\ 0 & 0 & 0 & 0 & 0 & s_{66} \end{bmatrix} \begin{bmatrix} \sigma_1 \\ \sigma_2 \\ \sigma_3 \\ \tau_4 \\ \tau_5 \\ \tau_6 \end{bmatrix} + \begin{bmatrix} 0 & 0 & d_{31}^* \\ 0 & 0 & d_{32}^* \\ 0 & 0 & d_{33}^* \\ 0 & d_{15}^* & 0 \\ d_{15}^* & 0 & 0 \\ 0 & 0 & 0 \end{bmatrix} \begin{bmatrix} E_1 \\ E_2 \\ E_3 \end{bmatrix} \quad (10)$$

where ε_i denotes the strain, s_{ij} - compliance coefficients, d_{ij}^* - effective, overall electromechanical coupling constants, E_j - electric field.

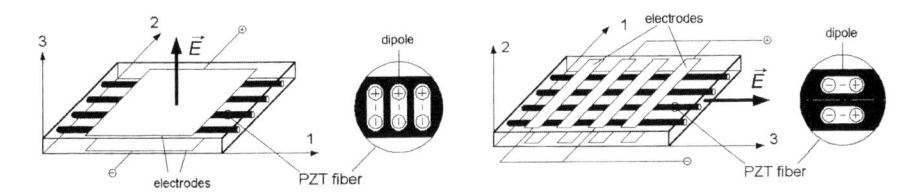

Figure 2. Transverse poling of active fibers in PFCs (left picture) and InterDigitated Electrode PFCs (on the right), see also Bent and Hagood [12]

In laminates with the interdigitated electrode pattern (see the right scheme in Fig. 2) two geometric arrangements of the electrode patterns can be predicted by a designer. They are shown Fig. 3.

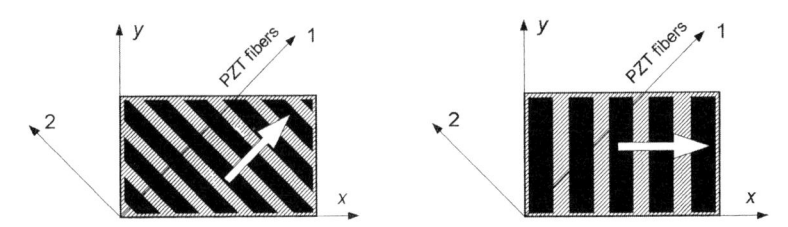

Figure 3. Interdigitated pattern of electrodes perpendicular to fibers "IDEPFC" (left scheme) and with electrodes perpendicular to the shaft symmetry axis "S-IDEPFC" (right scheme)

4. DYNAMIC CHARACTERISTICS

Consider a fixed-free composite shaft subject to pure torsion by an externally applied harmonic torque as shown in Fig. 4. The shaft is made of an active laminate containing piezoelectric fibers capable of generating mechanical strain and stress under an electric field.

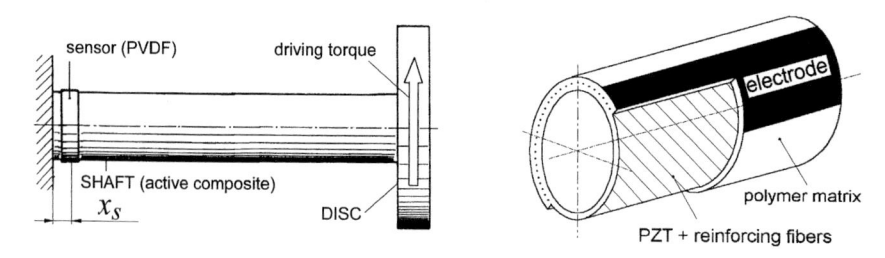

Figure 4. Considered model (left) and structural features of the shaft (right)

The equation of motion of torsional vibration has the following (linear) well-known form:

$$\rho J_0 \frac{\partial^2 \varphi}{\partial t^2} = \frac{\partial M_s}{\partial x} \quad \text{where} \quad M_s = J_0 \left(G_T \frac{\partial \varphi}{\partial x} - \frac{1}{r} \Xi_\tau E [H(x) - H(x-l)] \right) \quad (11)$$

where ρ is the mass density, M_s - the twisting moment, E - the applied electric field. In fact, the direction of the vector $\mathbf{E}$ is different in each case of the applied electrode pattern, i.e. $\mathbf{E}_{PFC} = [0,0,E_3]$, $\mathbf{E}_{IDE} = [E_1,0,0]$, $\mathbf{E}_{SID} = [E_x,0,0]$ but the magnitude remains the same: $E_3 = E_1 = E_x$. The electromechanical coupling coefficient, marked as Ξ_τ in (11), means:

$$\Xi_\tau = \begin{cases} \dfrac{1}{2} \overline{\Xi}_{13} \dfrac{\overline{Q}_{66}}{\overline{Q}_{11}} & \text{for anti-symmetric PFCs} \\[2mm] \dfrac{1}{2} \overline{\Xi}_{11} \dfrac{\overline{Q}_{66}}{\overline{Q}_{11}} & \text{for anti-symmetric IDEPFCs} \\[2mm] \dfrac{1}{2} \overline{\Xi}_{11}{}^{(S)} \dfrac{\overline{Q}_{66}}{\overline{Q}_{11}} & \text{for anti-symmetric S-IDEPFCs} \\[2mm] \dfrac{1}{N} \left(\overline{\Xi}_{33} - \dfrac{\overline{Q}_{16}}{\overline{Q}_{11}} \overline{\Xi}_{13} \right) & \text{for symmetric PFCs} \\[2mm] \dfrac{1}{N} \left(\overline{\Xi}_{31} - \dfrac{\overline{Q}_{16}}{\overline{Q}_{11}} \overline{\Xi}_{11} \right) & \text{for symmetric IDEPFCs} \\[2mm] \dfrac{1}{N} \left(\overline{\Xi}_{31}{}^{(S)} - \dfrac{\overline{Q}_{16}}{\overline{Q}_{11}} \overline{\Xi}_{11}{}^{(S)} \right) & \text{for symmetric S-IDEPFCs} \end{cases} \quad (12)$$

and where the overbars indicate that the quantities are given in the transformed (rotated by θ) co-ordinate system. Explicitly:

$$\overline{\Xi}_{11} = \Xi_{11}\cos^2\theta + \Xi_{21}\sin^2\theta , \quad \overline{\Xi}_{13} = \Xi_{13}\cos^2\theta + \Xi_{23}\sin^2\theta$$
$$\overline{\Xi}_{31} = (\Xi_{21} - \Xi_{11})\sin\theta\cos\theta , \quad \overline{\Xi}_{33} = (\Xi_{23} - \Xi_{13})\sin\theta\cos\theta$$
$$\overline{\Xi}_{11}^{(S)} = \Xi_{11}\cos^3\theta + (\Xi_{21} + 2\Xi_{32})\sin^2\theta\cos\theta \quad (13)$$
$$\overline{\Xi}_{31}^{(S)} = ((\Xi_{21} - \Xi_{11})\cos^2\theta + \Xi_{32}\cos 2\theta)\sin\theta$$

where Ξ_{ij} are

$$\Xi_{11} = d_{33}^* Q_{11} + d_{31}^* Q_{12} , \quad \Xi_{21} = d_{33}^* Q_{12} + d_{31}^* Q_{22} , \quad \Xi_{32} = d_{15}^* Q_{66}$$
$$\Xi_{13} = d_{31}^* Q_{11} + d_{32}^* Q_{12} , \quad \Xi_{23} = d_{31}^* Q_{12} + d_{32}^* Q_{22} \quad (14)$$

and $\overline{Q}_{ij}$ can be easily found in any primer on composite materials, see e.g. Ashton et al. [13] and Jones [14]. Applying a simple control strategy based on the velocity feedback:

$$E = c_d \frac{N}{h} \frac{dU_S}{dt} \quad (15)$$

where U_S is the voltage produced by the sensor (possibly PVDF film), c_d - gain factor, h - thickness of the shaft, and assuming that the measured voltage is directly proportional to the first derivative $\partial\varphi/\partial x$ of the torsional displacement in the point x_s the sensor is bonded at (which is true for very short sensing patches [7]) one writes down the equation of motion in the following form:

$$\frac{\partial^2\varphi}{\partial t^2} - \frac{G_T}{\rho}\frac{\partial^2\varphi}{\partial x^2} + \beta\frac{\partial^2\varphi(x_s)}{\partial x \partial t}\{\delta(x) - \delta(x-l)\} = m(x,t)\delta(x-l) \quad (16)$$

where $\beta = \beta(c_d,\theta)$ represents a coefficient of damping due to active control (β depends on the gain, lamination angle and electromechanical properties of the sensors and actuating fibers). The quantities $m(x,t)$ and $\delta(.)$ denote the torque excitation and Dirac's delta function, respectively. By predicting the solution to (16) in the form of an infinite series:

$$\varphi(x,t) = \sum_{i=1}^{\infty} \Phi_i(x) T_i(t) \quad (17)$$

one obtains an infinite set of ordinary differential equations:

$$\begin{cases} \ddot{T}_1 + \beta_{11}\dot{T}_1 + \beta_{12}\dot{T}_2 + \ldots\ldots\ldots + \beta_{1n}\dot{T}_n + \ldots\ldots\ldots + \omega_1^2 T_1 = \mu_1 \\ \ddot{T}_2 + \beta_{21}\dot{T}_1 + \beta_{22}\dot{T}_2 + \ldots\ldots\ldots + \beta_{2n}\dot{T}_n + \ldots\ldots\ldots + \omega_2^2 T_2 = \mu_2 \\ \ldots\ldots\ldots\ldots\ldots\ldots\ldots\ldots\ldots\ldots\ldots\ldots\ldots\ldots\ldots\ldots\ldots\ldots\ldots\ldots\ldots\ldots \\ \ddot{T}_n + \beta_{1n}\dot{T}_1 + \beta_{2n}\dot{T}_2 + \ldots\ldots\ldots + \beta_{nn}\dot{T}_n + \ldots\ldots\ldots + \omega_n^2 T_n = \mu_n \\ \ldots\ldots\ldots\ldots\ldots\ldots\ldots\ldots\ldots\ldots\ldots\ldots\ldots\ldots\ldots\ldots\ldots\ldots\ldots\ldots\ldots\ldots \end{cases} \quad (18)$$

where $\Phi_i(x)$ is the *i*-th eigenmode and $T_i(t)$ - arbitrary function of time, and:

$$\mu_i = m_0(t) \frac{\Phi_i(l)}{\int_0^l \Phi^2(x)\,dx}, \quad \beta_{ij} = -\beta \frac{\Phi_i(l)}{\int_0^l \Phi^2(x)\,dx} \frac{d\Phi_j(x_s)}{dx} \quad (19)$$

where ω_i is the *i*-th eigenfrequency corresponding to the assumed boundary (fixed-free) conditions.

The results of numerical simulations in the form of amplitude frequency characteristics are shown below. Each time the efficiency of three methods of developing the electric field (PFC, IDEPFC, S-IDEPFC) is compared in a symmetric (left diagrams) and anti-symmetric (right diagrams) laminates, see Figs 5- 8.

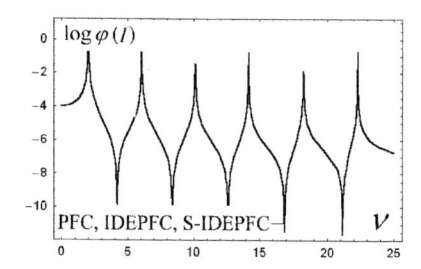

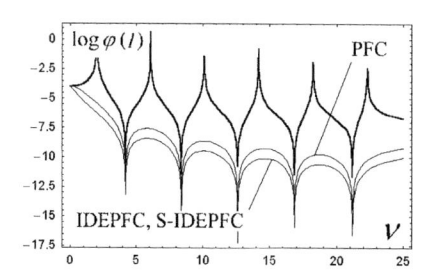

Figure 5. Amplitude frequency characteristics for $\theta = 0^0$

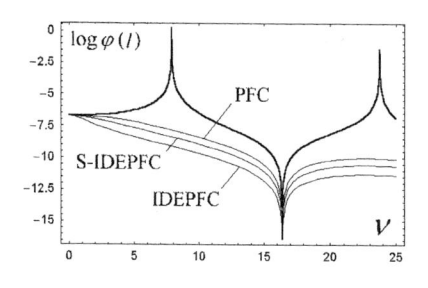

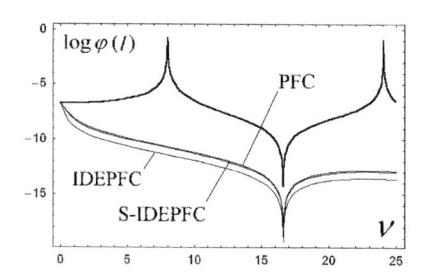

Figure 6. Amplitude frequency characteristics for $\theta = 45^0$

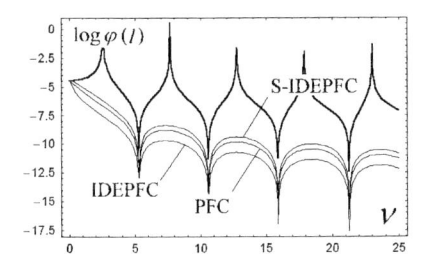

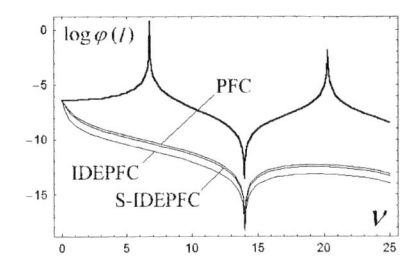

Figure 7. Amplitude frequency characteristics for $\theta = \theta_{\max ef}$

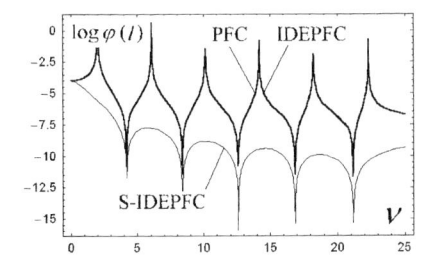

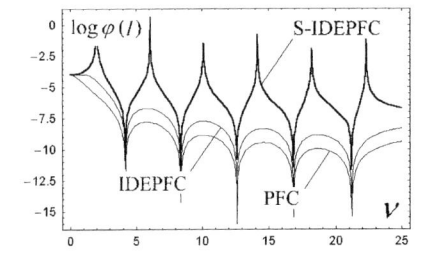

Figure 8. Amplitude frequency characteristics for $\theta = 90^0$

The thick lines in the above figures correspond to the system with disabled control. The thin lines present drops in the vibration amplitude due to the applied damping method for the three different ways of generating the electric field within the laminate. In some cases the curves cover one another or are in line with the thick curve, which implies inability to reduce the torsional vibration. In Fig. 7 the resonance characteristics are shown for a lamination angle denoted as $\theta_{\max ef}$, i.e. θ at which the efficiency of the control is the greatest. In anti-symmetric laminates $\theta_{\max ef}$ is about 80^0 and 70^0 in symmetric ones. These values can be conveniently found from static characteristics revealing static angular displacement under a constant electric field.

5. CONCLUDING REMARKS

In the paper a method of active damping of torsional vibration by making use of piezoelectric elements is presented. The method is based on the application of piezoelectric active composites – the state-of-the-art structural materials that have appeared in the field of mechatronics.

The integration of piezoceramic (PZT) fibers within composite materials represents a new type of material evolution. Tiny PZT fibers of 30 µm in diameter can be aligned in an array, electrodized with interdigital electrodes and then integrated into planar architectures. Such architectures are embedded within glass or graphite fiber-reinforced polymers and become piezoelectric after being poled. Matrix and ceramic combinations, volume fractions, and ply angles contribute to the tailorability of PFCs, which make them applicable to structures requiring highly distributed actuation and sensing. Manufacturing technologies of PFCs have been adopted from graphite/epoxy manufacturing methods. Today, PFCs are being equipped with an interdigitated electrode patterns. Regardless of the electrode arrangement the piezoelectric composites create a class of active materials that can cover or create entire structures – the actuators that are conformable to curved elements such as shafts, tubes or shells.

The paper is intended to show the applicability of piezoelectric fiber composites to control of torsional vibration and introduces different ways of supplying the controlling voltage to the structure. The extent to which the proposed method of the electric field development affects the damping capability of the considered system is discussed and illustrated on appropriate dynamic characteristics.

ACKNOWLEDGEMENT

The work presented in the paper was supported by the Polish State Committee for Scientific Research (KBN Grant No. 8 T07 A01 421), which is gratefully acknowledged.

6. REFERENCES

1. Meng-Kao Yeh and Chih-Yuan Chin (1994) Dynamic response of circular shaft with piezoelectric sensor, *Journal of Intelligent Material Systems and Structures* **5**(11) 833-840
2. Chia-Chi Sung, Vasundara V. Varadan, Xiao-Qi Bao and Vijay K. Varadan (1994) Active torsional vibration control experiments using shear-type piezoceramic sensors and actuators, *Journal of Intelligent Material Structures and Systems* **5**(3) 436-442
3. Spearritt, D.J. and Asokanthan, S.F. (1996) Torsional vibration control of a flexible beam using laminated PVDF actuators, *Journal of Sound and Vibration* **193**(5) 941-956
4. Lee, C.K. and Moon, F.C. (1989) Laminated piezopolymer plates for torsion and bending sensors and actuators, *Journal of Acoustical Society of America* **85**(6) 2432-2439
5. Büter, A. and Breitbach, E. (2000) Adaptive blade twist – calculations and experimental results, *Aerospace Science Technology* **4** 309-319
6. Kawiecki, G., Smith, W.P. and Hu, C. (1995) Feasibility study of a tosrional-bending piezoelectric actuator, *Journal of Intelligent Material Systems and Structures* **6**(7) 465-473
7. Przybyłowicz, P.M. (1995) Torsional vibration control by active piezoelectric system, *Journal of Theoretical and Applied Mechanics* **33**(4) 809-823
8. Newnham, R.E., Bowen, K.A., Klicker, K.A. and Cross, L.E. (1980) Composite piezoelectric transducers, *Material Engineering* **2** 93-106
9. Zhu, M.-L., Ricky Lee, S.-W., Li, H.-L., Zhang, T.-Y. and Tong., P. (2002) Modeling of torsional vibration induced by extension-twisting coupling of anisotropic composite laminates with piezoelectric actuators, *Smart Materials and Structures* **11** 55-62
10. Vidoli, S. and Batra, R.C. (2001) Coupled extensional and torsional deformations of a piezoelectric cylinder, *Smart Materials and Structures* **10** 300-304
11. Nye, J.F. (1985) *Physical Properties of Crystals*,:Clarendon, Oxford
12. Bent, A.A. and Hagood, N.W. (1997) Piezoelectric fiber composites with interdigitated electrodes, *Journal of Intelligent Material Systems and Structures* **8**(11) 903-919
13. Ashton, J.E., Halpin, J.C. and Petit, P.H. (1969*) Primer on Composite Materials: Analysis*, Technomic Publishing, Westport
14. Jones, R.M. (1975) *Mechanics of Composite Materials*, McGraw-Hill Scripta Book, Washington

High-Performance PZT and PNN-PZT Actuators

JINHAO QIU AND JUNJI TANI
Institute of Fluid Science, Tohoku University
2-1-1 Katahira, Aoba-ku, Sendai 980-8577, Japan
qiu@ifs.tohoku.ac.jp

HIROFUMI TAKAHASHI
Fuji Ceramics Corporation, 2320-11 Yamamiya
Fujinomiya-City, Shizuoka-Pref. 418-0111, Japan

1. INTRODUCTION

Piezoelectric materials have been widely used as sensor and actuators in smart materials and structural systems [1,2]. Piezoelectric materials have attracted special attention among the many functional materials because they possess both the sensor and actuator functions, which can even be realized in a single element. Applications of piezoelectric materials in smart material and structures, especially those for vibration and noise control require high performance, such as larger force and displacement output.

The microwave sintering process has been recognized to offer a number of advantages over the conventional process in the sintering of ceramics, such as uniform and rapid heating [3-7]. In order to improve the performance of piezoelectric actuators, a hybrid sintering process, which is the combination of 28 GHz microwave sintering and hot-press, was developed and applied to the sintering of a commercial PZT material C82 developed by Fuji Ceramics Corporation and a PNN-PZT (Pb($Ni_{1/3}Nb_{2/3}$)$_{0.5}$($Ti_{0.7}Zr_{0.3}$)$_{0.5}O_3$) material, with the optimal sintering conditions [2],[3]. Due the rapid and uniform heating of the samples in microwave sintering process, the sintering time was reduced by about 87% for PZT actuators and 70% for PNN-PZT actuators, compared to the time needed for the conventional process. On the other hand, shorter sintering time resulted in less evaporation of Pb in the sintering process and consequently improvement of properties. Pores were significantly reduced and the density was increased due to high pressure of hot-press. The density of the PZT and PNN-PZT specimens sintered with the hybrid sintering process was higher than 99% of the theoretical density. The measurement results showed that the electromechanical coupling factor was improved significantly, the piezoelectric constants of PZT increased by nearly 40% and that of PNN-PZT increased by 21%.

2. FABRICATION AND SINTERING PROCESS

2.1. Preparation of Sintering Specimens

The PZT specimens used in the study were made from PZT powder C82 developed by Fuji Ceramics Corporation. The powder was compacted into specimens of cylindrical shape which 17 mm in diameter and 15 mm thick, with cold press under 200 MPa. The binder of the specimens was burned out at 700°C for five hours.

In order to develop piezoelectric actuators with higher performance, a new material was chosen from the ternary system $Pb(Ni_{1/3}Nb_{2/3})O_3$-$PbTiO_3$-$PbZrO_3$. The ternary system has been recognized to possess large piezoelectric constant. Specimens of 40 compositions in the ternary system were fabricated and their piezoelectric constants, dielectric constants and electromechanical factors were measured. Among the 40 compositions, $Pb(Ni_{1/3}Nb_{2/3})_{0.5}(Ti_{0.7}Zr_{0.3})_{0.5}O_3$ (designated as PNN-PZT), which has relatively high performance, was used as the material in this study.

The PNN-PZT powder was prepared from chemical reagent-grade PbO, NiO, Nb_2O_5, TiO_2, and ZrO_2 (all 99.9% purity), using the traditional ceramic fabrication techniques for PbO-based ceramics. Weighed raw materials of the given composition were wet-milled in a plastic mill, with partially stabilized zirconia balls and pure water, for 12 hours and then dried. The powder was calcined at 750°C for 2 hours in a covered magnesia crucible and then cooled to room temperature. After the calcined material had been cooled, it was wet-ground for 12 hours in the plastic mill and then dried. The ground powder material was thoroughly mixed with poly(vinyl alcohol) (5 wt%) in an auto-mortar, and the powder was formed into a cylinder 17 mm in diameter and 15 mm thick, under a pressure of 200 MPa. The binder was burned out at 700°C for 3 hours. The sample was then sintered under the given temperature and pressure schedule for the hybrid sintering process.

2.2. Conventional and Hybrid Sintering Processes

The conventional sintering process uses resistance heating of electric furnace. Since the specimen is heated from the surface, internal temperature is always lower than the surface temperature. In order to keep the temperature difference between the surface and center as small as possible, the temperature must be raised very slowly.

The hybrid sintering process used for the sintering of the PZT and PNN-PZT ceramics in this study is a combination of microwave heating and hot-press. The hybrid sintering apparatus is schematically illustrated in Figure 1. The applicator of the microwave sintering furnace is connected to the 28 GHz gyrotron, the power of which is 10 kW, capable of heating specimens up to 2000°C. The typical dimensions of the applicator are 600 mm in diameter and 900 mm in length, which allows a volume of 300 mm (diameter) by 300 mm (height) in the applicator to offer spatially uniform distribution of wave field. Figure 2 shows photo of the hybrid sintering machine.

On the other hand, the hot-press accessory is mounted in applicator of the microwave sintering furnace and the hydraulic press subsystem can deliver the maximum force of 49 kN. The die used in the hot-press accessory was made from alumina. Alumina lagging is also placed around the die to avoid heat dissipation. The preliminarily formed samples were covered with zirconia powders to prevent them from sticking to the alumina die.

In the case of microwave heating, the internal temperature is usually higher than surface temperature, since they are theoretically uniformly heated and the surface temperature becomes lower due to heat dissipation. The temperature of the

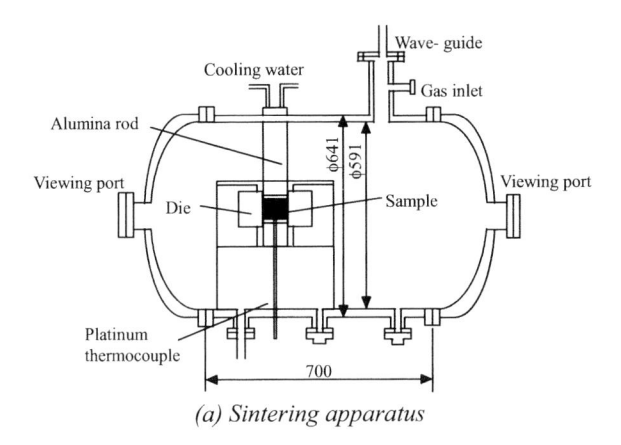

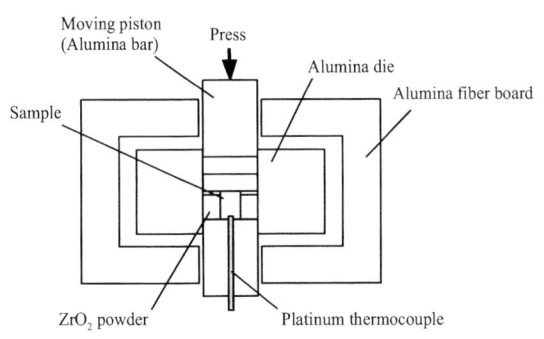

(a) Sintering apparatus

(b) Detailed view of the press system
Figure 1. Schematic diagram of microwave and hot-press hybrid sintering

Figure 2. Photo of the sintering apparatus

specimens was measured with a platinum thermo-couple, the measurement error of which was less than 5°C. In order to measure the internal temperature, a hole of 5 mm in diameter and 1.5 mm deep was made on the bottom of the specimen and the tip of the thermo-couple was inserted into the hole. The measured temperature was used in the feedback control of the output power of the microwave generator to ensure the temperature of the specimens in agreement with the scheduled temperature of the sintering process.

The temperature scheduling of microwave heating and the pressure scheduling of hot-press for the sintering of PZT are shown in Figure 3. In the hybrid sintering process, the temperature of specimens was raised from the room temperature to the sintering temperature at the rate of 1800°C/hour, maintained for 30 minutes and then cooled at the same rate as that of the heating phase. The whole cycle of the hybrid sintering process should have taken about two hours. However, the temperature of specimens decreases much more slowly than the scheduled rate due to the lack of forced cooling and the actual sintering process took about 7 hours. Sintering temperatures between 1100°C and 1250°C were tried at the step of 50°C in the experiment. The pressure of hot-press was raised to the predetermined value (20 MPa or 40 MPa) when the temperature of the PZT specimens was maintained at the maximum sintering temperature. Since the conventional process took about 55 hours, the hybrid process reduces the total sintering time by about 87%.

The temperature scheduling of conventional sintering process and temperature and pressure scheduling of hybrid sintering process for PNN-PZT are shown in Figure 4. In the conventional sintering process, the temperature was raised from room temperature to 300°C at the rate of °C/h, kept for 2 hours and then raised to the sintering temperature at the rate of 100°C/h. The specimens were kept at the sintering temperature for 2 hours, cooled at the rate of 100°C/h to 600°C, and then cooled naturally. In the hybrid sintering process, the temperature was raised at the rate of 400°C/h to 600°C, then at the rate of 600°C to the sintering temperature, kept for 30 min and then cooled at 200°C/h to the room temperature. A pressure of 20 MPa was applied when the specimens were kept at the sintering temperature. The conventional process took about 23 hours, but the hybrid process was only about 7 hours, about 70% reduction in total sintering time.

2.3. Evaluation of Properties

Specimens of 1.2 mm thick were cut from the hybrid-sintered PZT and PNN-PZT cylinders. The surfaces of the specimens were polished in a polishing machine and the thickness was reduced to 1 mm. The density of the specimens were measured with the Archimedes technique using water. The phase structure of the specimens was established by X-ray diffraction (XRD) using CuKα. Furthermore the PZT specimens were mirror-polished and chemically etched in 30% HNO_3 solution added with several drops of HF. The PNN-PZT specimens were thermally etched. The microstructure of etched surface of the specimens was observed under

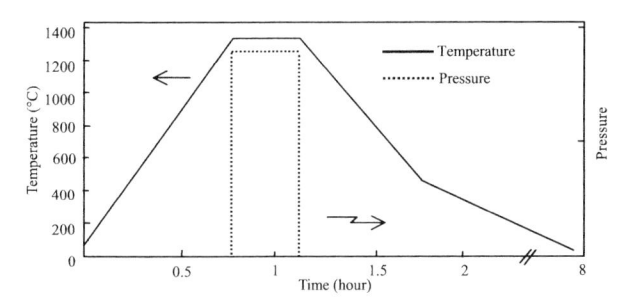

Figure 3. Temperature and pressure schedules of hybrid sintering process

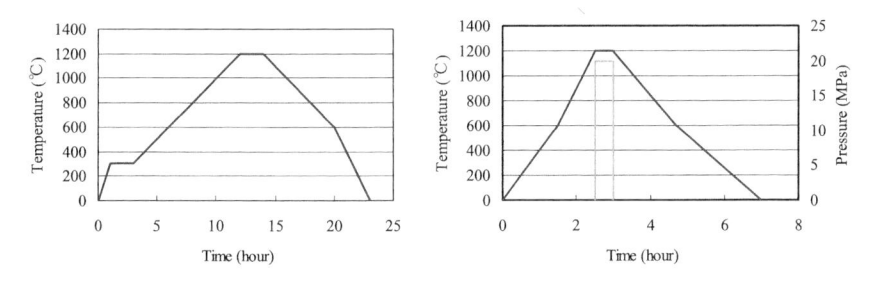

Figure 4. Sintering schedule of PNN-PZT, left: conventional, right: hybrid

scanning electron microscope (SEM), and the grain size was measured with linear intercept method.

The specimens for the measurement of k_p were disks of 15 mm in diameter and 1 mm in thickness, while the specimens for the measurement of k_{31} were plates of 12mm long and 3 mm wide cut from the disks. After sputtered with gold electrodes on both surfaces, specimens were polarized in a silicon oil bath with a DC filed of 2 kV/mm and oil temperature of 170°C for 30 minutes, and then cooled down to 50°C in the field. The dielectric constants of the polarized specimens were measured at the frequency of 1 kHz. The electromechanical coupling factors k_p, relative dielectric constant $\varepsilon_r = \varepsilon_{33}^T / \varepsilon_0$ and Young's modulus $1/S_{11}^E$ were determined by the resonant-antiresonant frequency method with a HP4194A impedance analyzer.

The same measurements were also performed for PZT and PNN-PZT specimens of the same shape fabricated with conventional sintering process. The measurement results are used in the comparison to show the improvement of performances due to the new sintering process.

3. PZT ACTUATORS AND THEIR PROPERTIES

3.1. Microstructure and Grain Size

The dependence of the grain size of the specimens on the sintering temperature was also investigated. The SEM microstructures are shown in Figure 5 (a) and (b) for specimens of the hybrid sintering process and in Figure 5 (c) and (d) for specimens of the conventional sintering process. It can obviously seen that the average grain size of specimens of the hybrid sintering process, which is about 4 μm for the sintering temperature of 1250°C, is larger than that of the specimens of conventional sintering process. It can also be seen from Figure 5(a) and (b) that the average grain size increases with the sintering temperature. For the conventional sintering process, only porous specimens can be obtained at 1150°C as shown in Figure 5(c), but pores were significantly reduced when sintered at 1250°C, as shown Figure 5(d). For the hybrid sintering process, specimens with few pores can be obtained even at 1150°C as shown in Figure 5(a).

Figure 6 illustrates the dependence of the density and grain size of the specimens on the sintering temperature of the hybrid sintering process and the conventional process. The theoretical density of the material is 7.75 g/cm³. The density of the specimens fabricated with hybrid sintering process was larger than that of conventionally sintered specimens when the sintering temperature was above 1100°C. For the conventionally sintered specimens, the density increases with sintering temperature below 1250°C but decreases slightly over 1250°C while the grain size increases with sintering temperature in the whole considered temperature range. For the specimens of hybrid sintering process, no strong correlation was observed between the density and grain size.

3.2. Piezoelectric Properties

Figures 7 and 8 show the electromechanical coupling factor k_p and the piezoelectric constant d_{31} as functions of the sintering temperature. The d_{31} and k_p of the specimens of the same sintering process exhibit very similar tendency of variation. Moreover, The d_{31} and k_p of the hybrid-sintered specimens are much larger than those of the conventionally sintered specimens. The maximum value of k_p of hybrid-sintered specimen, achieved at the sintering temperature of 1250°C, is approximately 75%, and the maximum value of d_{31} of the same specimen, achieved at the same sintering temperature, is about 360×10^{-12} m/V, nearly 40% larger than 260×10^{-12} m/V, the d_{31} of conventionally fabricated specimens. If the curves of d_{31} and k_p are compared with those of grain size in Figure 8, it can be concluded that the increase of both d_{31} and k_p is attributed to the increase of grain size in the specimens. In the conventional sintering, the percentage of pore usually increases with increasing grain size so that the piezoelectric constant d_{31} and coupling factor k_p decrease. However, the grain size was increased without increasing pores in hybrid sintering process due to the rapid and internal heating.

(a) 1150°C Hybrid sintering   *(c) 1150°C Conventional*

(b) 1250°C Hybrid sintering   *(d) 1250°C Conventional*

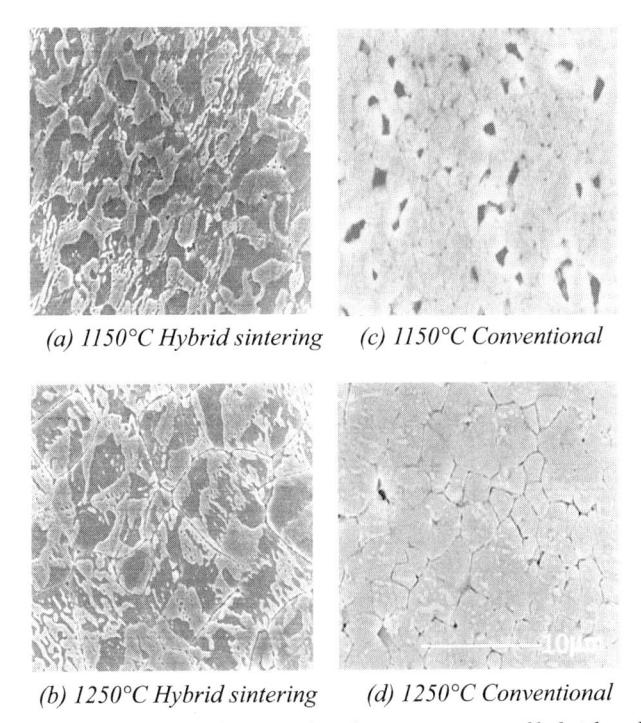

Figure 5. The SEM photographs of microstructure of hybrid and conventionally sintered specimens

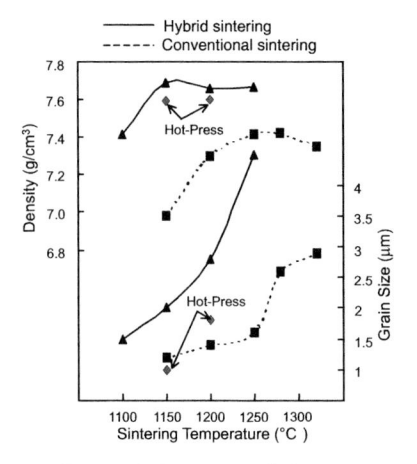

Figure 6. Density and grain size

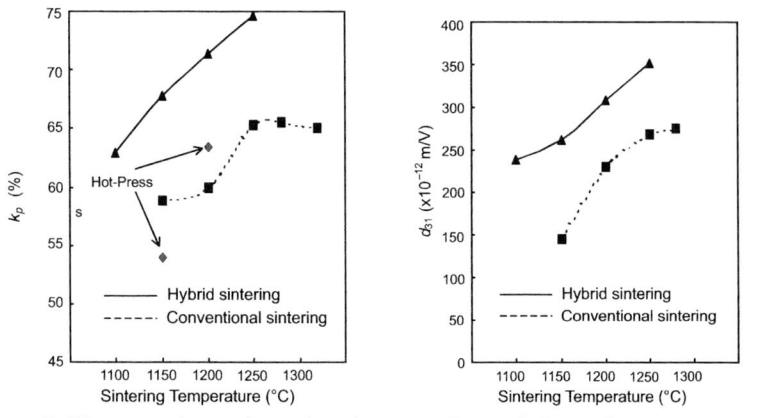

Figure 7. Electromechanical coupling factor   *Figure 8. Piezoelectric constants*

4. PNN-PZT ACTUATORS AND THEIR PROPERTIES

The planar electromechanical coupling factor k_p, dielectric constant ε_{33}^T (the superscript T means the parameter is measured under constant stress), and piezoelectric constant d_{31} of the specimens fabricated by the conventional and hybrid processes were measured. Figure 9 shows that the electromechanical coupling factor k_p increases with sintering temperature under 1230°C and decreases slightly above 1230°C for both processes. The maximum value of k_p at 1230°C was increased from 64.45% of the conventional process to 70.52% of the hybrid process. The dielectric

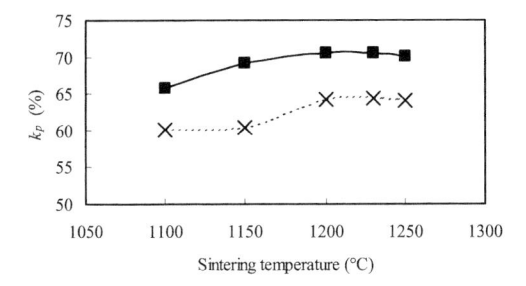

Figure9. Electromechanical coupling factor

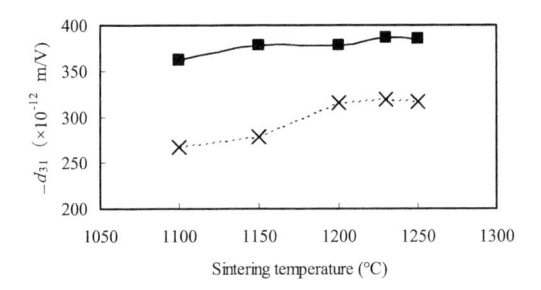

Figure10. Piezoelectric constant

constant versus sintering temperature relationship has the similar tendency as that of k_p.

Figure 10 shows the variation of dielectric constant with sintering temperature. It reaches maximum at the sintering temperature of 1230°C. The maximum value was raised from 319 ×10^{-12} m/V of the conventional process to 386 ×10^{-12} m/V, with about 21% increase. The increase of piezoelectric constant can be attributed to the improvement the microstructure, such as few pores, and suppression of Pb due to shorter sintering time.

5. SUMMARY

A hybrid sintering process, which is the combination of 28 GHz microwave heating and hot-press, was developed and applied to the sintering of a commercial PZT material and a PNN-PZT (Pb(Ni$_{1/3}$Nb$_{2/3}$)$_{0.5}$(Ti$_{0.7}$Zr$_{0.3}$)$_{0.5}$O$_3$) material, with the optimal sintering conditions. Due the rapid and uniform heating of the samples in microwave sintering process, the sintering time was reduced by about 87% for PZT actuators and 70% for PNN-PZT actuators, compared to the time needed for the conventional process. On the other hand, shorter sintering time suppressed less evaporation of Pb in the sintering process. Pores were significantly reduced and the density was increased. The density of hybrid-sintered samples was higher than 99% of the theoretical density due to high pressure of hot-press. The measurement results showed that the electromechanical coupling factor was increased significantly, the piezoelectric constants of PZT could be improved by nearly 40%, and that of PNN-PZT was improved by 21% with the hybrid process.

6. REFERENCE

1. J. Tani, T. Takagi and J. Qiu (1998), Intelligent Material Systems: Application of Functional Materials, *Applied Mechanics Reviews*, ASME., **51**, 505-521.
2. J. Qiu and J. Tani(1995), Vibration Control of a Cylindrical Shell Using Distributed Sensors and Actuators, *J. Inter. Mat. Sys. Struc.*, **6**, pp.474-481.
3. Y. Setuhara, M. Kanai, S. Kinoshita, N. Abe, S. Miyake and T. Saji (1996), Advance Ceramics Sintering using High-Power Millimeter-Wave Radiation, *Mat. Res. Soc. Symp.Proc.*, **430**, pp.533-538.
4. T. Saji (1996), Microwave Sintering of Large Products, Mat. Res. Soc. Symp. Proc., 430, 15-20.
5. H. Takahashi, K. Kato, J. Qiu, J. Tani (2001), of Lead Zirconate Titannate Actuator Manufactured with Microwave Sintering Process, *Jpn. J. Appl. Phys.*, **40**, pp.724-727.
6. H. Takahashi, K. Kato, J. Qiu, J. Tani, and K. Nagata (2001), Fabrication of High-Performance Lead Zirconate Titannate Actuators Using the Microwave and Hot-Press Hybrid Sintering Process, *Jpn. J. Appl. Phys.*, **40**, 4611-4614.
7. H. Takahashi, K. Kato, J. Qiu, J. Tani, and K. Nagata (2001), Properties of Lead Zirconate Titanate Ceramics Determined Using the Microwave and Hot-Press Hybrid Sintering Process, *Jpn. J. Appl. Phys.*, **40**, 5642-5646.

ENERGY RELEASE RATE CRITERIA FOR PIEZOELECTRIC SOLIDS

R.K.N.D. RAJAPAKSE
Department of Mechanical Engineering
University of British Columbia
Vancouver, Canada, V6T 1Z4
E-mail: rajapakse@mech.ubc.ca

S.X. XU
Department of Civil Engineering
University of Manitoba
Winnipeg, Canada R3T 5V6

1. INTRODUCTION

Criteria for crack propagation are fundamentally important in the study of fracture mechanics. Stress-based criteria and energy-based criteria result in quite similar fracture predictions for isotropic elastic materials. However, this is not the case for anisotropic materials. The prediction of crack propagation in piezoelectric materials is further complicated by the coupling between mechanical and electrical fields. Pak [1] studied an impermeable plane crack perpendicular to the poling direction and showed that both a positive and a negative electric field retard crack propagation based on the criterion of total energy release rate. Park and Sun [2] proposed strain energy release rate as the fracture criterion, which predicts that a positive electric field promotes crack propagation and a negative one retards propagation. The assumption of self-similar crack propagation was extensively used in the above studies.

A crack may deviate from a straight path in piezoelectrics due to material anisotropy. In an experimental study, McHenry and Koepke [3] noted that a straight crack branched off under a symmetrically applied electric field. To theoretically predict crack propagation path, the knowledge of energy release rate along an arbitrary direction at the crack tip is required. Analytical solutions for angular distribution of energy release rate have not appeared in the literature. Based on a finite element analysis, Kumar and Singh [4,5] calculated the angular distribution of energy release rate at the tip of an impermeable plane crack perpendicular to the poling direction. They discussed crack propagation using the criterion of maximum energy release rate and assumed isotropic fracture toughness.

Based on the experimental data, Pisarenko, et al [6] and Chen, et al [7], etc, reported that the fracture toughness in piezoelectrics is orientation dependent. In a recent study [8], a stress-based criterion incorporated with fracture toughness anisotropy was applied to predict crack propagation. It was noted that distinctly different propagation directions would be predicted if the assumption of isotropic fracture toughness was relaxed. The study showed that crack orientation with respect to the poling direction has a significant effect on fracture.

The main objective of this study is to examine the applicability of energy release rate criteria in the presence of fracture toughness anisotropy and non-self-similar crack propagation. Angular distribution of energy release rates at the tip of an arbitrarily oriented crack in a plane piezoelectric medium is first sought by using a branched crack model. A simple model of critical fracture energy is used to describe fracture toughness anisotropy. The criteria of modified strain energy release rate and modified total energy release rate are then applied to discuss crack propagation path. The suitability of the two energy-based criteria are discussed in the numerical study.

2. BASIC EQUATIONS

Consider a centre crack (impermeable) of length $2a$ in a piezoelectric plane, as shown in Figure 1a. Two Cartesian coordinate systems, i.e. systems (x, y, z) and (x', y', z'), are used with $y \equiv y'$. The poling direction of the medium is denoted by z'-axis, which makes angle β with respect to the z-axis. Uniform mechanical and electric loading σ_{xx}^∞, σ_{zz}^∞, σ_{xz}^∞ and D_x^∞, D_z^∞ (or E_x^∞, E_z^∞) are applied at far field.

Assuming plane stress or plane strain conditions, the constitutive equations in the xz system can be expressed as,

$$\begin{Bmatrix} \epsilon_{xx} \\ \epsilon_{zz} \\ 2\epsilon_{xz} \end{Bmatrix} = \begin{pmatrix} a_{11} & a_{12} & a_{13} \\ a_{12} & a_{22} & a_{23} \\ a_{13} & a_{23} & a_{33} \end{pmatrix} \begin{Bmatrix} \sigma_{xx} \\ \sigma_{zz} \\ \sigma_{xz} \end{Bmatrix} + \begin{pmatrix} b_{11} & b_{21} \\ b_{12} & b_{22} \\ b_{13} & b_{23} \end{pmatrix} \begin{Bmatrix} D_x \\ D_z \end{Bmatrix}$$

$$\begin{Bmatrix} E_x \\ E_z \end{Bmatrix} = -\begin{pmatrix} b_{11} & b_{12} & b_{13} \\ b_{21} & b_{22} & b_{23} \end{pmatrix} \begin{Bmatrix} \sigma_{xx} \\ \sigma_{zz} \\ \sigma_{xz} \end{Bmatrix} + \begin{pmatrix} d_{11} & d_{12} \\ d_{12} & d_{22} \end{pmatrix} \begin{Bmatrix} D_x \\ D_z \end{Bmatrix} \quad (1)$$

where ϵ_{ij}, σ_{ij}, D_i and E_i $(i, j = x, z)$ denote the strain tensor, stress tensor, electric displacement in the i-direction and electric field in the i-direction, respectively. a_{ij}, b_{ij} and d_{ij} are two-dimensional elastic, piezoelectric and dielectric constants, respectively [8].

The general solutions for plane piezoelectrics can be expressed as [8],

$$\{u_x, u_z, \phi\}^T = 2Re \sum_{n=1}^{3} \{p_n, q_n, s_n\}^T \varphi_n(z_n) \quad (2)$$

where u_x, u_z and ϕ are displacements in the $x-$, z-directions and electric potential, respectively; a superscript T denotes transpose of a vector; $z_n = x + \mu_n z$; $\varphi_n(z_n)$ are complex potential functions; Re denotes the real part of a complex-valued quantity; p_n, q_n, s_n and μ_n $(n = 1, 2, 3)$ are complex constants that depend on material properties and polarization angle β [8].

The boundary conditions on an impermeable crack faces are

$$\sigma_{xz} = \sigma_{zz} = 0, \quad D_z = 0; \quad \Delta u_z = u_z^+ - u_z^- \geq 0 \quad (3)$$

where the superscript $+$ and $-$ indicate the upper and lower crack surfaces, respectively.

Employing the approach of distributed dislocations [9], the potential functions $\varphi_n(z_n)$ in eqn (2) for an arbitrarily oriented crack are [8],

$$\varphi_n(z_n) = [t_{1n}\sigma_{zz}^\infty + t_{2n}\sigma_{xz}^\infty + t_{3n}D_z^\infty]\{z_n - \sqrt{z_n^2 - a^2}\} \quad (4)$$

where complex constants t_{jn} are functions of material properties and polarization angle [8].

Using the polar coordinate system (r, ω) defined at the right crack tip (Figure 1a), crack tip hoop stress, shear stress and hoop electric displacement can be expressed as,

$$\{\sigma_{\omega\omega}, \sigma_{r\omega}, D_\omega\}^T = \frac{1}{\sqrt{2r}} Re \sum_{n=1}^{3} \sqrt{T_n} \{T_n, -R_n, -\delta_n\}^T h_n \qquad (5)$$

where δ_n are defined in eqn (A3), $R_n = \mu_n \cos\omega - \sin\omega$, $T_n = \cos\omega + \mu_n \sin\omega$, $h_n = -\frac{2}{\sqrt{\pi}}(t_{1n}K_I + t_{2n}K_{II} + t_{3n}K_D)$, and K_I, K_{II} and K_D are conventional field intensity factors given by [10],

$$K_I = \sqrt{\pi a}\sigma_{zz}^\infty; \qquad K_{II} = \sqrt{\pi a}\sigma_{xz}^\infty; \qquad K_D = \sqrt{\pi a}D_z^\infty \qquad (6)$$

A set of generalized intensity factors (hoop stress intensity factor $K_{\omega\omega}$, shear stress intensity factor $K_{r\omega}$ and hoop electric displacement intensity factor $K_{D\omega}$) can be defined as

$$< K_{\omega\omega}, K_{r\omega}, K_{D\omega} > = \lim_{r \to 0} \sqrt{2\pi r} < \sigma_{\omega\omega}, \sigma_{r\omega}, D_\omega > \qquad (7)$$

3. ANGULAR DISTRIBUTION OF ENERGY RELEASE RATES

The energy release rate associated with crack extension in a given mode is equal to the work done in closing an infinitesimally extended crack back to its original length [11]. In the case of self-similar crack propagation, energy release rates at the crack tip can be readily obtained by using the straight crack model solution (eqn (4)) in the preceding section. The results are

$$G^M = Im \sum_{n=1}^{3} [q_n t_{1n}(K_I)^2 + p_n t_{2n}(K_{II})^2 + (q_n t_{2n} +$$

$$p_n t_{1n})K_I K_{II} + p_n t_{3n} K_{II} K_D + q_n t_{3n} K_I K_D]$$

$$G^E = Im \sum_{n=1}^{3} [s_n t_{1n} K_I K_D + s_n t_{2n} K_{II} K_D + s_n t_{3n}(K_D)^2] \qquad (8)$$

where G^M denotes mechanical energy release rate, G^E denotes electric energy release rate, and the total energy release rate $G = G^M + G^E$.

As noted by McHenry and Koepke [3], crack branching happens in piezoelectrics even under symmetric loading, which suggests limited application of eqn (8). In order to properly discuss crack propagation directions, the knowledge of energy release rates along an arbitrary direction at the crack tip is required. In the ensuing part of this section, angular distribution of energy release rates is formulated. Considering the extension of a small branch of length

L along the ω-direction (Figure 1b), the corresponding total energy release rate can be expressed as

$$G(\omega) = \lim_{L \to 0} \frac{1}{2L} \int_0^L [\sigma_{\omega\omega}(r,\omega)\Delta u_\eta + \sigma_{r\omega}(r,\omega)\Delta u_\xi + D_\omega(r,\omega)\Delta\phi]\, dr \quad (9)$$

where $\sigma_{\omega\omega}(r,\omega)$, $\sigma_{r\omega}(r,\omega)$ and $D_\omega(r,\omega)$ are given by eqn (5); Δu_η, Δu_ξ and $\Delta\phi$ are the crack branch opening displacement along the η-direction, ξ-direction, and the jump in the electric potential across the branch faces, respectively.

Xu and Rajapakse [8] presented a continuously distributed generalized dislocation model to study branched cracks. The branched crack problem was reduced to the solution of a system of three singular integral equations with generalized dislocation densities b_i ($i = 1, 2, 3$) along the branch line as unknowns. The dislocation density functions $b_i(s)$ are singular at the branch knee and the branch tip. Extract the singularity from the dislocation densities $b_i(s)$ by introducing $B_i(s)$ as

$$b_i(s) = B_i(s)/\sqrt{s(L-s)} \quad (10)$$

The branch opening displacements and jump in the electric potential are

$$\{\Delta u_x(s), \Delta u_z(s), \Delta\phi(s)\}^T = -\int_L^s \{b_1(\xi), b_2(\xi), b_3(\xi)\}^T\, d\xi \quad (11)$$

For an open branch, it is required that, $\Delta u_\eta(s) = \Delta u_x(s)\cos\omega + \Delta u_z(s)\sin\omega \geq 0$.

It can be shown that $B_i(\hat{s})$ is proportional to $\sqrt{L}$. Therefore Δu_η, Δu_ξ and $\Delta\phi$ are also proportional to $\sqrt{L}$ according to eqns (15) and (11). Performing a variable change from $[0, L]$ to $[0, 1]$ on eqn (9) and making use of eqn (7) yield

$$G(\omega) = \lim_{L \to 0} \frac{1}{2L} \int_0^1 \frac{1}{\sqrt{2\pi L\hat{s}}}[K_{\omega\omega}\Delta u_\eta(\hat{s}) + K_{r\omega}\Delta u_\xi(\hat{s}) + K_{D\omega}\Delta\phi(\hat{s})]L\, d\hat{s} \quad (12)$$

Due to the fact that Δu_η, Δu_ξ and $\Delta\phi$ are proportional to $\sqrt{L}$ for $L \to 0$, eqn (12) can be expressed as,

$$G(\omega) = \frac{1}{2\sqrt{2\pi}} \int_0^1 \frac{1}{\sqrt{\hat{s}}}\left[K_{\omega\omega}\frac{\Delta u_\eta(\hat{s})}{\sqrt{L}} + K_{r\omega}\frac{\Delta u_\xi(\hat{s})}{\sqrt{L}} + K_{D\omega}\frac{\Delta\phi(\hat{s})}{\sqrt{L}}\right] d\hat{s} \quad (13)$$

Therefore, energy release rates (strain, electrical and total) are independent of the branch length L. The numerical approach of piecewise quadratic polynomials proposed by Gerasoulis [12] is used in this study to evaluate the singular integral equation for $b_i(s)$ and eqns (10), (11) and (13).

Alternatively, energy release rates along an arbitrary direction at the crack tip can be obtained by following a common practice [13]. Transforming eqn (8) to $\xi\eta$ system yields,

$$G^M(\omega) = Im \sum_{n=1}^{3} [\tilde{q}_n \tilde{t}_{1n}(K_I^b)^2 + \tilde{p}_n \tilde{t}_{2n}(K_{II}^b)^2 + (\tilde{q}_n \tilde{t}_{2n} + \tilde{p}_n \tilde{t}_{1n}) K_I^b K_{II}^b + \tilde{p}_n \tilde{t}_{3n} K_{II}^b K_D^b + \tilde{q}_n \tilde{t}_{3n} K_I^b K_D^b]$$

$$G^E(\omega) = Im \sum_{n=1}^{3} [\tilde{s}_n \tilde{t}_{1n} K_I^b K_D^b + \tilde{s}_n \tilde{t}_{2n} K_{II}^b K_D^b + \tilde{s}_n \tilde{t}_{3n} (K_D^b)^2] \quad (14)$$

where variables $\tilde{p}_n$, $\tilde{q}_n$, $\tilde{s}_n$ and $\tilde{t}_{jn}$ are defined in the $\xi\eta$ system, corresponding to p_n, q_n, s_n and t_{jn} in the xz system; K_I^b, K_{II}^b and K_D^b are field intensity factors at the branch tip.

It is noted that eqn (14) gives the energy release rates based on an infinitesimal extension of an existing infinitesimal small branch, while eqn (13) gives the results due to the occurrence of branching. A comparison of results calculated by these two approaches is made in the numerical study.

4. MODIFIED ENERGY RELEASE RATE CRITERIA

Experimental data for critical energy release rate G_c is currently not available, whereas the critical stress intensity factor K_c has been measured (e.g.[7]). A simple model was used in a previous study to describe K_c [8]. Referring to Figure 1a, $K_c(\theta)$ is expressed by two principal toughness $K_c(\theta = 0°) = K_0$ and $K_c(\theta = 90°) = K_{90}$,

$$K_c(\theta) = K_0 \cos^2\theta + K_{90} \sin^2\theta \quad (15)$$

Note that $\theta = 90° - \omega - \beta$ in Figure 1a, therefore

$$K_c^\omega = K_c(\theta) = K_0 \sin^2(\omega + \beta) + K_{90} \cos^2(\omega + \beta) \quad (16)$$

In view of the fact that energy release rates are quadratic functions of field intensity factors (e.g. eqn (8)), it is assumed that

$$G_c^\omega = k \ (K_c^\omega)^2 \quad (17)$$

where $G_c^\omega = G_c(\theta)$ and $K_c^\omega = K_c(\theta)$ are the critical values of energy release rate and stress intensity factor, respectively; k is a uniform constant for a given piezoelectric material.

Modified strain energy release rate $H^M(\omega)$ and modified total energy release rate $H(\omega)$ are defined as [14]

$$H^M = G^M(\omega)/G_c^\omega; \quad H = G(\omega)/G_c^\omega \quad (18)$$

where $G^M(\omega)$ and $G(\omega)$ are angular distributions of strain and total energy release rates, respectively.

The criterion of modified strain energy release rate assumes that, at a pre-existing crack tip, crack growth takes place along the direction α which renders modified strain energy release rate $H^M(\omega)$ maximum, and a crack propagates when $G^M(\alpha) \geq G_c^\alpha$. Similarly, the criterion of modified total energy release rate assumes that crack growth takes place along the direction α which renders modified total energy release rate $H(\omega)$ maximum, and a crack propagates when $G(\alpha) \geq G_c^\alpha$

5. NUMERICAL RESULTS AND DISCUSSION

In this section, the suitability of the two energy-based criteria are examined through numerical examples. Plain strain conditions are assumed. The condition for an open main crack is satisfied in all cases, and the condition for an open branch after an infinitesimal crack extension is checked during the computation. PZT-4 is used in the numerical study and the material properties are [2]:

$c_{11} = 13.9 \times 10^{10} N/m^2$, $c_{12} = 7.78 \times 10^{10} N/m^2$, $c_{13} = 7.43 \times 10^{10} N/m^2$
$c_{33} = 11.3 \times 10^{10} N/m^2$, $c_{44} = 2.56 \times 10^{10} N/m^2$
$e_{31} = -6.98 C/m^2$, $e_{33} = 13.84 C/m^2$, $e_{15} = 13.44 C/m^2$
$\varepsilon_{11} = 6.00 \times 10^{-9} CV/m$, $\varepsilon_{33} = 5.47 \times 10^{-9} CV/m$

Strain and total energy release rates computed from eqns (13) and (14) are first compared. Consider a general polarization angle ($\beta = 30°$), and a set values of branch angle ω, far field tensile stress σ_{zz}^∞ and electric field E_z^∞. It is found that the total energy release rates from the two different approaches are identical. However, this is not the case for the strain energy release rate, especially when the ratio of electric to mechanical load is large. Therefore, eqn (14) based on the common practice can be applied to calculate total energy release rate, but not strain energy release rate. In view of this observation, eqn (13), corresponding to branch nucleation, is used in the ensuing part of this section.

Taking $\beta = 0°$ and $\omega = 0°$, it is found that the results of present scheme agree with Pak [1]. Figures 2 and 3 present modified strain and total energy-release rates in PZT-4 with varying crack branch angles (ω) under mechanical and electric loading. Based on [7], $K_0/K_{90} = 2$ is assumed in this study. Critical fracture energy release rate G_c^ω is calculated based on eqns (16) and (17). The corresponding results for modified hoop stress intensity factor $K^*(\omega) = K_{\omega\omega}/K_c^\omega$ given in [8] are also shown for the purpose of comparison. Only results corresponding to the ranges of branch angles which satisfy the condition of an open branch are shown.

Consider tensile stress $\sigma_{zz}^\infty = 0.6 MPa$ at far field. Figure 2a shows normalized modified strain energy release rate $H^M G_0/a$, where G_0 is the critical energy release rate along the poling direction ($\theta = 0°$). The results of modified total energy release rate are found to be virtually identical to those of strain energy release rate, and are therefore not shown. The results of normalized modified hoop stress intensity factor are shown in Figure 2b. It is seen that modified strain energy release rate and hoop stress intensity factor have very similar trends except for the case of $\beta = 90°$. The theoretical branching angles based on these two criteria are also close to each other when $\beta \neq 90°$. The potential branching directions, based on the modified strain energy release rate criterion, are $-25°$, $-37°$, $-48°$ and $\pm 64°$ for polarization angle $\beta = 30°$, $45°$, $60°$ and $90°$, respectively. A self-similar extension is expected for $\beta = 0°$. Therefore, in the case of mechanical loading, the criterion of modified strain energy release rate is virtually equivalent to the criterion of modified total energy release rate. The prediction results of branching angles based on the two energy-based criteria are close to those predicted by the stress-based criterion except for the case of $\beta = 90°$.

The results of normalized modified strain energy release rate $H^M G_0/a$ and modified total energy release rate HG_0/a are shown in Figures 3a and 3b, respectively for electric loading. Figure 3c shows the results of modified hoop stress intensity factor. In contrast to the case of mechnical loading, it is found that the results of total energy release rate are now totally different from those of strain energy release rate. Negative values of total energy release rate are observed for all values of ω and β, implying no crack propagation based on the criterion of modified total energy release rate. The modified strain energy release rate and hoop stress intensity factor have somewhat similar trends, but the theoretical branching angles based on these two criteria are quite different. The potential branching directions, based on the modified strain energy release rate criterion, are $\pm 30°$, $-7°$, $-48°$, $-61°$ and $-88°$ for polarization angles $\beta = 0°$, $30°$, $45°$, $60°$ and $90°$, respectively. Note that a crack tends to deviate from the straight extension path regardless of crack orientation with respect to the poling direction. This prediction qualitatively agrees with the experimental findings reported by McHenry and Koepke [3]. Therefore, in the case of electric loading, the predictions based on the criterion of modified strain energy release rate can qualitatively explain the experimental observation of crack branching, while the criterion of modified total energy release rate contradicts it. Distinctly different propagation directions would be predicted if isotropic fracture toughness was used in Figures 2 and 3. For example, based on the criterion of modified strain energy release rate, a self-similar crack extension would be predicted rather than branching along $\omega = \pm 30°$ for a crack under electric loading when $\beta = 0°$.

6. CONCLUSIONS

Analytical solutions for angular distribution of energy release rates at the crack tip in piezoelecrics are successfully obtained and applied to construct two new energy-based criteria, namely modified strain energy release rate criterion and modified total energy release rate criterion. The salient features of the two new criteria include the relaxation of self-similar crack extension assumption and the consideration of fracture toughness anisotropy. In the case of mechanical loading, the two energy-based criteria are found to be nearly equivalent. In the case of electric loading or combined electromechanical loading, the prediction results indicate that the criterion of modified total energy release rate is not suitable. The criterion of strain energy release rate can explain the experimental findings of crack branching under applied electric loading. The complex dependence of crack propagation on the direction of applied electric field, crack orientation with respect to the poling direction and crack branching angle is noted through the criterion of modified strain energy release rate. Coordinated experimental studies are needed to verify the accuracy of theoretical models and play a critical role in future advances of fracture mechanics of piezolectric materials.

Acknowledgment: The work presented in this paper was supported by the Natural Sciences and Engineering Research Council of Canada grant A-6507.

7. REFERENCES

1. Pak, Y. E. (1992), Linear electro-elastic fracture mechanics of piezoelectric materials, *International Journal of Fracture*, 54, 79-100.
2. Park, S. B. and Sun, C. T. (1995), Fracture criteria for piezoelectric ceramics, *Journal of the American Ceramic Society*, 78, 1475-1480.
3. McHenry, K.D. and Koepke, B.G. (1983), Electric fields effects on subcritical crack growth in PZT-4. In *Fracture Mechanics of Ceramics*, (Edited by R.C. Bradt, D.P. Hasselman and F.F. Lange), Vol. 5, 337-352.
4. Kumar, S. and Singh, R. N. (1997a), Energy release rate and crack propagation in piezoelectric materials. Part I: mechanical/electrical load, *Acta Materiala*, 45, 849-857.
5. Kumar, S. and Singh, R. N. (1997b), Energy release rate and crack propagation in piezoelectric materials. Part II: combined mechanical and electrical loads, *Acta Materiala*, 45, 859-868.
6. Pisarenko, G.G., Chushko, V.M. and Kovalev, S. P. (1985), *Journal of the American Ceramic Society*, 68, 259.
7. Chen, W., Lupascu, D. and Lynch, C.S. (1999), Fracture behaviour of ferroelectric ceramics, *Proc. SPIE Conference of Smart Structures and Materials*, 3667, 145-155.
8. Xu, X.-L. and Rajapakse, R. K. N. D. (2000), Theoretical study of branched cracks in piezoelectrics, *Acta Materialia*, 48, 1865-1882.
9. Gross, D. (1982), Spanungintensitatsfaktoren von ribsystemen (stress intensity factor of systems of cracks), *Ing Arch*, 51, 301-310.
10. Suo, Z., Kuo, C. M., Barnett, D. M. and Willis, J. R. (1992), Fracture mechanics for piezoelectric ceramics, *Journal of the Mechanics and Physics of Solids*, 40, 739-765.
11. Irwin, G. R. (1957), Analysis of stresses and strains near the end of a crack traversing a plate, *Journal of Applied Mechanics*, 24, 361-364.
12. Gerasoulis, A. (1982), The use of piecewise quadratic polynomials for the solution of singular integral equations of Cauchy type, *Computational Mathematics with Applications*, 8, 15.
13. Azhdari, A. and Nemat-Nasser, S. (1996), Energy-release rate and crack kinking in anisotropic brittle solids, *Journal of the Mechanics and Physics of Solids*, 44, 929-951.
14. Azhdari, A. and Nemat-Nasser, S. (1998), Experimental and computational study of fracturing in an anisotropic brittle solid, *Mechanics of Materials*, 28, 247-262.

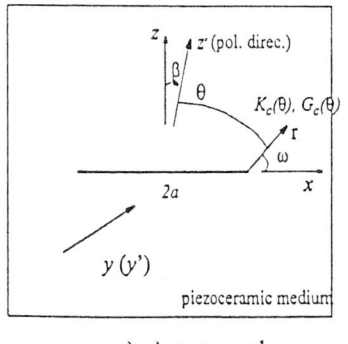

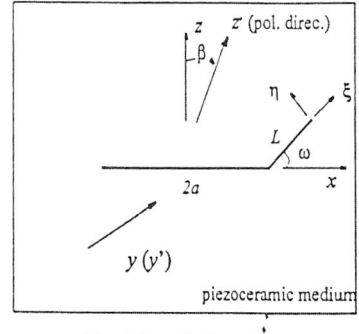

a). A centre crack   b). A branched crack

Figure 1. Crack configurations.

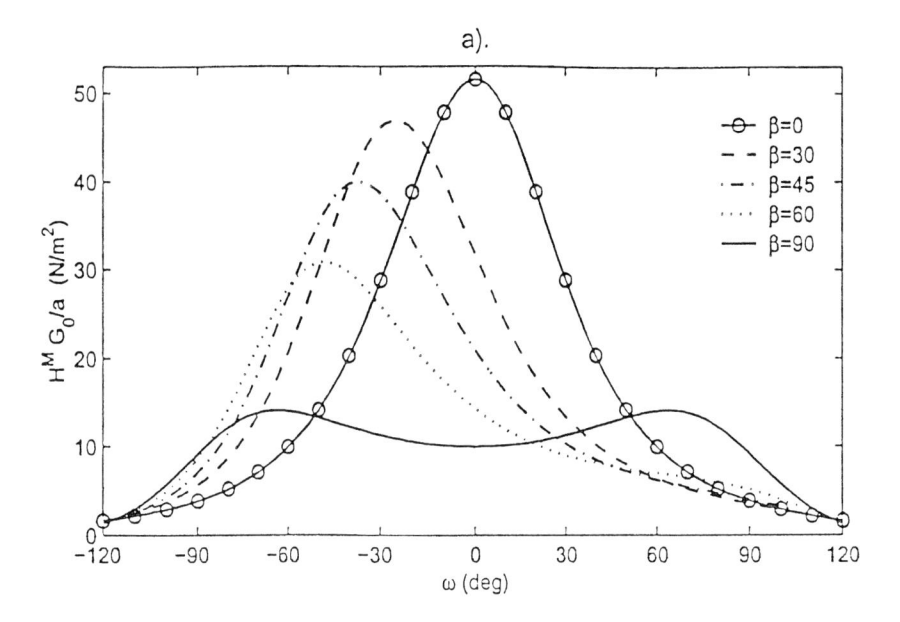

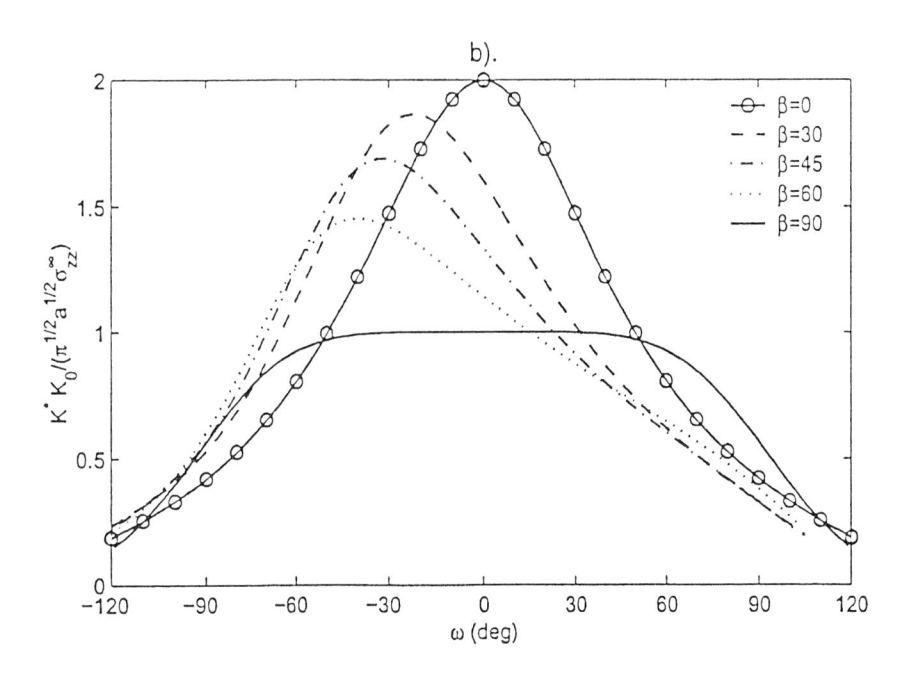

Figure 2. Modified strain energy release rate and modified hoop stress intensity factor under tension ($\sigma_{zz} = 0.6 MPa$)

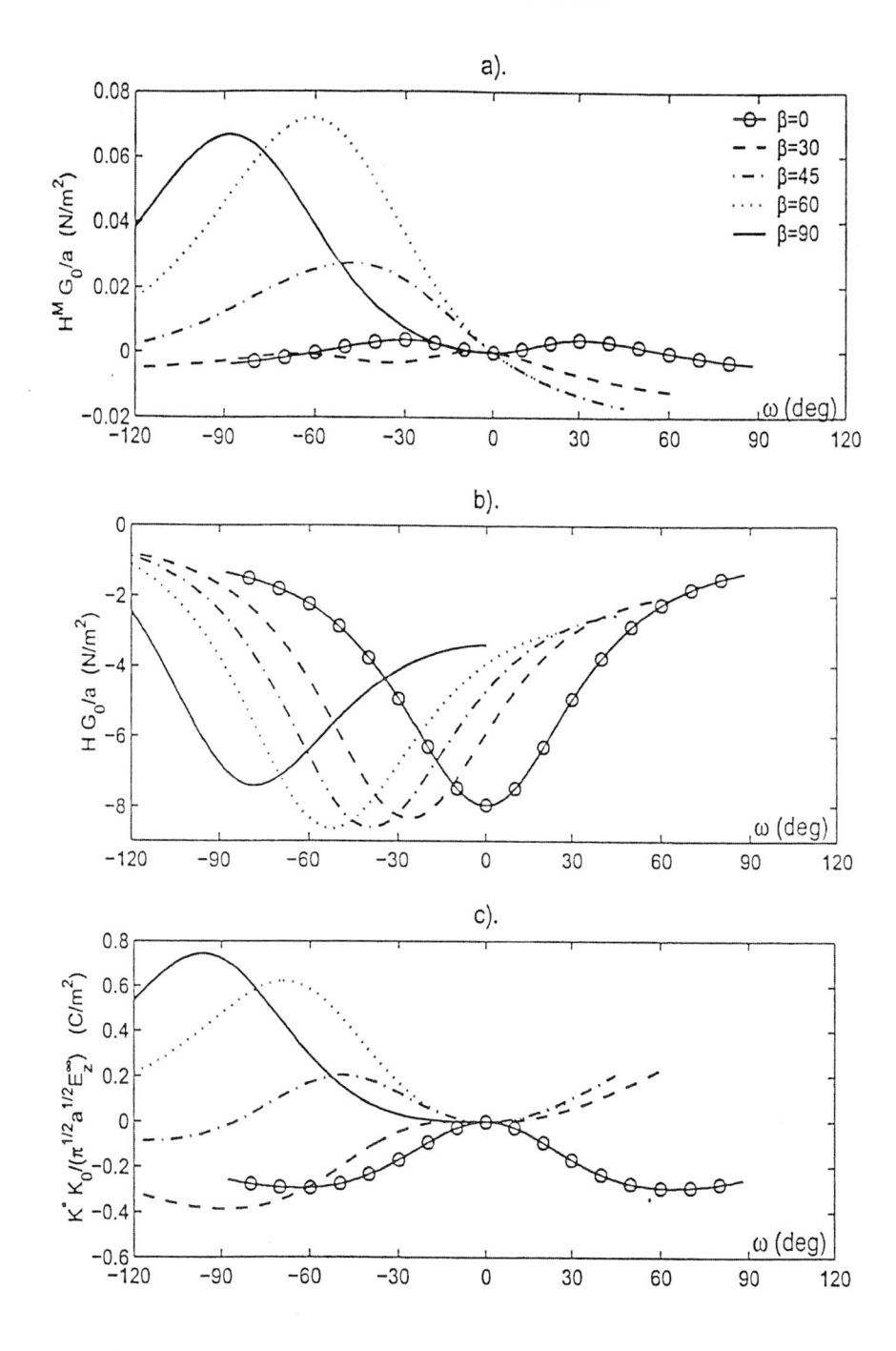

Figure 3. Modified strain energy release rate and modified hoop stress intensity factor under electric loading ($E_z = 12 KV/m$)

INTEGRAL APPROACH FOR VELOCITY FEEDBACK CONTROL IN A THIN PLATE WITH PIEZOELECTRIC PATCHES

IBRAHIM S. SADEK
Department of Computer Science, Mathematics and Statistics
American University of Sharjah, Sharjah, UAE
E-mail: sadek@aus.ac.ae

JOHN C. BRUCH, JR.
Department of Mechanical & Environmental Engineering, and
Department of Mathematics
University of California, Santa Barbara, CA, USA
E-mail: jcb@engineering.ucsb.edu

JAMES M. SLOSS
Department of Mathematics,
University of California, Santa Barbara, CA, USA
E-mail: jmsloss@silcom.com

SARP ADALI
School of Mechanical Engineering,
University of Natal, Durban, South Africa
E-mail: adali@nu.ac.za

1. INTRODUCTION

The developments of smart structures technology in recent years has provided numerous opportunities in applications. Examples where they have been effectively employed include acoustics for noise cancellations with applications to reduce interior noise in aircraft, aerodynamics to adjust wing surfaces and electronics where they are used in the reading heads in video cassette recorders and in compact discs as positioning devices.

Another example of use of piezo materials in adaptive structures is shape control by piezo-actuation. Adaptive materials and structures are presently being used in a variety of applications involving static control such as robotic end effectors and space structures. One of the important issues in the use of piezo actuators is their optimal deployment to minimize their weight and improve performance. In the present study the effectiveness of piezoelectric actuators, whose input is obtained by the sensor output, is studied by means of an integral equation formulation for varying actuator and sensor location. This type of control introduces a damping component.

Here, we consider the distributed vibration control of a thin plate. The plate is assumed to have a laminated patch piezoelectric sensor on the bottom of the plate and a laminated piezoelectric patch actuator on the top of the plate. The control will be a closed-loop velocity feedback control. The signal from the patch sensor is amplified and sent to the patch actuator. The formulation of this problem in terms of a differential equation is given in Tzou [1] and Banks et al. [2].

It will be shown that the solution of the differential equation formulation for the problem can be obtained by solving an integral equation in which the kernel is given explicitly. This will simplify finding a solution to the problem since the differential equation approach is non-standard, the equation itself contains distributions, whereas the integral equation approach is standard. Consequently the integral equation formulation facilitates the study of the effectiveness of the control. Eigensolutions of the integral equation are eigensolutions of the differential equation. In Bruch et al. [3] a similar integral approach was used for a displacement feedback control problem using patches on plates.

2. PROBLEM FORMULATION

Consider a rectangular thin plate of length a and width b occupying the area Ω which is reinforced with a sensor patch at $S^e \subset \Omega$ and actuator patch at $A^e \subset \Omega$, $\Omega = \{(x,y) | 0 < x < a, 0 < y < b\}$. For rectangular patches:

$S^e = \{(x,y) \in \Omega | x_{s_1} < x < x_{s_2}, y_{s_1} < y < y_{s_2}\} \subset \Omega$, $A^e = \{(x,y) \in \Omega | x_{a_1} < x < x_{a_2}, y_{a_1} < y < y_{a_2}\} \subset \Omega$

with $0 < x_{s_1} < x_{s_2} < a$, $0 < y_{s_1} < y_{s_2} < b$, and $0 < x_{a_1} < x_{a_2} < a$, $0 < y_{a_1} < y_{a_2} < b$.

In the absence of mechanical excitations, the equation of motion of the plate with externally applied control moments becomes (Tzou [1])

$$L[u] = \rho h u_{tt} + L_o[u] - G \frac{\partial}{\partial t}\{L_x[u] + L_y[u]\} \tag{1}$$

where the differential operators L_o, L_x, and L_y are defined by

$$L_o[u] = D \Delta^2 u = D\left[\partial_x^4 u + 2 \partial_x^2 \partial_y^2 u + \partial_y^4 u\right],$$

$$L_x[u] = \partial_x^2 M_{xx}^a,$$

and

$$L_y[u] = \partial_y^2 M_{yy}^a,$$

in which

$$\partial_z = \frac{\partial}{\partial z}, \ z = x \text{ or } y, \ M_{xx}^a = C_3 L_1[u] \chi_{A^e},$$

$$M_{yy}^a = C_4 L_1[u] \chi_{A^e}, L_1[u] = -\iint_{S^e} \left[C_1 \partial_x^2 u + C_2 \partial_y^2 u \right] dS^e,$$

$$C_1 = \frac{h^s}{S_s} h_{31} r_1^s, \quad C_2 = \frac{h^s}{S_s} h_{32} r_2^s, \quad C_3 = r_1^a d_{31} Y_p, \quad C_4 = r_2^a d_{32} Y_p,$$

$D = Y h^3 / 12(1-v^2)$ is the flexural rigidity of the plate, Y is Young's modulus of the plate, h^s is the thickness of the sensor layer, h_{31} and h_{32} are the piezoelectric stress constants of the sensor in the x and y directions, respectively, r_1^s and r_2^s are the distances from the neutral surface of the plate to the mid-surface of the sensor, S_s is the area of the sensor patch, h is the plate thickness, v is Poisson's ratio, r_1^a is the effective moment arm in the x – direction, d_{31} is the actuator piezoelectric constant in the x – direction, r_2^a is the effective moment arm in the y – direction, d_{32} is the actuator piezoelectric constant in the y – direction, Y_p is Young's modulus of the actuator, G is the gain, and χ_{A^e} is the characteristic function

$$\chi_{A^e} = \begin{cases} 1 & \text{on } A^e \\ 0 & \text{otherwise}. \end{cases}$$

Consider the case where the plate is simply supported. i.e.,

$$u \Big|_{\partial \Omega} = 0 \text{ and } \partial_n^2 u \Big|_{\partial \Omega} = 0, \text{ (n = normal)}. \tag{2}$$

3. FREE VIBRATION ANALYSIS

The deflection of the simply-supported plate is
$$u(x,y) = \Psi(x,y) e^{i \omega t},$$
where $\omega = \omega_1 + i \omega_2$ is the vibration frequency and $\Psi(x,y)$ is the corresponding deflection mode. Then equation (1) becomes when $\omega \neq 0$

$$L_{xy}[\Psi] = \sigma(\omega) \Psi \tag{3}$$

with $\sigma(\omega) = \omega^2 \rho h / D$ and
$$L_{xy}[\Psi] = (L_o[\Psi])/D - (i \omega G / D)\{ L_x[\Psi] + L_y[\Psi]\}$$
and the associated boundary conditions
$$\Psi \Big|_{\partial \Omega} = 0 \text{ and } \partial_n^2 \Psi \Big|_{\partial \Omega} = 0 \tag{4}$$

The normalized eigenfunctions of the eigenvalue problem, when the plate is simply supported and has no patches, are

$$\varphi_{mn}(x,y) = \frac{2}{\sqrt{ab}} \sin\left(\frac{m \pi x}{a}\right) \sin\left(\frac{n \pi y}{b}\right) \tag{5}$$

4. INTEGRAL EQUATION

Consider the integral equation

$$\sigma^{-1}(\omega)\Psi(x,y) = \iint_\Omega K(x,y;x_o,y_o)\Psi(x_o,y_o)\,dx_o\,dy_o, \qquad (6)$$

where

$$K(x,y;x_o,y_o;\omega) = g(x,y;x_o,y_o) + P(x,y)Q(x_0,y_o;\omega) \qquad (7)$$

and the auxiliary functions g, P and Q are given as follows:

(i). Green's function $g(x,y;x_o,y_o)$ satisfies the system

$$\Delta^2 g = \delta(x-x_o, y-y_o) \qquad (8)$$

$$g = 0 \quad \text{and} \quad \partial_n^2 g = 0 \quad \text{for } (x,y) \in \partial\Omega \qquad (9)$$

and

$$g(x,y;x_o,y_o) = \sum_{m,n=1}^{\infty} g_{mn}\varphi_{mn}(x,y)\varphi_{mn}(x_o,y_o), \qquad (10)$$

where

$$g_{mn} = \frac{1}{\omega_{mn}^2} \quad \text{and} \quad \omega_{mn} = \left(\frac{m\pi}{a}\right)^2 + \left(\frac{n\pi}{b}\right)^2.$$

(ii). $P(x,y)$ needs to satisfy

$$\Delta P(x,y) = (\partial_x^2 + \partial_y^2)P = \chi_{A^e}$$
$$= [H(x-x_{a_1}) - H(x-x_{a_2})][H(y-y_{a_1}) - H(y-y_{a_2})] \qquad (11)$$

and thus

$$P(x,y) = \sum_{m,n=1}^{\infty} p_{mn}\varphi_{mn}(x,y), \qquad (12)$$

where

$$p_{mn} = \frac{-2\sqrt{ab}}{mn\,\pi^2\,\omega_{mn}} A_m(x_{a1},x_{a2})B_n(y_{a1},y_{a2}),$$

in which

$$A_k(x_1,x_2) = \cos\frac{k\pi}{a}x_1 - \cos\frac{k\pi}{a}x_2 \quad \text{and} \quad B_k(y_1,y_2) = \cos\frac{k\pi}{b}y_1 - \cos\frac{k\pi}{b}y_2.$$

(iii).
$$Q(x_o,y_o;\omega) = \frac{i\omega GC_3 L_1[g](x_o,y_o)}{D - i\omega C_3 L_1[P]}, \qquad (13)$$

where the actuator is considered as isotropic, i.e., $C_3 = C_4$, and

$$L_1[g](x_o, y_o) = -\int_{y_{s1}}^{y_{s2}} \int_{x_{s1}}^{x_{s2}} g_s(x, y; x_o y_o) dx dy,$$

where

$$g_s(x, y; x_o, y_o) = \left[C_1 \partial_x^2 + C_2 \partial_y^2\right] g(x, y; x_o, y_o).$$

In view of equation (10), we have

$$L_1[g](x_o, y_o) = -\sum_{m,n=1}^{\infty} \alpha_{mn} g_{mn} \left(\frac{2\sqrt{ab}}{mn\pi^2}\right) A_m(x_{s2}, x_{s1}) B_n(y_{s2}, y_{s1}) \varphi_{mn}(x_o, y_o), \tag{14}$$

in which

$$\alpha_{mn} = C_1 \left(\frac{m\pi}{a}\right)^2 + C_2 \left(\frac{n\pi}{b}\right)^2$$

and

$$L_1[P] = -\sum_{m,n=1}^{\infty} \alpha_{mn} P_{mn} \left(\frac{2\sqrt{ab}}{mn\pi^2}\right) A_m(x_{s_2}, x_{s_2}) B_n(y_{s_2}, y_{s_1}) \tag{15}$$

Note that $K(x, y; x_o, y_o; \omega)$ is not necessarily symmetric in the sense that $K(x, y; x_o, y_o; \omega) \neq K(x_o, y_o; x, y; \omega)$.

5. EQUIVALENCE OF INTEGRAL EQUATION FORMULATION

Note that if the actuator is isotropic and thus $C_3 = C_4$ and if $\{\sigma(\omega), \Psi(x, y)\}$ is a solution of the integral equation (6), then $\{\sigma, \Psi\}$ is a solution of the differential equation

$$L_{xy}[\Psi; \omega] = \sigma(\omega)\Psi,$$

where

$$L_{xy}[\Psi] = (L_o[\Psi]/D - (i\omega G/D)\{L_x[\Psi] + L_y[\Psi]\}$$
$$= (L_o[\Psi])/D - (i\omega G/D) L_1[\Psi](C_1 \partial_x^2 + C_4 \partial_y^2) \chi_{A^e}$$
$$= (L_o[\Psi])/D - (i\omega G/D) L_1[\Psi] \Delta \chi_{A^e}$$

This can be seen by noting that

$$L_o[g(x, y; x_o, y_o)]/D = \Delta^2 g = \delta(x - x_o, y - y_o),$$

where L_o operating with respect to (x, y), and hence

$$L_{xy}[P(x, y)Q(x_o, y_o; \omega)] = (L_o[\Psi])/D - (i\omega G/D) L_1[g](x_o, y_o) \Delta \chi_{A^e}.$$

Also,

$$L_o[P] = D \Delta^2 P = D \Delta \chi_{A^e},$$

$$L_{xy}[P(x,y)Q(x_o,y_o;\omega)] = (L_o[\Psi]/D - (i\omega G/D)L_1[\Psi]\Delta\chi_{A^e}Q(x_o,y_o)$$

and using the definition of Q, equation (13), it is seen that

$$L_{xy}[K(x,y;x_o,y_o;\omega)] = \delta(x-x_o)\delta(y-y_o)$$

and the result follows.

6. METHOD OF SOLUTION OF THE INTEGRAL EQUATION

The method for solving the integral equation involves the following steps.

(i) Choose a complete orthonormal set of functions, e.g., the functions defined by equation (5).

(ii) Expand the kernel $K(x_o,y_o;x,y;\omega)$ in terms of the Fourier series of the orthonormal functions, i.e., using equations (10), and (12)-(15).

(iii) Reduce the integral equation to an infinite set of linear equations. The integral equation

$$u(x,y) = \sigma(\omega)\iint_\Omega K(x,y;x_o,y_o;\varpi)u(x_o,y_o)\,dx_o\,dy_o \qquad (16)$$

becomes

$$\mu(\omega)u(x,y) = \sum_{m,n=1}^{\infty} c_{mn}\,g_{mn}\,\varphi_{mn}(x,y) + P(x,y)\bar{c}(\omega) \qquad (17)$$

where $\mu(\omega) = \sigma^{-1}(\omega)$ and

$$c_{mn} = \iint_\Omega \varphi_{mn}(x_o,y_o)u(x_o,y_o)\,dx_o\,dy_o,$$

$$\bar{c}(\omega) = \iint_\Omega Q(x_o,y_o;\omega)u(x_o,y_o)\,dx_o\,dy_o$$

in which $u(x,y)$ will be determined if c_{mn} and $\bar{c}$ can be determined. A system of equations is derived when equation (17) is multiplied by φ_{jk} and integrated over Ω to obtain

$$\mu(\omega)c_{jk} = [g_{jk}\,c_{jk} + \bar{c}(\omega)p_{jk}], \quad j,k=1,2,3,\ldots\ldots \qquad (18)$$

where equation (12) was used and

$$p_{jk} = \iint_\Omega P(x,y)\varphi_{jk}(x,y)\,dx\,dy.$$

An additional equation is derived by multiplying equation (17) by $Q(x,y;\omega)$ and integrating the resulting equation over Ω and noting that

$$Q(x,y;\omega) = \sum_{m,n=1}^{\infty} q_{mn}(\omega)\varphi_{mn}(x,y),$$

where

$$q_{mn}(\omega) = \iint_\Omega Q(x,y;\omega) P(x,y) \, dx \, dy \ .$$

This computation gives

$$\mu(\omega) \bar{c}(\omega) = \sum_{m,n=1}^{\infty} g_{mn} c_{mn} q_{mn}(\omega) + \bar{c}(\omega) \bar{q}(\omega) \ . \tag{19}$$

Finding the eigensolutions $\{\sigma(\omega), \Psi(x,y)\}$ of the integral equation is then replaced by finding the eigensolutions $\{\mu(\omega), \vec{v}\}$ of the matrix B, i.e., solving the eigenvalue problem

$$B(\omega) \vec{v} = \begin{bmatrix} \bar{D} & C \\ R(\omega) & S(\omega) \end{bmatrix} \vec{v} = \mu(\omega) \vec{v} \tag{20}$$

where
$\bar{D}$ = diagonal matrix $[g_{11}, g_{12}, g_{21}, \ldots]$,

$R(\omega)$ = row vector $[g_{11} q_{11}(\omega), g_{12} q_{12}(\omega), g_{21} q_{21}(\omega), \ldots]$

C = column vector $[P_{11}, P_{12}, P_{21}, \ldots]$, and $S(\omega) = Matrix[\bar{q}]$.

(iv) Consider the integral equation in which the kernel is replaced by the MN – term Fourier Series expansion of the kernel. The kernel K can be approximated by

$$K(x, y; x_o, y_o; \omega) = \sum_{m=1}^{M} \sum_{n=1}^{N} K_{mn}(x, y; x_o, y_o; \omega), \tag{21}$$

where

$$K_{mn}(x, y; x_o, y_o; \omega) = g_{mn} \varphi_{mn}(x_o, y_o) + p_{mn} \varphi_{mn}(x,y) \bar{Q}(x,y) \tag{22}$$

with

$$Q(x_o, y_o; \omega) = \frac{i\omega G C_3 L_1[g]^*(x_o, y_o)}{D - i\omega C_3 L_1[P]^*}, \tag{23}$$

where $L_1[g]^*$ is a truncation of the series given in equation (14) and similarly for $L_1[P]^*$ and equation (15).

(v) This new integral equation has a degenerate kernel and its solution is reduced to solving the algebraic eigenvalue problem (20) with MN equations.

7. NUMERICAL RESULTS

The previous theoretical considerations will be applied to the example of a Plexiglas plate with PVDF rectangular patches. For this analysis a six term approximation will be used, i.e., $m + n \leq 4$. The relevant physical constants are [Tzou (1993)]:

Table 1. Material Properties.

Plexiglas Plate Patches	Plate PVDF
$Y = 3.0 \times 10^9 \text{ N/m}^2$	$Y_p = 2 \times 10^9 \text{ N/m}^2$
$h = 1.6 \times 10^{-3} \text{ m}$	$h^s = h^a = 4 \times 10^{-5} \text{ m}$
$a = 0.2 \text{ m}$	$\nu_p = 0.20$
$b = 0.2 \text{ m}$	$\rho_p = 1.80 \times 10^3 \text{ kg/m}^3$
$\nu = 0.35$	$h_{31} = h_{32} = 4.32 \times 10^8 \text{ V/m}$
$\rho = 1.19 \times 10^3 \text{ kg/m}^3$	$d_{31} = d_{32} = 1 \times 10^{-11} \text{ m/V}$
	$r_1^s = r_2^s = (h^s + h)/2$
	$r_1^a = r_2^a = (h^a + h)/2$

The first six natural frequencies which are defined as $\omega = \left(\sigma \dfrac{D}{\rho h}\right)^{\frac{1}{2}}$, are summarized in Table 2 for case 1: $x_{s_1} = x_{a_1} = 0.002 \text{ m}$, $x_{s_2} = x_{a_2} = 0.198 \text{ m}$, $y_{s_1} = y_{a_1} = 0.002 \text{ m}$, $y_{s_2} = y_{a_2} = 0.198 \text{ m}$.

Table 2. Natural frequencies (Hz) for case 1.

G	m = 1, n = 1	m = 1, n = 2	m = 2, n = 1	m = 2, n = 2	m = 1, n = 3	m = 3, n = 1
0	(61.49, 0.0)	(153.72, 0.0)	(153.72, 0.0)	(245.95, 0.0)	(307.43, 0.0)	(307.43, 0.0)
0.01	(61.49, 0.02)	(153.72, 0.0)	(153.72, 0.0)	(245.95, 0.0)	$(307.43, 0.30 \times 10^{-2})$	(307.43, 0.09)
0.1	(61.49, 0.16)	(153.72, 0.0)	(153.72, 0.0)	(245.95, 0.0)	$(307.43, 0.29 \times 10^{-3})$	(307.43, 0.89)
0.5	(61.49, 0.81)	(153.72, 0.0)	(153.72, 0.0)	(245.95, 0.0)	$(307.43, 0.35 \times 10^{-3})$	(307.38, 4.45)
1.0	(61.49, 1.62)	(153.72, 0.0)	(153.72, 0.0)	(245.95, 0.0)	$(307.43, 0.29 \times 10^{-4})$	(307.43, 8.90)
5.0	(61.44, 8.18)	(153.72, 0.0)	(153.72, 0.0)	(245.95, 0.0)	$(307.43, 0.12 \times 10^{-5})$	(301.71, 44.40)
10.0	(61.31, 17.07)	(153.72, 0.0)	(153.72, 0.0)	(245.95, 0.0)	$(307.43, 0.29 \times 10^{-5})$	(283.68, 88.09)

Table 3 presents the natural frequencies for case 2 where $x_{s_1} = x_{a_1} = 0.02$ m, $x_{s_2} = x_{a_2} = 0.08$ m, $y_{s_1} = y_{a_1} = 0.002$ m, $y_{s_2} = y_{a_2} = 0.198$ m.

Table 3. Natural Frequencies (Hz) for case 2.

G	m = 1, n = 1	m = 1, n = 2	m = 2, n = 1	m = 2, n = 2	m = 1, n = 3	m = 3, n = 1
0	(61.49, 0.0)	(153.72, 0.0)	(153.72, 0.0)	(245.95, 0.0)	(307.85, 0.0)	(307.01, 0.0)
0.01	(61.49, 0.01)	(153.72, 0.0)	(153.72, 0.05)	(245.95, 0.0)	(307.85, .04)	(307.01, 0.04)
0.1	(61.49, 0.05)	(153.72, 0.0)	(153.72, 0.54)	(245.95, 0.0)	$(307.43, 0.35 \times 10^{-5})$	(307.43, 0.86)
0.5	(61.50, 0.27)	(153.72, 0.0)	(153.70, 2.70)	(245.95, 0.0)	$(307.43, 0.36 \times 10^{-5})$	(307.30, 4.32)
1.0	(61.51, 0.54)	(153.72, 0.0)	(153.78, 5.42)	(245.95, 0.0)	$(307.43, 0.35 \times 10^{-5})$	(306.87, 8.63)
5.0	(62.04, 2.64)	(153.72, 0.0)	(155.49, 29.13)	(245.95, 0.0)	$(307.43, 0.37 \times 10^{-6})$	(292.91, 41.22)
10.0	(63.63, 4.76)	(153.72, 0.0)	(160.31, 84.52)	(245.95, 0.0)	$(307.43, 0.35 \times 10^{-7})$	(244.82, 56.70)

In both tables small values of G are taken. It appears that the G value can't be very large and the number of digits used in the MAPLE program also affects the results. When G is large, the system will be unstable.

As seen in Table 2 the natural frequencies in columns 2-5 and those in columns 2,4 and 5 in Table 3 are unaffected by the control being applied for any value of G. Thus there is no controllability of those modes for the shapes of the piezoelectrics used. Tzou (1993) observed the same phenomenon in his study of distributed control of plates. His theoretical derivation suggested that there were observability and controllability problems for antisymmetrical modes if single-piece symmetrically distributed sensor and actuator layers were used. His method for overcoming this problem was to segment both the distributed piezoelectric sensor and actuator into collocated subsections or sub-areas. The theoretical approach presented herein can be extended to include his latter case.

8. CONCLUSIONS

An integral equation formulation and solution were presented for the vibration control of thin plates using finite size piezo sensor and actuator patches. The approach involves converting the piezo control problem to an integral equation one by constructing a kernel in terms of a suitable Green's function and other functions satisfying certain criteria to take the discontinuities of the problem into account. The solution of the integral equation is obtained by introducing an orthonormal set of functions based on the free vibration of the uncontrolled plate and expanding the kernel in terms of these eigenfunctions. The resulting integral equation is reduced to a system of linear equations, the solutions of which yield the eigenfrequencies and the eigenfunctions of the full problem. Numerical results are given for several gain factors and the eigenfrequencies of the uncontrolled and controlled plates are compared.

ACKNOWLEDGEMENT

This material is based upon work supported by the National Science Foundation under award No. INT-9906092.

REFERENCES

1. Tzou H.S. (1993), *Piezoelectric Shells*, Kluwer Academic Publishers, Dordrecht, The Netherlands.
2. Banks H.T., Smith R.C., and Wang Y. (1995), The modeling of piezoceramic patch interactions with shells, plates and beams, *Quarterly of Applied Mathematics* LIII(2), pp. 353-381.
3. Bruch J.C., Jr., Sadek I.S., Sloss J.M., and Adali S. (2002), Analytic solution of a plate vibration problem controlled by piezoelectric patches, *Proceedings of the SPIE, Smart Structures and Materials, Modeling, Signal Processing, and Control in Smart Structures*, edited by S. Rao, San Diego, CA, March 17-21, 2002, Vol. 4693, (in press).

NON-PARAMETRIC REPRESENTATIONS OF MR LINEAR DAMPER BEHAVIOUR

BOGDAN SAPIŃSKI
Department of Process Control, University of Mining and Metallurgy
30 – 059 Cracow, al. Mickiewicza 30, Poland
E-mail: deep@uci.agh.edu.pl

1. INTRODUCTION

Magnetorheological (MR) dampers are semi-active control devices that include promising features such as fast response to variable control signals, full reflexivity of MR fluid transformation and low power demand, which creates particular interest due to its potential application in vibration control. Unfortunately, the highly non-linear dynamic nature of MR dampers (the inherent hysteresis and jump-type behaviour depending on velocity), resulting from specific properties of MR fluids, reduces their applicability.

Two general methods for the modelling of MR damper behaviour exist, called parametric and non-parametric modelling. The parametric method characterizes the MR damper as a system of springs, dashpots and other physical elements and derives its dynamics from the laws of physics. The non-parametric method employs analytical expressions or artificial intelligence techniques that enable us to describe the characteristics of the MR damper. These models demonstrate that complex relationships governing MR damper behaviour can be captured by employing the following: appropriately selected functions (polynomial, shape, delay and offset), polynomial with the power of piston velocity, fuzzy logic as well as neural network, and next fitting the model's parameters to the experimental data obtained for the damper under investigation.

Data for training and checking models used in this study was acquired from experimental tests of the MR linear damper (the RD-1005 series manufactured by Lord Corporation) [7], contrary to the models from references that used data produced by the parametric model developed by Spencer [4]. The effectiveness of the discussed non-parametric models is shown by comparing the measured results with the predicted ones, under various operating conditions.

2. DESCRIPTION OF A MR LINEAR DAMPER

The diagram of a structure of a medium-sized MR linear damper is shown in Fig. 1a (longitudinal section) and in Fig. 1b (cross-section). The damper of cylindrical shape is filled with a type of MR fluid. The magnetic field excited in the damper is generated by the control current (I) in the coil incorporated in the piston. The design of a piston with an integrated coil ensures that the magnetic field is focused within a gap, inside a volume of the active portion of MR fluid.

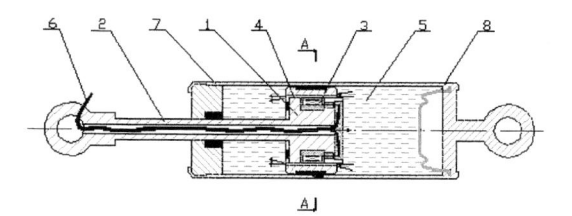

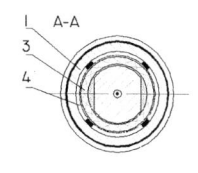

Figure 1a. Longitudinal section: 1 – piston, 2 – rod, 3 – coil, 4 – gap, 5 – MR fluid, 6 - wires, 7 – housing, 8 – accumulator.

Figure 1b. Cross-section: 1 – piston, 3 – coil, 4 – gap.

The MR damper operation is based on the MR effect [6], whose most significant feature is the change of fluid viscosity within a time of milliseconds. As a result, MR fluid flow through the gap is restricted, and in consequence hydraulic resistance against piston displacement increases. Finally, a damping force of appropriate magnitude is generated.

In the absence of a magnetic field ($I=0 \Rightarrow H=0$), the aggregates of ferromagnetic particles are suspended in the carrier fluid (Fig. 2a), their magnetic moments are without any ordered structure and the resultant magnetic moment is equal to zero. In the presence of an external field ($I \neq 0 \Rightarrow H \neq 0$), the aggregates are polarized and their magnetic moments are arranged along field lines. These aggregates form chain-like structures parallel to field lines (perpendicular to fluid flow direction), thereby increasing the shear stress and the fluid viscosity and restricting its motion (Fig. 2b). The magnetic energy required to form such structures increases due to the increase in magnetic field strength.

The behaviour of MR fluid is often described by using the Bingham model. In this model, in the post-yield region the MR fluid exhibits non-Newtonian behaviour (visco-plastic properties) and in the pre-yield one, it exhibits Newtonian-like behaviour (visco-elastic properties), [5].

Bearing in mind that the height of the gap is far less than its width and length, the MR fluid flow in the gap can be approximated by the model with parallel plates. This assumes that the damping force generated by the MR damper has three components that are: the force resulting from static friction (dependent on the seal applied in the damper), the force dependent on the fluid viscosity, and the force dependent on magnetic field strength (particularly on magnetic flux density in the gap due to the current intensity in the coil) [8].

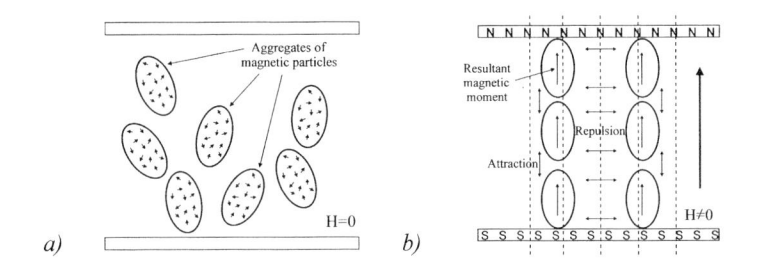

Figure 2. Behaviour of MR fluid: a) H=0, b) H≠0.

3. EXPERIMENTAL TESTS FOR THE EVALUATION OF FORCE CHARCTERISTICS

The goal of laboratory tests was to identify the non-linear force characteristics of the MR linear damper in a broad range of stable operating conditions.

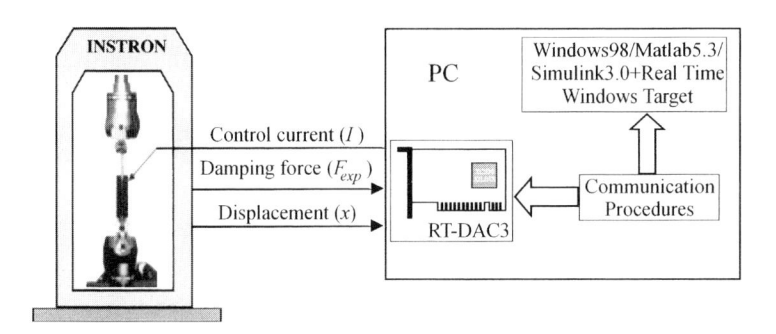

Figure 3. Schematic diagram of experimental setup damper tests.

The tests were performed at the experimental setup, see Fig. 3, with a computer-controlled INSTRON test machine and data acquisition system with the I/O board of RTDAC-3 type and in the software environment of Matlab/Simulink and Real Time Windows Target [10]. The machine was programmed to move up and down in a sinusoidal wave for seven levels of frequency (f) (0.5, 1, 2.5, 4, 6, 8, 10)Hz. The responses were measured for eight levels of control current (I) (0.0, 0.2, 0.4, 0.6, 0.8, 1.0, 1.2, 1.6)A. Data for the time histories of piston displacement (x) and damping force (F_{exp}) for each applied current were recorded with a sampling frequency of 1000Hz, and, together with the computed velocity data ($\dot{x}$), used to replicate the input and output relationship of the non-parametric models discussed below.

4. DEVELOPMENT OF NON-PARAMETRIC REPRESENTATIONS

The general idea of a non-parametric model construction of a particular device is associated with finding the best approximation of the function, which enables us to represent its dynamic behaviour. In the case of MR dampers this idea is presented in Fig. 4 and is outlined by the following steps:
- collect samples of input and output data as measured by the damper tests,
- use an appropiate learning algorithm for a non-parametric model to be created that relates displacement, velocity and current to the damping force,
- validate a new model of the damper by comparison of the predicted damping force to the measured one by the same input signals.

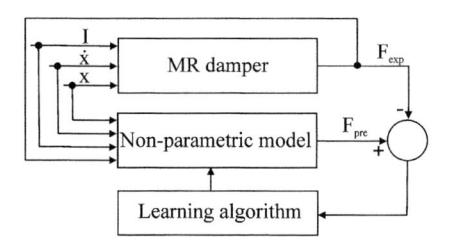

Figure 4. Construction of a non-parametric model of a MR damper.

In order to identify the non-parametric models of the damper, the following algorithms were employed: *fmincon* ⇒ model with selected functions, *polyfit* ⇒ model with polynomial function, *anfis* ⇒ fuzzy model, *Levenberg-Marquardt* ⇒ neural network model. The identified model will fully represent the damper behaviour if the acquired data is high quality and contains information in the whole range of operating conditions. Knowledge of the models was acquired using experimentally obtained displacement and damping force data and then velocity computed at various levels of applied current. The learning was completed when the predicted damping force (F_{pre}) was "close" to that measured (F_{exp}). The force-velocity measured data was considered for lower and upper loop separately.

4.1. Model with selected functions

The model is based on an approximation of the force-velocity characteristics by the use of selected functions [1]. It was developed by:
- separate evaluation of different aspects of MR damper behaviour using force traces: force vs. time, force vs. displacement and force vs. velocity at various levels of current,
- combination of a series of equations associated with appropriately selected functions to capture trends characterizing this behaviour,
- use of constrained nonlinear optimization to obtain the coefficients of these equations that fit the experimental data.

In order to capture the highly nonlinear dynamic nature of the damper, a combination of the functions which are presented in Fig.5, was employed:

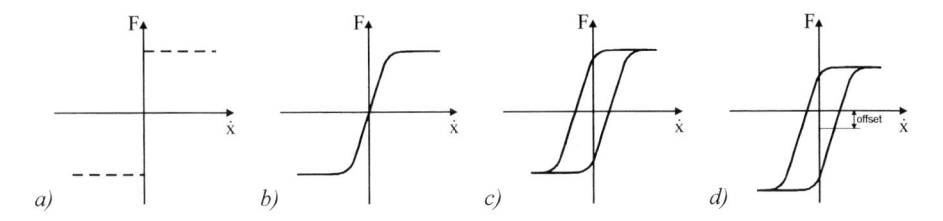

*Figure 5. Shaping of force-velocity relationship by use of the selected functions:
a) saturation, b) shape, c) delay, d) offset.*

For the estimation of model parameters the subroutine *fmincon* (which finds the minimum value of a function with several variables under some constraints imposed on the variables) available in Optimization Toolbox of Matlab was used. The objective function was assumed as:

$$J = \sum_{k=1}^{N} [F_{pre}(k) - F_{exp}(k)]^2 \qquad (1)$$

where N is total number of the experimental data for each level of applied current within a range of (0.0-1.6)A.

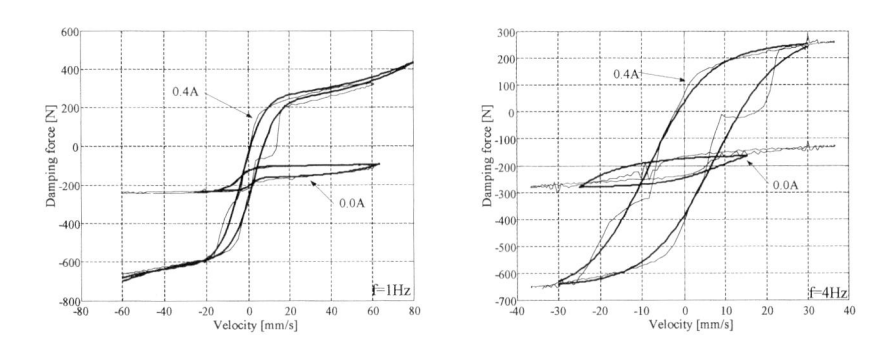

Figure 6. Model with selected functions: comparison of force-velocity curves between measurement (—) and prediction (—).

The comparison of the predicted and measured force-velocity curves is shown in Fig. 6 and refers to the values of function parameters obtained for I=0.0A; I=0.4A and for f=1Hz; f=4Hz. These values were computed as the local minima of a function, so they cannot be treated as parameters which enable us to predict the damping force for other applied currents with similar accuracy, as was mentioned in [1].

4.2. Polynomial model

This polynomial model is based on the assumption that the hysteresis loop is divided into two regions according to the acceleration sign $(\ddot{x})$, see Fig.7.

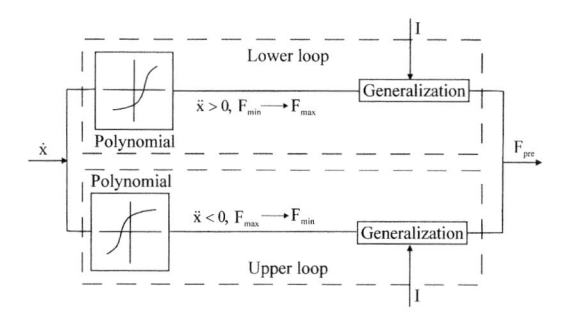

Figure 7. Model for damping force prediction by polynomial with the power of piston velocity.

Then, each loop can be fitted by the polynomial with the power of damper piston velocity $(\dot{x})$ [3]. Thus, the damping force can be expressed by

$$F_{pre} = \sum_{i=0}^{n} a_i \dot{x}^i \qquad (2)$$

where a_i are the coefficients to be determined from the force-velocity data obtained in the experiment by the objective function, as defined in formula (1).

In order to obtain the force-velocity relationship in the form as written in (2) displacement data handling was necessary by employing smoothing spline functions. For this purpose the subroutine *spaps*, from Splines Toolbox of Matlab was used. Then the velocity data was calculated by numerical differentiation. Afterwards, the lower and upper loops of hysteresis were approximated by the polynomial with the power of velocity using the *polyfit* subroutine, a standard function of Matlab. The polynomial degrees determined by the restricted velocity range are: n=5 for I=0.0A and n=4 for I=0.4A for both f=1Hz and f=4Hz. The comparison of the predicted and measured force-velocity curves is shown in Fig. 8.

The value of n=6 mentioned in [3] made computation difficult numerically from the available experimental data for both the lower and upper loop. Both the polynomial form of the force-velocity relationship and the statement that coefficients a_i can be linearized with respect to the applied current [3] provide promising features for the real time control of the damping force.

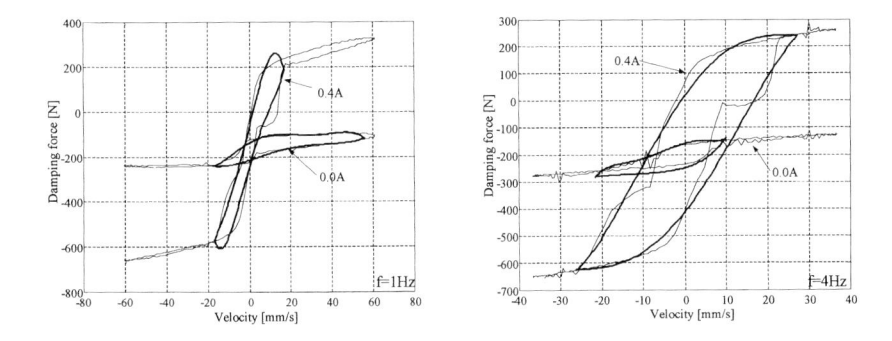

Figure 8. Polynomial model: comparison of force-velocity curves between measurement(—) and prediction (—).

4.3. Fuzzy model

The fuzzy model was assumed to be based on first order Takagi-Sugeno-Kang architecture [9], which seems to be well suited to capture nonlinearities inherent in the damper, see Fig.9.

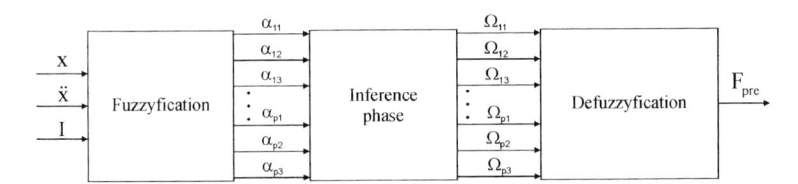

Figure 9. The fuzzy model structure of a MR damper.

The inference system was designed with three inputs and single output. It was assumed that the initial inference system contained no knowledge of the target behaviour. In order to have clear displacement and velocity data for the training and checking models, an analytical formula for the displacement signal was estimated. Then, this signal was differentiated and the obtained data served to estimate the analytical formula for the new velocity signal. Thus, displacement and velocity data became continuous and differentiable functions of time. This data, together with the measured force data, was used to create the fuzzy model by the use of Adaptive Neuro-Fuzzy Inference System (*anfis*), available in Fuzzy Logic Toolbox of Matlab. This neuro-fuzzy application uses a hybrid learning algorithm to create a fuzzy inference system, whose membership functions are iteratively adjusted to a given set of input and output data.

The number of input membership functions was determined during simulation tests, bearing in mind a relatively small learning error, no fuzzy model overfitting and computing efficiency. As a result, nine input memberships from 100 samples for f=1Hz and eleven input membership functions from 125 samples for f=4Hz in both bell or gaussian type were assumed. Thus, it has turned out that the designed three input and single output fuzzy inference system was able to represent the MR damper behaviour with satisfactory accuracy. The predicted force-velocity characteristics, see Fig.10, resemble the experimental data more closely than the two previously considered models. It can be observed especially at higher velocities. It was also demonstrated in [9] that the speed of execution for the fuzzy model was far less than for parametric ones. Therefore this model can become a useful tool for semi-active control.

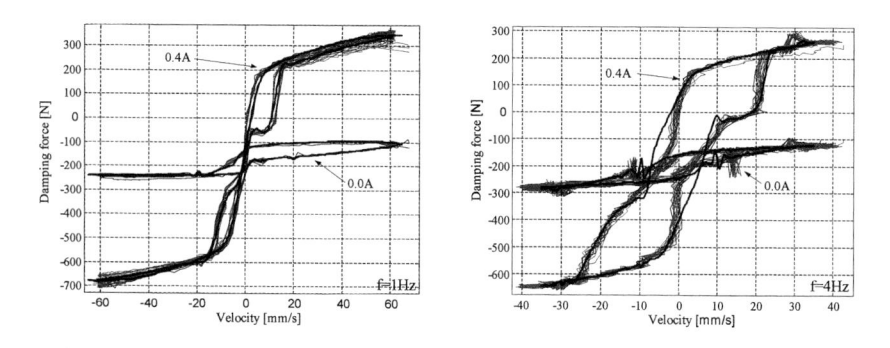

Figure 10. Fuzzy model: comparison of force-velocity curves between measurement (—) and prediction (—).

4.4. Neural network model

The key aim in the construction of a neural network model was to find one of the best approximations of the function that is able to represent the behaviour of the damper with desired accuracy [2]. This involves two main problems to be considered, i. e., how many hidden layers should be chosen and how many neurons should be used in each layer to achieve such accuracy. The first one is usually solved based on research that for most problems one or two hidden layers are sufficient (there are no methods for a priori approximation of the appropriate structure of the network). The second is associated with determination of an optimal network structure, and refers to the sufficient number of neurons characterizing the capacity of the network to emulate highly nonlinear MR damper behaviour (there is no relation to the size of the neural network that could clearly explain this problem). Moreover, there is no universal solution for the problem of optimal structure of the neural network model for the damper. This problem was particulary significant because the quality of the neural network model of the damper was identified by use of training data selected from data acquired experimentally. It was contrary to the model, that used data generated from the parametric model developed by Spencer.

According to the above, data sets acquired in measurements (from which training and checking data was selected) and the structure of multi-layer feed-forward networks were determined a priori. The networks were trained and learned using the standard approach for minimization of mean-square error criteria. For this purpose the Gauss-Newton based *Levenberg-Marquardt* algorithm, available in Neural Networks Toolbox of Matlab, was employed. For the initialization of a layer's weights and biases the Nguyen-Widraw algorithm was used.

The input data was divided into sets according to frequency, current and character of input signal i.e. displacement and velocity. Unfortunately, because of high nonlinearities inherent in the damper and not high enough quality of the acquired data, no universal structure of the neural network that can represent the behaviour of the damper was found. Moreover, the different network structures (not optimized) for a decreasing and increasing input signal were obtained. An example of neural network structure for velocity data is given in Fig. 11. The comparison of the predicted and measured force-velocity curves for this structure is shown in Fig. 12.

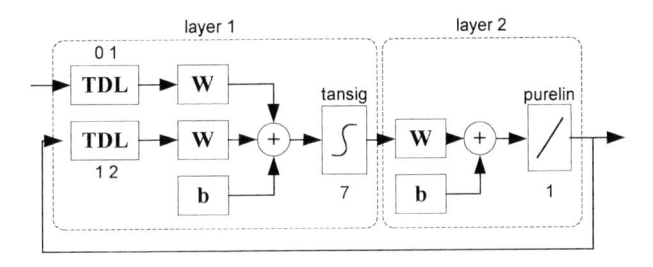

Figure 11. The neural network structure for velocity data.

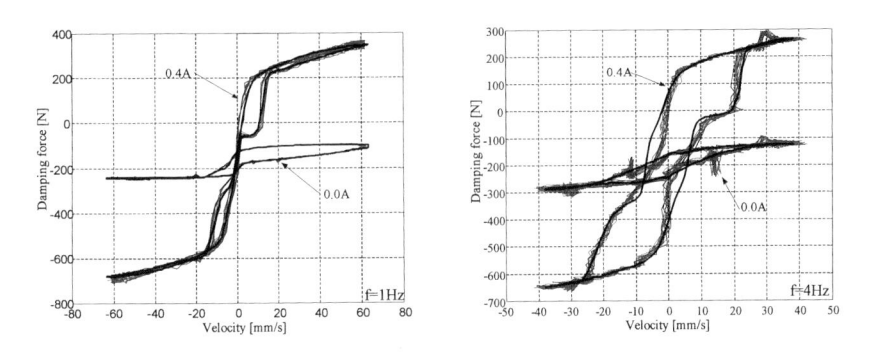

Figure 12. Neural network model: comparison of force-velocity curves between measurement (—) and prediction (—).

It is clearly seen that the trained and learned network structure enable us to represent the MR damper behaviour with comparable accuracy to that achieved for the fuzzy model. So the neural network model also provides promising features for MR damper evaluation and semi-control development. The value of mean square error (*rms*) of the network learning was tightly linked to the type of learning sets (diplacement or velocity) and quality of data acquired in measurement.

5. CONCLUSIONS

Non-parametric models employing different approximation methods of the function representing the behaviour of MR linear dampers were discussed. Contrary to previously presented papers, the non-parametric models were identified using data acquired experimentally. It was demonstrated how they can alternatively portray the behaviour of the damper in a broad operating range and what the shortcomings of non-parametric models are. These models feature reasonable fidelity and therefore they are able to predict MR damper force characteristics with an accuracy satisfactory for system evaluation. This is a big challenge to undertake further investigations for the semi-active control communitty either to improve the MR damper models or assume its less precise control. However, at present it is difficult to recommend, unequivocally, a particular non-parametric model for the purpose of control algorithm development. The improvement of predicted results will be possible by collection of high quality input and output data sets used for model identification.

6. REFERENCES

1. Ahmadian, M. (1999) A Non Parametric Model for Magnetorheological Dampers, *Proc. ASME Design Engineering Technical Conferences* 1-10.
2. Chang, C.C. and Roschke, P. (1998) Neural Network Modeling of a Magnetorheological Damper, *Journal of Intelligent Material Systems and Structures* **9** 755-764.
3. Choi, S.B. and Lee, S. K. and Park, Y. P. (2001) A hysteresis model for the field-dependent damping force of a magnetorheological damper, *Journal of Sound and Vibration* **245** 375-383.
4. Dyke, S. and Spencer, B. and Sain, M. and Carlson, J. (1996) Phenomenological Model of a Magnetorheological Damper, *Journal of Engineering Mechanics.*
5. Jolly, M. and Bender, J. W. and Carlson, J. D. (1999) Properties and Applications of Commercial Magnetorheological Fluids, *Journal of Intelligent Material Systems and Structures* **10** 5-13.
6. Kordonski, W. (1993) Elements and Devices Based on Magnetorheological Effect, *Journal of Intelligent Systems and Structures* **4** 65-69.
7. Sapiński, B. (2002) Parametric Iidentification of MR Linear Automotive Size Damper, *Journal of Theoretical and Applied Mechanics*, (to be published).
8. Sapiński, B. (2001) Influence of Fluctuations of Magnetic Field on Mechanical Characteristics of MR Damper, *Mechanics* **83** 243-252.
9. Schurter, K. C. and Roschke, P. N. (2000) Fuzzy Modeling of a Magnetorheological Damper Using ANFIS, *Proc. 9 th IEEE International Conference on Fuzzy Systems* 122-127.
10. Using MATLAB (1999) The Math Works Inc. Natick, Massachusetts.

This research is supported by the National Science Council: Grant No. 8T07B 03520

ACTIVE CONTROL OF SMART STRUCTURES USING PORT CONTROLLED HAMILTONIAN SYSTEMS

KURT SCHLACHER and KURT ZEHETLEITNER
Department of Automatic Control and Control Systems Technology
Christian Doppler Laboratory for Automatic Control in Steel Industries
Johannes Kepler University of Linz, Altenbergerstaße 69, A-4040, Austria
E-mail: kurt.schlacher@jku.at

1. INTRODUCTION

Smart structures based on piezoelectric composites have turned out to be excellent actuators and sensors for active and passive damping in vibration control. In the case of small displacements a linear approach suffices [8], if hysteresis or depolarization of the active material are negligible [7]. This contribution presents a unifying way for the mathematical modeling of smart structures based on **P**ort **C**ontrolled **H**amiltonian **S**ystems, see [3].

It is well known that the generalized Hamiltonian formulation is extremely useful for lumped parameter control systems, where the mathematical model includes nonlinear algebraic and ordinary differential equations. As a result of this development the interest in the PCH formulation of distributed parameter control systems, the mathematical model includes also nonlinear partial differential equations, is strongly increasing. Some successfully solved examples lead us to suppose that in the latter case the Hamiltonian approach is even more important than in the first one [6]. Recently, a PCH formulation of Maxwell's equations has been presented [3].

The present contribution is organized as follows. In the next section we draw together the mathematical notation required for the subsequent investigations, see [1], [5]. The third section presents an introductory example of a PCH system together with its rigorous geometric picture. The fourth section generalizes this approach to lumped and distributed parameter systems of the Lagrangian and Hamiltonian type. The general mathematical model of a piezoelectric structure is presented in the fifth section. In the sixth section we give some remarks concerning the controller design and close this contribution with some final remarks.

2. SOME NOTATIONS

A smooth bundle with the $\dim \mathcal{E} = (p+q)$-dimensional total manifold $\mathcal{E}$ the $\dim \mathcal{B} = p$-dimensional base $\mathcal{B}$ and projection $\pi : \mathcal{E} \to \mathcal{B}$ is denoted by $(\mathcal{E}, \pi, \mathcal{B})$. The set $\pi^{-1}(x) = \mathcal{E}_x$ is called the fiber over x. We use locally adapted coordinates (x^i) for $\mathcal{B}$ and (x^i, u^α) for $\mathcal{E}$, such that x^i, $i = 1, \ldots, p$ denote the independent variables and u^α, $\alpha = 1, \ldots, q$ denote the dependent variables. Latin indices are employed for the independent and Greek indices for the dependent variables, and we use the Einstein convention for sums.

Let $\mathcal{M}$ be a smooth finite-dimensional manifold, then $\mathcal{T}(\mathcal{M})$ denotes the tangent bundle of $\mathcal{M}$, $\mathcal{T}^*(\mathcal{M})$ the cotangent bundle of $\mathcal{M}$, $\wedge_k^*(\mathcal{M})$ the corresponding exterior k bundles and $\wedge^*(\mathcal{M})$ the exterior algebras over $\mathcal{M}$. $d : \wedge_k^*(\mathcal{M}) \to \wedge_{k+1}^*(\mathcal{M})$ is the exterior derivative and $i : \mathcal{T}(\mathcal{M}) \times \wedge_{k+1}^*(\mathcal{M}) \to \wedge_k^*(\mathcal{M})$ is the interior product written as $i_X(\omega)$ with $X \in \mathcal{T}(\mathcal{M})$ and $\omega \in \wedge_{k+1}^*(\mathcal{M})$. $\wedge$ denotes the exterior product of the exterior algebra $\wedge^*(\mathcal{M})$. The Lie derivative of $\omega \in \wedge^*(\mathcal{M})$ along the tangent vector field $f \in \mathcal{T}(\mathcal{M})$ is written as $f(\omega)$. Furthermore, we set $\wedge_0^*(\mathcal{M}) = C^\infty(\mathcal{M})$, $\wedge_1^*(\mathcal{M}) = \mathcal{T}^*(\mathcal{M})$.

Let $\sigma = (x^i, f^\alpha(x))$ be a section $\sigma : \mathcal{B} \to \mathcal{E}$ of $(\mathcal{E}, \pi, \mathcal{B})$, such that $\pi \circ \sigma = \mathrm{id}_\mathcal{B}$, $u = f(x)$ is met for points of its domain, then we write

$$\frac{\partial^k}{\partial_1^{j_1} \ldots \partial_p^{j_p}} f^\alpha = \partial_J f^\alpha , \quad \partial_i = \frac{\partial}{\partial x^i}$$

for a k-th order partial derivative of f^α with respect to the independent coordinates with the ordered multi-index $J = j_1, \ldots, j_p$ and $k = \#J = \sum_{i=1}^p j_i$. We use the abbreviations $1_k = j_1, \ldots, j_p$, $j_i = \delta_{ik}$ and $J + \bar{J} = j_1 + \bar{j}_1, \ldots j_p + \bar{j}_p$. The first prolongation of σ, the map $x \to (x^i, f^\alpha(x), \partial_i f^\alpha(x))$ is denoted by $j^1\sigma$, it is an element of the first jet bundle $(J^1(\mathcal{E}), \pi, \mathcal{B})$ of $\mathcal{E}$. Furthermore, adapted coordinates (x^i, u^α) of $\mathcal{E}$ induce an adapted coordinate system of $(J^1(\mathcal{E}), \pi, \mathcal{B})$ of $\mathcal{E}$, which is given by $(x^i, u^\alpha, u_{1_i}^\alpha)$ with the pq new coordinates $u_{1_i}^\alpha$. From $(J^1(\mathcal{E}), \pi, \mathcal{B})$ one derives another important structure, the bundle $(J^1(\mathcal{E}), \pi_0^1, \mathcal{E})$. The nth-prolongation $j^n\sigma$ of sections of $\mathcal{E}$ to higher order jet bundles $(J^n(\mathcal{E}), \pi, \mathcal{B})$ are defined analogously, as well as the bundles $(J^n(\mathcal{E}), \pi_m^n, J^m(\mathcal{E}))$, $n > m \geq 0$.

Let $(\mathcal{E}, \pi, \mathcal{B})$ be a bundle with adapted coordinates (x^i, u^α), then we introduce special tangent vector fields d_i on $\mathcal{T}(J^\infty(\mathcal{E}))$, the so called total derivatives with respect to the independent coordinates x^i, given by

$$d_i = \partial_i + u_{J+1_i}^\alpha \partial_\alpha^J , \quad \partial_\alpha^J = \frac{\partial}{\partial u_J^\alpha} . \tag{1}$$

They are maps $C^\infty(J^n(\mathcal{E})) \to C^\infty(J^{n+1}(\mathcal{E}))$, which obey the rule

$$d(f) \circ j^{n+1}\sigma = \partial_i (f \circ j^n \sigma) \tag{2}$$

for any smooth function $f \in C^\infty(J^n(\mathcal{E}))$ and section σ of $\mathcal{E}$. The dual objects to the fields $\pi_{n,*}^\infty d_i \in J^{n+1}(\mathcal{E})$ are called *contact forms*. The special 1-forms

$\omega_J^\alpha \in \mathcal{T}^* J^{n+1}(\mathcal{E})$,
$$\omega_J^\alpha = \mathrm{d} u_J^\alpha - u_{J+1_i}^\alpha \mathrm{d} x^i \, , \quad \#J \leq n \tag{3}$$
form a basis for the contact forms.

3. A FIRST EXAMPLE

Let us consider the following lumped parameter model
$$\tfrac{\mathrm{d}}{\mathrm{d}t} x^\alpha = x_1^\alpha = \left(A^{\alpha\beta} - S^{\alpha\beta} \right) \partial_\beta V + b_r^\alpha u^r \, , \quad y_r = b_r^\alpha \partial_\alpha V \tag{4}$$

with the n-dimensional state x^α, $\alpha, \beta = 1, \ldots, n$, the m-dimensional input u^r and the output y_r, $r = 1, \ldots, m$. We model the system (4) on the bundle $(\mathcal{E}, \mathrm{pr}_1, \mathcal{B})$, $\mathcal{E} = \mathcal{B} \times \mathcal{M}$, $\mathcal{B} \subseteq \mathbb{R}$, $\mathcal{M} \subseteq \mathbb{R}^n$ with local coordinates (t, x^α) and assume that $[A^{\alpha\beta}]$ is a skew symmetric matrix and $[S^{\alpha\beta}]$ is a positive semi definite matrix with functions $A^{\alpha\beta}, S^{\alpha\beta}, b_r^\alpha, V \in C^\infty(\mathcal{E})$. Let us choose $u^r \in C^\infty(\mathcal{B})$, then we can construct the vector field $f_e = \partial_1 + x_1^\alpha \partial_\alpha \in \mathcal{T}(\mathcal{E})$ with x_1^α from (4), which leads to one of the key relations for the control loop design

$$f_e(V) = \mathrm{i}_{f_e}(\mathrm{d}V) = -\partial_\alpha V S^{\alpha\beta} \partial_\beta V + \partial_\alpha V b_r^\alpha u^k + \partial_1 V \leq y_k u^k + \partial_1 V \tag{5}$$

with $\mathrm{d}V \in \mathcal{T}^*(\mathcal{E})$. In many cases V corresponds to the energy of the system. Then it follows from (5) that one can control the time change of V by u^k based on the measurement of y_k only. If also the whole state or some part is available, then one gains an influence on the internal energy transport by changing A, one may change S or even the energy function V itself (e.g. see [4]).

To develop further the geometric picture, we consider the vertical bundle $\mathcal{T}(\mathcal{E}) \supset \mathcal{V}(\mathcal{E}) = \{ X \in \mathcal{T}(\mathcal{E}) \mid \pi_*(X) = 0 \}$ together with the horizontal 1-form bundle $\wedge_1^*(\mathcal{E}) \supset \mathcal{H}_1^*(\mathcal{E}) = \{ \eta \in \wedge_1^*(\mathcal{E}) \mid \mathrm{i}_X(\eta) = 0, X \in \mathcal{V}(\mathcal{E}) \}$. It is well known that neither $\mathcal{V}(\mathcal{E})$ nor $\mathcal{H}_1^*(\mathcal{E})$ have a distinguished complement. Surprisingly, they have a distinguished complement, if we pull them back to $J^1(\mathcal{E})$, [5]. Taking the decomposition of any $\eta \in \wedge_1^*(\mathcal{E})$ by

$$\eta_\alpha \mathrm{d} x^\alpha + \eta_1 \mathrm{d}t = \eta_\alpha \omega^\alpha + (\eta_\alpha x_1^\alpha + \eta_1) \, \mathrm{d}t$$

with ω^α from (3) into account, we see that $C\mathcal{H}_1^*(\mathcal{E}) = \{ \eta \in \wedge_1^*(\mathcal{E}) \mid \eta = \eta_\alpha \omega^\alpha \}$ is the required complement of $\mathcal{H}_1^*$. Obviously, the complement $C\mathcal{V}(\mathcal{E})$ of $\mathcal{V}(\mathcal{E})$ follows from $C\mathcal{V}(\mathcal{E}) = \{ X \in \mathcal{T}(\mathcal{E}) \mid \mathrm{i}_X(\eta) = 0, \eta \in C\mathcal{H}_1^*(\mathcal{E}) \}$. Now, we are ready to present the geometric picture for (4) as $x_1^\alpha \partial_\alpha \in \mathcal{V}(\mathcal{E})$, $b_r^\alpha \partial_\alpha = b_r \in \mathcal{V}(\mathcal{E})$, $\partial_\alpha V \mathrm{d} x^\alpha \in C\mathcal{H}_1^*(\mathcal{E})$ together with two linear maps $C\mathcal{H}_1^*(\mathcal{E}) \to \mathcal{V}(\mathcal{E})$ given by the tensors $A^{\alpha\beta} \partial_\alpha \otimes \partial_\beta$, $S^{\alpha\beta} \partial_\alpha \otimes \partial_\beta$.

4. THE LAGRANGIAN AND HAMILTONIAN PICTURE

Let us consider the lumped parameter case, where we take the trivial bundle $(\mathcal{E}, \mathrm{pr}_1, \mathcal{B})$, $\mathcal{E} = \mathcal{B} \times \mathcal{M}$, $\mathcal{B} \subseteq \mathbb{R}$ with the q-dimensional configuration manifold $\mathcal{M} \subseteq \mathbb{R}^q$ and local coordinates (t, u^α). We restrict our consideration to the simple Lagrangian L,

$$\int_\mathcal{D} L \, dt \, , \quad L \in C^\infty \left(J^1 \left(\mathcal{E} \right) \right) , \tag{6}$$

where $\mathcal{D}$ denotes the interval $[t_1, t_2]$. Let φ be a section of $\mathcal{E}$, and X a vector field on $\mathcal{E}$ with flow ψ_ε, $\varepsilon \in \mathbb{R}$, then the variation of φ induced by X is the 1-parameter family of sections $\psi_\varepsilon \circ \varphi$. We admit only fiber preserving variations and get $X \in \mathcal{V}(\mathcal{E})$. Now, a section φ is called *extremal*, iff it meets

$$\tfrac{\partial}{\partial \varepsilon} \int_\mathcal{D} \left(j^1 \left(\psi_\varepsilon \circ \varphi \right) \right)^* (L dt) = \int_\mathcal{D} \left(j^1 \left(\varphi \right) \right)^* \left(j^1 (X) (L dt) \right) = 0 \tag{7}$$

for any X. Following Cartan we introduce the 1-form

$$C = L dt + p_\alpha \left(du^\alpha - u_1^\alpha dt \right) , \quad p_\alpha = \partial_\alpha^1 L ,$$

which we call the *Cartan form* of L. We already see that the *Legendre transform* $\mathcal{L}$, given by $p_\alpha = \partial_\alpha^1 L$ is here a map $\mathcal{V}(\mathcal{E}) \to C\mathcal{H}_1^* (\mathcal{E})$. The celebrated Euler-Lagrange equations follow from

$$\int_\mathcal{D} \left(j^2 (\varphi) \right)^* \left(i_{d_1} \circ i_X (dC) \right) dt + \left(j^1 (\varphi) \right)^* \left(i_X (C) \right) = 0$$

as

$$i_{d_1} (dC) \wedge dt = 0 \, , \quad d_1 = \partial_1 + u_1^\alpha \partial_\alpha + u_2^\alpha \partial_\alpha^1 . \tag{8}$$

From now on we assume that $\mathcal{L}$ is invertible and we set $d_1 \in \mathcal{T} \left(J^1 (\mathcal{E}) \right)$ for d_1 from (8). If we introduce the Hamiltonian H by

$$H dt = (p_\alpha u_1^\alpha - L) \circ \mathcal{L}^{-1} (p) \, dt . \tag{9}$$

then the famous Hamilton equations follow as

$$i_{\bar{d}_1} \left(d \left(p_\alpha du^\alpha - H dt \right) \right) = 0 \, , \quad \bar{d}_1 = \partial_1 + u_1^\alpha \partial_\alpha + p_\alpha^1 \bar{\partial}^\alpha \, , \quad \bar{\partial}^\alpha = \tfrac{\partial}{\partial p_\alpha} . \tag{10}$$

From (10) it follows

$$\bar{d}_1 (H) - \partial_1 H = \mathrm{v} \left(d_1 \right) (H) = 0 \, , \tag{11}$$

where $\mathrm{v}(d_1) = u_1^\alpha \partial_\alpha + p_\alpha^1 \bar{\partial}^{\bar\alpha}$ denotes the vertical part of d_1.

Let us consider the special Lagrangian

$$L = \left(L_0 + L_r U^r + L_\alpha^1 u_1^\alpha V^s \right) dt \, , \quad L_0 \in C^\infty \left(J^1 (\mathcal{E}) \right) \, , \quad L_r, L_{\alpha,r} \in C^\infty (\mathcal{E})$$

with the generalized forces U^r, V^s, then from

$$p_\alpha = \partial_\alpha^1 L_0 + L_\alpha^1 V^s \, , \quad H = H_0 - L_r U^r \, , \quad H_0 = L_0 - \partial_\alpha^1 L_0 u_1^\alpha$$

and (11) we get
$$\mathrm{v}\left(\bar{d}_1\right)(\mathrm{d}H_0) = Y_r U^r \,, \quad Y_r = \mathrm{v}\left(\bar{d}_1\right)(L_r) \,. \tag{12}$$

If we compare (12) with (5) then a natural choice for the output is Y_r.

Now we consider the distributed parameter case, where we take the trivial bundle $(\mathcal{E}, \mathrm{pr}_1, \mathcal{B})$, $\mathcal{E} = \mathcal{B} \times \mathcal{M}$, with the p-dimensional base $\mathcal{B} \subseteq \mathbb{R}^p$ and the q-dimensional configuration manifold $\mathcal{M}$ and local coordinates (x^i, u^α) and consider the Lagrangian
$$\int_{\mathcal{D}} l\Omega \,, \quad l \in C^\infty \left(J^1\left(\mathcal{E} \right) \right) \tag{13}$$

with Lagrangian density l, where $\mathcal{D}$ denotes the domain of integration and $\Omega = \mathrm{d}x^1 \wedge \cdots \wedge \mathrm{d}x^p$ is the volume form on $\mathcal{B}$, which induces an orientation. To find the *Legendre transform* $\mathcal{L}$, we proceed in the following way. There is a natural map between $J^1(\mathcal{E})$ and the space of maps $\mathcal{T}(\mathcal{B}) \to \mathcal{V}(\mathcal{E})$, whose sections are tensors of the type $v_\alpha \otimes \mathrm{d}x^i$, $v_\alpha \in \mathcal{V}(\mathcal{E})$. Their dual objects are given by $\omega^\beta \otimes \partial_j$, $\omega^\beta \in C\mathcal{H}_1^*(\mathcal{E})$ together with the pairing $\mathrm{i}_{v_\alpha}\left(\omega^\beta\right) \mathrm{i}_{\partial_j}\left(\mathrm{d}x^i\right) = \mathrm{i}_{v_\alpha}(\omega^\alpha)$. Since $\mathcal{B}$ has an orientation, we can use the volume form Ω to construct an isomorphism between these tensors and tensors of the type $\omega^\beta \wedge \mathrm{i}_{\partial_j}(\Omega)$. We choose the space of these tensors as the dual space together with the pairing $\mathrm{i}_{v_\alpha}\left(\omega^\beta\right) \mathrm{d}x^i \wedge \mathrm{i}_{\partial_j}(\Omega) = \mathrm{i}_{v_\alpha}\left(\omega^\beta\right) \Omega$. Now, we are ready to present the Cartan form c for (13)
$$c = l\Omega + p_\alpha^j \omega^\alpha \wedge \mathrm{i}_{\partial_j}(\Omega) \,, \quad p_\alpha^j = \partial_\alpha^{1_j} l \,,$$

see (3), as well as the Euler-Lagrange equations
$$0 = \mathrm{I}_d(\mathrm{d}c) \wedge \Omega \,, \quad \mathrm{I}_d = \mathrm{i}_{d_p} \circ \cdots \circ \mathrm{i}_{d_1} \,, \quad d_i = \partial_i + u_{1_i}^\alpha \partial_\alpha + u_{1_i+1_j}^\alpha \partial_\alpha^{1_j}$$
$$0 = \left(\partial_\alpha l - d_i \left(p_\alpha^i \right) \right) \mathrm{d}u^\alpha \wedge \Omega \tag{14}$$

where the boundary conditions follow from $\mathrm{i}_X(c) = 0$ with the vertical field X that induces the variation.

To construct the Hamilton equations we assume that the Legendre transform $\mathcal{L}$ with respect to p^1 is invertible and choose x^1 as the distinguished variable. Then the Hamiltonian density h is given by
$$h\Omega = \left(p_\alpha^1 u_{1_1}^\alpha - l \right) \circ \mathcal{L}^{-1}\left(p^1\right) \Omega \,. \tag{15}$$

Now, we rewrite the relations (14) and derive the Hamiltonian equations in the following form
$$\mathrm{I}_{\bar{d}} \left(\mathrm{d} \left(p_\alpha^1 \mathrm{d}u^\alpha \wedge \mathrm{i}_{\partial_1}(\Omega) + p_\alpha^k \omega^\alpha \wedge \mathrm{i}_{\partial_k}(\Omega) - H\Omega \right) \right) \wedge \mathrm{i}_{\partial_1}(\Omega)$$
$$= \left(\left(u_{1_1}^\alpha - \bar{\partial}_1^\alpha H \right) \mathrm{d}p_\alpha^1 - \left(p_{\alpha,1_1}^1 + \bar{d}_k\left(p_\alpha^k\right) + \partial_\alpha H \right) \mathrm{d}u^\alpha \right) \wedge \mathrm{i}_{\partial_1}(\Omega)$$
$$+ \left(\bar{d}_1(H) - \partial_1 H \right) \Omega = 0 \,, \tag{16}$$
$$\bar{d}_i = \partial_i + u_{1_i}^\alpha \partial_\alpha + p_{\alpha,1_i}^1 \bar{\partial}_1^\alpha + u_{1_i+1_k}^\alpha \partial_\alpha^{1_k} \,, \quad k = 2, \ldots, p \,, \quad \bar{\partial}_i^\alpha = \frac{\partial}{\partial p_\alpha^i} \,.$$

The relations analogous to (11) follow directly from (16) as
$$\left(\bar{d}_1(H) - \partial_1 H \right) \mathrm{i}_{\partial_1}(\Omega) = \mathrm{v}(d_1)(H) \mathrm{i}_{\partial_1}(\Omega) = 0 \,.$$

It is worth mentioning that the pairing of the tensors $v \otimes dt$, $\omega^1 \wedge i_{\partial_1}(\Omega)$ form a Dirac structure, e.g. see [3], for our variational problem. Since $\mathcal{B}$ has an orientation, this pairing induces also a pairing on the boundary $\partial \mathcal{D}$.

5. THE MATHEMATICAL MODEL

To derive the mathematical model of a piezoelectric structure, we choose a Riemannian manifold $\mathcal{M}$ with local coordinates x^i, $i = 1, \ldots, p = 3$ and metric $g = g_{ij} dx^i \otimes dx^j$ for the reference manifold and another Riemannian manifold $\bar{\mathcal{M}}$ with local coordinates u^α, $i = 1, \ldots, q = 3$ and metric $\bar{g} = \bar{g}_{\alpha\beta} du^\alpha \otimes du^\beta$ for the configuration manifold, see [1]. We introduce the base manifold $\mathcal{B}$ by $\mathcal{B} = T \times \mathcal{M}$, $T \subseteq \mathbb{R}$ with coordinates (x^0, x^i), $x^0 = t$ and construct the bundle $(\mathcal{E}, \text{pr}_1, \mathcal{B})$, $\mathcal{E} = \mathcal{B} \times \bar{\mathcal{M}}$ then a motion is section σ of $\mathcal{E}$. Following the considerations above we use the isomorphism between $J^1(\mathcal{E})$ and tensors of the type $u^\alpha_{1_i} \partial_\alpha \otimes dx^i + u^\alpha_{1_0} \partial_\alpha \otimes dx^0$ and define the tensor

$$f^\alpha_i \partial_\alpha \otimes dx^i + v^\alpha \partial_\alpha \otimes dx^0 \in \mathcal{V}(\mathcal{E}) \otimes T^*(\mathcal{B}) \tag{17}$$

where the spatial part is the *deformation gradient* and the time part is the *material velocity*. To simplify the following we take the trivial metric $g_{ij} = \delta_{ij}$, $\bar{g}_{\alpha\beta} = \delta_{\alpha\beta}$ for both Riemannian manifolds.

5.1. The Quasi Electrostatic Field

The simplest way to derive the field equations starts with the Maxwell's equations on the configuration manifold $\bar{\mathcal{M}}_e = T \times \bar{\mathcal{M}}$ extended by the time t, where we use coordinates (t, u^α). Let du denote the canonical volume form on the Riemannian manifold $\bar{\mathcal{M}}$, then we may introduce the magnetic flux density $\bar{B}$ by $\bar{B} = \bar{B}^\alpha i_{\partial_\alpha}(du)$. Together with the electrical field strength $\bar{E} = \bar{E}_\alpha du^\alpha$ the first set of Maxwell's equations is given by

$$d(\bar{B} + \bar{E} \wedge dt) = 0 . \tag{18}$$

With the electric flux density $\bar{D} = \bar{D}^\alpha i_{\partial_\alpha}(du)$ and the magnetic field strength $\bar{H} = \bar{H}_\alpha du^\alpha$ the second set of Maxwell's equations follows as

$$d(\bar{D} - \bar{H} \wedge dt) = \bar{\rho} du + \bar{j} \wedge dt \tag{19}$$

with the electric charge density $\bar{\rho}$ and the electrical current flux density $\bar{j} = \bar{j}^\alpha i_{\partial_\alpha}(du)$, see [1]. For a static electrical field we set $\bar{B} = 0$, $\bar{H} = 0$, $\bar{j} = 0$. Then (18) implies the local existence of a potential $\bar{U}$ for $\bar{E}$ and we derive the final set of equations for the quasi static case in the form

$$(d\bar{U} + \bar{E}) \wedge dt = 0 , \quad (d\bar{D} - \bar{\rho} du) \wedge dt , \tag{20}$$

as well as

$$\bar{E} \wedge d\bar{D} + d(\bar{U} d\bar{D}) + \bar{U} d(\bar{\rho} du) = 0 , \tag{21}$$

where $\bar{E} \wedge d\bar{D}$ is the electrical power density and we set $d\left(\bar{\rho}\bar{U}du\right) = 0$, since $\bar{\rho}$ vanishes on the boundary of interest.

Since we use a Lagrangian description of the piezoelectric structure, we have to pull back these equations from the configuration onto the reference manifold by $\bar{E}_\alpha u^\alpha_{1_i} dx^i = E_i dx^i$, $\bar{D}^\alpha i_{f^{-1}(\partial_\alpha)}\left(u^\alpha_{1_i} dx^i\right) J = D^i i_{\partial_i}(dx)$, $J\bar{\rho} = \rho$ with the functions $U, E^\alpha, D^\alpha, \rho, J \in C^\infty\left(J^1(\mathcal{E})\right)$ and $J = \det([f^\alpha_i])$ with f^α_i from (17). The result is

$$(d_H U + E) \wedge dt = 0 , \quad (d_H D - \rho dx) \wedge dt = 0 . \qquad (22)$$

Here we have applied the horizontal exterior derivative $d_H \eta$, which we define for any horizontal k-form η, $i_v(\eta) = 0$, $\forall v \in \mathcal{V}(\mathcal{E})$ by

$$d\eta = d_H \eta + \lambda_{\alpha_1,\ldots\alpha_{k+1}} \bigwedge_{i=1}^{k+1} \omega^{\alpha_i} , \quad \alpha_1 < \cdots < \alpha_{k+1}$$

with the contact forms ω^{α_i} from (3) and the unique horizontal $(k+1)$-form $d_H \eta$.

5.2. The Mechanical Model

We use the standard equations for the mechanical model in Lagrangian coordinates based on the conservation of mass and the symmetry of stress given by

$$\nabla_{\partial_0}(v^\alpha \partial_\alpha) \otimes \rho dx = a^\alpha \partial_\alpha \otimes \rho dx - \partial_\alpha \otimes i_{\partial_0}\left(dt \wedge d_H\left(P^{\alpha i} i_{\partial_i}(dx)\right)\right) = 0 \quad (23)$$

with the mass density ρ on $\mathcal{M}$, [2]. The tensor P is the Piola stress tensor $P^{\alpha i} \partial_\alpha \otimes \partial_i$ and $a^\alpha \rho \partial_\alpha \otimes dx$ describes the body forces. We use the exterior product with dt in the third term to cancel terms which contain the form dt. Since we have chosen the trivial metric for $\mathcal{M}$, we may simplify the covariant derivative as $\nabla_{\partial_0}(v^\alpha \partial_\alpha) = \partial_0(v^\alpha) \partial_\alpha$. Applying the metric on $\bar{\mathcal{M}}$ to (23) we derive the dual version

$$d_0\left(p^0_\alpha\right) \rho du^\alpha \wedge dx = \delta_{\alpha\beta} a^\beta \rho du^\alpha \wedge dx + \delta_{\alpha\beta} du^\beta \wedge i_{\partial_0}\left(dt \wedge d_H\left(P^{\beta i} i_{\partial_i}(dx)\right)\right) ,$$

with $p^0_\alpha = \delta_{\alpha\beta} v^\beta$. We use the pairing between the tensors $p^0_\alpha \omega^\alpha \wedge dx$ and $v^\alpha \partial_\alpha \otimes dt$ and get the energy relation

$$p^0_\alpha v^\alpha \rho dt \wedge dx = v^\alpha \delta_{\alpha\beta} a^\beta \rho dt \wedge dx + v^\alpha \delta_{\alpha\beta} dt \wedge d_H\left(P^{i\beta} i_{\partial_i}(dx)\right) \qquad (24)$$

or

$$\begin{aligned}\left(\tfrac{1}{2} d_0\left(v^\alpha \delta_{\alpha\beta} v^\beta\right) \rho + D_{ij} S^{ij}\right) dt \wedge dx = \\ a^\alpha \delta_{\alpha\beta} v^\alpha dt \wedge dx + dt \wedge d_H\left(v^\alpha \delta_{\alpha\beta} P^{i\beta} i_{\partial_i}(dx)\right)\end{aligned} \qquad (25)$$

with the *deformation tensor* $C = C_{ij} dx^i \otimes dx^j = \delta_{\alpha\beta}\left(u^\alpha_{1_i} + u^\beta_{1_j}\right) dx^i \otimes dx^j$, which is the pull back of the metric, and the tensor $2D = d_0(C)$. The equation

(25) shows also the induced pairing between the tensor $\delta_{\alpha\beta}P^{i\beta}\omega^\alpha \wedge i_{\partial_i}(dx)$ and $v^\alpha\partial_\alpha \otimes dt$ on the boundary. If in addition, the relation

$$d(2w(C,t,x)\rho dt \wedge dx) = S^{ij}dC_{ij} \wedge dt \wedge dx \qquad (26)$$

is met, then we may set $D_{ij}S^{ij} = d_0(w)$ in (25). Furthermore, we call w the *stored energy function*. In this case, the system (24) can directly be modelled as a PCH system.

5.3. Piezoelastic Coupling

The simplest way to study the coupling of the electrical field with the mechanical one starts with the stored energy function w, which now may also depend on D in order to take the electrical contribution into account, see (21). We assume that the coupling is energy preserving and that the charge density vanishes (see (21), (22). Extending the stored energy function (26) by the electrical part we get from

$$d(w(C,D,x)\rho dx) = \tfrac{1}{2}S^{ij}dC_{ij} \wedge dx + E \wedge dD = \tfrac{1}{2}S^{ij}dC_{ij} \wedge dx - E_i dD^i \wedge dx$$

the required equations

$$E_i = -\partial_{D^i}w(C,D,x)\rho, \quad S^{ij} = 2\partial_{C_{ij}}w(C,D,x)\rho. \qquad (27)$$

Obviously, the energy conserving coupling also preserves the PCH character of the piezolectric structure, since we consider the quasi static electrical case only. In a linearized scenario, one takes the Taylor expansion of (27) and derives the relations

$$\Delta E_k = a^{ij}_k \Delta C_{ij} + b_{ki}\Delta D^i, \quad \Delta S^{ij} = c^{ijkl}\Delta C_{kl} + d^{ij}_k \Delta D^k,$$

where the coupling constants meet the symmetries $c^{ijkl} = c^{jikl} = c^{ijlk}$, $b_{ki} = b_{ik}$, $-d^{ij}_k = a^{ij}_k = a^{ji}_k$. In addition to the piezoelectric active material and the substrate, a piezoelectric structure is also built up by metallic electrodes. This approach does not include the stress caused by the electrical surface charge densities on the electrodes. But it is worth mentioning that one can take the temperature in a straightforward manner into account by an extension of the stored energy function.

6. THE CONTROLLER DESIGN

The piezoelectric smart structures under consideration are composite beams and plates consisting of a large number of thin layers of laminae with and without piezoelectric properties. The various layers are considered to be perfectly bonded together. Apart from the possibility of varying the voltage applied to the electrodes of an piezoelectric actuator layer or of measuring the electrical charge (current) of a piezoelectric sensor layer we have the additional degree of freedom of shaping the surface electrode. The effect of shaping can be well expressed in the stored energy function of the piezoelectric laminae, see [6] and the citations therein.

To apply the presented theory, we choose the coordinates $x^i, u^\alpha, i = 1, \ldots, p$, $\alpha = 1, \ldots, q$, $x_0 = t$ and assume that the piezoelectric structure can be described by a Hamiltonian like

$$H = H_0 - H_l U^l, \qquad H_0 = \int_\mathcal{B} h_0 dx, \qquad H_i = \int_\mathcal{B} h_i dx, \qquad (28)$$

where $H_0 = H_0(x, p, u, u_{1_i})$ denotes the Hamiltonian of the free system. The control input u^l, $l = 1, \ldots, m$ enters the equation by the control Hamiltonians $H_i = H_i(x, u, u_{1_i})$. The integrals are taken over the body $\mathcal{B}$ in the reference configuration. This assumption can be justified for special stored energy functions (27) which meet $\partial_{C_{ji}}(\partial_{D^k} w)^2 = 0$. The input U^l is the voltage applied to the electrodes and $\mathrm{v}(\bar{d}_0) H_l$, see (16), is the current, therefore the pairing of the input U^l with the output $Y_l = \mathrm{v}(\bar{d}_0) H_l$ forms already the port in the sense of a PCH system. One of the advantages of piezoelectric structures is that one can construct sensor layers that allow to measure the required output Y_l. Sometimes it is even possible to measure the corresponding electrical charge. In this case, an additional output is available, which we denote by Iy_l.

For the following we assume that the function H_0 is positive besides at the equilibrium, where $H_0 = 0$ is met. As a first and simple controller we choose

$$U^m = -K^{lr} Iy_r - D^{lr} Y_r, \qquad r = 1, \ldots, m \qquad (29)$$

the PD-law with positive definite matrices $[K^{lr}]$ and $[D^{lr}]$. From the relations

$$d_0 \left(H_0 + \tfrac{1}{2} Iy_l K^{lr} Iy_r \right) = -Y_l D^{lr} Y_r,$$

we see that the control law (29) preserves stability. Next, let us try to minimizes the objective function J_2,

$$J_2 = \sup\nolimits_{T \in [0, \infty)} \inf\nolimits_{u \in L_2^m[0,T]} \tfrac{1}{2} \int_0^T \left(\|Y\|^2 + \|U\|^2 \right) dt \qquad (30)$$

with euclidean norm $\| \ \|$. The Hamilton-Jacobi-Belman inequality of this H_2-problem is given by

$$\inf\nolimits_u \left(d_0(V) + \tfrac{1}{2} \left(\|Y\|^2 + \|U\|^2 \right) \right) \leq 0. \qquad (31)$$

A short calculation shows that the ansatz $V = \rho E$, $\mathbb{R} \ni \rho > 0$ leads to the simple control law $U^i = -\rho Y^i$, $i = 1, \ldots, m$ and converts (31) to $\tfrac{1-\rho^2}{2} \|Y\|^2 \leq 0$. The choice $\rho = 1$ solves the problem exactly and the objective function J_2,

$$J_2 = -\int_0^\infty Y_l U^l dt$$

is equal to the energy dissipated by the controller. This optimization problem (30) offers an interesting extension. We solve the problem $J_2(\varepsilon)$, $\varepsilon \geq 0$,

$$J_2(\varepsilon) = \tfrac{1}{2} \int_0^\infty \left(\varepsilon \|Iy\|^2 + \|Y\|^2 + \|U\|^2 \right) dt$$

for a linearized lumped parameter approximation of the distributed parameter system. Since the solutions of $J_2(0)$ with output feedback is a simple strictly passive D-law and J_2 depends continuously on ε, we may expect that these properties are preserved for the solutions of the problem $J_2(\varepsilon)$ for ε sufficiently small. Furthermore, the passivity of the closed loop is preserved and therefore, it is possible to use this control law for the distributed system. Since we derive an output feedback law, we avoid the common problem to measure the state of an finite state approximation, which has no physical meaning in general.

7. FINAL REMARKS

The goal of this contribution was to show that piezoelectric smart structures can be modelled as distributed parameter PCH systems. The main problem was the construction of the Dirac structure, which turned out to be a pairing between a suitable bundle of vertical vectors and its dual. The key to the solution was to start with a variational problem, where the Dirac structure already appears in the Cartan form. This approach requires the mathematical machinery of jet bundles, which was presented at the beginning. The profit of this approach is that now several powerful methods can be applied to the controller design, of which we were able to presented a very small part only.

8. ACKNOWLEDGEMENT

This work has been done in the context of the European sponsored project GeoPlex with reference code IST-2001-34166. Further information is available at http://www.geoplex.cc. It has also been partially supported by the LCM at the Johannes Kepler University as part of the strategic project 4.4.

References

1. Frankel T. The Geometry of Physics, An Introduction. Cambridge University Press, Cambridge, 1997.
2. Marsden J.E. and Hughes T.J.R. Mathematical Foundations of Elasticity. Dover Publications, 1993.
3. Maschke B.M., van der Schaft A.J. Port controlled Hamiltonian representation of distributed parameter systems. In N. E. Leonard and R. Ortega, editors, *Proceedings of the IFAC Workshop on Lagrangian and Hamiltonian methods for nonlinear control*, Princeton University, pp. 28-38, 2000.
4. Ortega R., van der Schaft A., Mareels I., Maschke B., Putting energy back in control. *IEEE Control Systems Magazine*, vol. 21, no. 2, pp. 18-33 , 2001.
5. Saunders D. J. The Geometry of Jet Bundles. Cambridge University Press, Cambridge, 1989.
6. Schlacher K., Kugi A. Control of elastic systems, a Hamiltonian approach. In N. E. Leonard and R. Ortega, editors, *Proceedings of the IFAC Workshop on Lagrangian and Hamiltonian methods for nonlinear control*, Princeton University, pp. 80-85, 2000.
7. Tzou H. S. Active Piezoelectric Shell Continua, Intelligent Structural Systems. In H. S. Tzou, G. L. Anderson, editors, Kluwer Academic Publishers, 1992.
8. Ziegler F., Mechanics of Solids and Fluids. Springer-Verlag, New York Vienna, 1991.

Numerical Simulation for Control of Progressive Buckling with Defects on Axisymmetric Shell Structure

YOTSUGI SHIBUYA
Department of Mechanical Engineering, Akita University,
Akita, Akita 010-8502 Japan
E-mail: shibuya@ipc.akita-u.ac.jp

SUMITAKA WATANABE
Graduate School of Mining and Engineering, Akita University

1. INTRODUCTION

Behavior of buckling for shell structure has been wide attention for absorption of impact energy[1]. Both high strength and the ability to absorb energy during structural collapse are important factors for structural design in crush problems. The collapse analyses have been carried out analytically and experimentally[2]-[4]. Recently, numerical techniques have become available and computer simulation for collapse analysis has become very attractive[5],[6]. Buckling and initial post-buckling behavior of cylindrical shells under axial compression have been studied. The behavior of thin-walled shells is sensitive to structural imperfection. A progressive plastic buckling has been presented for the cylindrical structure of thick-walled shell[7] and has been classified[8]. Usually, the first peak load becomes maximum during the collapse process. It is necessary to decrease the first peak load in order to soften an impact.

In this paper, suitable arrangement of defects for an axisymmetric shell is considered to control the progressive plastic buckling. After first peak load appears on the load-displacement diagram, following formation of collapse folds is developed due to the inward and outward buckling during axial compression. For the numerical analysis, a nonlinear finite element method based on the updated Lagrange approach can be used to analyze the structure. Large displacement, frictional contact, localization of buckling pattern and elastic-plastic analysis are included in the quasi-static impact problem. The material nonlinearity is modeled as a power plastic hardening. Load displacement diagram and progressive behavior of buckling are simulated on the axisymmetric cylindrical shell. Defects to control the buckling are modeled by grooves at the inner surface of the cylindrical shell. The artificial defects decrease the first peak load and control the progressive plastic buckling. A cylindrical shell subjected to axial compression is placed on the rigid flat body with frictional contact.

In the numerical calculation, the effect of arrangement of defects on the behavior of the buckling is considered. First peak load, average load and their ratio are chosen as suitable parameters to evaluate the absorption of impact energy. Optimal arrangement of defects in the structure is discussed to control the progressive plastic buckling.

2. PROGRESSUVE BUCKLING OF CYLINDRICAL SHELL

A cylindrical shell under compression is considered for evaluation of absorption of impact load. Figure 1 is a photograph for typical axisymmetric progressive plastic buckling of a cylindrical shell. The material is aluminum. During the compression, stable buckling with plastic deformation occurs progressively. As the buckling proceeds continuously, the cylinder can absorb a large amount of impact energy.

Figure 1. Photograph of typical progressive plastic buckling.

3. FINITE ELEMENT ANALYSIS

Nonlinear finite element analysis based on the updated Lagrange method is employed to perform large deformation analysis of plastic buckling. Commercial finite element code [MARC K7.1] is used in the analysis where effects of elasto-plastic and frictional contact are included.

As the constitutive equation can be expressed in the objective rate for the large deformation analysis. We introduce the Truesdell rate $\dot{\sigma}_{ij}^T$ as

$$\dot{\sigma}_{ij}^T = \dot{\sigma}_{ij} - \frac{\partial v_i}{\partial x_k}\sigma_{kj} - \sigma_{ik}\frac{\partial v_j}{\partial x_k} + \sigma_{ij}\frac{\partial v_k}{\partial x_k} \tag{1}$$

where σ_{ij} and $\dot{\sigma}_{ij}$ are the Cauchy stress and its rate, respectively, and v_j is rate of displacement. $x(x_1, x_2, x_3)$ is defined in the deformed body. The constitutive equations is formulated in terms of the Truesdell rate of Cauchy stress as

$$\dot\sigma_{ij}^T = D_{ijkl}(\sigma_{mn})\dot\varepsilon_{kl} \qquad (2)$$

where $\dot\varepsilon_{ij}$ is the strain rate. For isotropic material, the modulus is written in general expression as

$$D_{ijkl} = 2\mu[\frac{\delta_{ik}\delta_{jl}+\delta_{il}\delta_{ij}}{2} + \frac{\upsilon}{1-2\upsilon}\delta_{ij}\delta_{kl}] \qquad (3)$$

in the elastic regions, or

$$D_{ijkl} = 2\mu\{\frac{\delta_{ik}\delta_{jl}+\delta_{il}\delta_{ij}}{2} + \frac{\upsilon}{1-2\upsilon}\delta_{ij}\delta_{kl} - \frac{3\sigma'_{ij}\sigma'_{kl}}{2\bar\sigma^2[1+H'/(3\mu)]}\} \qquad (4)$$

in the elasto-plastic regions. μ is shear modulus; υ, Poisson's ratio; δ_{ij}, Kronecker's delta; and $\bar\sigma$, equivalent stress. The plastic modulus is defined as $H' = d\bar\sigma/d\varepsilon^p$ where $d\varepsilon^p$ is plastic strain.

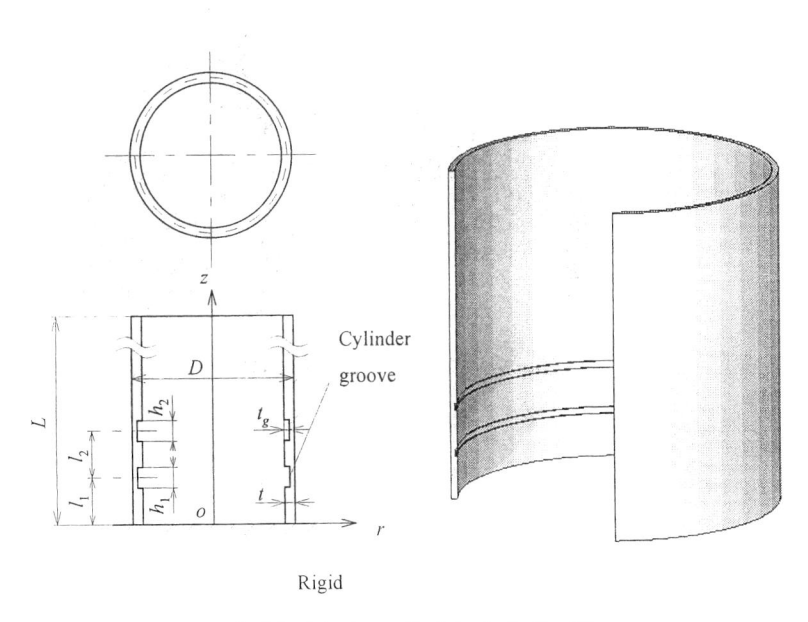

Figure 2. Model of a cylindrical shell with grooves.

4. MODEL OF CYLINDRICAL SHELL WITH DEFECTS

A cylindrical shell structure under axial compression is considered for evaluation of energy absorption as shown Figure 2. The shell is placed on the rigid flat body with Coulomb frictional contact. The length of the shell is L, outer diameter of the shell is D, and the thickness is t_c. To control the buckling load, grooves are made in the shell with the distance l_i, the width and depth are h_i and t_g, respectively.

A numerical analysis of a cylindrical shell is made as axisymmetric problem by using nonlinear finite element analysis based on updated Lagrange method. The axisymmetric 8-node elements are used in the analysis. The bottom end of the shell is contacted under Coulomb friction, and the top end is compressed at constant displacement rate.

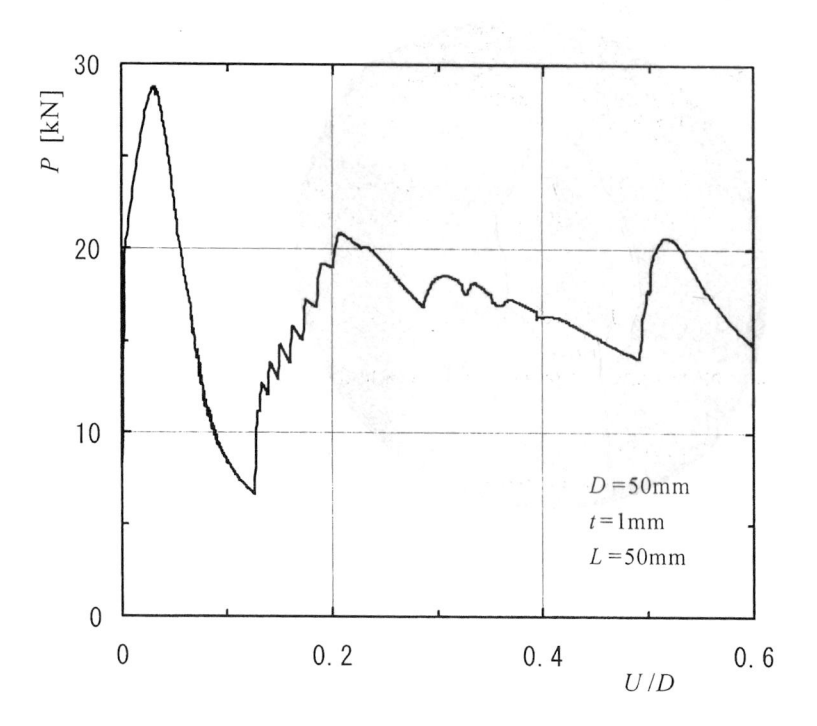

Figure 3. Load displacement diagram of a cylindrical shell without groove.

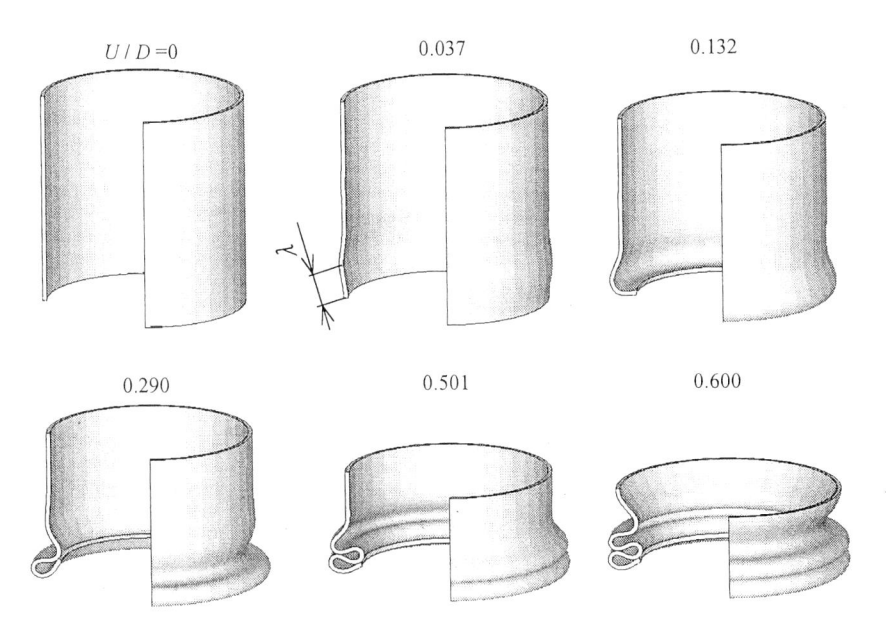

Figure 4. Buckling process of a cylindrical shell without grooves.
U / D : normalized displacement at the top end.

5. NUMERICAL RESULT AND DISCUSSIONS

The material obeys a power plastic hardening law. The constitutive relation is written as

$$\sigma = \sigma_Y(1 + H\varepsilon^p)^n \tag{5}$$

where σ_Y is initial yield stress; H the plastic coefficient; and n the work-hardening exponent. The material treated in this study is aluminum. The following properties of the material are used: Young's modulus $E = 66.8$ GPa, Poisson's ratio $\nu = 0.3$, initial yield stress $\sigma_Y = 123$ MPa, plastic coefficient $H = 95.4$, and work-hardening exponent $n = 0.32$. The coefficient of Coulomb friction at contact surface is 0.3 and the coefficient of friction on the contact of the cylinder itself is also 0.3. Dimensions of the cylinder are $L = 50$ mm, $D = 50$ mm and $t_c = 1$ mm.

Typical load-displacement diagram and progressive buckling process of a cylinder without grooves are shown in Figures 3 and 4, respectively. U indicates displacement of the shell at the top end. The first peak occurs in the diagram when the shell starts to stretch outward. The load oscillates with compression due to progressive buckling as shown in Figure 3. The buckling starts at the contact side, and the first peak load P_{f0} becomes highest and the value is 28.7 kN. The average load P_{a0} for the displacement $U/D = 0.6$ is 17.1 kN and the ratio $r_d (=P_{a0}/P_{f0})$ is 0.596. These values will be used as controlled data to evaluate ability of impact energy in the cylindrical shell. In the figure, the large oscillation corresponds to the repetition of buckling and the small waves in the curve occur due to frictional contact.

Figure 5 shows the buckling process of the cylinder with one inner groove for $l_1 = 9$ mm. The distance $l_1 = 9$ mm is similar value with λ in Figure 4. Dimensions of the groove are $h_1 = 1$ mm and $t_g = 0.25$ mm. The fold starts to expand outside at the groove and continues into the buckling. Figure 6 shows the load-displacement diagram for the cylinder with the groove. Since the fold occurs at the groove, the first peak of the load significantly decreases in comparison with Figure 3, however the load diagram after second peak becomes similar with Figure 3.

Figure 7 shows effect of distance l_1 on the buckling process with first peak load P_f, average load P_a and their ratio r_d normalized by P_{f0}, P_{a0} and r_{d0}, respectively. If each value is 1.0, the case corresponds to the cylinder without groove. In the cylinder with one groove, behavior of the buckling is classified in two types. The buckling mode changes around $l_1 = 4$ mm and the first peak load and average load tends to increase around the distance. For the arrangement of the groove, the distance l_1 is small, the fold expands outwards at the end of the cylinder. The decrease of the average load is small in comparison with the first peak load on the small distance l_1 and ratio of their loads r_d becomes large.

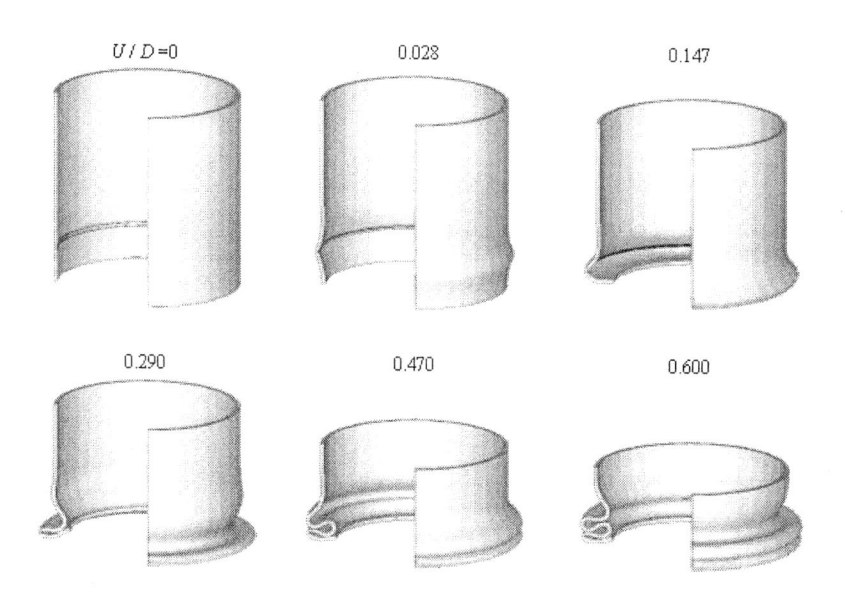

Figure 5. Buckling process of a cylindrical shell with one groove.

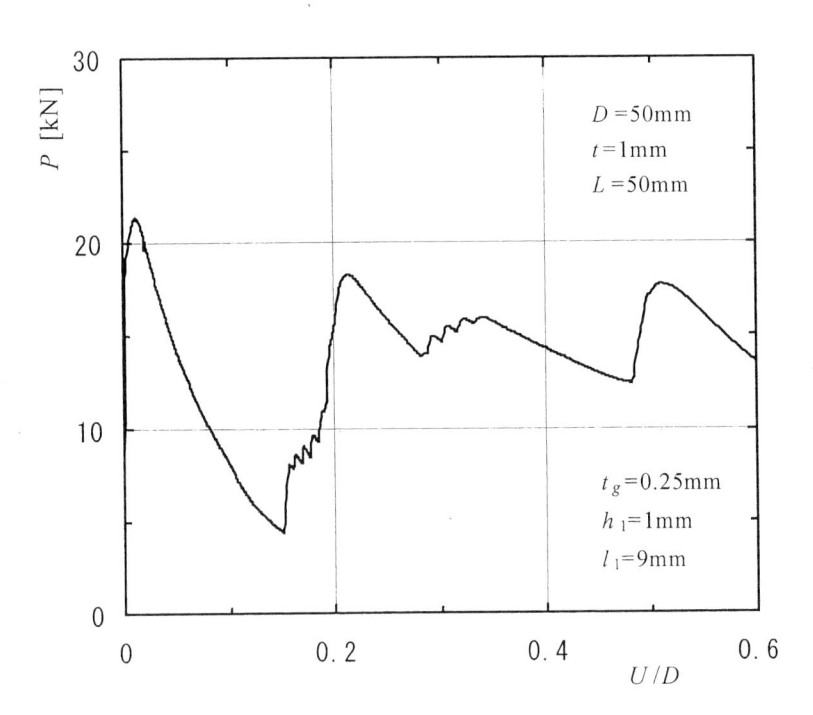

Figure 6. Load displacement diagram of a cylindrical shell with one groove.

Figure 8 shows the buckling process of the cylinder with two inner grooves for $l_1 = l_2 = 8$ mm. Dimensions of the groove are $h_1 = h_2 = 1$ mm and $t_g = 0.25$ mm. The cylinder is folded outside and inside at the grooves and continues to be crushed. Figure 9 shows the load-displacement diagram for the cylinder with two grooves. In this case, first and second peak loads have similar value.

Figure 10 shows effect of distance $l_1 = l_2$ on the buckling process with first peak load P_f, average load P_a and their ratio r_d normalized by P_{f0}, P_{a0} and r_{d0}, respectively, in similar with Figure 7. Three types of buckling mode are simulated as shown in the figure. The first peak and average loads change according to the buckling modes. For the arrangement of the groove, where the distance $l_1 = l_2$ is small, the fold expands outwards at the end of the cylinder. The ratio of their loads r_d becomes large due to relation of the average and the first peak loads on the small distance $l_1 = l_2$.

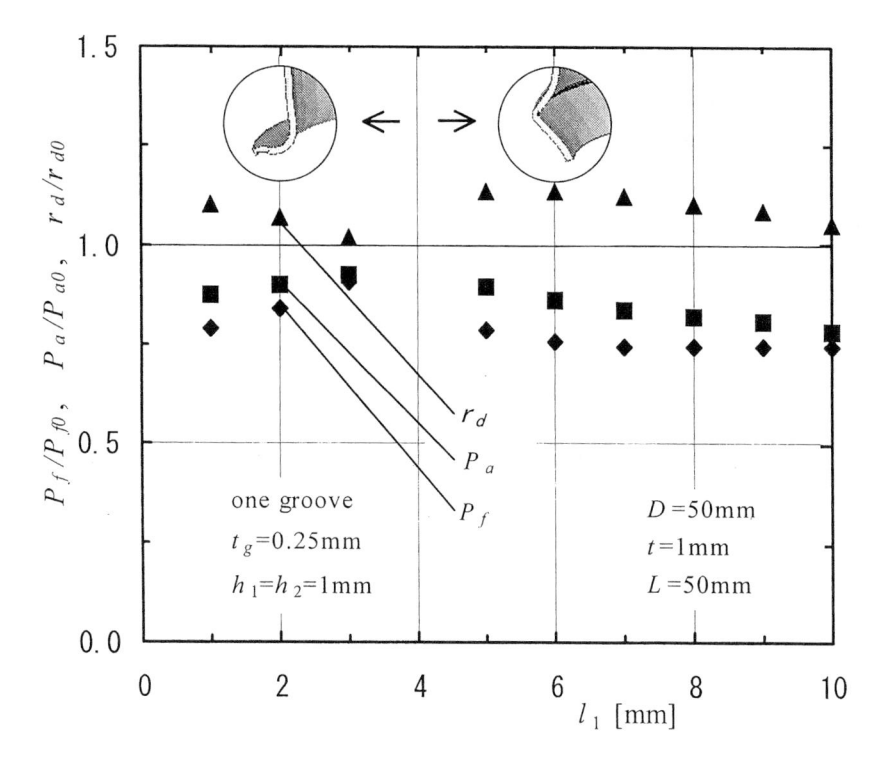

Figure 7. Evaluation of buckling load on a cylindrical shell with one groove.

6. CONCLUSION

Effect of arrangement of defects in the structure on the absorption of impact energy is considered. The energy is evaluated by using average load over the stroke within $U/D=0.6$. The buckling starts from the contact side of the cylinder due to the friction. Grooves are made in the cylinder as defects to control the buckling. The buckling is classified in two types for one groove and is classified in three types for two grooves. When the distance between rigid body and groove is small, the initial fold expands outwards. As the result, first peak load becomes small in comparison with decreasing of average load. The ratio of the average load and the first peak load becomes large.

$U/D=0$ 0.028   0.126

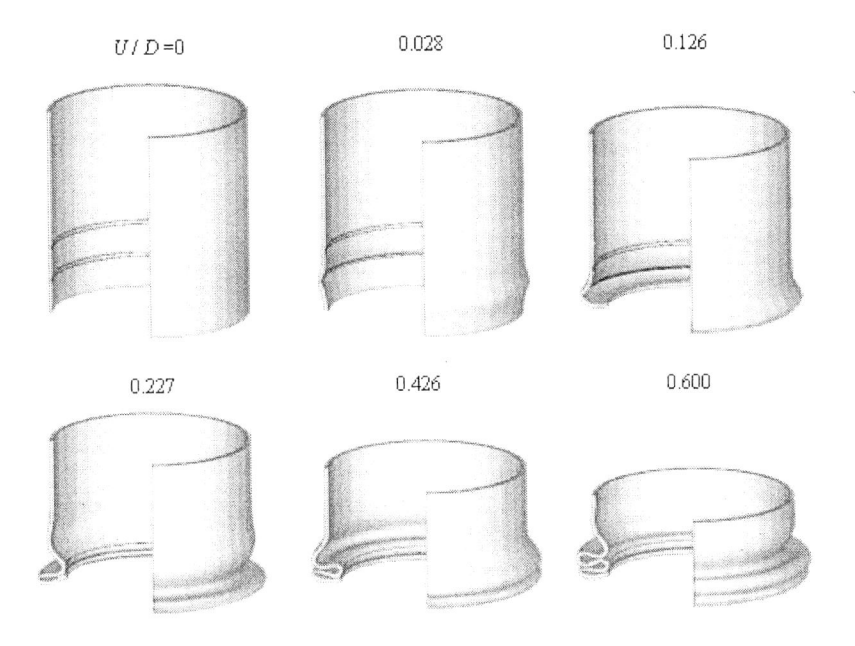

0.227   0.426   0.600

Figure 8. Buckling process of a cylindrical shell with two grooves.

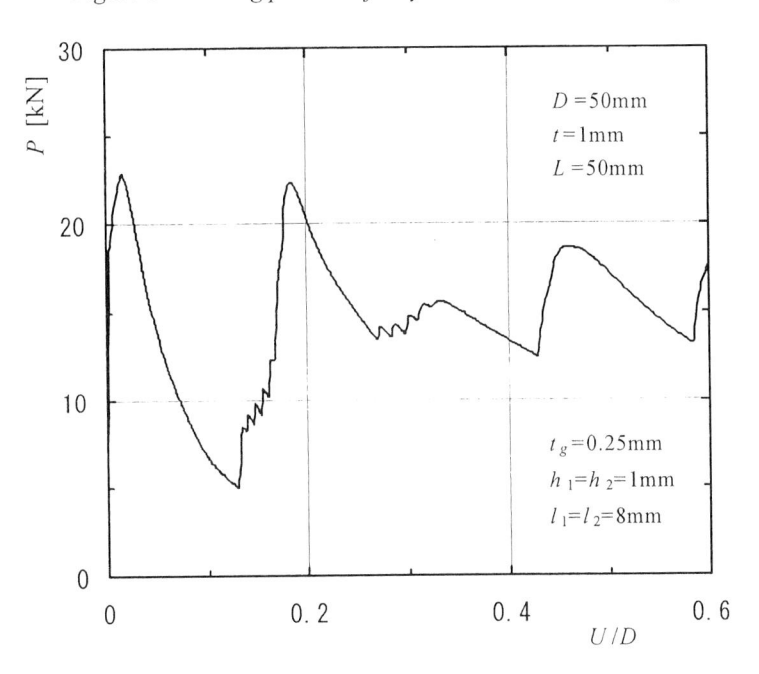

$D = 50$mm
$t = 1$mm
$L = 50$mm

$t_g = 0.25$mm
$h_1 = h_2 = 1$mm
$l_1 = l_2 = 8$mm

Figure 9. Load displacement diagram of a cylindrical shell with two grooves.

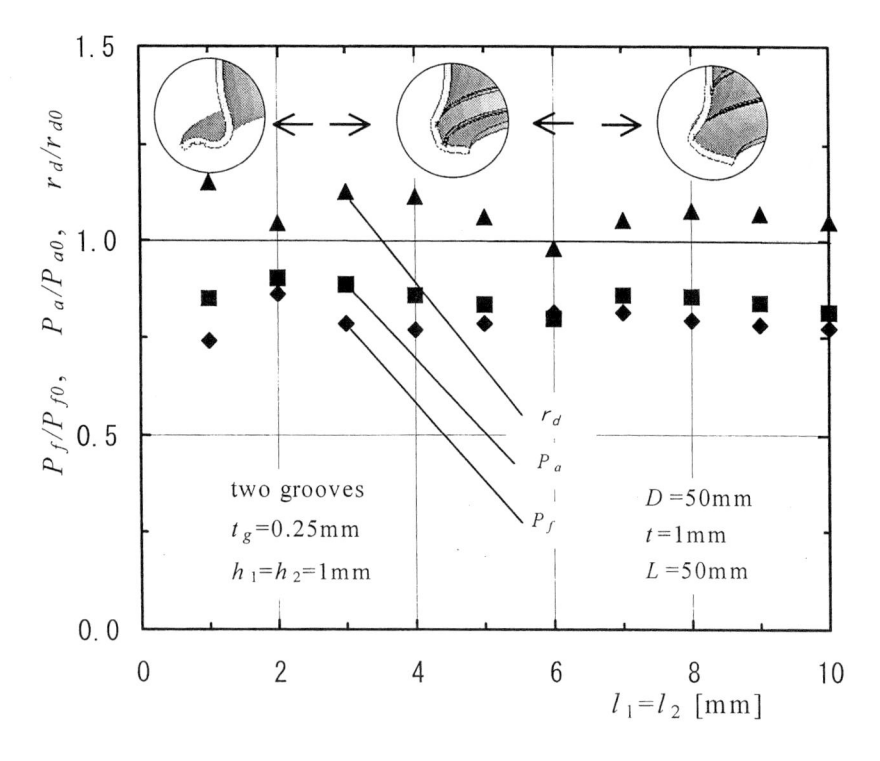

Figure 10. Evaluation of buckling load on a cylindrical shell with two grooves.

7. REFERENCES

1. Wierzbicki, T. and Abramowicz, W. (1983) On the crushing mechanics of thin-walled structure, *ASME J. Applied Mechanics*, 50 727-734.
2. Allan, T. (1968) Experimental and analytical investigation of the behaviour of cylindrical tubes subject to axial compressive forces, *J. Mechanical Engineering Science*, 10 182-197.
3. Singace, A.A., Elsobky, H. and Reddy, T.Y. (1995) On the eccentricity factor in the progressive crushing of tubes, *Int. J. Solids Structures*, 32 3589-3602.
4. Singace, A.A. and Elsobky, H. (1996) Further experimental investigation on the eccentricity factor in the progressive crushing of tubes, *Int. J. Solids Structures*, 33 3517-3538.
5. Ine, T. and Toi, Y. (1991) Basic Studies on the Crashworthiness of Structural Elements (Part 6) – Axisymmetric progressive collapse analysis of circular cylinders using finite element method -, *J. Soc. Nav. Arch. Japan*, 170 525-537.
6. Mikkelsen, L.P. (1999) A numerical axisymmetric collapse analysis of viscoplastic cylindrical shells under axial compression, *Int. J. Solids Structures*, 36 643-668.
7. Sobel, L.H. and Newman, S. Z. (1980) Plastic buckling of cylindrical shells under axial compression, *ASME J. Pressure Vessel Technology*, 102 40-44.
8. Andrews, K.R., England, G.L. and Ghani, E. (1983) Classification of the axial collapse of cylindrical tubes under quasi-static loading, *Int. J. Mech. Sci.*, 25 687-696.

WAVE PROPAGATION IN PIEZOELECTRIC CIRCULAR PLATE UNDER THERMO-ELECTRO-MECHANICAL LORDING

NAOBUMI SUMI
Faculty of Education, Shizuoka University,
Shizuoka, Shizuoka 422-8529 Japan
E-mail : einsumi@ipc.shizuoka.ac.jp

1. INTRODUCTION

In recent years, increasing attentions have been given to the smart structures with piezoelectric sensors and actuators for they can sense and alter the mechanical response during in-service operation. The literature on the piezothermoelectric behavior of materials is extensive. To author's knowledge, however, a majority of investigations have focused entirely on the stationary or vibration problems. Few attempts have been made to investigate the wave propagation in piezoelectric materials under thermo-electro-mechanical lording.

The object of this paper is to study the numerical treatment of wave propagation behavior of a piezoelectric circular plate under axisymmetric thermal, electric, and mechanical lording. In this paper, the wave response is treated as a one-spatial dimensional one. Moreover, a coupling between thermal and mechanical fields is not taken into account. The general dynamic equations, which include thermal, electric, and mechanical fields, are based on the first-order partial differential equations with stresses, particle velocity, electric displacement, and electric field intensity as dependent variables. This system of equations is analyzed by the method of characteristics. The method transforms the basic hyperbolic partial differential equations to the ordinary differential equations, each of which is valid along a different family of characteristic lines. A numerical procedure, which integrates the characteristic equations along characteristics via a step-by-step integration technique, is established.

The numerical examples are shown for two different cases of wave propagation in a traction free piezoelectric plate : Case (1) is for an infinite plate exposed to impulsive surface heating, and Case (2) is for a thin hollow circular plate subjected to impulsive inner surface heating in the radial direction. Graphical displays are utilized to present the influences of piezothermoelastic constants on the wave profiles.

2. BASIC EQUATIONS

The equations governing the propagation of axisymmetric waves in a piezothermoelastic solid possessing hexagonal material symmetry of class 6mm, expressed in cylindrical coordinates (r, z), are given as follows [1-4] :

(a) Equations of motion

$$\frac{\partial \sigma_{rr}}{\partial r} + \frac{\partial \sigma_{zr}}{\partial z} + \frac{\sigma_{rr} - \sigma_{\theta\theta}}{r} = \rho \frac{\partial^2 u_r}{\partial t^2}$$
$$\frac{\partial \sigma_{zr}}{\partial r} + \frac{\partial \sigma_{zz}}{\partial z} + \frac{\sigma_{zr}}{r} = \rho \frac{\partial^2 u_z}{\partial t^2} \tag{1}$$

(b) Constitutive equations

$$\sigma_{rr} = c_{11}\epsilon_{rr} + c_{12}\epsilon_{\theta\theta} + c_{13}\epsilon_{zz} - e_1 E_z - \beta_1 T$$
$$\sigma_{\theta\theta} = c_{12}\epsilon_{rr} + c_{11}\epsilon_{\theta\theta} + c_{13}\epsilon_{zz} - e_1 E_z - \beta_1 T$$
$$\sigma_{zz} = c_{13}\epsilon_{rr} + c_{13}\epsilon_{\theta\theta} + c_{33}\epsilon_{zz} - e_3 E_z - \beta_3 T \tag{2}$$
$$\sigma_{zr} = c_{44}\epsilon_{zr} - e_4 E_r$$

(c) Constitutive equations for the electric field

$$D_r = e_4 \epsilon_{zr} + \eta_1 E_r$$
$$D_z = e_1 \epsilon_{rr} + e_1 \epsilon_{\theta\theta} + e_3 \epsilon_{zz} + \eta_3 E_z + p_3 T \tag{3}$$

(d) Charge equilibrium equation of electrostatics

$$\frac{\partial D_r}{\partial r} + \frac{\partial D_z}{\partial z} + \frac{D_r}{r} = 0 \tag{4}$$

where σ_{ij} and ϵ_{ij} are the stress and strain components, u_i are the displacements components, D_i and E_i are the electric displacement and electric field components, T is the temperature rise from stress-free state, ρ denotes the density, c_{ij} are elastic stiffnesses, e_i are piezoelectric constants, β_i are stress-temperature coefficients, η_i are dielectric permitivities, and p_3 is the pyroelectric constant.

The strains are related to the displacements by

$$\epsilon_{rr} = \frac{\partial u_r}{\partial r}, \quad \epsilon_{\theta\theta} = \frac{u_r}{r}, \quad \epsilon_{zz} = \frac{\partial u_z}{\partial z}, \quad \epsilon_{zr} = \frac{\partial u_z}{\partial r} + \frac{\partial u_r}{\partial z} \tag{5}$$

The temperature T is governed by the Fourier's heat conduction equation

$$k_1 \left(\frac{\partial^2 T}{\partial r^2} + \frac{1}{r} \frac{\partial T}{\partial r} \right) + k_3 \frac{\partial^2 T}{\partial z^2} = \rho c_v \frac{\partial T}{\partial t} \tag{6}$$

in which k_i are coefficients of thermal conductivity, and c_v is the specific heat. For the uncoupled theory of piezothermoelasticity, the temperature state can be found independently of piezomechanical state of the body.

3. WAVE PROPAGATION IN INFINITE PLATE

3.1 Statement of Problem

As the first example, we consider the one-spatial dimensional wave propagation behavior involving an infinite piezothermoelastic plate of thickness h as shown in Fig.1. The plate is referred to a cylindrical coordinate system in which z means the perpendicular distance from a boundary surface. We will assume that the temperature on the surface $z = 0$ increases in a small but a finite time interval t_p from its initial value 0, linearly to its final value T_c, while the other surface $z = h$ is kept to its initial temperature 0. Then the boundary conditions for the temperature are

$$T(0,t) = \begin{cases} T_c \dfrac{t}{t_p}, & t \leq t_p \\ T_c, & t \geq t_p \end{cases}, \quad T(h,t) = 0 \quad (7)$$

Under these conditions, the heat flow occurs in the z-direction only, and the heat conduction equation (6) reduces to

$$\frac{\partial T}{\partial t} = \frac{1}{\kappa} \frac{\partial^2 T}{\partial z^2} \quad (8)$$

in which $\kappa = k_3/\rho c_v$ denotes the thermal diffusivity. For the method of characteristics, the temperature state can be obtained by writing Eq.(8) in the explicit finite-difference form by using the "leap-frog" method in the characteristic $(z - t)$ plane [6].

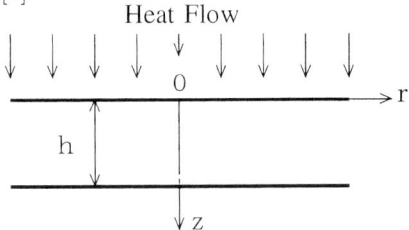

Figure 1. Infinite plate subjected to ramp-type surface heating.

For the elastic and electric fields, the surfaces of the plate are considered to be free of both traction and electric charge, i.e.

$$\sigma_{zz} = \sigma_{zr} = D_z = 0 \quad \text{on} \quad z = 0, h \quad (9)$$

The symmetry conditions of the problem imply that all the field variables depend on the distance z and time t only, and these assumptions lead to

$$u_r = E_r = D_r = D_z = 0 \quad (10)$$

3.2 Characteristics and Characteristic Equations

Substituting Eqs.(10) into Eqs.(1) $\sim$ (4), and differentiating these equations with respect to the time t, we have basic equations for one-spatial dimensional wave response in a piezothermoelastic infinite plate

$$\frac{\partial \sigma_{zz}}{\partial z} = \rho \frac{\partial v}{\partial t} \tag{11}$$

$$\frac{\partial \sigma_{zz}}{\partial t} = c_{33} \frac{\partial v}{\partial z} - e_3 \frac{\partial E_z}{\partial t} - \beta_3 \frac{\partial T}{\partial t} \tag{12}$$

$$\frac{\partial \sigma_{rr}}{\partial t} = c_{13} \frac{\partial v}{\partial z} - e_1 \frac{\partial E_z}{\partial t} - \beta_1 \frac{\partial T}{\partial t} \tag{13}$$

$$0 = e_3 \frac{\partial v}{\partial z} + \eta_3 \frac{\partial E_z}{\partial t} + p_3 \frac{\partial T}{\partial t} \tag{14}$$

in which $v = \partial u_z / \partial t$ represents the particle velocity in the z-direction. This is a system of four linear first-order partial differential equations with $\sigma_{zz}, \sigma_{rr} (= \sigma_{\theta\theta}), v$, and E_z as the four dependent variables. These equations are hyperbolic in nature. In the $(z-t)$ plane, certain curves may exist, along which these variables are continuous, but their derivatives may be discontinuous. These curves will be called the characteristics. The relations governing the variations of variables along characteristics will be called the characteristic equations. The characteristics and the characteristic equations may be derived by the conventional directional derivative approach [5-7]. For the system of Eqs. (11) $\sim$ (14), the characteristics are found to be composed of three families of characteristic lines $I_i, (i = 1, 2, 3)$:

$$I_i \quad : \quad \frac{dz}{dt} = V_i = (c_1, -c_1, 0), \quad (i = 1, 2, 3) \tag{15}$$

where c_1 is the wave velocity defined by

$$c_1 = \sqrt{\frac{e_3^2 + c_{33}\eta_3}{\rho \eta_3}} \tag{16}$$

The I_1 and I_2 characteristics describe two characteristic families of lines with slopes c_1 and $-c_1$, respectively, in the $(z-t)$ plane. The third characteristic I_3 consists of straight lines parallel to the t-axis as shown in Fig.2.

The characteristic equations along $I_j : dz/dt = V_j, (j = 1, 2)$ are given by

$$d\sigma_{zz} - \rho V_j dv = \left(\frac{e_3 p_3}{\eta_3} - \beta_3 \right) \frac{\partial T}{\partial t} dt, \quad (j = 1, 2) \tag{17}$$

The characteristic equations along $I_3 : dz/dt = V_3$ are

$$d\sigma_{zz} - \frac{c_{33}}{c_{13}}d\sigma_{rr} - \left(\frac{c_{33}e_1}{c_{13}} - e_3\right)dE_z = \left(\frac{c_{33}\beta_1}{c_{13}} - \beta_3\right)dT \qquad (18)$$

$$-\frac{e_3}{c_{13}}d\sigma_{rr} - \left(\frac{e_3 e_1}{c_{13}} + \eta_3\right)dE_z = \left(\frac{e_3\beta_1}{c_{13}} + p_3\right)dT \qquad (19)$$

The equations (17) ~ (19) are applicable for continuous fields with possible discontinuities in the derivatives of the variables $\sigma_{zz}, \sigma_{rr}, v$, and E_z. In this paper, a numerical procedure involving stepwise integration of characteristic equations along characteristics is employed to solve the problem. The $(z - t)$ plane is subdivided into a network constructed by three families of characteristic lines I_i, as shown in Fig.2. At a typical interior point D, the four quantities $\sigma_{zz}, \sigma_{rr}, v$, and E_z can be calculated if all the quantities at three neighboring points A, C and E are known from the previous calculations. Between points A and D along a I_1 characteristic, and between points C and D along a I_2 characteristic, Eqs.(17) are written in finite-difference form using the trapezoidal integration rule. Along a I_3 characteristic from point E to D, Eqs.(18) and (19) are also expressed in finite-difference form. The four unknowns $\sigma_{zz}, \sigma_{rr}, v$, and E_z at point D can be determined from these four equations.

Along the boundary points where one of the four variables is prescribed, the remaining three unknown quantities may be determined from the same procedure for which a characteristic equation along a characteristic extending outside of the region is eliminated.

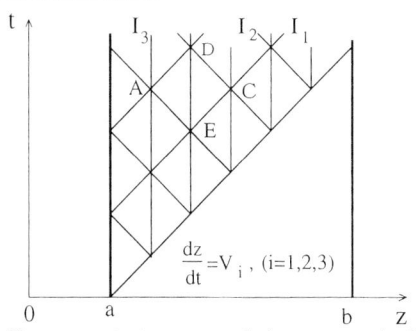

Figure 2. Characteristic network for numerical procedure.

3.3 Numerical Example

For illustrative purposes, the plate is considered to be cadmium selenide with the following properties [1-4]:

$$\rho = 5655 \, [kg/m^3], \quad c_v = 274 \, [J/KgK], \quad k = 9 \, [W/mK]$$
$$c_{11} = 74.1 \times 10^9 \, [Nm^{-2}], \quad c_{12} = 45.2 \times 10^9 \, [Nm^{-2}]$$
$$c_{13} = 39.3 \times 10^9 \, [Nm^{-2}], \quad c_{33} = 83.6 \times 10^9 \, [Nm^{-2}]$$
$$\beta_1 = 0.621 \times 10^6 \, [NK^{-1}m^{-2}], \quad \beta_3 = 0.551 \times 10^6 \, [NK^{-1}m^{-2}] \qquad (20)$$
$$e_1 = -0.160 \, [Cm^{-2}], \quad e_3 = 0.347 \, [Cm^{-2}], \quad d_1 = -3.92 \times 10^{-12} \, [CN^{-1}]$$
$$\eta_3 = 90.3 \times 10^{-12} \, [C^2N^{-1}m^{-2}], \quad p_3 = -2.94 \times 10^{-6} \, [CK^{-1}m^{-2}]$$
$$Y_r = 42.8 \times 10^9 \, [Nm^{-2}], \quad \alpha_r = 4.4 \times 10^{-6} \, [K^{-1}], \quad c_r = 3367 \, [ms^{-1}]$$

where Y_r is Young's modulus, α_r is the coefficient of linear thermal expansion, c_r is dilatational wave speed, and d_1 is the piezoelectric coefficient. Since the values of the coefficients of heat conduction for cadmium selenide cannot be found in the literatures, the value $k = k_i$ is assumed.

In the presentation of the numerical results, the following dimensionless quantities are introduced:

$$z^* = \frac{z}{l}, \quad t^* = \frac{c_r t}{l}, \quad \sigma_{ij}^* = \frac{\sigma_{ij}}{Y_r \alpha_r T_c}, \quad v^* = \frac{v}{\alpha_r T_c c_r}, \quad E_z^* = \frac{|d_1| E_z}{\alpha_r T_c}$$
$$T^* = \frac{T}{T_c}, \quad c_{13}^* = \frac{c_{13}}{Y_r}, \quad c_{33}^* = \frac{c_{33}}{Y_r}, \quad \beta_1^* = \frac{\beta_1}{Y_r \alpha_r}, \quad \beta_3^* = \frac{\beta_3}{Y_r \alpha_r} \qquad (21)$$
$$e_1^* = \frac{e_1}{Y_r |d_1|}, \quad e_3^* = \frac{e_3}{Y_r |d_1|}, \quad \eta_3^* = \frac{\eta_3}{Y_r |d_1|^2}, \quad p_3^* = \frac{p_3}{Y_r |d_1| \alpha_r}$$

in which l is the characteristic length of the problem.

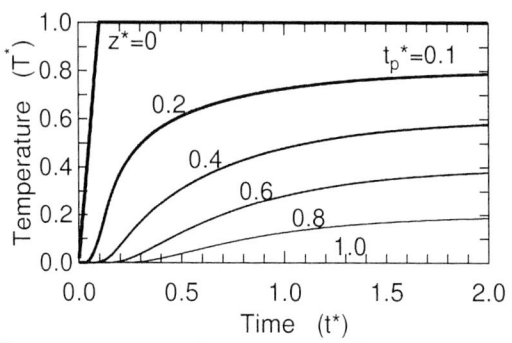

Figure 3. Time variations of temperature T^* at various positions z^*.

Calculations are carried out for $h^* = h/l = 1, t_p^* = 0.1$ by taking the characteristic length $l = 10^{-8} [m]$. In this case, responses show wave fronts traveling with speed $c_1^* = c_1/c_r = 1.15$. A mesh size $\Delta z^* = h^*/500$ is used in the solution of the majority of numerical examples. Figures 3 ∼ 5 exhibit the time variations of temperature T^*, normal stress σ_{zz}^*, and electric field E_z^* at various positions z^* of the plate. The temperature T^* increases monotonously by the diffusion phenomenon. At various positions z^*, the abrupt wave front arrives at time z^*/c_1^*, and propagates toward inside of the plate at velocity c_1^*. Therefore, the normal stress σ_{zz}^* and electric field E_z^* retain sharp wave fronts, and undergo a frequency response as a result of repeated reflections of waves from the upper and lower surfaces of the plate.

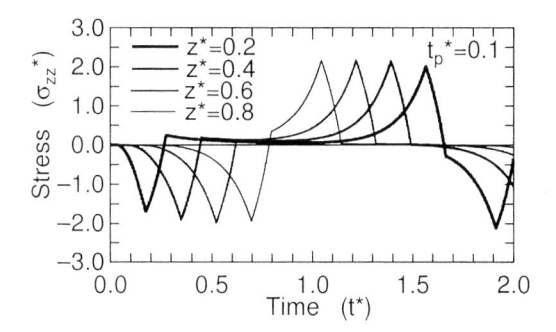

Figure 4. Time variations of normal stress σ_{zz}^* at various positions z^*.

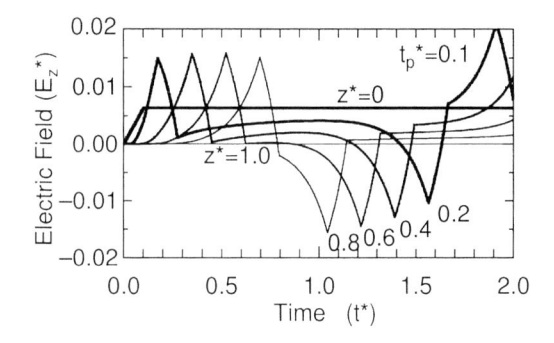

Figure 5. Time variations of electric field E_z^* at various positions z^*.

4. WAVE PROPAGATION IN HOLLOW CIRCULAR PLATE

4.1 Statement of Problem

As a second example, we consider the one-spatial dimensional wave propagation in a thin hollow circular plate of inner radius a and outer radius b as shown in Fig.6. The plate is subjected to ramp-type heating on the inner surface $r = a$, zero temperature rise on the outer surface $r = b$, and thermal insulation on the top and bottom surfaces of the plate. For this case, the temperature varies in the radial direction only, and the boundary conditions take the form

$$T(a,t) = \begin{cases} T_c \dfrac{t}{t_p}, & t \leq t_p \\ T_c, & t \geq t_p \end{cases}, \quad T(b,t) = 0 \tag{22}$$

The heat conduction equation (6) reduces to

$$\frac{\partial^2 T}{\partial r^2} + \frac{1}{r}\frac{\partial T}{\partial r} = \frac{1}{\kappa}\frac{\partial T}{\partial t} \tag{23}$$

in which $\kappa = k_1/\rho c_v$ denotes the thermal diffusivity.

For the elastic and electric fields, the plate is considered to be in a plane-stress state, and the top and bottom surfaces are taken to be free from traction and electric charge. Under these conditions, the field variables depend on the coordinate r only, and we have

$$\sigma_{zz} = \sigma_{zr} = D_z = 0 \tag{24}$$

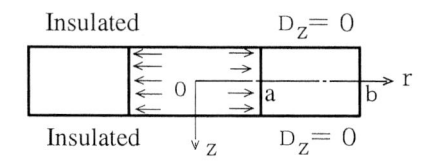

Figure 6. Hollow circular plate subjected to ramp-type inner surface heating.

4.2 Characteristics and Characteristic Equations

Following the procedure similar to that introduced in the previous section, the basic equations are given by

$$\frac{\partial \sigma_{rr}}{\partial r} + \frac{\sigma_{\theta\theta} - \sigma_{rr}}{r} = \rho \frac{\partial v}{\partial t} \tag{25}$$

$$\frac{\partial \sigma_{rr}}{\partial t} = \bar{c}_{11}\frac{\partial v}{\partial r} + \bar{c}_{12}\frac{v}{r} - \bar{e}_1\frac{\partial E_z}{\partial t} - \bar{\beta}_1\frac{\partial T}{\partial t} \tag{26}$$

$$\frac{\partial \sigma_{\theta\theta}}{\partial t} = \bar{c}_{12}\frac{\partial v}{\partial r} + \bar{c}_{11}\frac{v}{r} - \bar{e}_1\frac{\partial E_z}{\partial t} - \bar{\beta}_1\frac{\partial T}{\partial t} \tag{27}$$

$$0 = \bar{e}_1\frac{\partial v}{\partial r} + \bar{e}_1\frac{v}{r} + \bar{\eta}_3\frac{\partial E_z}{\partial t} + \bar{p}_3\frac{\partial T}{\partial t} \tag{28}$$

where $v = \partial u_r/\partial t$, and

$$\bar{c}_{11} = c_{11} - \frac{c_{13}^2}{c_{33}}, \quad \bar{c}_{12} = c_{12} - \frac{c_{13}^2}{c_{33}}, \quad \bar{e}_1 = e_1 - \frac{c_{13}e_3}{c_{33}}$$

$$\bar{\beta}_1 = \beta_1 - \frac{c_{13}\beta_3}{c_{33}}, \quad \bar{\eta}_3 = \eta_3 + \frac{e_3^2}{c_{33}}, \quad \bar{p}_3 = p_3 + \frac{e_3\beta_3}{c_{33}} \tag{29}$$

Equations (25) ~ (28) constitute a system of four linear first-order partial differential equations with $\sigma_{rr}, \sigma_{\theta\theta}, v$ and E_z as the four dependent variables. For the system of Eqs.(25)-(28), the characteristics are

$$I_i \; : \; \frac{dr}{dt} = V_i = (\bar{c}_1, -\bar{c}_1, 0), \quad (i = 1, 2, 3) \tag{30}$$

where $\bar{c}_1$ is the wave velocity defined by

$$\bar{c}_1 = \sqrt{\frac{\bar{e}_1^2 + \bar{c}_{11}\bar{\eta}_3}{\rho\bar{\eta}_3}} \tag{31}$$

The characteristic equations along $I_j : dr/dt = V_j, (j = 1, 2)$ are

$$d\sigma_{rr} - \rho V_j dv = \left(\bar{c}_{12} + \frac{\bar{e}_1^2}{\bar{\eta}_3}\right)\frac{v}{r}dt - \frac{\sigma_{rr} - \sigma_{\theta\theta}}{r}V_i dt - \left(\bar{\beta}_1 - \frac{\bar{p}_3 \bar{e}_1}{\bar{\eta}_3}\right)\frac{\partial T}{\partial t}dt \tag{32}$$

The characteristic equations along $I_3 : dr/dt = V_3 = 0$ are

$$d\sigma_{rr} + \left(\bar{e}_1 + \frac{\bar{c}_{11}\bar{\eta}_3}{\bar{e}_1}\right) dE_z = (\bar{c}_{12} - \bar{c}_{11})\frac{v}{r}dt - \left(\bar{\beta}_1 + \frac{\bar{c}_{11}\bar{p}_3}{\bar{e}_1}\right)\frac{\partial T}{\partial t}dt \tag{33}$$

$$d\sigma_{\theta\theta} + \left(\bar{e}_1 + \frac{\bar{c}_{12}\bar{\eta}_3}{\bar{e}_1}\right) dE_z = (\bar{c}_{11} - \bar{c}_{12})\frac{v}{r}dt - \left(\bar{\beta}_1 + \frac{\bar{c}_{12}\bar{p}_3}{\bar{e}_1}\right)\frac{\partial T}{\partial t}dt \tag{34}$$

Finally, the values of $\sigma_{rr}, \sigma_{\theta\theta}, v$, and E_z may be found by integrating numerically the characteristic equations (32) ~ (34) along characteristics.

4.3 Numerical Example

Calculations are carried out for thin hollow circular plate of $a^* = 1$, $b^* = 2$, $t_p^* = 0.1$ by taking $l = 10^{-8} [m]$. Figures 6 and 7 exhibit the time variations of hoop stress $\sigma_{\theta\theta}^*$ and electric field E_z^* at various positions z^*, respectively.

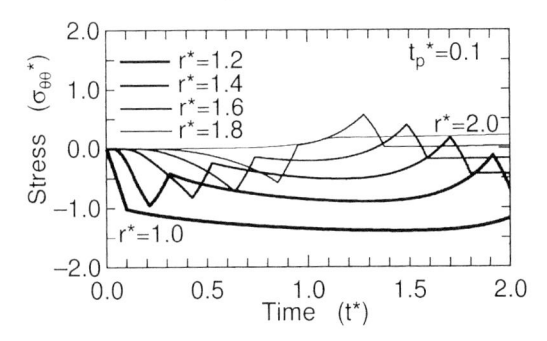

Figure 6. Time variations of hoop stress $\sigma_{\theta\theta}^*$ at various positions z^*.

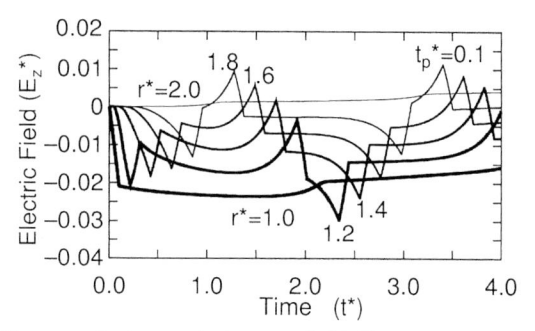

Figure 7. Time variations of electric field E_z^* at various positions z^*.

5. CONCLUSION

The major accomplishment of this paper is to present the numerical method for the solution of one-spatial dimensional wave propagation in a piezothermoelastic material under thermo-electro-mechanical lording. The solutions is are obtained by the method of characteristics, yielding characteristics and characteristic equations. Numerical examples are shown for two cases of wave propagation : Case (1) is for an infinite plate, and case (2) is for a thin hollow circular plate. The present study is valuable for the understanding of the piezothermoelastic behavior of smart structures. The method presented here is easily adaptable to computer calculations, and is believed to be used successfully for many problems involving multiple wave reflections.

6. REFERENCES

1. Ashida, F. and Tauchert, T. R., An inverse problem for determination of transient surface temperature from piezoelectric sensor measurement, *Journal of Applied Mechanics,***65**(1998), pp.367-373.
2. Ashida, F. and Tauchert, T. R., Transient response of a piezothermoelastic circular disk under axisymmertic heating, *Acta Mechanica,***128**(1998), pp.1-14.
3. Ashida, F. and Tauchert, T. R., Plane stress problem of a piezothermoelastic plate, *Acta Mechanica,***145**(2000), pp.127-134.
4. Ashida, F. and Tauchert, T. R., A general plane-stress solution in cylindrical coordinates for a piezothermoelastic plate, *International Journal of Solids and Structures,***38**(2001), pp.4960-4985.
5. Chou, P. C. and Koenig, H. A., A unified approach to cylindrical and spherical elastic waves by method of characteristics, *Journal of Applied Mechanics,*(1966), pp.159-167.
6. Sumi, N. and Noda, N., Dynamic thermal stresses in laminated media by the method of characteristics, *Journal of Thermal stresses,***15**(1992), pp.379-392.
7. Sumi, N., Numerical solutions of thermoelastic wave problems by the method of characteristics, *Journal of Thermal Stresses,***24**(2001), pp.509-530.

MODELING OF PIEZOELECTRIC / MAGNETOSTRICTIVE MATERIALS FOR SMART STRUCTURES

MEHMET SUNAR
Mechanical Engineering Department, King Fahd University of Petroleum and Minerals, Dhahran 31261, Saudi Arabia
E-mail: mehmets@kfupm.edu.sa

1. INTRODUCTION

There have been many research activities in the last two decades in system sensing and control involving smart materials and structures. Smart structures contain smart materials such as piezoelectric and magnetostrictive materials, and shape memory alloys for highly adaptive structural behavior.

The piezoelectric phenomenon is referred to as the relation between mechanical and electrical fields in certain crystals, ceramics and polymeric films. The uses of piezoelectric materials as sensors and actuators initiated a remarkable number of research studies in recent years. Some of the recent research work in piezoelectricity include the finite element modeling of transient response of micro-electromechanical system (MEMS) sensors by Lim et al [1], and the piezoelectric actuators for dual bending/twisting control of wings by Tzou et al [2].

The phenomenon of magnetostriction or piezomagnetism is defined as the interaction between mechanical and magnetic fields in a body. To a lesser degree than the piezoelectric materials, the magnetostrictive materials have also attracted a good number of research studies in sensing and control. They were used by Shaw [3] in active vibration isolation, and by Monaco et al [4] on damage detection.

The piezoelectric and magnetostrictive layers can be bonded together to form a composite with a magnetoelectric effect. A magnetoelectric coefficient was defined by Harshe et al [5] as the coefficient relating the electrical field to the magnetic field at open circuit conditions. This effect enables the coupling and interaction of magnetic and electrical fields. Various piezoelectric/magnetostrictive composite combinations were formed by Avellaneda and Harshe [6] and the magnetoelectric coefficient and parameters which characterize the efficiency of energy conversion among the layers were calculated. Closed-form solutions for the magnetoelectric coefficients in composites with piezoelectric and piezomagnetic phases were found by Wu and Huang [7].

In this paper, the constitutive and finite element equations are presented for a medium where mechanical, electrical, thermal and magnetic fields interact with each other. The equations for piezoelectric and magnetostrictive media are extracted from these equations. An illustrative example is then taken to study the performance of a piezoelectric and magnetostrictive sensor and actuator, whose locations are varied along a cantilever beam structure.

2. GENERAL CONSTITUTIVE AND DIFFERENTIAL EQUATIONS

To derive the constitutive equations governing the quasi-static behavior of a medium where mechanical, electrical, thermal and magnetic fields interact with each other, an approach similar to the one by Mindlin [8] is taken. A quadratic thermodynamic potential G is defined such that

$$G = \frac{1}{2} S^T cS - \frac{1}{2} E^T \varepsilon E + \frac{1}{2} B^T \mu^{-1} B - \frac{1}{2} \alpha \theta^2 - S^T eE - E^T P\theta - S^T \lambda \theta$$
$$- S^T \ell B - B^T r\theta - B^T bE. \qquad (1)$$

where S, E and B are the vectors of strain, electrical field and magnetic flux density, respectively, and θ is a small temperature change, c, ε, μ, e, ℓ and b are the matrices of constitutive coefficients, P, λ and r are the vectors of constitutive coefficients, α is the scalar constitutive coefficient given as $\alpha = \rho c_v \Theta_o^{-1}$, where ρ is the mass density, c_v is specific heat and Θ_o is reference temperature. From Equation (1), the constitutive equations are derived as follows

$$T = \frac{\partial G}{\partial S} = cS - eE - \ell B - \lambda \theta, \quad D = -\frac{\partial G}{\partial E} = e^T S + \varepsilon E + b^T B + P\theta$$

$$H = \frac{\partial G}{\partial B} = -\ell^T S - bE + \mu^{-1} B - r\theta, \quad \eta = -\frac{\partial G}{\partial \theta} = \lambda^T S + P^T E + r^T B + \alpha \theta \qquad (2)$$

which contain the vectors of mechanical stress T, electrical displacement D and magnetic field intensity H, and the entropy η.

To obtain the differential equations governing the dynamic behavior, two energy functionals Π and Ψ are defined as

$$\Pi = \int_V (G + \eta\Theta) dV - \int_V u^T P_b dV - \int_S u^T P_s dV + \int_V \phi \rho_v dV + \int_S \phi \sigma dS -$$
$$\int_V A^T J dV + \int_S A^T H'_E n dS, \quad \Psi = \int_V (\Gamma - \mu \Theta \dot{\Theta} - W\Theta) dV + \int_S h^T n dS \qquad (3)$$

where Θ is the absolute temperature given as $\Theta = \Theta_o + \theta$, P_b and P_s are the vectors of body and surface forces; u, A, J and h are the vectors of mechanical displacement, magnetic potential, volume current density and external heat flux; n is the vector normal to the surface; ϕ, ρ_v, σ and W are the scalars of electrical potential, volume charge density, surface charge and heat source density; and H'_E is the matrix of applied magnetic field intensity defined in cartesian coordinate system as

$$H'_E = \begin{bmatrix} 0 & H_z & -H_y \\ -H_z & 0 & H_x \\ H_y & -H_x & 0 \end{bmatrix}_E. \qquad (4)$$

Γ in Equation (3) is given as

$$\Gamma = \frac{1}{2}\nabla^T \Theta \, K \nabla \Theta \tag{5}$$

where K is the matrix of heat conduction coefficients and ∇ is the gradient vector. The generalized Hamilton's principle has the following forms

$$\delta \int_{t_1}^{t_2}(Ki - \Pi)\,dt = 0, \quad \delta \int_{t_1}^{t_2} \Psi \,dt = 0 \tag{6}$$

where the kinetic energy Ki is defined as

$$Ki = \int_V \frac{1}{2}\rho \dot{u}^T \dot{u}\, dt \tag{7}$$

whose variation is computed as

$$\delta \int_{t_1}^{t_2} Ki\, dt = \delta \int_{t_1}^{t_2}\int_V \frac{1}{2}\rho \dot{u}^T \dot{u}\, dt\, dV = -\int_{t_1}^{t_2} dt \int_V \rho\, \delta u^T \ddot{u}\, dt. \tag{8}$$

The variation of the thermodynamic potential is given as

$$\delta G = \delta S^T T - \delta E^T D + \delta B^T H - \delta \theta\, \eta. \tag{9}$$

Recall the relations

$$S = L_u u, \quad E = -\nabla\phi - \dot{A}, \quad B = \nabla \times A = L_A A \tag{10}$$

where L_u and L_A are differential operators, and substitute them along with Equations (8) and (9) into the first form of Hamilton's principle to obtain

$$\int_{t_1}^{t_2} dt \Big[\int_V \delta u^T \left(-\rho \ddot{u} + L_u^T T + P_b\right) dV + \int_V \delta \phi^T \left(\nabla^T D - \rho_v\right) dV$$
$$+ \int_V \delta A^T \left(\dot{D} - \nabla \times H + J\right) dV + \int_S \delta u^T (P_s - N T)\, dS \tag{11}$$
$$- \int_S \delta \phi^T \left(\sigma + D^T n\right) dS - \int_S \delta A^T \left(H_E' - H'\right) n\, dS \Big] = 0.$$

where H' is the internal magnetic field intensity defined as in Equation (4) and N is the matrix containing appropriate surface normals. Therefore, the following equations must be satisfied with appropriate boundary conditions on the corresponding surfaces

$$-\rho \ddot{u} + L_u^T T + P_b = 0, \quad \nabla^T D - \rho_v = 0, \quad \dot{D} - \nabla \times H + J = 0. \tag{12}$$

The first equation in Equation (12) is the equations of motion for the mechanical field, and the second and third equations are the Maxwell's equations for the electrical and magnetic fields [9]. It can be shown [10] that the Hamilton's principle on Ψ given in Equation (6) yields

$$\int_{t_1}^{t_2} dt \Big[\int_V \delta \theta \left(\nabla^T q - W + \Theta \dot{\eta}\right) dV + \int_S \delta \theta \left(h^T - q^T\right) n\, dS \Big] = 0 \tag{13}$$

where q is the vector of heat flux. Hence the Hamilton's principle yields the generalized heat equation given as

$$\nabla^T q - W + \Theta \dot{\eta} = 0. \tag{14}$$

3. FINITE ELEMENT EQUATIONS

For the finite element formulation, u, ϕ, A and θ are written for each finite element as

$$\boldsymbol{u}_e = N_u \boldsymbol{u}_i, \quad \phi_e = N_\phi \phi_i, \quad \boldsymbol{A}_e = N_A \boldsymbol{A}_i, \quad \theta_e = N_\theta \theta_i \tag{15}$$

where the subscript e and i respectively stand for the element and nodes of the element, and N's are the shape function matrices. The following relations are also written

$$\boldsymbol{E}_e = -\nabla \phi_e - \dot{\boldsymbol{A}}_e = -[\nabla N_\phi] \, \phi_i - N_A \dot{\boldsymbol{A}}_i = -B_\phi \phi_i - N_A \dot{\boldsymbol{A}}_i, \quad \nabla \theta_e = [\nabla N_\theta] \, \theta_i = B_\theta \theta_i$$

$$\boldsymbol{S}_e = L_u \boldsymbol{u}_e = [L_u N_u] \boldsymbol{u}_i = B_u \boldsymbol{u}_i, \quad \boldsymbol{B}_e = L_A \boldsymbol{A}_e = [L_A N_A] \boldsymbol{A}_i = B_A \boldsymbol{A}_i. \tag{16}$$

Substituting Equations (2), (15) and (16) into Equations (11) and (13) yields the following finite element equations (after assemblage)

$$M_{uu} \ddot{\boldsymbol{u}} + C_{uA} \dot{\boldsymbol{A}} + K_{uu} \boldsymbol{u} + K_{u\phi} \phi - K_{uA} \boldsymbol{A} - K_{u\theta} \theta = \boldsymbol{F}, \quad -C_{\phi A} \dot{\boldsymbol{A}} + K_{\phi u} \boldsymbol{u} - K_{\phi\phi} \phi + K_{\phi A} \boldsymbol{A} + K_{\phi\theta} \theta = \boldsymbol{G}$$

$$M_{AA} \ddot{\boldsymbol{A}} - C_{Au} \dot{\boldsymbol{u}} + C_{A\phi} \dot{\phi} + C_{AA} \dot{\boldsymbol{A}} - C_{A\theta} \dot{\theta} - K_{Au} \boldsymbol{u} + K_{A\phi} \phi + K_{AA} \boldsymbol{A} - K_{A\theta} \theta = \boldsymbol{M} \tag{17}$$

$$M_{\theta A} \ddot{\boldsymbol{A}} - C_{\theta u} \dot{\boldsymbol{u}} + C_{\theta\phi} \dot{\phi} - C_{\theta A} \dot{\boldsymbol{A}} - C_{\theta\theta} \dot{\theta} - K_{\theta\theta} \theta = \boldsymbol{Q}$$

where $\boldsymbol{u}$, ϕ, $\boldsymbol{A}$ and θ are the global vectors of displacement, electrical and magnetic potentials, and small temperature variation, respectively. The element matrices in Equation (17) are computed as

$$[M_{uu}]_e = \int_{V_e} \rho_e N_u^T N_u \, dV, \quad [C_{uA}]_e = \int_{V_e} B_u^T e_e N_A \, dV, \quad [K_{uu}]_e = \int_{V_e} B_u^T c_e B_u \, dV$$

$$[K_{u\phi}]_e = \int_{V_e} B_u^T e_e B_\phi \, dV, \quad [K_{uA}]_e = \int_{V_e} B_u^T \ell_e B_A \, dV, \quad [K_{u\theta}]_e = \int_{V_e} B_u^T \lambda_e N_\theta \, dV$$

$$[C_{\phi A}]_e = \int_{V_e} B_\phi^T \varepsilon_e N_A \, dV, \quad [K_{\phi\phi}]_e = \int_{V_e} B_\phi^T \varepsilon_e B_\phi \, dV, \quad [K_{\phi A}]_e = \int_{V_e} B_\phi^T b_e^T B_A \, dV$$

$$[K_{\phi\theta}]_e = \int_{V_e} B_\phi^T P_e N_\theta \, dV, \quad [C_{AA}]_e = \int_{V_e} \left(B_A^T b_e N_A - N_A^T b_e^T B_A \right) dV \tag{18}$$

$$[M_{AA}]_e = \int_{V_e} N_A^T \varepsilon_e N_A \, dV, \quad [C_{A\theta}]_e = \int_{V_e} N_A^T P_e N_\theta \, dV, \quad [K_{AA}]_e = \int_{V_e} B_A^T \mu_e^{-1} B_A \, dV$$

$$[K_{A\theta}]_e = \int_{V_e} B_A^T r_e N_\theta \, dV, \, [M_{\theta A}]_e = \int_{V_e} \Theta_{oe} N_\theta^T P_e^T N_A \, dV, \, [C_{\theta u}]_e = \int_{V_e} \Theta_{oe} N_\theta^T \lambda_e^T B_u \, dV$$

$$[C_{\theta\phi}]_e = \int_{V_e} \Theta_{oe} N_\theta^T P_e^T B_\phi \, dV, \quad [C_{\theta A}]_e = \int_{V_e} \Theta_{oe} N_\theta^T r_e^T B_A \, dV$$

$$[C_{\theta\theta}]_e = \int_{V_e} \Theta_{oe} N_\theta^T \alpha_e N_\theta \, dV, \quad [K_{\theta\theta}]_e = \int_{V_e} B_\theta^T K_e B_\theta \, dV.$$

Furthermore, $K_{\phi u} = K_{u\phi}^T$, $K_{Au} = K_{uA}^T$, $K_{A\phi} = K_{\phi A}^T$, $C_{Au} = C_{uA}^T$ and $C_{A\phi} = C_{\phi A}^T$. The elemental applied mechanical force, electrical charge, magnetic current and heat vectors ($\boldsymbol{F}$, $\boldsymbol{G}$, $\boldsymbol{M}$ and $\boldsymbol{Q}$) in Equation (17) are found as

$$\boldsymbol{F}_e = \int_{V_e} N_u^T \boldsymbol{P}_{be} \, dV + \int_{S_e} N_u^T \boldsymbol{P}_{se} \, dS + N_u^T \boldsymbol{P}_{ce}, \quad \boldsymbol{G}_e = -\int_{V_e} N_\phi^T \rho_{ve} dV - \int_{S_e} N_\phi^T \sigma_e \, dS$$

$$\boldsymbol{M}_e = \int_{V_e} N_A^T \boldsymbol{J}_e \, dV - \int_{S_e} B_A^T \boldsymbol{H}'_{Ae} \boldsymbol{n} \, dS, \quad \boldsymbol{Q}_e = -\int_{V_e} N_\theta^T W_e \, dV + \int_{S_e} N_\theta^T \boldsymbol{h}_e^T \boldsymbol{n} \, dS \tag{19}$$

where the concentrated force $\boldsymbol{P}_{ce}$ is also added in the formulation of $\boldsymbol{F}_e$.

4. EQUATIONS OF PIEZOELECTRICITY AND MAGNETOSTRICTION

The constitutive equations for piezoelectricity can easily be obtained from Equation (2) by only considering the mechanical and electrical fields. These equations are given as follows

$$T = cS - eE$$
$$D = e^T S + \varepsilon E. \tag{20}$$

The finite element equations for a piezoelectric medium are obtained from Equation (17) as

$$M_{uu}\ddot{u} + C_{uu}\dot{u} \; K_{uu}u + K_{u\phi}\phi = F$$
$$K_{\phi u}u - K_{\phi\phi}\phi = G \tag{21}$$

where the proportional damping matrix C_{uu} is added into the equations. The form of C_{uu} is given as

$$C_{uu} = \alpha M_{uu} + \beta K_{uu} \tag{22}$$

where α and β are positive constants.

Keeping only the terms for the mechanical and magnetic fields in Equations (2) and (17) yields the constitutive and finite element equations for a magnetostrictive medium as

$$T = cS - \ell B$$
$$H = -\ell^T S + \mu^{-1} B \tag{23}$$

and

$$M_{uu}\ddot{u} + C_{uu}\dot{u} + C_{uA}\dot{A} + K_{uu}u - K_{uA}A = F$$
$$M_{AA}\ddot{A} - C_{Au}\dot{u} + C_{AA}\dot{A} - K_{Au}u + K_{AA}A = M. \tag{24}$$

5. ILLUSTRATIVE EXAMPLE

A cantilever beam bonded with two smart material layers is taken as an example (Figure 1). The beam is composed of a host beam structure made of structural steel, a piezoelectric or magnetostrictive sensor at the top, and actuator at the bottom. The piezoelectric and magnetostrictive materials are tested separately to observe their performances in structural sensing and control. The piezoelectric material is assumed to be BM500 of Sensor Technology of Canada, which has properties similar to PZT-5A of Morgan Matroc Limited of UK [11] and the magnetostrictive material is taken as the magnetostrictive ceramic $CoFe_2O_4$. The material properties of these materials are listed in Table 1 [5, 6, 11]. The dimensions (length×width×depth) for the beam, and for both the sensor and actuator are taken as 0.12m×0.00115m×0.0254m, and 0.005m×0.005m×0.0254m, respectively. For the finite element analysis, the beam is modeled according to the Euler beam theory and the smart layers are modeled using two dimensional rectangular elements with pseudo internal degrees of freedom [12]. The constants in Equation (24) are assumed to be $\alpha = 8$ rad/s and $\beta = 0$. The LQG/LQR control technique is used for the closed-loop system [13].

The locations of both the sensor and actuator is varied along the beam through the change in the parameter of d shown in Figure 1. Two values of d are tested in simulations, one corresponding to the case when both the sensor and actuator are at the fixed end of the beam (d=0.0025 m), and the other one for the sensor and actuator being in the middle of the beam (d=0.06 m). A step force of magnitude 2 N is applied at the tip of the beam for the open-loop and closed-loop structure simulations.

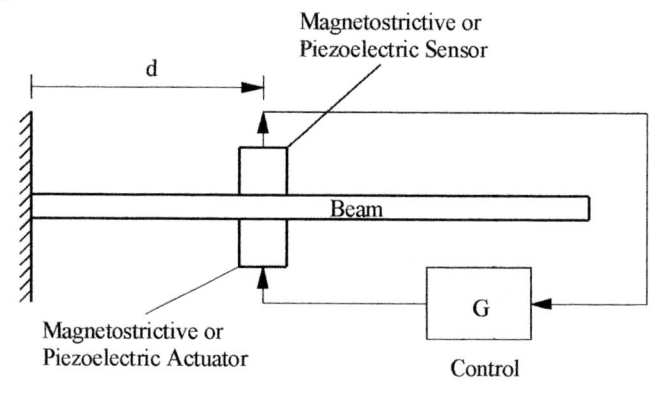

Figure 1. System consisting of beam and smart materials.

Table 1. Properties of materials.

Piezoceramic (PZT-5A, BM500)	
$c_{11}, c_{12}, c_{22}, c_{33}$ (Pa × 10^{10})	11.1, 7.52, 12.1, 2.26
e_{31}, e_{33} (C/m^2)	-5.4, 15.8
$\varepsilon_{11}, \varepsilon_{33}$ (F/m × 10^{-3})	1.151, 1.043
ρ (kg/m^3)	7800
Magnetoceramic (CoFe$_2$O$_4$)	
c_{11} (Pa)	15.432 × 10^{10}
ℓ (N/Wb or A/m)	2.86 × 10^8
μ_{11} or μ_{33} (H/m)	1.885 × 10^{-7}

The results when the smart materials are piezoelectric are shown in Figures 2 through 4. Figure 2 corresponds to the case of d=0.0025 m, and Figures 3 and 4 are for d=0.06 m. The open-loop Fast Fourier Transform (FFT) system responses in Figure 3 show clearly the first natural frequency at around 70 Hz, and the second natural frequency is slightly seen for d=0.06 m for the piezoelectric voltage output from node 9 (top right corner, Figure 1) at about 350 Hz. The closed-loop system response damps out quickly when the piezoelectric actuator is placed at the fixed end of the beam (Figures 2 and 4).

The results of the magnetostrictive material are shown in Figures 5 to 7. Note that the magnetic vector potential **A** has horizontal and vertical components, which

MODELING OF MATERIALS FOR SMART STRUCTURES

are shown in the FFT responses of the open-loop system (Figure 6). The open-loop responses again clearly show the first natural frequency, but the second natural frequency is barely noticeable at around 350 Hz. Furthermore, for the closed-loop system response, as for the piezoelectric case, structural vibrations are attenuated faster when the magnetostrictive actuator is placed at the fixed end (Figures 5 and 7).

6. CONCLUSION

The general constitutive equations of a medium where mechanical, electrical, thermal and magnetic fields interact with each other are obtained through a thermodynamic potential. Two energy functionals are used in Hamilton's principle and the finite element method is employed to obtain the general finite element equations describing the dynamic behavior of the medium. Then the constitutive and finite element equations for piezoelectricity and magnetostriction are extracted from these general constitutive and finite element equations.

Finally, the performance of a piezoelectric/magnetostrictive sensor and actuator is tested in monitoring and eliminating structural vibrations of a cantilever beam structure. It is shown that both the piezoelectric and magnetostrictive sensors can monitor structural vibrations at the transient stage. For the closed-loop system response, based on the numerical modeling and results, it is concluded that both the piezoelectric and magnetostrictive actuators perform better when they are at the fixed end of the beam.

Acknowledgment: The author gratefully acknowledges the support of King Fahd University of Petroleum & Minerals in carrying out this research.

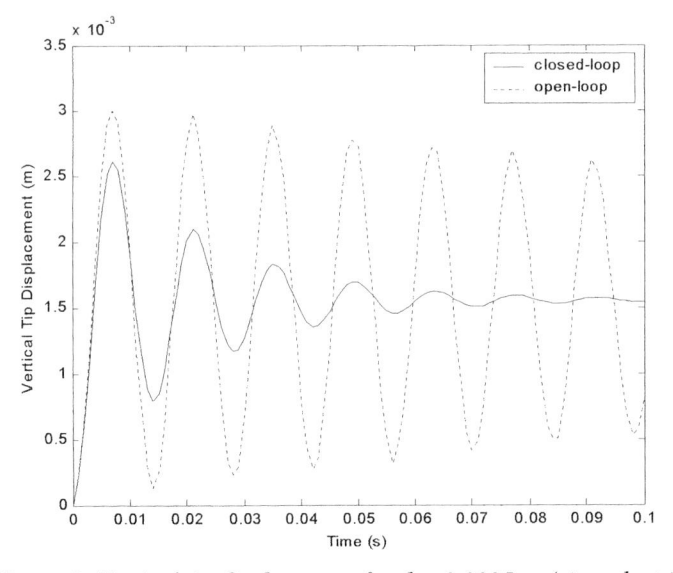

Figure 2. Vertical tip displacement for d = 0.0025 m (piezoelectric).

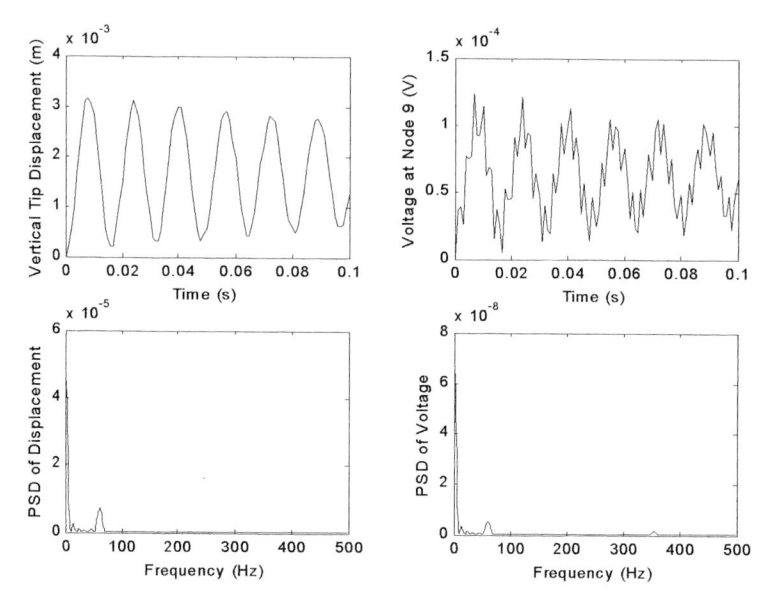

Figure 3. Vertical tip displacement, voltage at node 9 and FFT's for d = 0.06 m (piezoelectric sensor).

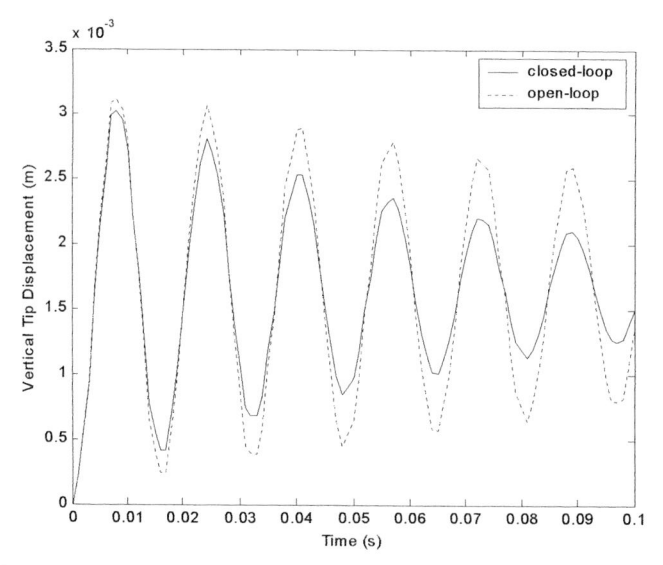

Figure 4. Vertical tip displacement for d = 0.06 m (piezoelectric).

MODELING OF MATERIALS FOR SMART STRUCTURES 395

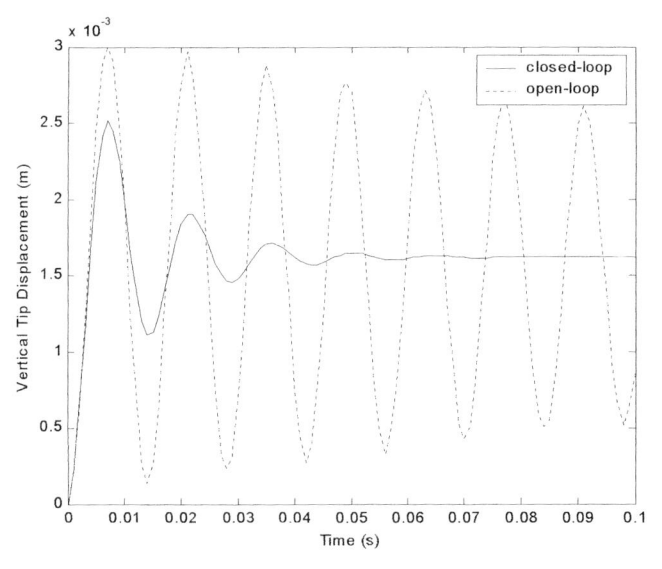

Figure 5. Vertical tip displacement for d = 0.0025 m (magnetostrictive).

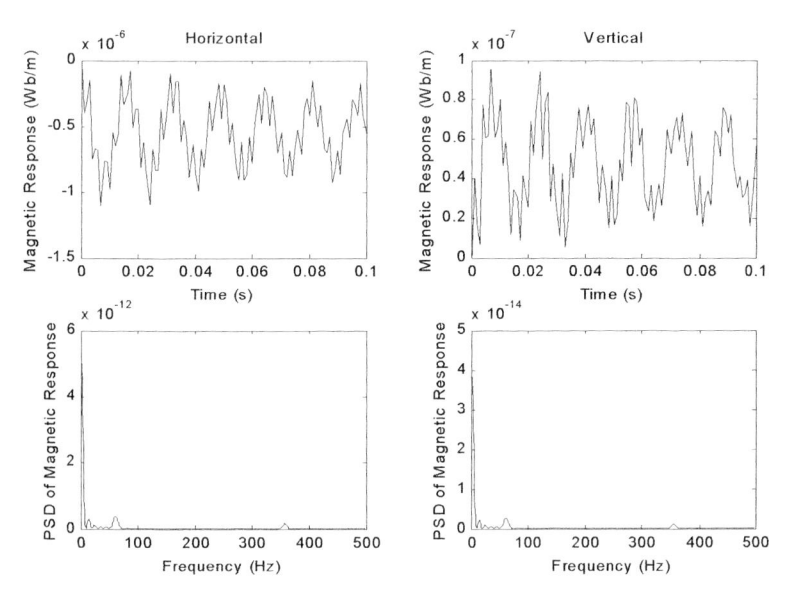

Figure 6. Magnetic vector potential at node 9 and FFT's for d = 0.06 m (magnetostrictive sensor).

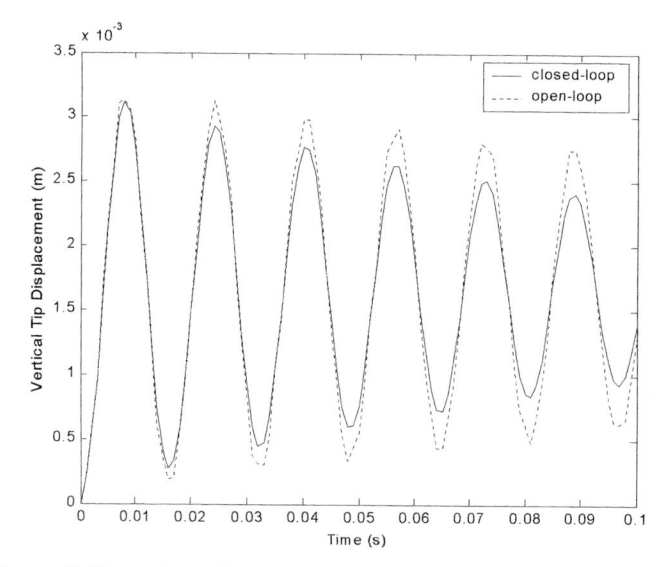

Figure 7. Vertical tip displacement for d = 0.06 m (magnetostrictive).

7. REFERENCES

1. Lim, Y.-H., Varadan, V. V. and Varadan, V. K. (1997) Finite element modeling of the transient response of MEMS sensors, *Smart Materials and Structures* **6** 53-61.
2. Tzou, H. S., Ye, R. and Ding, J. H. (2001) A new X-actuator design for dual bending/twisting control of wings, *Journal of Sound and Vibration* **241** 271-281.
3. Shaw, J. (2001) Active vibration isolation by adaptive control, *JVC/Journal of Vibration and Control* **7** 19-31.
4. Monaco, E., Calandra, G. and Lecce, L. (2000) Experimental activities on damage detection using magnetostrictive actuators and statistical analysis, *Proceedings of SPIE* **3985** 186-196.
5. Harshe, G., Dougherty, J. P. and Newnham, R. E. (1993) Theoretical modeling of multilayer magnetoelectric composites, *International Journal of Applied Electromagnetics in Materials* **4** 145-159.
6. Avellaneda, M. and Harshe, G. (1994) Magnetoelectric effect in piezoelectric/magnetostrictive multilayer (2-2) composites, *Journal of Intelligent Material Systems and Structures* **5** 501-513.
7. Wu, T.-L. and Huang, J. H. (2000) Closed-form solutions for the magnetoelectric coupling coefficients in fibrous composites with piezoelectric and piezomagnetic phases, *International Journal of Solids & Structures* **37** 2981-3009.
8. Mindlin, R. D. (1974) Equations of high frequency vibrations of thermopiezoelectric crystal plates, *International Journal of Solids & Structures*, **10** 625-637.
9. Sadiku, M. N. O. (1995) *Elements of Electromagnetics*, 2nd Edition, Oxford University Press, New York, 428.
10. Nowacki, W. (1978) Some general theorems of thermopiezoelectricity, *Journal of Thermal Stresses* **1** 171-182.
11. *Piezoelectric Ceramics Data Book*, Morgan Matroc Limited, Transducer Products Division, Thornhill, Southampton, Hampshire, England.
12. Tzou, H. S. and Tseng, C. I. (1990) Distributed piezoelectric sensor/actuator design for dynamic measurement/control of distributed parameter systems: a piezoelectric finite element approach, *Journal of Sound and Vibration* **137** 1-18.
13. Maciejowski, J. M. (1989) *Multivariable Feedback Design*, Addison-Wesley, Wokingham, England, Chapter 5.

CONTROL OF THERMALLY-INDUCED STRUCTURAL VIBRATIONS VIA PIEZOELECTRIC PULSES

THEODORE R. TAUCHERT
Department of Mechanical Engineering, University of Kentucky
Lexington, KY 40506-0046 USA
E-mail: tauchert@engr.uky.edu

FUMIHIRO ASHIDA
Department of Electronic and Control Systems Engineering
Shimane University, Matsue, Shimane 690-8504 Japan
E-mail: ashida@ecs.shimane-u.ac.jp

1. INTRODUCTION

Numerous investigations have demonstrated the feasibility of employing piezoelectric elements for active control of structural systems. Attenuation of structural vibrations, for example, can be achieved through application of an appropriate electric field to piezoelectric actuators incorporated within the structure. The dynamic behavior of composite beams having piezoelectric layers was examined by Abramovich and Livshits [1]. Bruch et al. [2] investigated the problem of actively controlling the transient vibrations of a simply supported beam reinforced with piezoelectric actuators. In the case of a beam exposed to a thermal environment, Rao and Sunar [3] analyzed vibration sensing and control of a cantilever thermoelastic beam having two PVDF layers functioning as the distributed sensor and actuator. Tzou and Ye [4] investigated distributed control of a PZT/steel laminated beam subjected to a steady-state temperature gradient. Lee and Saravanos [5] considered the response of a graphite/epoxy beam with piezoceramic actuators; they demonstrated the capability of suppressing deflections caused by thermal gradients. Tauchert et al. [6] studied piezo-control of forced vibrations of a thermoelastic beam. Chandrashekhara and Tenneti [7] developed a finite-element model for active control of thermally-induced vibrations of laminated plates; they presented numerical results for the case of a temperature field having a linear variation in the thickness direction. Other studies on the control of composite structures via piezoelectric actuation are reviewed in [8].

The focus of the present research is the control of forced vibrations caused by a sudden temperature rise over a surface of a layered structure. It will be shown that the induced deflections throughout the structure can be significantly reduced if constant amplitude electric pulses are applied when the uncontrolled deflections reach peak values and discontinued after one-half the damped fundamental period of vibration.

2. STRUCTURAL CONFIGURATIONS AND LOADINGS

2.1 Configurations

Three different structural configurations are considered, namely a beam, a rectangular plate, and an axisymmetrically-loaded circular plate. In each case the edges of the structure are taken to be simply-supported, and the system consists of symmetrically placed isotropic thermoelastic structural layers and transversely isotropic piezothermoelastic actuator layers. Strain-rate damping in each material is included.

2.2 Thermal-Shock Loading

The structures considered are assumed to be at rest initially. At time $t = 0$ the lower surface $(z = c)$ of the structure experiences a sudden uniform temperature rise T_0. Assuming that the edges and upper surface $(z = -c)$ of the structure are thermally insulated, and that the structural and piezoelectric materials have similar thermal diffusivity κ, the temperature distribution is given by [9]

$$T(z,t) = T_0 \left\{ 1 + \sum_{n=1}^{\infty} \frac{2}{j\pi} (-1)^n e^{-j^2 \psi t} \cos[\frac{j\pi}{2c}(z+c)] \right\} \quad (1)$$

in which $j = n - 1/2$ and $\psi = \pi^2 \kappa / 4c^2$.

2.3 Electric-Pulse Loading

In order to simplify the analysis, it is assumed that the electric field resulting from variations in stress or temperature (the *direct piezoelectric effect*) is insignificant compared with that produced by the electric loading. In this case, if a constant amplitude electric potential V_k is applied to the lower surface $(z = z_k)$ of a piezoelectric layer k for a time duration $t : t_i \to t_f$, while the upper surface $(z = z_{k-1})$ is grounded, the electric field intensity $\mathcal{E}_z$ in the thickness direction within the layer may be approximated as

$$[\mathcal{E}_z(t)]_k = \frac{-V_k}{(z_k - z_{k-1})}[\Phi(t-t_i) - \Phi(t-t_f)] \qquad (2)$$

where $\Phi(t-t_j)$ denotes a unit step at $t = t_j$.

3. CONTROL OF BEAM VIBRATIONS

3.1 Formulation

Under the assumptions of classical bending theory, the bending stress σ_x in layer k of the composite beam is expressed as

$$(\sigma_x)_k = (-Ezw_{,xx} - \eta z\dot{w}_{,xx} - E\alpha T - Ed\mathcal{E}_z)_k \qquad (3)$$

where w is the transverse displacement, E is Young's modulus, η is the coefficient of strain-rate damping, α is the coefficient of thermal expansion, and d is a piezoelectric coefficient (taken to be zero if the layer is non-piezoelectric). The resulting bending moment in the beam is then given by

$$M = -(Dw_{,xx} + D'\dot{w}_{,xx} + M^T + M^{\mathcal{E}}) \qquad (4)$$

in which

$$D = b\sum_k E_k \int_{z_{k-1}}^{z_k} z^2 dz, \quad D' = b\sum_k \eta_k \int_{z_{k-1}}^{z_k} z^2 dz \qquad (5)$$

$$M^T = b\sum_k E_k \alpha_k \int_{z_{k-1}}^{z_k} Tz\,dz, \quad M^{\mathcal{E}} = b\sum_k E_k d_k \int_{z_{k-1}}^{z_k} (\mathcal{E}_z)_k z\,dz \qquad (6)$$

where b is the beam's width, and M^T and $M^{\mathcal{E}}$ define thermal and electric moments. The equation of motion governing the transverse displacement w, namely

$$M_{,xx} + q(x,t) = \mu\ddot{w} \qquad (7)$$

yields

$$D\nabla^4(w + 2\gamma\dot{w}) + \mu\ddot{w} = q - \nabla^2(M^T + M^{\mathcal{E}}) \qquad (8)$$

where q is an applied force-per-unit-length, μ is mass-per-unit-length, and

$$\nabla^2 = \frac{\partial^2}{\partial x^2}, \quad 2\gamma = \frac{D'}{D}, \quad \mu = b\sum_k \rho_k(z_k - z_{k-1}) \qquad (9)$$

in which ρ is density.

To obtain a solution to Eq. (8) in the case of simply-supported edges, we express the loads in Fourier sine series form as

$$(q, M^T, M^\varepsilon) = \sum_{m=1}^{\infty}(q_m(t), M_m^T(t), M_m^\varepsilon(t))\sin(m\pi x/L) \qquad (10)$$

where the Fourier coefficients are given by

$$(q_m, M_m^T, M_m^\varepsilon) = \frac{2}{L}\int_0^L (q, M^T, M^\varepsilon)\sin(m\pi x/L)dx \qquad (11)$$

and L is the beam length. For distributions of temperature and electric fields described by Eqs. (1) and (2), the thermal and electric moments are, from Eq. (6),

$$M^T(t) = bT_0 \sum_k E_k \alpha_k \int_{z_{k-1}}^{z_k} \left[\sum_{n=1}^{\infty} \frac{2}{j\pi}(-1)^n e^{-j^2\psi t}\cos[\frac{j\pi}{2c}(z+c)]\right]zdz \qquad (12)$$

$$M^\varepsilon(t) = -b\sum_k E_k d_k V_k \left[\frac{z_k + z_{k-1}}{2}\right][\Phi(t-t_i) - \Phi(t-t_f)] \qquad (13)$$

Taking

$$w(x,t) = \sum_{m=1}^{\infty} \zeta_m(t)\sin(m\pi x/L) \qquad (14)$$

and substituting Eqs. (12)-(14) into (8), yields the equation governing normal mode m:

$$\ddot{\zeta}_m + 2\gamma\Omega_m^2\dot{\zeta}_m + \Omega_m^2\zeta_m = Z_m(t) \qquad (15)$$

in which circular frequency Ω_m and generalized force $Z_m(t)$ are given by

$$\Omega_m^2 = \frac{D}{\mu}\left(\frac{m\pi}{L}\right)^4, \quad Z_m(t) = \frac{1}{\mu}\left[q_m + \left(\frac{m\pi}{L}\right)^2(M_m^T + M_m^\varepsilon)\right] \qquad (16)$$

For modal damping less than critical ($\gamma\Omega_m < 1$), the solution to Eq. (15) can be expressed as

$$\zeta_m(t) = e^{-\gamma\Omega_m^2 t}[a_m\cos(\bar{\Omega}_m t) + b_m\sin(\bar{\Omega}_m t)] + \frac{1}{\bar{\Omega}_m}\int_0^t Z_m(\tau)e^{-\gamma\Omega_m^2(t-\tau)}\sin[\bar{\Omega}_m(t-\tau)]d\tau \qquad (17)$$

where $\bar{\Omega}_m = \Omega_m\sqrt{1-\gamma^2\Omega_m^2}$ are the damped frequencies, and coefficients a_m, b_m are

$$a_m = \frac{2}{L}\int_0^L w(x,0)\sin(m\pi x/L)dx, \; b_m = \frac{2}{L\bar{\Omega}_m}\int_0^L [\dot{w}(x,0)+\gamma\Omega_m^2 w(x,0)]\sin(m\pi x/L)dx \quad (18)$$

In the case of zero initial conditions, the beam deflection is

$$w(x,t) = \sum_{m=1}^\infty \frac{1}{\bar{\Omega}_m}\left\{\int_0^t Z_m(\tau)e^{-\gamma\Omega_m^2(t-\tau)}\sin[\bar{\Omega}_m(t-\tau)]d\tau\right\}\sin(m\pi x/L) \quad (19)$$

3.2 Numerical Results

Consider a three-layer beam of length $L = 1m$, width $b = 50mm$ and depth $2c = 7mm$. The central layer ($k = 2$) is aluminum, having thickness $h_a = 5mm$ and material properties: $E_a = 69.9GPa$, $\rho_a = 2720 kg/m^3$, $\alpha_a = 23\cdot 10^{-6}/^0K$, $\kappa_a = 4.84\cdot 10^{-5}m^2/s$. The outer layers ($k = 1,3$) are piezoceramic, each having thickness $h_p = 1mm$ and properties: $E_p = 63GPa$, $\rho_p = 7650kg/m^3$, $\alpha_p = 4.4\cdot 10^{-6}/^0K$, $d = -166\cdot 10^{-12} m/V$.

In order to simplify determination of thermally-induced beam displacements, the thermal moment $M^T(t)$ given by Eq. (12) was curve-fit with the expression $M^T(t) = c_1 e^{-c_2 t} + c_3 e^{-9c_2 t}$; this simplification was found to have little effect on the calculated results. Superposition of thermally-induced (uncontrolled) displacements and displacements induced by the electric moment M^ε gives the corresponding controlled displacements. Figure 1 shows how the uncontrolled and controlled displacements (given in meters) vary with time (seconds) at the center of the beam, resulting from a surface temperature rise $T_0 = 1K$. The control voltages for this case were single pulses of amplitude $V_1 = -8.5V$ and $V_3 = 8.5V$, applied to the upper and

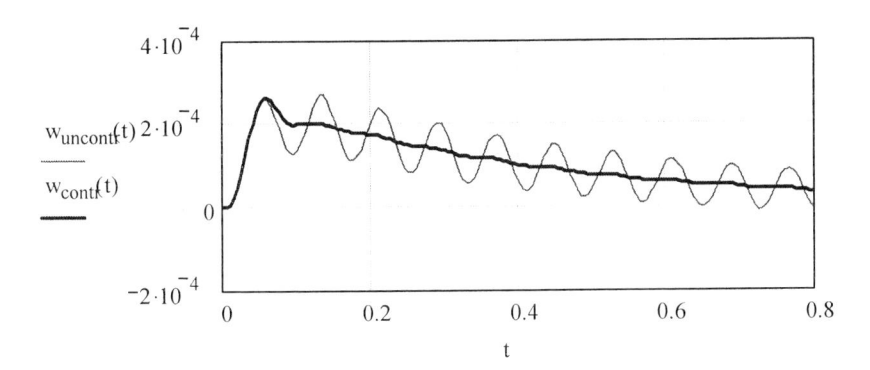

Figure 1. Uncontrolled and controlled displacements in a beam.

lower beam surfaces when the uncontrolled damped displacement reached its first maximum, and discontinued after one-half the damped fundamental period $\bar{\tau}_1 = 2\pi/\bar{\Omega}_1$. A strain-rate damping coefficient $\gamma = 1 \cdot 10^{-4} s$ was assumed in this case.

4. CONTROL OF RECTANGULAR PLATE VIBRATIONS

4.1 Formulation

The stresses in layer k of a rectangular composite plate are expressed as

$$(\sigma_x)_k = \left[\frac{1}{1-\nu^2}(E\varepsilon_x + \eta\dot{\varepsilon}_x - E\alpha T - Ed\mathcal{E}_z + \nu E\varepsilon_y + \nu\eta\dot{\varepsilon}_y - \nu E\alpha T - \nu Ed\mathcal{E}_z)\right]_k$$

$$(\sigma_y)_k = \left[\frac{1}{1-\nu^2}(E\varepsilon_y + \eta\dot{\varepsilon}_y - E\alpha T - Ed\mathcal{E}_z + \nu E\varepsilon_x + \nu\eta\dot{\varepsilon}_x - \nu E\alpha T - \nu Ed\mathcal{E}_z)\right]_k \quad (20)$$

$$(\sigma_{xy})_k = \left[\frac{1}{2(1+\nu)}(E\gamma_{xy} + \eta\dot{\gamma}_{xy})\right]_k$$

where ν denotes Poisson's ratio. The corresponding strains due to bending are

$$\varepsilon_x = -zw_{,xx}, \; \varepsilon_y = -zw_{,yy}, \; \gamma_{xy} = -2zw_{,xy} \quad (21)$$

The equation of motion in terms of bending moment resultants, namely

$$M_{x,xx} + 2M_{xy,xy} + M_{y,yy} + q(x,y,t) = \mu\ddot{w} \quad (22)$$

then becomes

$$D\nabla^4(w + 2\gamma\dot{w}) + \mu\ddot{w} = q - \nabla^2(M^T + M^\varepsilon) \quad (23)$$

in which

$$\nabla^2 = \frac{\partial^2}{\partial x^2} + \frac{\partial^2}{\partial y^2}, \quad 2\gamma = \frac{D'}{D}, \quad \mu = \sum_k \rho_k(z_k - z_{k-1})$$

$$D = \sum_k \frac{E_k}{1-\nu_k^2}\int_{z_{k-1}}^{z_k} z^2 dz, \quad D' = \sum_k \frac{\eta_k}{1-\nu_k^2}\int_{z_{k-1}}^{z_k} z^2 dz \quad (24)$$

$$M^T = \sum_k \frac{E_k\alpha_k}{1-\nu_k}\int_{z_{k-1}}^{z_k} Tz\,dz, \quad M^\varepsilon = \sum_k \frac{E_k d_k}{1-\nu_k}\int_{z_{k-1}}^{z_k} (\mathcal{E}_z)_k z\,dz$$

CONTROL OF THERMALLY INDUCED STRUCTURAL VIBRATIONS   403

where q now denotes force-per-unit-area, and μ is mass-per-unit-area.

For a solution to Eq. (23), we express the loads as

$$(q, M^T, M^\varepsilon) = \sum_{m=1}^{\infty}\sum_{n=1}^{\infty} (q_{mn}(t), M^T_{mn}(t), M^\varepsilon_{mn}(t))\sin(\frac{m\pi x}{a})\sin(\frac{n\pi y}{b}) \qquad (25)$$

where a and b are the plate dimensions in the x and y directions, and

$$(q_{mn}, M^T_{mn}, M^\varepsilon_{mn}) = \frac{4}{ab}\int_0^a\int_0^b (q, M^T, M^\varepsilon)\sin(\frac{m\pi x}{a})\sin(\frac{n\pi y}{b})dxdy \qquad (26)$$

and we take

$$w(x, y, t) = \sum_{m=1}^{\infty}\sum_{n=1}^{\infty} \zeta_{mn}(t)\sin(\frac{m\pi x}{a})\sin(\frac{n\pi y}{b}) \qquad (27)$$

The equation governing normal mode mn then becomes

$$\ddot{\zeta}_{mn} + 2\gamma\Omega^2_{mn}\dot{\zeta}_{mn} + \Omega^2_{mn}\zeta_{mn} = Z_{mn}(t) \qquad (28)$$

in which

$$\Omega^2_{mn} = \frac{D}{\mu}\left[\left(\frac{m\pi}{a}\right)^4 + 2\left(\frac{m\pi}{a}\right)^2\left(\frac{n\pi}{b}\right)^2 + \left(\frac{n\pi}{b}\right)^4\right] \qquad (29)$$

$$Z_{mn}(t) = \frac{1}{\mu}\left\{q_{mn} + \left[\left(\frac{m\pi}{a}\right)^2 + \left(\frac{n\pi}{b}\right)^2\right](M^T_{mn} + M^\varepsilon_{mn})\right\} \qquad (30)$$

For modal damping less than critical ($\gamma\Omega_{mn} < 1$), the solution to Eq. (28) is

$$\zeta_{mn}(t) = e^{-\gamma\Omega^2_{mn}t}[a_{mn}\cos(\overline{\Omega}_{mn}t) + b_{mn}\sin(\overline{\Omega}_{mn}t)]$$
$$+ \frac{1}{\overline{\Omega}_{mn}}\int_0^t Z_{mn}(\tau)e^{-\gamma\Omega^2_{mn}(t-\tau)}\sin[\overline{\Omega}_{mn}(t-\tau)]d\tau \qquad (31)$$

where $\overline{\Omega}_{mn} = \Omega_{mn}\sqrt{1-\gamma^2\Omega^2_{mn}}$ are the damped frequencies, and

$$a_{mn} = \frac{4}{ab}\int_0^a\int_0^b w(x, y, 0)\sin(\frac{m\pi x}{a})\sin(\frac{n\pi y}{b})dxdy \qquad (32)$$

$$b_{mn} = \frac{4}{ab\overline{\Omega}_{mn}}\int_0^a\int_0^b [\dot{w}(x, y, 0) + \gamma\Omega^2_{mn}w(x, y, 0)]\sin(\frac{m\pi x}{a})\sin(\frac{n\pi y}{b})dxdy \qquad (33)$$

In the case of zero initial conditions, the plate deflection is given by

$$w(x,y,t) = \sum_{m=1}^{\infty}\sum_{n=1}^{\infty}\frac{1}{\overline{\Omega}_{mn}}\left\{\int_0^t Z_{mn}(\tau)e^{-\gamma\Omega_{mn}^2(t-\tau)}\sin[\overline{\Omega}_{mn}(t-\tau)]d\tau\right\}\sin(\frac{m\pi x}{a})\sin(\frac{n\pi y}{b}) \quad (34)$$

4.2 Numerical Results

Now consider a three-layer plate having in-plane dimensions $a = b = 2m$, geometric and material properties in the z-direction identical to those of the beam considered in Section 3.2, Poisson's ratios $v_a = 0.33$ and $v_p = 0.481$, and damping coefficient $\gamma = 2 \cdot 10^{-4} s$. Figures 2 and 3 show the displacements at the center of the plate due to a surface temperature rise $T_0 = 1K$. For Fig. 2 the control voltages applied to the top

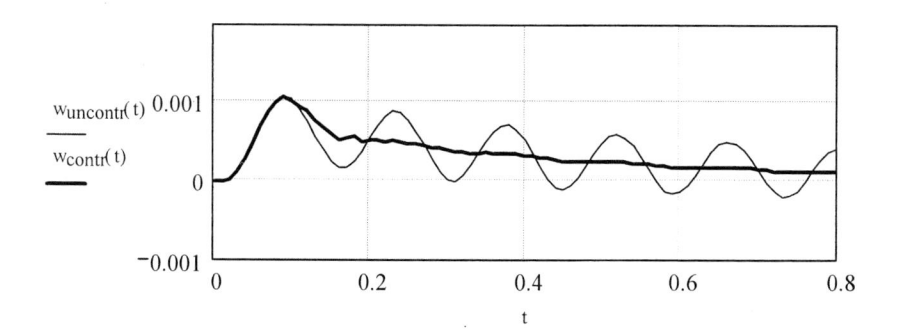

Figure 2. Vibration control in a rectangular plate via a single piezoelectric pulse.

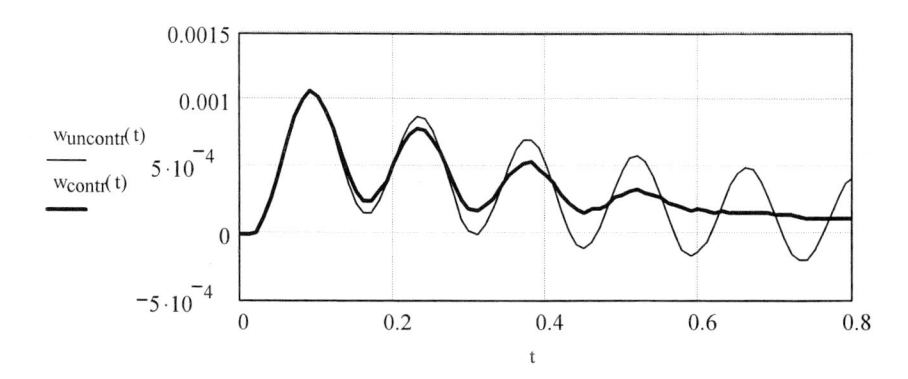

Figure 3. Vibration control in a rectangular plate via four piezoelectric pulses.

and bottom plate surfaces consisted of single pulses of amplitude $\mp 11.8V$, introduced when the uncontrolled displacement reached its first peak, and terminated after one-half the damped fundamental period $\bar{\tau}_{11} = 2\pi/\bar{\Omega}_{11}$. Fig. 3 shows results for the case of four consecutive voltage pulses spaced one period apart, with amplitude $\mp 2.8V$ and duration $0.5\bar{\tau}_{11}$.

5. CONTROL OF CIRCULAR PLATE VIBRATIONS

5.1 Formulation

In the case of a circular plate under axisymmetric loading, the equation of motion is

$$D\nabla^4(w+2\gamma\dot{w}) + \mu\ddot{w} = q - \nabla^2(M^T + M^\varepsilon) \tag{35}$$

where now

$$\nabla^2 = \frac{\partial^2}{\partial r^2} + \frac{1}{r}\frac{\partial}{\partial r} \tag{36}$$

and with expressions for all other quantities identical to those given earlier for plates of rectangular planform.

The simply-supported edge conditions for a circular plate of radius a are

$$\text{On } r = a: \quad w = 0, \quad M_r = 0 \tag{37}$$

In order to simplify the analysis, we modify the second condition by taking the edge conditions to be

$$\text{On } r = a: \quad w = 0, \quad \nabla^2 w = 0 \tag{38}$$

This modification was shown in Ref. [10] to have negligible influence upon the forced vibrations of a circular plate. For a solution satisfying the modified edge conditions, the loads are expressed as

$$(q, M^T, M^\varepsilon) = \sum_{m=1}^{\infty} (q_m(t), M_m^T(t), M_m^\varepsilon(t)) J_0(\alpha_m r) \tag{39}$$

in which α_m are taken as the roots of the equation $J_0(\alpha_m a) = 0$ so that Eqs. (38) are satisfied. The Fourier coefficients are

$$(q_m, M_m^T, M_m^\varepsilon) = \frac{2}{a^2 J_1^2(\alpha_m a)} \int_0^a (q, M^T, M^\varepsilon) r J_0(\alpha_m r) dr \tag{40}$$

We now take

$$w(r,t) = \sum_{m=1}^{\infty} \zeta_m(t) J_0(\alpha_m r) \tag{41}$$

The equation governing normal mode m then becomes

$$\ddot{\zeta}_m + 2\gamma\Omega_m^2 \dot{\zeta}_m + \Omega_m^2 \zeta_m = Z_m(t) \tag{42}$$

where

$$\Omega_m^2 = \frac{D}{\mu}\alpha_m^4, \quad Z_m(t) = \frac{1}{\mu}\left[q_m + \alpha_m^2(M_m^T + M_m^\varepsilon)\right] \tag{43}$$

For modal damping less than critical ($\gamma\Omega_m < 1$), the solution to Eq. (42) is

$$\zeta_m(t) = e^{-\gamma\Omega_m^2 t}[a_m \cos(\bar{\Omega}_m t) + b_m \sin(\bar{\Omega}_m t)] + \frac{1}{\bar{\Omega}_m}\int_0^t Z_m(\tau)e^{-\gamma\Omega_m^2(t-\tau)}\sin[\bar{\Omega}_m(t-\tau)]d\tau \tag{44}$$

in which $\bar{\Omega}_m = \Omega_m\sqrt{1-\gamma^2\Omega_m^2}$ are the damped frequencies, and a_m, b_m are

$$a_m = \frac{2}{a^2 J_1^2(\alpha_m a)} \int_0^a w(r,0) r J_0(\alpha_m r) dr \tag{45}$$

$$b_m = \frac{2}{a^2 J_1^2(\alpha_m a)\bar{\Omega}_m} \int_0^a [\dot{w}(r,0) + \gamma\Omega_m^2 w(r,0)] r J_0(\alpha_m r) dr \tag{46}$$

Assuming zero initial conditions, the deflection of the circular plate is

$$w(r,t) = \sum_{m=1}^{\infty} \frac{1}{\bar{\Omega}_m}\left\{\int_0^t Z_m(\tau) e^{-\gamma\Omega_m^2(t-\tau)} \sin[\bar{\Omega}_m(t-\tau)]d\tau\right\} J_0(\alpha_m r) \tag{47}$$

5.2 Numerical Results

Equation (47) was used to calculate the response of a circular plate of radius $a = 1m$, with cross-sectional properties identical to those of the beam and plate considered earlier. The damping coefficient was taken to be $\gamma = 1 \cdot 10^{-4} s$. Figure 4 shows the displacements at the center of the plate for single-pulse control voltages $\mp 10.5V$ applied when the uncontrolled displacement reached its first peak, and discontinued after one-half the damped fundamental period. Results for the case in which single-pulse voltages $\pm 10.5V$ were applied when the uncontrolled displacement reached its next minimum are plotted in Fig. 5. Figure 6 presents the radial variation of the

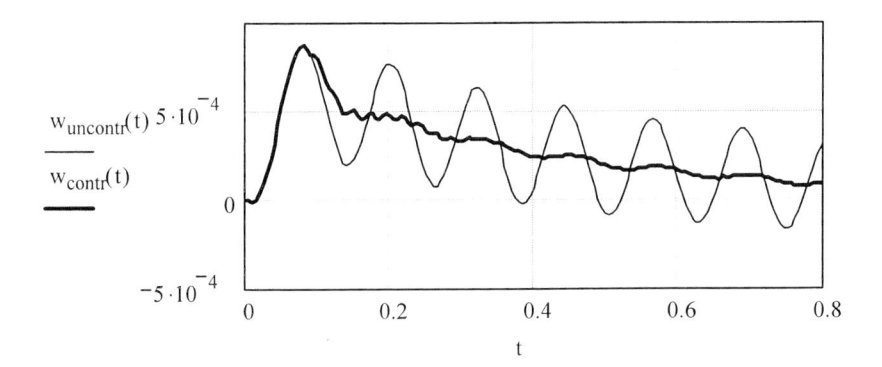

Figure 4. Vibration control in a circular plate via a single piezoelectric pulse, applied when the displacement reaches a maximum.

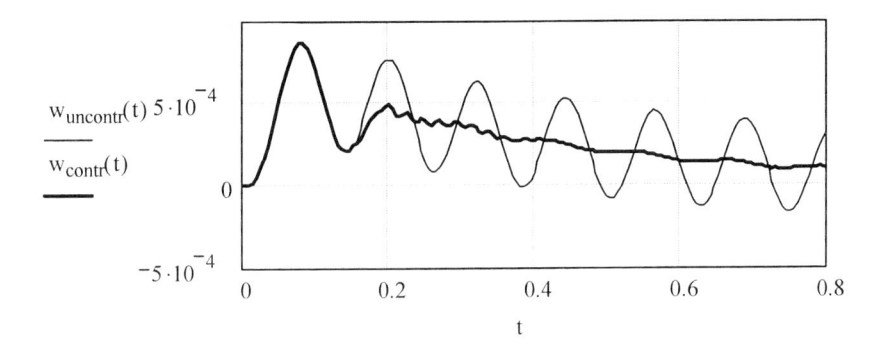

Figure 5. Vibration control in a circular plate via a single piezoelectric pulse, applied when the displacement reaches a minimum.

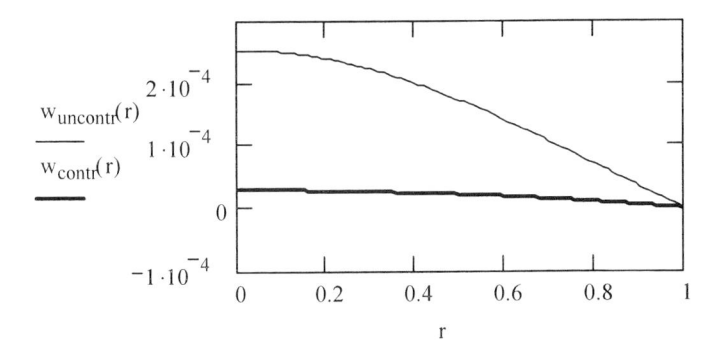

Figure 6. Radial variation of displacements at time $t = 1.3s$.

uncontrolled and controlled displacements after ten cycles of vibration (time $t = 1.3s$), with control voltages taken to be the same as those used for Fig. 5.

6. CONCLUDING REMARKS

The present results, plus those for additional cases not reported here, indicate that vibrations induced in composite structures by a sudden temperature rise can be controlled effectively by means of constant-amplitude electric potential pulses applied to piezoelectric layers. The pulses are introduced when the thermally-induced (uncontrolled) deflection at the center of a structure reaches a peak value, and discontinued after one-half the damped fundamental period of vibration. When multiple piezoelectric pulses rather than a single pulse are employed, a similar degree of vibration suppression can be achieved using correspondingly smaller voltage amplitudes.

Further research is underway to develop an intelligent structural system that incorporates the present piezoelectric-pulse concept together with piezoelectric sensors, for active control of composite structures exposed to thermal shock.

7. REFERENCES

1. Abramovich, H. & Livshits, A. (1993) Dynamic behavior of cross-ply laminated beams with piezoelectric layers, *Composite Structures*, 25, 371-379.

2. Bruch, J.C., Sloss, J.M., Adali, S. and Sadek, S. (1999) Modified bang-bang piezo-control of vibrating beams, *Smart Materials and Structures*, 8, 647-653.

3. Rao, S.S. and Sunar, M. (1993) Analysis of distributed thermopiezoelectric sensors and actuators in advanced intelligent structures, *AIAA J.*, 31(7), 1280-1286.

4. Tzou, H.S. and Ye, R. (1994) Piezothermoelasticity and precision control of piezoelectric systems: theory and finite element analysis, *J. Vibration and Acoustics*, 116, 489-495.

5. Lee, H.-J. and Saravanos, A. (1996) Coupled layerwise analysis of thermopiezoelectric composite beams, *AIAA J.*, 34(6), 1231-1237.

6. Tauchert, T.R., Adali, S. and Verijenko, V. (2001) Piezo-control of forced vibrations of a thermoelastic beam, *Proc. 4th Int. Congr. Thermal Stresses*, Osaka, Japan, 477-480.

7. Chandrashekhara, K. and Tenneti, R. (1995) Thermally induced vibration suppression of laminated plates with piezoelectric sensors and actuators, *Smart Materials and Structures*, 4, 281-290.

8. Tauchert, T.R., Ashida, F., Noda, N., Adali, S. and Verijenko, V. (2000) Developments in thermopiezoelasticity with relevance to smart composite structures, *Composite Structures*, 48, 31-38.

9. Ozisik, M.N. (1968) *Boundary Value Problems of Heat Conduction*, International Textbook, Scranton, PA.

10. Nowacki, W. (1963) *Dynamics of Elastic Systems*, Wiley, New York, p.221.

ACTIVE DAMPING OF PARAMETRIC VIBRATIONS OF MECHANICAL DISTRIBUTED SYSTEMS

ANDRZEJ TYLIKOWSKI
Warsaw University of Technology
Institute of Machine Design Fundamentals
Narbutta 84, 02-524 Warsaw, Poland
E-mail: aty@simr.pw.edu.pl

1. INTRODUCTION

A theoretical investigation of dynamic stability for linear elastic structures due to time dependent harmonic and stochastic inner forces is presented. A viscous model of external damping with a constant proportionality coefficient is assumed to describe a dissipation of the structure energy. The concept of intelligent structure is used to insure the active damping. The study is based on the application of distributed sensors, actuators, and an appropriate feedback and is adopted for stability problems of system consisting of plate with control part governed by uniform partial differential equations with time-dependent coefficients. To estimate deviations of solutions from the equilibrium state (the distance between a solution with nontrivial initial conditions and the trivial solution) a scalar measure of distance equal to the square root of the functional is introduced. The Liapunov method is used to derive a velocity feedback implying non increasing of the functional along an arbitrary beam motion and in consequence to balance the supplied energy by the parametric excitation and the dissipated energy by the inner and control damping. In order to calculate the energetic norm of disturbed solution as a function of time the partial differential equation is solved numerically. The numerical tests performed for the simply supported beam with surface bonded actuators and sensors show the influence of the feedback constant on the vibration decrease.

Piezoelectric sensors and actuators have been applied successfully in the closed-loop control (Bailey and Hubbard [1], Newman [2]). A comprehensive static model for piezoelectric actuators glued to a beam has been presented by Crawley and de Luis [3] with detail analysis of bonding layers. This one-dimensional theory was extended by Dimitriadis et al. [4] on thin plates with two-dimensional piezoelectric patches. A dynamic model for a simply supported beam with a piezoelectric actuator perfectly bonded to each of its upper and lower surfaces was developed by Jie Pan and Hansen [5]. The collocated sensor/actuator system was analysed by Dosch, Inman, and Garcia [6]. The general dynamic coupling model of the beam with bonding sensors and actuators was used to derive the Liapunov control strategy especially useful in the collocated sensor-actuator systems (Tylikowski [7]). The dynamic extensional strain on the beam surface was calculated by considering the dynamic coupling between the actuator and the beam, and by taking into account a finite bonding layer with the finite stiffness (Tylikowski, [8]). Lee and Moon [9] introduced novel distributed sensors/actuators, which sense and actuate the particular modal co-ordinate without extensive computational requirements. Tzou and Fu [10] analysed models of a plate with segmented distributed piezoelectric sensors and actuators, and showed that segmenting improves the observability and the controllability of the system. Designed by Kumar et al. [11] the piezoelectric actuators with constant properties over a wide-band frequencies seems to be appropriate for applying in systems subjected to wide-band stochastic excitations. Chow and Maestrello [12] examined the exponential stabilization of panels with time-dependent parametric excitations and concluded that a stronger mode of control, such as a distributed control, should be used to stabilize time-dependent systems. Dynamic stability analysis of a composite beam with piezolayers subjected to axial harmonic parametric loading was investigated by Chen, L.W., et al. [13]. Using unimodal approximation Shen and Yang [14] determined instability regions of laminated piezoelectric shells subjected to uniformly distributed harmonic axial force. The direct Liapunov method was applied to the stabilization problem of the beam subjected to a wide-band axial time-dependent force (Tylikowski [7]).

The purpose of the present paper is to solve an active control problem of plate parametric vibrations excited by in-plane time-dependent forces. The problem is solved using the concept of distributed piezoelectric sensors and actuators with a sufficiently large value of velocity feedback. Real mechanical systems are subjected not only to nontrivial initial conditions but also to permanently acting excitations and the active vibration control should be modify in order to balance the supplied energy by external parametric excitation. The applicability of active vibration control is extended to distributed systems with stochastic parametric excitation. The effective estimation of the feedback constant stabilizing the beam parametric vibration is derived analytically. The analysis is based on assumption of an appropriate two-dimensional sinusoidal polarization profile of sensors and actuators [15]. The minimal value of the feedback constant is effectively expressed by the constant component of axial force, intensity of stochastic component, geometry, mechanical and piezoelectric properties of actuators and sensors. A study of stability of non-deflected middle plane or surface w=0 splits into two branches. First, under the assumption that in-plane forces are harmonic functions of time conditions of asymptotic stability of the trivial solution are derived from the Mathieu equation. Our second purpose is to discuss almost sure asymptotic stability of the trivial solution, if the forces are stochastic physically realizable (non-white) processes. The almost sure stochastic stability domains are obtained using the Liapunov direct method.

2. DYNAMIC EQUATIONS

2.1 Basic Assumptions

Consider the Kirchhoff plate of length a, width b, and thickness h, loaded by in-plane time-dependent forces with piezoelectric layers mounted on each of two opposite sides (Fig. 1). The plate is simply supported on all four edges. The piezoelectric layers are assumed to be bonded on the plate surfaces and the mechanical properties of the bonding material are represented by the effective retardation time λ calculated from the rule of mixture. The effective retardation time is a linear function of both the plate and bonding layer retardation times. It is assumed that the transverse motion dominates the in-plane vibrations.

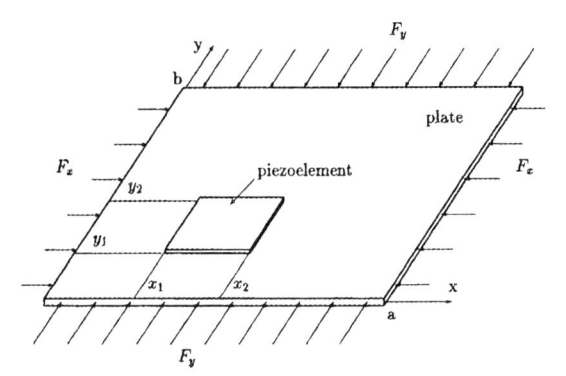

Figure 1. Geometry of plate with piezoactuators

The thickness of the actuator and the sensor is denoted by h_a and h_s, respectively. Assuming a negligible stiffness of the piezolayer in comparison with that of the plate and the influence of the piezoelectric actuator on the plate can be reduced to bending moments M_x, M_y distributed along the actuator edges (Fig. 2)..

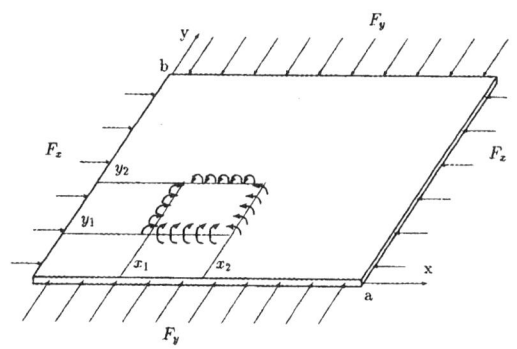

Figure 2. Geometry of plate with equivalent loading

2.2 Sensor and Actuator Equations

Sensor electric displacement in direction perpendicular to the plate is given by

$$D_3 = -e_{s31} \in_{s1} - e_{s32} \in_{s2} \tag{1}$$

where e_{s31} and e_{s32} are the piezoelectric stress/charge coefficients, and $\in_{s1}$, $\in_{s2}$ are sensor strains. Expressing strains by the plate curvatures and the distance from the neutral plane we have

$$D_3 = \frac{(h_s + h_p)E_s}{2(1-v_s^2)}\left[(d_{s31} + v_s d_{s32})w_{,xx} + (d_{s32} + v_s d_{s31})w_{,yy}\right] \tag{2}$$

where d_{s31} and d_{s32} are the piezoelectric strain/charge coefficients of sensor, and v_s denotes the sensor Poisson's ratio. Electric charge Q is obtained by integrating dielectric displacement over the sensor area taking into account the sensor polarization profile Φ^s. Finally, the sensor voltage is calculated using the formula for a flat capacitor

$$V_s = \frac{1}{C}\int_S D_3 \Phi^s dS \tag{3}$$

The voltage applying to the electrodes of the actuaror is obtained by differentiating the voltage produced by the sensor

$$V_a = \frac{K_a}{C}\frac{dV_s}{dt} \tag{4}$$

Substituting Eq (2) and Eq (3) to Eq (4) we have

$$V_a = \frac{(h_s + h)E_s K_a}{2A \in_{33}(1-v_s^2)}\int_S \left[(d_{s31} + v_s d_{s32})w_{,xxt} + (d_{s32} + v_s d_{s31})\right]\Phi^s dS \tag{5}$$

where A is the effective sensor area, $\in_{33}$ is the permittivity coefficient.
Actuator stresses are

$$\sigma_{a1} = \frac{d_{a31}V_a}{h_a}\Phi^a \tag{6}$$

$$\sigma_{a2} = \frac{d_{a32}V_a}{h_a}\Phi^a \tag{7}$$

where d_{a31}, d_{a32} are the piezoelectric strain/charge coefficients of actuator, Φ^a is the actuator polarization profile. Integrating the actuator stresses with respect to the co-ordinate perpendicular to the plate yields the equivalent electric moments (Fig. 2)

$$M^e = \sigma_a h_a \frac{h+h_a}{2} \Phi^a \qquad (8)$$

2.3 Plate Dynamic Equation With Closed Loop Control

The linear differential equation that describes the transverse plate motion with the electric moments has the form

$$\rho h w_{,TT} + D\Delta^2 w + \varepsilon D\Delta^2 w_{,T} + N_X w_{,xx} + N_Y w_{,yy} +$$
$$+ M^p_{X,XX} + M^p_{Y,YY} = 0 \quad (X,Y) \in (0,a) \times (0,b) \qquad (9)$$

where $D=Eh^3/12(1-v^2)$ is the flexural rigidity, v is the Poisson's ratio, ρ is the uniform density of plate material, ε is the retardation time of the Voigt-Kelvin model, T, X, and Y denote time, and the plate co-ordinates, respectively. The time-dependent in-plane forces are denoted by N_x and N_y, respectively. M_x and M_y are distributed moments of electric origin described by the closed loop control equation. Introducing the dimensionless time $t = T(D/\rho h a^4)^{1/2}$, the dimensionless co-ordinates $x=X/a$, $y=Y/b$, the transverse velocity $v=w_{,t}$, the dimensionless retardation time $\lambda=0.5\varepsilon(\rho h a^4/D)^{1/2}$, the reduced in-plane loads $f_{ox}+f_x(t)=N_X(T)a^2/D$, $f_{oy}+f_y(t)=N_Y(T)a^2/D$ we obtain the dynamic equation of plate with closed loop control

$$w_{,tt} + \Delta^2 w + 2\lambda\Delta^2 v + f_x w_{,xx} + f_y w_{,yy} + m^e_{x,xx} + m^e_{y,yy} = 0, (x, y) \in (0,1) \times (0,1) \quad (10)$$

The transverse motion of the plate is described by the uniform equation with time-dependent coefficients. Its trivial solution $w=v=0$ corresponds to the undisturbed plane state. The trivial solution is called stable in Liapunov sense if the following definition is satisfied

$$\bigwedge_{\varepsilon>0} \bigvee_{\delta>0} \|w(0,.)\| < \delta \Rightarrow \bigwedge_{t>0} \|w(t,.)\| < \varepsilon \qquad (11)$$

where ||.|| is a measure of distance of disturbed solution w from the equilibrium state. When the in-plane forces are stochastic processes we call the trivial solution almost sure asymptotically stable if

$$P\{\lim_{t\to\infty} \|w(t,.)\| = 0\} = 1 \qquad (12)$$

3. STABILITY ANALYSIS

3.1 Harmonic Parametric Excitation

Assume that the plate is subjected to the equal harmonic forces in x and y directions. Consider the frequency p of parametric excitation in the range corresponding to the lowest resonance ω. Determine the polarization profiles of the sensor and actuator as follows

$$\Phi_{11}^s(x,y) = \alpha_{11} \sin \pi x \sin \pi y$$
$$\Phi_{11}^a(x,y) = \beta_{11} \sin \pi x \sin \pi y \qquad (13)$$

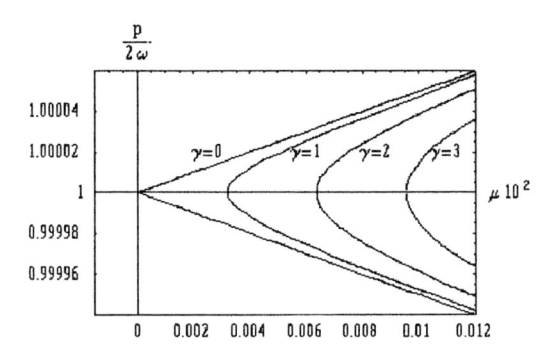

Figure 3. The first instability regions under deterministic harmonic in-plane forces

Therefore, the stability problem is reduced to the following Mathieu equation

$$\frac{d^2W}{dt^2} + 2\varepsilon \frac{dW}{dt} + \omega_o^2(1 + \mu \cos pt)W = 0 \qquad (14)$$

where $\omega_o^2 = 4D\pi^4/\rho h a^4$, $\mu = N_x a^4 \rho h / 2D\pi^4$
and $\varepsilon/\omega_o = \lambda/2 + ab(\beta_x + \beta_y)(d_{s31} + d_{s32})(1 + v_s)/2\pi^2 D$, β_x and β_y are the active damping coefficients. The instability regions are shown in Fig. 3 and 4.
In both cases increase of the total gain factor γ decrease instability regions. It is seen that the larger the control gain is, the greater the critical amplitude of the in-plane forces.

2.2 Stochastic Non-White Noise Stationary Excitations

The crucial point of the method is a construction of a suitable Liapunov functional, which is positive for any motion of the analysed plate. The energy-like Liapunov functional has the form

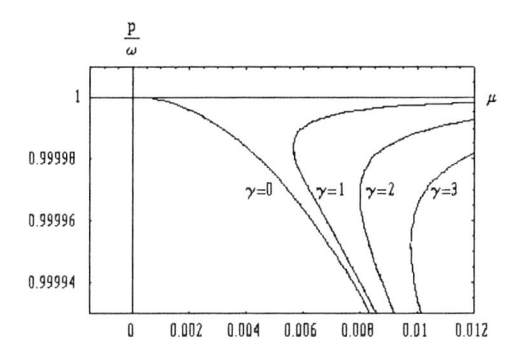

Figure 4. The second instability regions under deterministic harmonic in-plane forces

$$V = \frac{1}{2}\int_D [w_{,t}^2 + 2\lambda w_{,t} \Delta^2 w_{,xx} + 2\lambda^2 \Delta^2 w + (\Delta w)^2 - f_{ox} w_{,x}^2 - f_{oy} w_{,y}^2]dD \quad (15)$$

If the classical condition for static buckling is fulfilled, the functional (15) satisfies positive-definiteness condition, and the measure of distance can be chosen as the square root of the functional

$$\| w(.,t) \| = \sqrt{V} \quad (16)$$

If the trajectories of the forces are physically realizable ergodic processes the classical calculus is applied to calculation of the time-derivative of Eq (15). Upon differentiation with respect to time, substituting dynamic equation (10) and using plate boundary conditions we obtain the time-derivative of functional in the form

$$\frac{dV}{dt} = -2\lambda V + 2U \quad (17)$$

where the auxiliary functional U is defined. We look for a function χ defined as a maximum over all admissible functions w and v of the ratio U/V

$$\chi = \max_{w,v} \frac{U}{V} \quad (18)$$

As a maximum is a particular case of a stationary point we put to zero a variation of U/V

$$\delta(U - \chi V) = 0 \quad (19)$$

In order to effectively solve the associated Euler equations we have to determine the polarization profiles of sensor and actuators. We assume that they correspond to the modal shape function of the chosen ω_{MN}

$$\Phi^s_{MN}(x) = \alpha_{MN} \sin M\pi x \sin N\pi y$$
$$\Phi^a_{MN}(x) = \beta_{MN} \sin M\pi x \sin N\pi y \qquad (20)$$

The feedback gain factor of modal control is denoted by γ_{MN}. Upon solving the Euler equations we obtain the function χ

$$\chi = \max_{m,n=1,2,..} \left\{ \lambda(1-\omega_{mn}) - \frac{1}{2}\omega_{MN}\delta_{Mm}\delta_{Nn} + \right. \qquad (21)$$

$$\left. + \frac{1}{2}\left(\gamma^2_{MN}\delta_{Mm}\delta_{Nn} + \frac{[\omega_{mn}\lambda(2\omega_{mn}\lambda + \omega_{MN}\delta_{Mm}\delta_{Nn}) + m^2\pi^2 f_x + n^2\pi^2 f_y]^2}{(\lambda^2\omega_{mn}+1)\omega_{mn} - m^2\pi^2 f_{ox} - n^2\pi^2 f_{oy}}\right)^{1/2} \right\}$$

where ω_{mn} is the plate natural frequency. Using the property of function χ in Eq (17) leads to the first order differential inequality

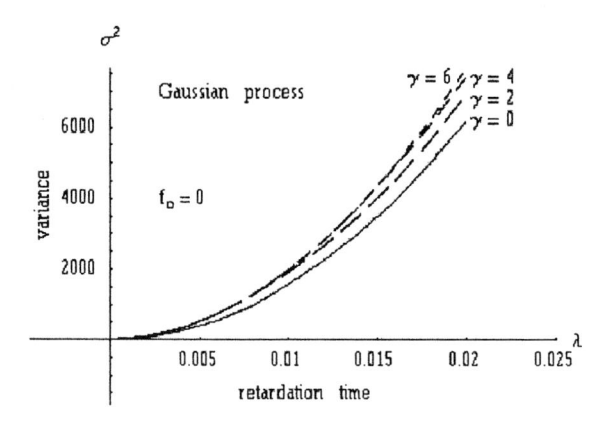

Figure 5. Stability domains of plate with the zero mean Gaussian force with the single mode control (M=N=1)

$$\frac{dV}{dt} \leq -2(\lambda - \chi(t))V \qquad (22)$$

The trivial solution of Eq (10) is almost sure asymptotically stable if the dimensionless retardation time is greater than the expectation of function χ

$$\lambda \geq E(\chi) \qquad (23)$$

It is assumed that the square plate is subjected to equal forces in both directions. The stability regions as functions of constant component of the axial force, in-plane loading variance σ^2, effective retardation time λ and gain factor γ are calculated numerically and shown in Fig. 5-7.

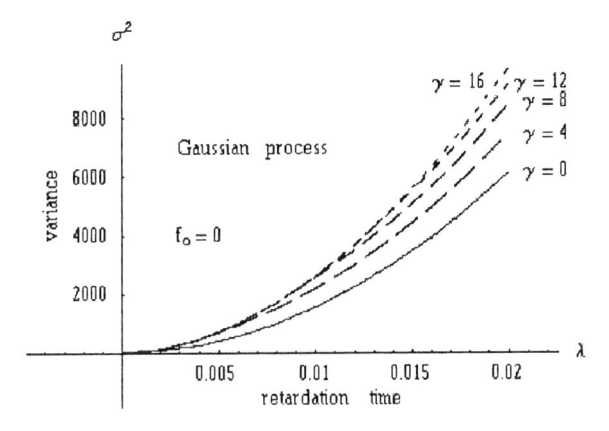

Figure 6. Stability domains of plate with the zero mean Gaussian force with the double modes control (M=N=1 and M=1, N=2)

4. CONCLUSIONS

The stabilization of vibrating plate with distributed piezoelectric sensor, actuator, and velocity feedback has been studied. The stabilization of stochastic parametric vibrations needs sufficiently large active damping. Admissible variances of loading depend strongly on the feedback gain factor. The saturation effect is observed for.

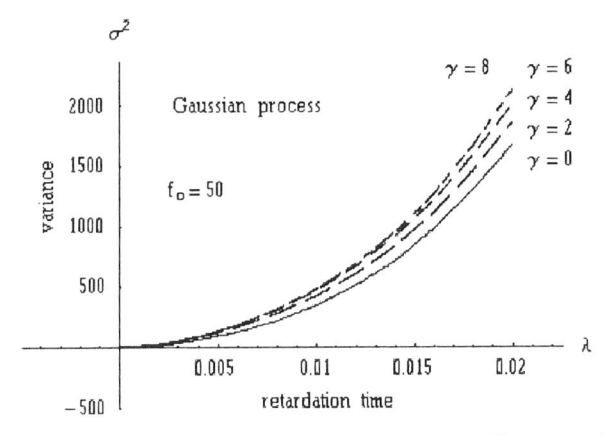

Figure 7. Stability domains of plate with the compressive Gaussian force with the single mode control (M=N=1)

large values of gain factor. Double-modes control enlarges the almost sure stability regions. Increase of constant components of in-plane forces decreases stability regions

ACKNOWLEDGEMENT

The work presented in the paper was supported by the Polish State Committee for Scientific Research (BS Grant No. 504/G/1152/0818/001), which is gratefully acknowledged.

5. REFERENCES

1. Bailey, T. and Hubbard, J. E. (1985) Distributed piezoelectric – polymer active vibration control of a cantilever beam, *Journal of Guidance, Control and Dynamic* **8** 605-611.
2. Newman, M. J. (1991) Distributed active vibration controlers, in *Recent Advances in Active Control of Sound and Vibration*, edited by Rogers, C. A., and Fuller, C. R., Technomic Publishing, Lancaster-Basel, pp.579-592.
3. Crawley, E. F. and de Luis, J. (1987) Use of piezoelectric actuators as elements of intelligent structures, *AIAA J.* **25** 1373-1385.
4. Dimitriadis, E., Fuller, C. R. and Rogers, C. A. (1991) Piezoelectric actuators for distributed vibration excitations of thin plates, *J. Appl. Mech.* **113** 100-107.
5. Jie Pan and Hansen, C. H. (1991) A study of the response of a simply supported beam to excitation by piezoelectric actuator, in *Recent Advances in Active Control of Sound and Vibration*, edited by Rogers, C. A., and Fuller, C. R., Technomic Publishing, Lancaster-Basel, pp.39-49.
6. Dosch, J. J., Inman, D. J. and Garcia, E. (1992) A self piezoelectric actuator for collocated control, *J. Intelligent Material Systems and Structures* **3** 166-185.
7. Tylikowski, A. (1995) Active Stabilization of Beam Vibrations Parametrically Excited by Wide-Band Gaussian Force, in *Proceedings of ACTIVE 95*, edited by Sommerfeld, S., and Hamada, H., Newport Beach, CA., pp. 91-102.
8. Tylikowski, A. (1993) Stabilization of beam parametric vibration, *Journal of Theoretical and Applied Mechanics* **32** 657-670.
9. Lee, C. K. and Moon, F. (1990) Modal sensors/actuators, *J. Appl. Mech.* **57** 434-441.
10. Tzou, H. S. and Fu, H. Q. (1992) A study on segmentation of distributed piezoelectric sensors and actuators; Part 1 – Theoretical analysis, *Active Control of Noise and Vibrations*, DSC – **38**, ASME, pp.239-246.
11. Kumar, S., Bhalla, A. S. and Cross, L. E. (1994) Smart ceramics for broad-band vibration control, *J. Intelligent Material Systems and Structures* **5** 673-677.
12. Chow, P. L. and Maestrello, L. (1993) Stabilization of the nonlinear vibration of an elastic panel by boundary control, in *Recent Advances in Active Control of Sound and Vibration*, edited by Burdisso, R. A., Technomic Publishing, Lancaster-Basel, pp.660-667.
13. Chen, L. W., Lin, C. Y. and Wang, C. C. (2002) Dynamic stability analysis and control of a composite beam with piezoelectric layers, Composite Structures **56** 97-109.
14. Yang, X. M. and Shen, Y. P. (2001) Dynamic instability of laminated piezoelectric shell, *Int. J. Solids & Structures* **38** 2291-2303.
15. Sulivan, J. M., Hubbard, J. E. and Burke, S. E. (1994) Distributed transducer design for plates: spatial shape and shading as design parameters, in *Smart Structures and Materials 1994, Mathematics and Control in Smart Structures*, edited by Banks, H. T., The International Society for Optical Engineering, **2192**, 132-144.

FINITE ELEMENT MODELS FOR LINEAR ELECTROELASTIC DYNAMICS

FRANCESCO UBERTINI
DISTART, Università di Bologna,
Viale Risorgimento 2, 40136 Bologna, Italy
E-mail: francesco.ubertini@mail.ing.unibo.it

1. INTRODUCTION

In the last decades, much attention has been focused on finite element modeling of electroelastic structural elements (see for example the survey in [1]). For most contributions in the covered literature, the reference variational framework is the extended Hamilton's principle and the finite element formulations involve mechanical displacement and electric potential as independent variables. Although these conventional finite element models are the most used in practice, they are often too stiff and susceptible to mesh distortion. To improve element accuracy, bubble or incompatible modes have been used for displacement and electric potential. An attractive alternative to enhance element performance is offered by mixed variational formulations [2]-[4]. Recently, a mixed approach has been successfully employed by Cannarozzi and Ubertini [5] to develop hybrid finite elements for static analysis, which involve stress and electric flux density as additional independent variables. These elements have been classified as hybrid in the sense of the original hybrid model proposed by Pian [6] for elasticity. In fact, the additional variables must satisfy locally the field balance equations and the corresponding primary variables (displacement and electric potential) turn out to be defined on the element boundary only. At the interelements, continuity requirements on the additional variables are relaxed, so that the parameters of their approximations can be condensed out at the element level. This leads to finite element equations involving only nodal displacements and electric potentials, which can be handled using standard finite element procedures.

A coherent extension of hybrid models to dynamics is not straightforward. In fact, the element displacement approximation needed to define inertia forces enters into dynamic equilibrium equations. Thus, satisfying a priori dynamic equilibrium implies that stress and displacement can not be independently approximated and the resulting finite element equations can not be handled using conventional procedures. For this reason, the most common extension consists of assuming the approximations used for statics (stresses equilibrating a priori the applied loads only) to provide for stiffness matrix and introducing a compatible element interior displacement which serves only to provide for a mass matrix. Although generally used, this approach can not be considered as consistent because equilibrium between inner and inertia forces cannot be satisfied within the element, but only at its boundary.

In this work, the consistent approach recently proposed in [7][8] to extend hybrid models to dynamics is considered. A mixed finite element formulation based on a Hellinger-Reissner-type functional is presented and a rational criterion to select consistent stress approximations is discussed. The rationale consists simply of decomposing the stress approximation into two parts. The first part is required to satisfy pointwise the statical equilibrium equations, as in the static case. The second part is introduced to balance inertia forces in a weak sense and is approximated consistently with the assumed displacements, although independently from them. In particular, the approximation space for the second stress part is assumed as the space which the inertia forces (resulting from the assumed displacements) implicitly define via dynamic equilibrium equations. Following this idea, the proposed formulation leads to element equations which can be expressed in terms of nodal displacements and electric potentials only. Thus, conventional finite element procedures can be applied and the resulting finite elements can be readily implemented in existing codes.

A four-node element is implemented based on the present model and its performance is assessed through four benchmark problems. Numerical results confirm that the proposed criterion to select consistent stresses is really effective in improving element accuracy. In fact, the present element is shown to be markedly superior to the conventional four-node element, notwithstanding the higher computational cost it involves at the element level. In addition, the present element is found to be more accurate than hybrid elements based on inconsistent stresses and only weakly sensitive to element geometry distortions.

2. BASIC EQUATIONS

Consider an electroelastic body that occupies a closed and bounded region, whose boundary and open set are $\partial\Omega$ and Ω. The state of the body is described by the displacement vector $\mathbf{u}$, the electric potential ϕ, the strain tensor $\boldsymbol{\epsilon}$, the stress tensor $\boldsymbol{\sigma}$, the electric field vector $\mathbf{e}$ and the electric flux density vector $\mathbf{d}$. The fundamental relations governing linear electroelastic dynamics are the balance equations, the compatibility equations and the constitutive equations:

$$\mathbf{D}^*\boldsymbol{\sigma} = \mathbf{b} - \rho\ddot{\mathbf{u}}, \quad \operatorname{div}\mathbf{d} = \gamma, \qquad (1)$$

$$\boldsymbol{\epsilon} = \mathbf{D}\mathbf{u}, \quad \mathbf{e} = -\nabla\phi, \qquad (2)$$

$$\epsilon = -\partial \pi/\partial \sigma = \mathbf{H}\sigma + \mathbf{h}^T \mathbf{d}, \qquad \mathbf{e} = \partial \pi/\partial \mathbf{d} = -\mathbf{h}\sigma + \mathbf{X}\mathbf{d}, \qquad (3)$$

with the boundary conditions

$$\sigma \mathbf{n} = \bar{\mathbf{t}} \quad \text{on } \partial\Omega_t, \qquad \mathbf{d} \cdot \mathbf{n} = -\bar{d} \quad \text{on } \partial\Omega_d, \qquad (4)$$
$$\mathbf{u} = \bar{\mathbf{u}} \quad \text{on } \partial\Omega_u, \qquad \phi = \bar{\phi} \quad \text{on } \partial\Omega_\phi, \qquad (5)$$

and the initial conditions

$$\mathbf{u}|_0 = \mathbf{u}_0, \qquad \dot{\mathbf{u}}|_0 = \mathbf{v}_0, \qquad (6)$$

where a superposed dot denotes differentiation with respect to time t, $\partial\Omega_u$ and $\partial\Omega_t$ are the portions of $\partial\Omega$ where displacements and tractions are prescribed, $\partial\Omega_\phi$ and $\partial\Omega_d$ are the portions of $\partial\Omega$ where electric potential and surface charge are prescribed. In the equations above, $\mathbf{D}$ and $\mathbf{D}^*$ are the differential compatibility and equilibrium operators, $\mathbf{b}$ and γ the prescribed load vector and electric charge, $\bar{\mathbf{u}}$ and $\bar{\phi}$ the prescribed boundary displacement and electric potential, $\bar{\mathbf{t}}$ and $\bar{d}$ the prescribed boundary traction and surface charge, $\mathbf{u}_0$ and $\mathbf{v}_0$ the initial displacement and velocity vectors, $\mathbf{H}$, $\mathbf{X}$ and $\mathbf{h}$ the constitutive tensors, ρ is the material density, π the mechanical enthalpy function, $\mathbf{n}$ the outward unit normal on $\partial\Omega$. Operators $\mathbf{D}$ and $\mathbf{D}^*$ are linear and adjoint.

The most common variational formulation of the above problem is based on the extended Hamilton's principle and involves displacement and electric potential as independent variables. Recently, an alternative formulation based on a Hellinger-Reissner-type functional, involving stress and electric flux density as additional independent variables, has been presented in [7]. Applying the standard procedure of space domain discretization, the Hellinger-Reissner-type functional is modified by introducing an independent displacement field $\tilde{\mathbf{u}}$ over interelement boundaries. Continuity between interior displacements $\mathbf{u}$ and interelement displacements $\tilde{\mathbf{u}}$ is imposed by means of traction $\mathbf{t}$ as a Lagrangian multiplier, which is assumed to be independent of stress σ. The modified Hellinger-Reissner-type functional for an individual element Ω_e is given by

$$\tilde{\Pi}_e^{SF}(\sigma, \mathbf{t}, \mathbf{d}, \mathbf{u}, \tilde{\mathbf{u}}, \phi) = \int_{\Omega_e} [\pi(\sigma, \mathbf{d}) + \sigma \cdot \mathbf{D}\mathbf{u} - \mathbf{b} \cdot \mathbf{u}] \, dV +$$
$$+ \int_{\partial\Omega_e} [\mathbf{t} \cdot (\tilde{\mathbf{u}} - \mathbf{u}) + \mathbf{d} \cdot \mathbf{n}\, \phi] \, dS - \int_{\partial\Omega_{te}} \bar{\mathbf{t}} \cdot \tilde{\mathbf{u}} \, dS + \int_{\partial\Omega_{de}} \bar{d}\, \phi \, dS, \quad (7)$$

subjected to the constraints that $\tilde{\mathbf{u}}$ and ϕ satisfy the prescribed boundary conditions and $\mathbf{d}$ satisfies the electrical balance equations within the element domain. In the above expression, $\partial\Omega_e$ is the element boundary, $\partial\Omega_{te}$ is the intersection between $\partial\Omega_e$ and $\partial\Omega_t$, $\partial\Omega_{de}$ is the intersection between $\partial\Omega_e$ and $\partial\Omega_d$. Notice that only $\tilde{\mathbf{u}}$ and ϕ must be continuous across interelement boundaries. Thus σ, $\mathbf{t}$, $\mathbf{d}$ and $\mathbf{u}$ are inner variables for each element.

Then, the semidiscrete version of the proposed formulation reads as

$$\sum_e \left\{ \int_I \delta \left(\tilde{\Pi}_e^{SF} - \Pi_e^K \right) dt - I_e^D \right\} = 0 \quad \forall (\delta\sigma, \delta\mathbf{d}, \delta\mathbf{t}, \delta\mathbf{u}, \delta\tilde{\mathbf{u}}, \delta\phi), \quad (8)$$

$$\Pi_e^K(\mathbf{u}) = \frac{1}{2} \int_{\Omega_e} \rho \dot{\mathbf{u}} \cdot \dot{\mathbf{u}} \, dV, \quad (9)$$

$$I_e^D(\mathbf{u}, \delta\mathbf{u}) = \int_{\Omega_e} \rho \left[\delta\mathbf{u}|_0 \cdot \mathbf{v}_0 - (\delta\mathbf{u} \cdot \dot{\mathbf{u}})|_{t_f} + \delta\dot{\mathbf{u}}|_0 \cdot (\mathbf{u}|_0 - \mathbf{u}_0) \right] dV, \quad (10)$$

where kinetic energy Π_e^K and term I_e^D serve to introduce the inertia forces and relax initial conditions, and $I = (0, t_f)$ denotes the time interval.

3. FINITE ELEMENT FORMULATION

Stress σ within each element is decomposed into two parts:

$$\sigma = \sigma_s + \sigma_d. \quad (11)$$

Term σ_s is taken as in the hybrid model for statics [5], that is as the sum of an undeterminate self-equilibrated field σ_h and a field σ_b in equilibrium with body forces:

$$\mathbf{D}^*\sigma_b = \mathbf{b}, \quad \mathbf{D}^*\sigma_h = \mathbf{0} \quad \text{in } \Omega_e. \quad (12)$$

Notice that σ_b is determined as a particular integral of the above equation and is not an undeterminate field. Term σ_d is an additional undeterminate field which is introduced upon the need to ensure equilibrium with the inertia forces. In fact, substituting eq. (11) into eq. (1) yields

$$\mathbf{D}^*\sigma_d = -\rho\ddot{\mathbf{u}} \quad \text{in } \Omega_e. \quad (13)$$

Since σ_d and $\mathbf{u}$ are assumed as independent fields, the above condition is weakly enforced in the present formulation. Analogously, $\mathbf{t}$ is separated into two parts:

$$\mathbf{t} = \mathbf{t}_s + \mathbf{t}_d \quad \text{such that} \quad \mathbf{t}_s = \sigma_s \mathbf{n} \quad \text{on } \partial\Omega_e. \quad (14)$$

It follows that $\mathbf{t}_s$ is no longer an independent field. Electric flux density $\mathbf{d}$ within each element is expressed as in the hybrid model [5]:

$$\mathbf{d} = \mathbf{d}_\gamma + \mathbf{d}_h \quad \text{such that} \quad \text{div}\,\mathbf{d}_\gamma = \gamma, \quad \text{div}\,\mathbf{d}_h = 0 \quad \text{in } \Omega_e, \quad (15)$$

where $\mathbf{d}_\gamma$ is determined as a particular integral of the above equation and $\mathbf{d}_h$ is an undeterminate null-divergence field. Note that in the static case σ_d and $\mathbf{t}_d$ do not make sense and can be dropped. As a result, the model coalesces into the fully hybrid model [5] retaining only σ_s, $\mathbf{d}$, ϕ and $\tilde{\mathbf{u}}$ as independent fields.

Each undeterminate field is expressed as product of a function of space variable x alone and a function of time variable t alone. The standard matrix-vector notation is employed hereinafter and the previous relations are intended as adjusted in accordance. The following approximations are introduced for inner element variables:

$$\sigma_h + \sigma_d = [\ \mathbf{P}_h(\mathbf{x})\ \ \mathbf{P}_d(\mathbf{x})\ ] \begin{bmatrix} \boldsymbol{\beta}_h(t) \\ \boldsymbol{\beta}_d(t) \end{bmatrix} = \mathbf{P}(\mathbf{x})\boldsymbol{\beta}(t), \tag{16}$$

$$\mathbf{d}_h = \mathbf{L}(\mathbf{x})\boldsymbol{\alpha}(t), \quad \mathbf{t}_d = \mathbf{T}(\mathbf{x})\boldsymbol{\tau}(t), \quad \mathbf{u} = \mathbf{U}(\mathbf{x})\mathbf{q}(t), \tag{17}$$

where $\mathbf{P}_h, \mathbf{P}_d, \mathbf{L}, \mathbf{T}$ and $\mathbf{U}$ are matrices of basis functions and $\boldsymbol{\beta}_h, \boldsymbol{\beta}_d, \boldsymbol{\alpha}, \boldsymbol{\tau}$ and $\mathbf{q}$ are the vectors of unknown parameters. Notice that self-equilibrated stress modes $\mathbf{P}_h$ and additional stress modes $\mathbf{P}_d$ are to be linearly independent, so to form a basis for $\boldsymbol{\sigma}$. The boundary displacement and the electric potential are interpolated by nodal values $\tilde{\mathbf{q}}$ and $\boldsymbol{\phi}$ as

$$\tilde{\mathbf{u}} = \tilde{\mathbf{U}}(\mathbf{x})\tilde{\mathbf{q}}(t), \quad \phi = \Phi(\mathbf{x})\boldsymbol{\phi}(t), \tag{18}$$

where $\tilde{\mathbf{U}}$ and Φ are the matrices of shape functions. In this way, identifying nodal values at common boundaries of adjacent elements ensures that $\tilde{\mathbf{u}}$ and ϕ are continuous across element boundaries.

The assumed approximations are substituted into $\tilde{\Pi}_e^{SF}$, Π_e^K and I_e^D to obtain the space-discretized form of eq. (8). Making the resultant expression satisfied for every $\delta\boldsymbol{\beta}$ and $\delta\boldsymbol{\alpha}$ leads to the compatibility equations of the model, which can be used to eliminate $\boldsymbol{\beta}$ and $\boldsymbol{\alpha}$. In order to condense out also the remaining internal variables $\boldsymbol{\tau}$ and $\mathbf{q}$ by making the statement satisfied for every $\delta\boldsymbol{\tau}$ and $\delta\mathbf{q}$, it can be easily verified that the following matrix

$$\mathbf{G}_u = \int_{\partial\Omega_e} \mathbf{T}^\mathrm{T}\mathbf{U}\,\mathrm{d}S \tag{19}$$

should be a square nonsingular matrix. This can be achieved by carefully selecting the approximations for $\mathbf{u}$ and $\mathbf{t}_d$, as shown in the next section. By doing so, the following variational form of semidiscrete equations involving only nodal values of displacement and electric potential is obtained

$$\int_I \begin{vmatrix} \delta\tilde{\mathbf{q}} \\ \delta\boldsymbol{\phi} \end{vmatrix}^\mathrm{T} \left(\begin{vmatrix} \mathbf{M}_u & 0 \\ 0 & 0 \end{vmatrix} \begin{vmatrix} \ddot{\tilde{\mathbf{q}}} \\ \ddot{\boldsymbol{\phi}} \end{vmatrix} + \begin{vmatrix} \mathbf{K}_u & \mathbf{K}_{u\phi} \\ \mathbf{K}_{u\phi}^\mathrm{T} & \mathbf{K}_\phi \end{vmatrix} \begin{vmatrix} \tilde{\mathbf{q}} \\ \boldsymbol{\phi} \end{vmatrix} + \begin{vmatrix} \mathbf{f}_u \\ \mathbf{f}_\phi \end{vmatrix} \right) \mathrm{d}t +$$

$$- \int_I (\delta\tilde{\mathbf{q}}^\mathrm{T}\mathbf{h}_u + \delta\boldsymbol{\phi}^\mathrm{T}\mathbf{h}_\phi)\mathrm{d}t + \delta\tilde{\mathbf{q}}|_0^\mathrm{T}(\mathbf{M}_u\dot{\tilde{\mathbf{q}}}|_0 - \tilde{\mathbf{q}}_{v0}) - \delta\dot{\tilde{\mathbf{q}}}|_0^\mathrm{T}(\mathbf{M}_u\tilde{\mathbf{q}}|_0 - \tilde{\mathbf{q}}_{u0}) = 0,$$

which should hold for every $\delta\tilde{\mathbf{q}}$ and $\delta\boldsymbol{\phi}$, where $\mathbf{M}_u$ (symmetric and positive definite) is the element mass matrix, $\mathbf{K}_u$ (symmetric and positive semidefinite) the element stiffness matrix, K_ϕ (symmetric and negative semidefinite) the element dielectric matrix, $K_{u\phi}$ the element electroelastic coupling matrix, h_u and h_ϕ are the vectors of generalized nodal forces and electric charges, f_u and f_ϕ are the vectors of equivalent nodal forces and electric charges, $\tilde{q}_{u0}$ and $\tilde{q}_{v0}$ the vectors of initial conditions for boundary displacements and velocities.

From the above statement, the standard form of semidiscrete equations can be obtained with the associated initial conditions:

$$\begin{vmatrix} h_u \\ h_\phi \end{vmatrix} = \begin{vmatrix} M_u & 0 \\ 0 & 0 \end{vmatrix} \begin{vmatrix} \ddot{\tilde{q}} \\ \ddot{\phi} \end{vmatrix} + \begin{vmatrix} K_u & K_{u\phi} \\ K_{u\phi}^T & K_\phi \end{vmatrix} \begin{vmatrix} \tilde{q} \\ \phi \end{vmatrix} + \begin{vmatrix} f_u \\ f_\phi \end{vmatrix}, \qquad (20)$$

$$M_u \tilde{q}|_0 - \tilde{q}_{u0} = 0, \qquad M_u \dot{\tilde{q}}|_0 - \tilde{q}_{v0} = 0. \qquad (21)$$

The conventional finite element procedure can be used to assemble the above equations and standard time integration algorithms can be applied. Note that the finite element equations in the variational form can be used to develop time integration methods by discretizing nodal values in the time domain. Time continuous or discontinuous methods are obtained depending on whether the time approximation for $\tilde{q}$ satisfies a priori initial conditions or not.

4. FINITE ELEMENT APPROXIMATIONS

Similar to conventional finite elements, $\tilde{u}$ and ϕ are uniquely defined when the nodal values are chosen. However, many choices are still possible for the approximations of inner variables. In this section, a rational approach for selecting consistent approximations is presented.

A local reference system is used to compute the element matrices and vectors. Boundary displacements $\tilde{u}$ and electric potential ϕ are interpolated using natural (parametric) coordinates, while interior displacements u using local coordinates x. To select $\tilde{u}$ and u in proper balance, the approximation for u is taken as the extension of the basis functions for $\tilde{u}$ over a regular element shape. In this way, u and $\tilde{u}$ have the same number of freedoms and, when the element geometry is regular, the same polynomial basis over $\partial\Omega_e$. In this case continuity turns out to be satisfied pointwise. Note that this perfect matching is not achieved when the element geometry is distorted, since the basis functions for $\tilde{u}$ (in terms of x) change in consequence of the transformation from natural to local coordinates. In this case continuity is weakly satisfied.

As mentioned in the previous section, t_d should be approximated such that eq. (19) yields a square nonsingular matrix. This is obtained by selecting the basis functions for t_d as the restriction of those for u over the element boundary:

$$T(x) = U(x)|_{\text{on } \partial\Omega_e},$$

since u vanishes on $\partial\Omega_e$ if and only if q is zero.

In the spirit of hybrid models, $\boldsymbol{\sigma}_d$ should be selected aiming at satisfying eq. (13). Taking into account for the assumed approximations, eq. (13) becomes

$$\mathbf{D}^*\mathbf{P}_d(\mathbf{x})\boldsymbol{\beta}_d(t) = -\rho\mathbf{U}(\mathbf{x})\ddot{\mathbf{q}}(t).$$

This equation suggests that a proper balance between assumed stresses $\boldsymbol{\sigma}_d$ and interior displacements $\mathbf{u}$ is achieved by taking $\mathbf{P}_d$ such that

$$\mathrm{span}(\mathbf{D}^*\mathbf{P}_d) = \mathrm{span}(\rho\mathbf{U}). \tag{22}$$

Using the above condition, the basis functions for $\boldsymbol{\sigma}_d$ can be easily constructed from the ones assumed for $\mathbf{u}$, since $\mathbf{u}$ is expressed in local coordinates. Otherwise, it could be an hard task because of the coordinate transformation. It follows that the number of $\boldsymbol{\beta}_d$-parameters is equal to that of $\mathbf{q}$-parameters and linear independence between $\mathbf{P}_h$ and $\mathbf{P}_d$ is ensured. In the plane case, the above criterion can be satisfied by determining the approximation for $\boldsymbol{\sigma}_d^T = |\ \sigma_{d11}\ \sigma_{d22}\ \sigma_{d12}\ |$ as follows:

$$\mathbf{P}_{dii}(\mathbf{x}) = \int_0^{x_i} \rho U_i(\mathbf{x})\,dx_i, \qquad \mathbf{P}_{d12}(\mathbf{x}) = 0, \tag{23}$$

where $\mathbf{P}_{dii}$ and $\mathbf{P}_{d12}$ are the basis functions of σ_{dii} and σ_{d12} (rows of $\mathbf{P}_d$), U_i are the basis functions of the interior displacement components (rows of $\mathbf{U}$).

Finally, the approximations for the null-divergence fields $\boldsymbol{\sigma}_h$ and $\mathbf{d}_h$ can be derived as usual for hybrid models [5][6]. For the purpose of element stability, these approximations are to be selected such as to provide rank-sufficient element matrices. Indeed, it is desirable to keep the number of assumed modes to a minimum in order to improve element accuracy and computational effectiveness. As regards the stress modes, this issue obviously refers to the total number of β-parameters and should be pursued by carefully selecting $\boldsymbol{\sigma}_h$.

A plane quadrilateral element has been implemented by interpolating $\tilde{\mathbf{u}}$ and ϕ using standard four-node shape functions. Then, bilinear expansions are taken for $\mathbf{u}$, the 4α-approximation labelled J2 in [9] is assumed for $\mathbf{d}$, an $8\beta_d$-approximation is derived for $\boldsymbol{\sigma}_d$ from eq. (23), a constant stress field is assumed for $\boldsymbol{\sigma}_h$. Notice that this ensures a proper balance between the strains corresponding to $\boldsymbol{\sigma}_h$ and the strains corresponding to $\mathbf{u}$. The resultant element has been checked to be stable. Hereinafter, this element is referred to as MD4.

5. NUMERICAL RESULTS AND CONCLUSIONS

In this section, four test cases are examined in order to assess the performance of element MD4. Numerical results in terms of natural frequencies are compared with those predicted by two four-node elements: the conventional element (C4) based on assumed displacements and electric potential and a hybrid element

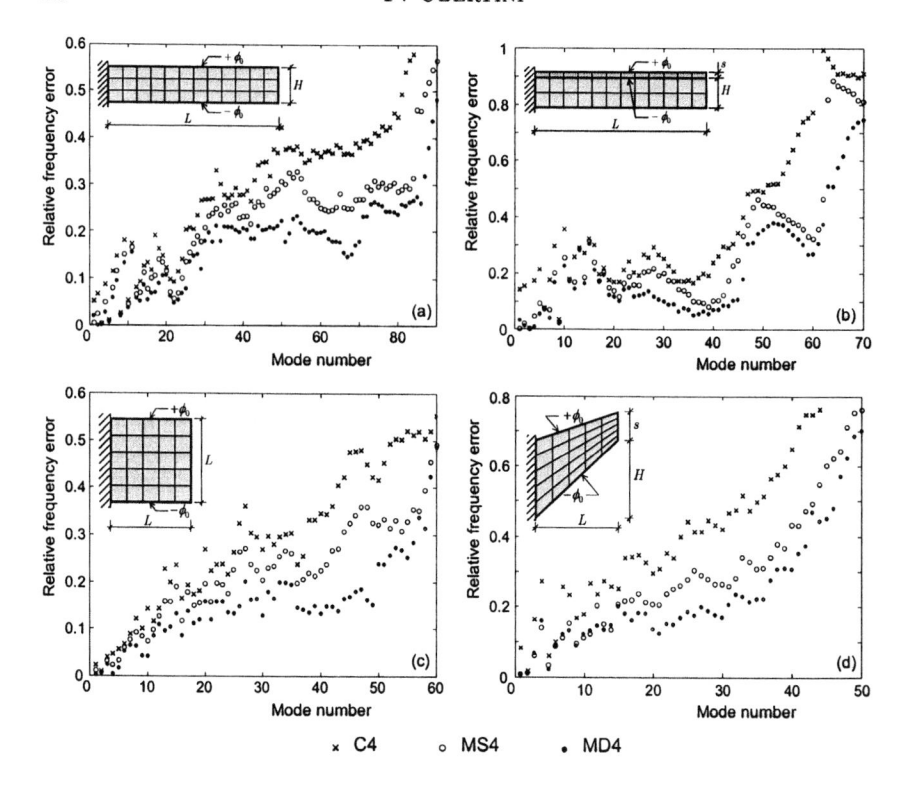

Figure 1: Relative errors on the frequency spectra.

(MS4) extended to dynamics by the usual (inconsistent) approach. This hybrid element is obtained by taking for stress the approximation of element $5\beta C$ in [10], while for electric flux the same approximation of MD4. The test case 1 is a cantilever beam of VIBRIT 420 [11], subjected to a prescribed electric potential on the top and bottom surfaces (Fig. 1a). The data are: $L = 1$, $H = 0.2$ and $\phi_0 = 1$. The test case 2 is a laminated cantilever beam made of a thick layer of aluminium and a thin layer of VIBRIT 420 adhered together (Fig. 1b). A potential difference is applied across the piezoelectric layer. The data are: $L = 1$, $H = 0.16$, $s = 0.04$ and $\phi_0 = 1$. The test case 3 is a square plate of VIBRIT 420, restrained on the left side and subjected to a prescribed electric potential on the top and bottom sides (Fig. 1c). The data are: $L = 1$ and $\phi_0 = 1$. The test case 4 is a swept plate of VIBRIT 420, subjected to the same conditions of the previous case (Fig. 1d). The data are: $L = 0.48$, $H = 0.44$, $s = 0.16$ and $\phi_0 = 1$.

The relative errors on the frequency spectra are shown in Fig. 1, together with the mesh used for the analysis. These pictures reveal the overall accuracy of the semidiscretizations. As it can be observed, the present element is

Table I: Percentage errors on the first six natural frequencies.

	Test case 1						Test case 2				
	C4			MD4			C4			MD4	
8×2	12×3	24×6	8×2	12×3	24×6	12×3	24×6	36×9	12×3	24×6	36×9
11.04	5.133	1.337	4.687	2.006	0.488	4.509	1.201	0.547	2.472	0.588	0.260
14.01	6.546	1.716	0.051	0.089	0.041	5.785	1.550	0.706	0.268	0.023	0.009
0.564	0.297	0.095	0.347	0.179	0.048	0.285	0.087	0.043	0.165	0.046	0.020
18.64	8.663	2.261	6.141	2.753	0.694	7.845	2.089	0.947	2.365	0.630	0.280
2.209	1.033	0.281	1.809	0.826	0.209	9.920	2.757	1.249	5.374	1.372	0.608
24.85	11.47	2.968	13.86	5.986	1.473	1.748	0.343	0.155	1.061	0.264	0.117

	Test case 3						Test case 4				
	C4			MD4			C4			MD4	
2×2	5×5	10×10	2×2	5×5	10×10	2×2	5×5	10×10	2×2	5×5	10×10
10.14	2.300	0.683	1.172	0.508	0.162	36.17	8.489	2.427	2.876	0.819	0.223
4.369	1.020	0.318	3.531	0.690	0.190	5.088	2.030	0.962	4.283	1.574	0.559
20.13	4.014	1.085	14.71	2.580	0.663	69.77	16.59	4.472	42.52	7.137	0.690
18.26	4.700	1.309	2.539	1.005	0.281	50.67	27.37	8.621	36.69	16.37	1.896
29.29	5.721	1.410	10.57	2.045	0.518	29.16	6.240	1.319	15.99	2.588	0.339
30.19	6.822	1.849	27.57	5.526	1.391	76.30	10.24	5.937	22.79	8.899	1.806

markedly superior to the conventional C4 element. All the frequencies are better estimated and the error tends to grow slower as the mode number increases. Notice that element MD4 is also more accurate than MS4 all over the frequency spectra. The results of convergence analysis on the first six frequencies are reported in Tab. 1. The sensitivity to element geometry distortions is investigated by solving the test case 3 using a series of progressively distorted meshes (Fig. 2). The relative errors on the first six frequencies are plotted versus the distortion parameter d/L in Fig. 3. It can be noticed that element MD4 is much less sensitive to element distortions than C4 element.

The above results demonstrate the effectiveness of the present approach to improve accuracy of conventional finite elements, as well as the validity of the proposed criterion to select consistent stress approximations.

ACKNOWLEDGMENTS: The author is deeply grateful to Prof. A.A. Cannarozzi, untimely died, for his valuable suggestions and stimulating discussions from which this work arose. The financial support by MURST is acknowledged. The computing facilities were provided by the Laboratory of Computational Mechanics (LAMC), DISTART, University of Bologna.

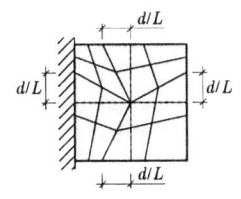

Figure 2: Distorted mesh scheme.

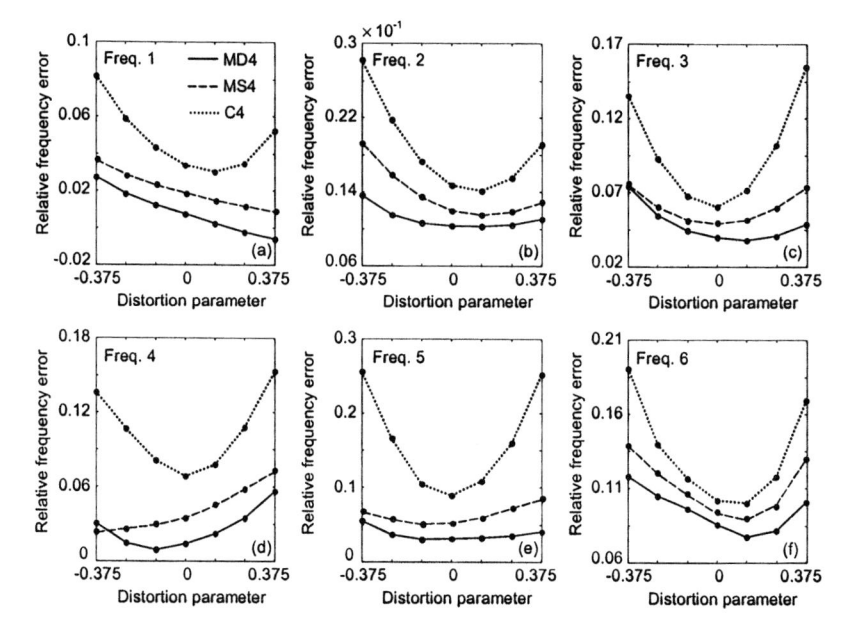

Figure 3: Sensitivity to element geometry distortions.

6. REFERENCES

1. Benjeddou, A. (2000) Advances in piezoelectric finite element modeling of adaptive structural elements: a survey, *Comp. Struct.* **76** 347-363.
2. Shen, M.H.H. (1995) A new modeling technique for piezoelectrically actuated beams, *Comp. Struct.* **57** 361-366.
3. Ghandi, K. and Hagood, N.W. (1997) A hybrid finite element model for phase transition in nonlinear electro-mechanically coupled material, in *Smart Struct. Mater.*, edited by Varadan, V.V. and Chandra, J., SPIE, 3039, Washington, pp. 97-112.
4. Sze, K.Y. and Pan, Y.S. (1999) Hybrid finite element models for piezoelectric materials, *J. Sound Vibrat.* **226** 519-547.
5. Cannarozzi, A.A. and Ubertini, F. (2001) Some hybrid variational methods for linear electroelasticity problems, *Int. J. Solids Structures* **38** 2573-2596.
6. Pian, T.H.H. (1973) Hybrid models, in *Numerical and Computer Methods in Structural Mechanics*, edited by Fenves, S.J., Perrone, R., Robinson, R. and Schonbrich, W.C., Academic Press, New York, pp. 50-80.
7. Cannarozzi, A.A. and Ubertini, F. (2001) Mixed stress variational approaches for linear dynamics of electroelastic solids, in *Proc. 2nd Europ. Conf. Comput. Mech.*, Cracow, Poland.
8. Ubertini, F. (2002) A consistent mixed stress approach for linear elastodynamic analysis, in *Proc. 5th Europ. Conf. Struct. Dyn.*, Munich, Germany.
9. de Miranda, S. and Ubertini, F. (2002) A mixed variational method for coupled deformation-diffusion analysis in elastic solids, in *Proc. 5th World Cong. Comput. Mech.*, Vienna, Austria.
10. Yuan, K.Y., Huang, Y.S. and Pian, T.H.H. (1993) New strategy for assumed stress for 4-node hybrid stress membrane element, *Int. J. Num. Meth. Engrg.* **36** 1747-1763.
11. Lerch, R. (1990) Simulation of piezoelectric devices by two- and three-dimensional finite elements, *IEEE Trans. Ultrasonics Ferroelectrics Frequency Control* **37** 233-247.

EXACT THERMOELASTICITY SOLUTION FOR CYLINDRICAL BENDING DEFORMATIONS OF FUNCTIONALLY GRADED PLATES

Senthil S. Vel
Department of Mechanical Engineering
University of Maine
Orono, ME 04469, U.S.A.
E-mail: vel@umeme.maine.edu

R. C. Batra
Department of Engineering Science and Mechanics, M/C 0219
Virginia Polytechnic Institute and State University
Blacksburg, VA 24061, U.S.A.
E-mail: rbatra@vt.edu

1. INTRODUCTION

The abrupt change in material properties across the interface between discrete layers in composite structures can result in large interlaminar stresses leading to delamination. One way to overcome these adverse effects is to use "functionally graded materials" which are inhomogeneous materials with continuously varying material properties. There are several three-dimensional (3D) solutions available for the thermoelastic analysis of inhomogeneous plates. Most of these studies have been conducted for laminated plates that have piecewise constant material properties in the thickness direction. Rogers et al. [1] have employed the method of asymptotic expansion to analyze 3D deformations of inhomogeneous plates. However, the boundary conditions on the edges of the plate in their theory are applied in an average sense like those in 2D plate theories and the plate is assumed to be only moderately thick. Tarn and Wang [2] have also presented an asymptotic solution that may be carried out to any order, but the manipulations become more and more involved as one considers higher order terms. Cheng and Batra [3] have also used the method of asymptotic expansion to study the 3D thermoelastic deformations of a functionally graded elliptic plate. Tanaka et al. [4] designed property profiles for functionally graded materials to reduce the thermal stresses. Reddy [5] has presented solutions for rectangular plates based on the third-order shear deformation plate theory. Reiter and Dvorak [6, 7] performed detailed finite element studies of discrete models containing simulated skeletal and particulate microstructures and compared results with those computed from homogenized models in which effective properties were derived by the Mori-Tanaka and the self-consistent methods. Cheng and Batra [8] have related the deflections of a simply supported functionally graded polygonal plate given by the first-order shear deformation theory (FSDT) and a third-order shear deformation theory (TSDT) to that of an equivalent homogeneous Kirchhoff plate.

The objective of this investigation is to present an exact solution to the thermoelastic cylindrical bending deformations of a simply supported functionally graded thick plate. By using suitable temperature and displacement functions, the governing partial differential equations are reduced to a set of coupled ordinary differential equations in the thickness coordinate, which are then solved by the power series method. We consider a two-phase graded material with a power-law variation of the volume fractions of the constituents through the thickness. The effective material properties at a point are determined in terms of the local volume fractions and the material properties of the two phases either by the Mori-Tanaka [9] or the self-consistent [10] scheme. Results are presented for an Al/SiC graded plate. We compare the exact results with those obtained from the classical plate theory [11] (CPT), the FSDT [12] and the TSDT [13].

2. PROBLEM FORMULATION

We use rectangular Cartesian coordinates x_i ($i = 1, 2, 3$) to describe the infinitesimal static thermoelastic deformations of an N-layer laminated plate occupying the region $[0, L] \times [-H/2, H/2] \times (-\infty, \infty)$ in the unstressed reference configuration. Each layer of the laminated plate is made of an isotropic material with material properties varying smoothly in the x_2 (thickness) direction only. The vertical positions of the bottom and the top surfaces, and the $N-1$ interfaces between the layers are denoted by $H^{(1)} = -H/2, H^{(2)}, \ldots, H^{(n)}, \ldots, H^{(N)}, H^{(N+1)} = H/2$. The equations of mechanical and thermal equilibrium in the absence of body forces and internal heat sources are

$$\sigma_{ij,j} = 0, \quad q_{j,j} = 0, \tag{1}$$

where σ_{ij} and q_j are, respectively, the components of the Cauchy stress tensor and the heat flux. A comma followed by index j denotes partial differentiation with respect to the position x_j of a material particle, and a repeated index implies summation over the range of the index. The constitutive equations for a linear isotropic thermoelastic material are

$$\sigma_{ij} = \lambda \varepsilon_{kk} \delta_{ij} + 2\mu \varepsilon_{ij} - \beta \delta_{ij} T, \quad q_j = -\kappa T_{,j}, \tag{2}$$

where λ and μ are the Lamé constants, β is the stress-temperature modulus, κ is the thermal conductivity, ε_{ij} are components of the infinitesimal strain tensor and T is the change in temperature of a material particle from that in the stress-free reference configuration. The material properties λ, μ, β and κ are functions of x_2. The infinitesimal strain tensor is related to the mechanical displacements u_i by $\varepsilon_{ij} = (u_{i,j} + u_{j,i})/2$. The edges of the plate are assumed to be simply supported and maintained at the reference temperature. That is,

$$\sigma_{11} = 0, \quad u_2 = 0, \quad T = 0 \quad \text{at} \quad x_1 = 0, L. \tag{3}$$

The mechanical boundary conditions prescribed on the top and the bottom surfaces can be either a displacement component u_j or the corresponding traction component σ_{3j}. However, typically non-zero normal and zero tangential tractions are prescribed on these two surfaces. Since the normal load can be expanded as a Fourier series in x_1, it suffices to consider loads of the form

$$\sigma_{12} = 0, \quad \sigma_{22} = p^{\pm} \sin r x_1 \text{ at } x_2 = \pm H/2, \tag{4}$$

where p^+ and p^- are known constants, $r = k\pi/L$ and k is a positive integer. The thermal boundary conditions on the top and the bottom surfaces are specified as

$$\vartheta^{\pm} T(x_1, \pm H/2) + \xi^{\pm} q_2(x_1, \pm H/2) = \varphi^{\pm} \sin r x_1. \tag{5}$$

By appropriately choosing values of constants $\vartheta^{\pm}$ and $\xi^{\pm}$, various boundary conditions corresponding to either a prescribed temperature, a prescribed heat flux or exposure to an ambient temperature through a boundary conductance can be specified. The interfaces between adjoining layers are assumed to be perfectly bonded together and in ideal thermal contact so that

$$[\![u_i]\!] = 0, \quad [\![\sigma_{i2}]\!] = 0, \quad [\![T]\!] = 0, \quad [\![q_2]\!] = 0 \text{ on } x_2 = H^{(2)}, H^{(3)}, \ldots, H^{(N)}. \tag{6}$$

Here, $[\![u_i]\!]$ denotes the jump in the value of u_i across an interface. Since the applied loads and material properties are independent of x_3 and the body is of infinite extent in the x_3 direction, we postulate that the displacements $\mathbf{u}$ and temperature T are functions of x_1 and x_2 only, and thus correspond to plane strain deformation.

3. EXACT SOLUTION

We construct a local rectangular Cartesian coordinate system $x_1^{(n)}, x_2^{(n)}, x_3^{(n)}$ with local axes parallel to the global axes and the origin at the point where the global x_2-axis intersects the mid-surface of the nth lamina. In the local coordinate system, the nth lamina occupies the region $[0, L] \times [-h^{(n)}/2, h^{(n)}/2] \times (-\infty, \infty)$, where $h^{(n)} = H^{(n+1)} - H^{(n)}$. We drop the superscript n for convenience with the understanding that all material constants and variables belong to this layer. We assume that within each layer, the Lamé constants λ and μ, the stress-temperature modulus β and the thermal conductivity κ are analytic functions of x_2 and thus can be represented by a Taylor series expansion about its midsurface as

$$[\lambda, \mu, \beta, \kappa] = \sum_{\alpha=0}^{\infty} [\tilde{\lambda}^{(\alpha)}, \tilde{\mu}^{(\alpha)}, \tilde{\beta}^{(\alpha)}, \tilde{\kappa}^{(\alpha)}] x_2^{\alpha}. \tag{7}$$

It should be noted that λ, μ, β and κ are positive quantities for all x_2, and therefore $\tilde{\lambda}^{(0)}, \tilde{\mu}^{(0)}, \tilde{\beta}^{(0)}$ and $\tilde{\kappa}^{(0)}$ are positive.

3.1 The Temperature Field

A solution for the change in temperature of points in the nth layer is sought in the form

$$T = \theta(x_2) \sin r x_1. \tag{8}$$

The assumed temperature function T identically satisfies the boundary conditions $(3)_3$ at the edges of the plate. We assume a solution for θ in the form of a power series

$$\theta(x_2) = \sum_{\gamma=0}^{\infty} \tilde{\theta}^{(\gamma)} x_2^{\gamma}. \tag{9}$$

Substitution for T from (8) and (9) into $(2)_2$ and the result into $(1)_2$, and equating coefficients of x_2^{γ} to zero gives the following recursive relation

$$\sum_{\gamma=0}^{\alpha} \left\{ \tilde{\kappa}^{(\gamma)} [\tilde{\theta}^{(\alpha-\gamma+2)}(\alpha-\gamma+2)(\alpha-\gamma+1) - \tilde{\theta}^{(\alpha-\gamma)} r^2] \right. \\ \left. + \tilde{\kappa}^{(\gamma+1)} \tilde{\theta}^{(\alpha-\gamma+1)}(\gamma+1)(\alpha-\gamma+1) \right\} = 0, \tag{10}$$

for $\alpha = 0, 1, 2, \ldots$. Evaluation of the recursion formula (10) successively for $\alpha = 1, 2, \ldots$, gives $\tilde{\theta}^{(\alpha+2)}$ in terms of arbitrary constants $\tilde{\theta}^{(0)}$ and $\tilde{\theta}^{(1)}$. There are two unknown constants for each layer, resulting in a total of $2N$ unknowns for an N-layer plate. The constants are determined by satisfying the thermal boundary conditions (5) on the top and the bottom surfaces of the plate and the interface continuity conditions $(6)_{3,4}$ for the thermal quantities between adjoining layers.

3.2 The Displacement Field

A solution for the displacement field in the nth layer is sought in the form

$$u_1 = U_1(x_2) \cos r x_1, \quad u_3 = U_2(x_2) \sin r x_1, \tag{11}$$

which identically satisfies the homogeneous boundary conditions $(3)_{1-2}$ at the simply supported edges. We assume a power series solution for the displacements as

$$U_i(x_2) = \sum_{\gamma=0}^{\infty} \tilde{U}_i^{(\gamma)} x_2^{\gamma}, \quad (i=1,2). \tag{12}$$

Substitution for $\mathbf{u}$ from (11) and (12) into $(2)_1$, for $\boldsymbol{\sigma}$ into $(1)_1$, and equating the coefficients of x_2^{γ} to zero, we obtain two coupled recurrence algebraic relations for every non-negative integer α. The recurrence relations are evaluated successively for $\alpha = 0, 1, \ldots$, to obtain $\tilde{U}_1^{(\alpha+2)}$ and $\tilde{U}_2^{(\alpha+2)}$ in terms of arbitrary constants $\tilde{U}_1^{(0)}$, $\tilde{U}_1^{(1)}$, $\tilde{U}_2^{(0)}$ and $\tilde{U}_2^{(1)}$. Thus, there are four unknown constants for each layer which are determined by satisfying the mechanical boundary conditions (4) on the top and the bottom surfaces of the plate and the interface continuity conditions $(6)_{1,2}$ for the mechanical quantities between adjoining layers.

4. EFFECTIVE MODULI OF TWO-PHASE COMPOSITES

We summarize the Mori-Tanaka and the self-consistent methods for estimating the effective properties of two-phase composite materials, and use them to analyze functionally graded materials.

4.1 The Mori-Tanaka Estimate

The Mori-Tanaka [9] scheme for estimating the effective moduli is applicable to regions of the graded microstructure which have a well-defined continuous matrix and a discontinuous particulate phase. It is assumed that the matrix phase, denoted by the subscript 1, is reinforced by spherical particles of a particulate phase, denoted by the subscript 2. The following estimates for the effective local bulk modulus K and shear modulus μ are useful for a random distribution of isotropic particles in an isotropic matrix,

$$\frac{K - K_1}{K_2 - K_1} = V_2 / (1 + (1 - V_2) \frac{K_2 - K_1}{K_1 + (4/3)\mu_1}),$$
$$\frac{\mu - \mu_1}{\mu_2 - \mu_1} = V_2 / (1 + (1 - V_2) \frac{\mu_2 - \mu_1}{\mu_1 + f_1}), \quad (13)$$

where V denotes the volume fraction and $f_1 = \mu_1(9K_1 + 8\mu_1)/6(K_1 + 2\mu_1)$. The effective thermal conductivity κ is given by [14]

$$\frac{\kappa - \kappa_1}{\kappa_2 - \kappa_1} = \frac{V_2}{1 + (1 - V_2)(\kappa_2 - \kappa_1)/3\kappa_1}, \quad (14)$$

and the coefficient of thermal expansion α by the correspondence relation [15]

$$\frac{\alpha - \alpha_1}{\alpha_2 - \alpha_1} = (\frac{1}{K} - \frac{1}{K_1}) / (\frac{1}{K_2} - \frac{1}{K_1}). \quad (15)$$

4.2 Self-consistent Estimate

The self-consistent method [10] is particularly suitable for determining the effective moduli in those regions which have an interconnected skeletal microstructure. The effective moduli for the self-consistent method are given by

$$\delta/K = V_1/(K - K_2) + V_2/(K - K_1),$$
$$\eta/\mu = V_1/(\mu - \mu_2) + V_2/(\mu - \mu_1), \quad (16)$$

where $\delta = 3 - 5\eta = K/(K + 4\mu/3)$. These are implicit expressions for the unknowns K and μ. The first equation in (16) can be solved for K in terms of μ to obtain

$$K = 1/(V_1/(K_1 + 4\mu/3) + V_2/(K_2 + 4\mu/3)) - 4\mu/3, \quad (17)$$

and μ is obtained by solving the following quartic equation

$$[V_1 K_1/(K_1 + 4\mu/3) + V_2 K_2/(K_2 + 4\mu/3)] + 5[V_1\mu_2/(\mu - \mu_2) \\ + V_2\mu_1/(\mu - \mu_1)] + 2 = 0. \quad (18)$$

The self-consistent estimate of the thermal conductivity coefficient [16] is in the implicit form

$$V_1(\kappa_1 - \kappa)/(\kappa_1 + 2\kappa) + V_2(\kappa_2 - \kappa)/(\kappa_2 + 2\kappa) = 0. \quad (19)$$

The self-consistent estimate of α is obtained by substitution of the self-consistent estimate of the bulk modulus K from (17) into the correspondence relation (15).

5. RESULTS AND DISCUSSION

Here we present exact results for a representative simply supported plate with its top surface subjected to either a mechanical load or a thermal load:

$$[\sigma_{33}(x_1, H/2), T(x_1, H/2)] = [p^+, T^+] \sin(\pi x_1/L). \tag{20}$$

The bottom surface is traction free and held at the reference temperature, that is, $\sigma_{i2}(x_1, -H/2) = 0$ and $T(x_1, -H/2) = 0$. We first consider a plate made of a single layer, and in Section 5.1 a plate with five layers. Since it is common in

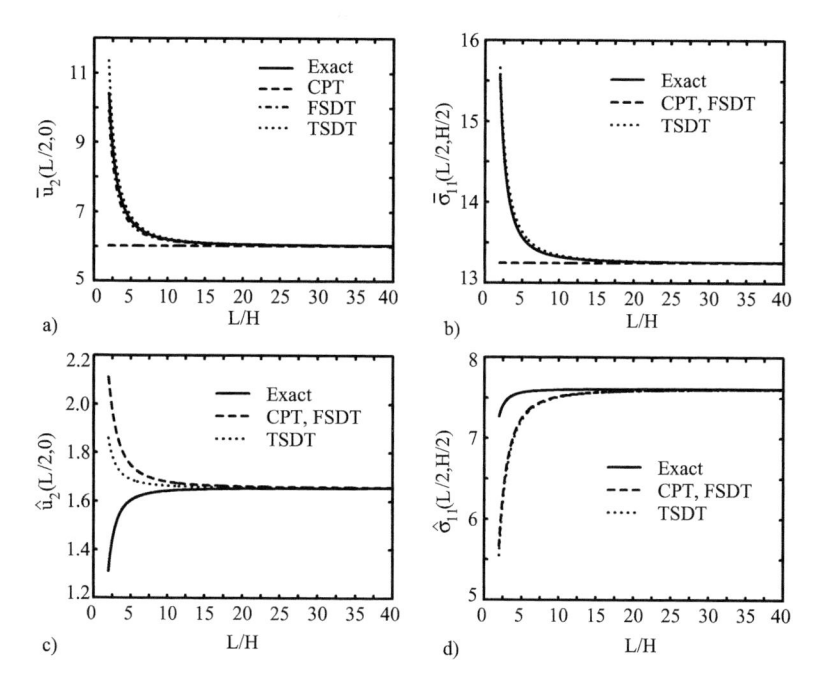

Figure 1: Transverse deflection and longitudinal stress versus length-to-thickness ratio for the Al/SiC functionally graded plate computed with the Mori-Tanaka homogenization scheme and $V_c^- = 0$, $V_c^+ = 1$, $n = 2$: (a,b) mechanical load and (c,d) thermal load.

high-temperature applications to employ a ceramic top layer as a thermal barrier to a metallic structure, we choose the constituent materials of the functionally graded plate to be Al with material properties $E_m = 70$ GPa, $\nu_m = 0.3$, $\alpha_m = 23.4 \times 10^{-6}$/K, $\kappa_m = 233$ W/mK and SiC with $E_c = 427$ GPa, $\nu_c = 0.17$, $\alpha_c = 4.3 \times 10^{-6}$/K and $\kappa_c = 65$ W/mK. We assume that the volume fraction of the ceramic phase is given by the power-law type function $V_c = V_c^- + (V_c^+ - V_c^-)(1/2 + x_2/H)^n$. Here V_c^+ and V_c^- are, respectively, the volume fractions of the ceramic phase on the top and the bottom surfaces of the plate, and n is a parameter that dictates the volume fraction profile through the thickness. The physical quantities are non-dimensionalized by relations

$$\bar{u}_1 = \frac{100 E_m H^2 u_1}{p^+ L^3}, \quad \bar{u}_2 = \frac{100 E_m H^3 u_2}{p^+ L^4}, \quad \bar{\sigma}_{11} = \frac{10 H^2 \sigma_{11}}{p^+ L^2}, \quad \bar{\sigma}_{12} = \frac{10 H \sigma_{12}}{p^+ L},$$

for the applied mechanical load, and by

$$\hat{u}_1 = \frac{10 u_1}{\alpha_m T^+ L}, \quad \hat{u}_2 = \frac{100 H u_2}{\alpha_m T^+ L^2}, \quad \hat{\sigma}_{11} = \frac{10 \sigma_{11}}{E_m \alpha_m T^+},$$
$$\hat{\sigma}_{12} = \frac{100 L \sigma_{12}}{E_m \alpha_m T^+ H}, \quad \hat{\sigma}_{22} = \frac{100 L^2 \sigma_{22}}{E_m \alpha_m T^+ H^2},$$

for the thermal load. Consider a simply supported metal-ceramic plate with the metal (Al) taken as the matrix phase and the ceramic (SiC) taken as the particulate phase. The exact solution for displacements and stresses at specific points in the plate is compared with the CPT, the FSDT and the TSDT results in Fig. 1 for length-to-thickness ratio, L/H, ranging from 2 to 40. The effective material properties are obtained by the Mori-Tanaka scheme. The transverse deflection $\bar{u}_2(L/2, 0)$ of the plate centroid for the mechanical load predicted by the FSDT and the TSDT is in excellent agreement with the exact solution even for thick plates with $L/H < 10$, but the CPT solution exhibits significant error for thick plates due to shear deformation. The CPT and the FSDT give identical values of the longitudinal stress $\bar{\sigma}_{11}$ which deviates from the exact solution as the length-to-thickness ratio decreases. The TSDT gives accurate results for $\bar{\sigma}_{11}$ even for thick plates. When the plate is subjected to the thermal load, the deflection $\hat{u}_2$ and the longitudinal stress $\hat{\sigma}_{11}$ given by each one of the three plate theories are inaccurate for thick plates with $L/H < 10$. The through-the-thickness variations of the displacements and longitudinal stresses at points on the centroidal axis and the shear stress variation at an edge are depicted in Fig. 2 for a thick plate ($L/H = 5$). When subjected to the mechanical load, the TSDT overestimates the transverse deflection $\bar{u}_2$ at all points within the plate, the CPT underestimates it at all points, and the FSDT value of $\bar{u}_2$ is close to the average value given by the exact solution. It is clear that the thickness of the plate changes and the transverse normal strain is not uniform through the plate thickness. It should be noted that all three plate theories give very good values of the transverse shear stress. When the functionally graded thick plate with $L/H = 5$ is subjected to the temperature load, all three plate theories exhibit large errors due to the assumption of inextensibility of the normals to the midsurface.

In the above examples, we used a single homogenization scheme to estimate the effective properties for the entire plate. This approach is appropriate only for functionally graded plates that have the same microstructure everywhere. Reiter and Dvorak [7] performed detailed finite element studies of the response of simulated discrete models containing both skeletal and particulate microstructures and concluded that homogenized models of combined microstructures which employ only a single averaging method do not provide reliable agreements with the

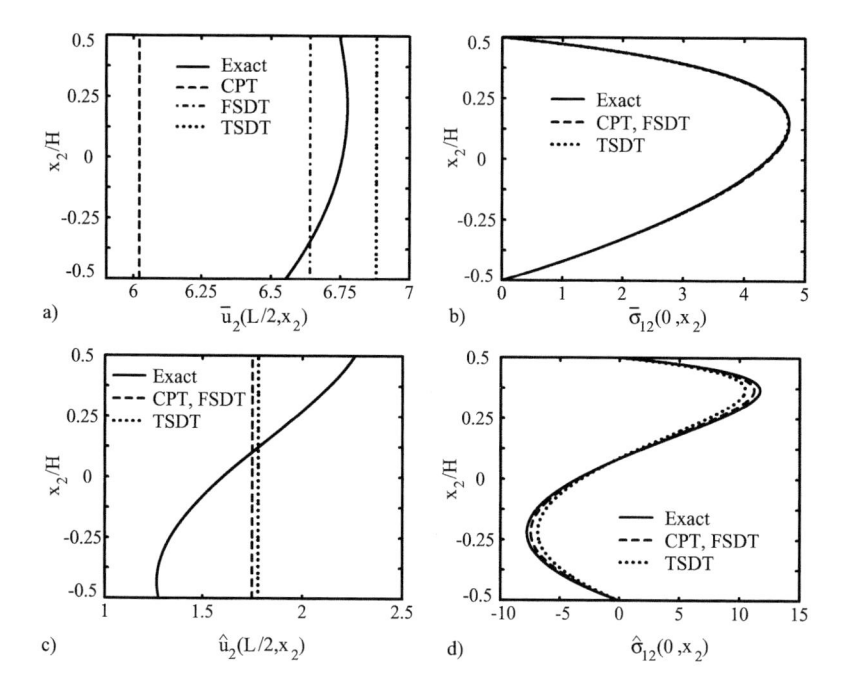

Figure 2: Through-the-thickness variation of the transverse deflection and stresses in the Al/SiC functionally graded plate computed with the Mori-Tanaka homogenization scheme and $V_c^- = 0$, $V_c^+ = 1$, $L/H = 5$, $n = 2$: (a,b) mechanical load and (c,d) thermal load.

discrete model predictions. However, close agreement with the discrete model was shown by homogenized models which employ different effective property estimates for regions of the plate that have different microstructures. We consider a functionally graded plate that has an affine variation of the ceramic volume fraction given by $V_c = 1/2 + x_2/H$. It is assumed to have a well-defined continuous metallic matrix with discontinuous ceramic particles in the metal-rich region $-0.5H \leq x_2 \leq -0.2H$ adjacent to the bottom surface and a well-defined continuous ceramic matrix with discontinuous metallic particles in the ceramic-rich region $0.2H \leq x_2 \leq 0.5H$ adjacent to the top surface. The plate is assumed to have a skeletal microstructure in the central region $-0.2H \leq x_2 \leq 0.2H$. We use a *combined model*, wherein the effective properties in the metal-rich region adjacent to the bottom surface are obtained by the Mori-Tanaka scheme with a metallic matrix phase (MTM), the effective properties in the ceramic-rich region adjacent to the top surface are obtained by the Mori-Tanaka scheme with a ceramic matrix phase (MTC) and the effective material properties in the central region are obtained by the self-consistent scheme (SC). To accommodate the

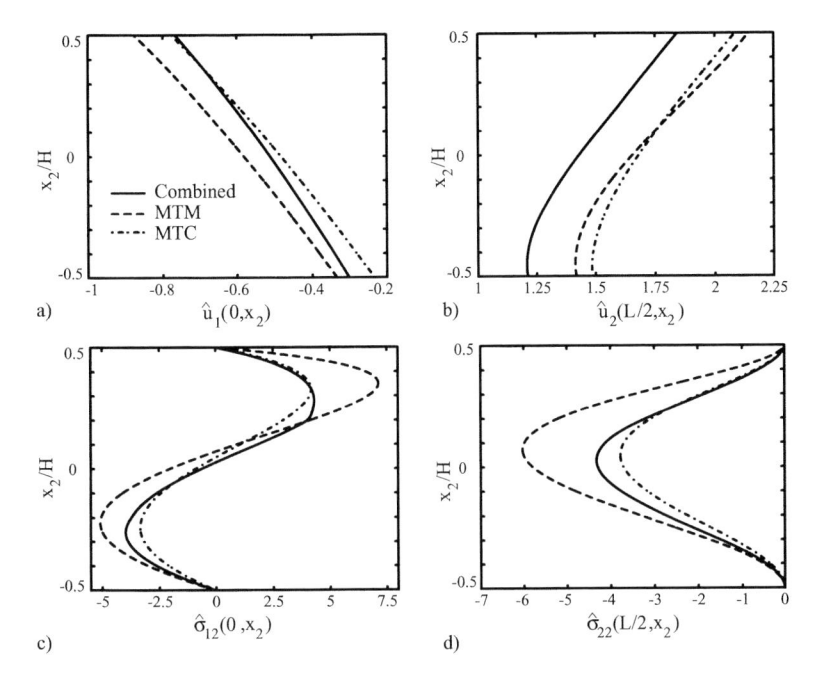

Figure 3: Through-the-thickness variation of the transverse deflection and stresses in the Al/SiC functionally graded plate obtained by various homogenization schemes; thermal load, $L/H = 5$.

discontinuities in homogenized material properties predicted at the boundaries between the different regions, we employ the third-order transition functions used by Reiter and Dvorak [7] in transition regions of width $0.05H$ centered at $x_2 = -0.2H$ and $x_2 = 0.2H$. A comparison of the through-the-thickness variation of the deflection and stresses obtained by the combined model and the two single averaging methods, namely MTM and MTC, are shown in Fig. 3 for a thick plate ($L/H = 5$) subjected to the thermal load. The MTM, the MTC and the combined model all give significantly different values for the transverse shear stress $\hat{\sigma}_{12}$ and the transverse normal stress $\hat{\sigma}_{22}$ for the thermal load.

6. CONCLUSIONS

We have analyzed thermomechanical deformations of a simply supported functionally graded Al/SiC plate subjected to either a sinusoidal pressure or a sinusoidal temperature field on the top surface. For thick functionally graded plates, there are significant differences between the exact solution and that obtained with any one of these three plate theories. It is found that the displacements, stresses, and temperatures computed with either the Mori-Tanaka scheme or the self-consistent method or their combination agree qualitatively but differ quantitatively.

7. REFERENCES

1. Rogers, T. G., Watson, P. and Spencer, A. J. M. (1995) Exact three-dimensional elasticity solutions for bending of moderately thick inhomogeneous and laminated strips under normal pressure, *International Journal of Solids and Structures* **32** 1659–1673.

2. Tarn, J. Q., and Wang, Y. M. (1995) Asymptotic thermoelastic analysis of anisotropic inhomogeneous and laminated plates, *Journal of Thermal Stresses* **18** 35–58.

3. Cheng, Z.Q., and Batra, R. C. (2000) Three-dimensional thermoelastic deformations of a functionally graded elliptic plate, *Composites: Part B* **31** 97–106.

4. Tanaka,K., Tanaka, Y., Watanabe, H., Poterasu, V. F., and Sugano, Y. (1993) An improved solution to thermoelastic material design in functionally gradient materials: Scheme to reduce thermal stresses, *Computer Methods in Applied Mechanics and Engineering* **109** 377–389.

5. Reddy, J. N. (2000) Analysis of functionally graded plates, *International Journal for Numerical Methods in Engineering* **47** 663–684.

6. Reiter, T., and Dvorak, G. J. (1997) Micromechanical modelling of functionally graded materials, in *IUTAM Symposium on Transformation Problems in Composite and Active Materials*, edited by Bahei-El-Din, Y., and Dvorak, G. J., Kluwer academic publishers, London, pp. 173-184.

7. Reiter, T., and Dvorak, G. J. (1998) Micromechanical models for graded composite materials: II.Thermomechanical loading, *Journal of the Mechanics and Physics of Solids* **46** 1655–1673.

8. Cheng, Z. Q., and Batra, R. C. (2000) Deflection relationships between the homogeneous Kirchhoff plate theory and different functionally graded plate theories, *Archives of Mechanics* **52** 143–158.

9. Mori, T. and Tanaka, K. (1973) Average stress in matrix and average elastic energy of materials with misfitting inclusions, *Acta Metallurgica* **21** 571–574.

10. Hill, R. (1965) A self-consistent mechanics of composite materials, *Journal of the Mechanics and Physics of Solids* **13** 213–222.

11. Hyer, M. W. (1998) *Stress Analysis of Fiber-Reinforced Composite Materials*, McGraw-Hill Higher Education.

12. Whitney, J. M., and Pagano, N. J. (1970) Shear deformation in heterogeneous anisotropic plates, *Journal of Applied Mechanics* **37** 1031–1036.

13. Reddy, J. N. (1997) *Mechanics of Laminated Composite Plates: Theory and Analysis*, CRC Press, Inc.

14. Hatta, H., and Taya, M. (1985) Effective thermal conductivity of a misoriented short fiber composite, *Journal of Applied Physics* **58** 2478–2486.

15. Rosen, B. W., and Hashin, Z. (1970) Effective thermal expansion coefficients and specific heats of composite materials, *International Journal of Engineering Science* **8** 157–173.

16. Hashin, Z. (1968) Assessment of the self consistent scheme approximation: Conductivity of particulate composites, *Journal of Composite Materials* **2** 284–300.

SHAPE MEMORY: HETEROGENEITY AND THERMODYNAMICS

DAVID VOKOUN AND VRATISLAV KAFKA
Institute of Theoretical and Applied Mechanics,
Academy of Sciences of the Czech Republic
E-mail: davidvokoun@yahoo.com.tw

1. INTRODUCTION

Shape memory (SM) of some mechanical systems is a very useful and widely used property. In a broader sense, this term relates not only to the special 'shape memory materials', but also to all elastic bodies, springs, bi-metals, and generally to many mechanical actuators, such as thermostats and others. The aim of our contribution is to show that all these systems or materials have one common factor: at least one elastic or thermoelastic continuous substructure. Some of these systems are composed of only thermoelastic substructures, e.g. of two thermoelastic materials in the case of bi-metals. In such a case the system works without dissipation of mechanical energy. On the other hand, if taking into consideration the spongy bone, tendons or cartilage, the viscous constituent, that ensures very useful dissipation of energy in cases of dangerous impacts, is present in the material. But even here, the material contains the elastic substructures composed of collagen and elastin that ensure SM. Similar situation is found in some shock-absorbers.

In the cases mentioned above the factor of an elastic substructure is clear enough. It is not so clear in the special SM materials. But even here – as will be shown in what follows – an elastic substructure is present as the factor ensuring the property of memory of shape.

2. NANOSCOPIC HETEROGENEITY IN METALLIC BINARY SM-ALLOYS

2.1 General outline

It is taken for granted that it is not by chance that significant shape memory (SM) is observed in binary and ternary alloys. Their nanoscopic heterogeneity seems to play the essential role in their SM manifestation. There are two scales of heterogeneity in polycrystals of these materials: the nanoscopic heterogeneity of the two or three different sets of atoms of the elements forming the alloy, and that of the subvolumes of the parent phase and martensite. But it is clear enough that it is the nanoscopic heterogeneity which is essential for the SM phenomena, as they are observed also in single crystals. The mechanism of the nanoscopic heterogeneity leading to SM effects has been described by Kafka [1,2]. In the bcc–B2 parent phase (that is typical for these alloys – e.g. for nitinol) the first strongly bound atomic neighbors are Ti-Ni, the second weekly bound neighbors are Ni-Ni and Ti-Ti. The changes of the interatomic distances between the second neighbors with decreasing temperature and with increasing deformation are schematically shown in Fig.1. In the austenitic state, the situation with the respective interatomic forces resulting from different interatomic potentials is shown in Fig.2. Fig.3 shows the analogous situation in the martensitic state. Transition from one state to the other one as a consequence of different shifts of the interatomic potentials due to changing temperature (different thermal dilatations) is straightforward. It is clear from these figures that in the course of the transition, the changes are assumed to be only elastic in the Ti-substructure, whereas in the Ni-substructure the energetic barriers are overcome and the changes are also inelastic, accompanied by dissipation of energy of the interatomic forces. The characteristic features of this scheme are higher Young's modulus, lower strength and higher thermal dilatation of Ni compared with Ti.

Making full use of the results comprised in the monograph [2], we are going to outline the possibilities of our nanoscopic heterogeneous scheme.

2.2 Mathematical model

On the basis of the considerations shortly presented above, the nanoscopic atomic structure of SM binary alloys was described as a heterogeneous one in [1]. The resulting mathematical form of the respective model follows:

$$d\bar{e}_{ij} = \mu d\bar{s}_{ij} + v_n s_{ijn} d\lambda_n \quad , \qquad d\bar{\varepsilon} = \rho d\bar{\sigma} \tag{1}$$

$$ds_{ijn} = d\bar{s}_{ij} - \frac{[(v_t + v_n \omega \eta \tau)s_{ijn} - v_t s'_{ijn}]d\lambda_n/\mu - \omega \eta s'_{ijt} d\tau}{1 + \eta + v_n \omega \eta \tau / v_t} \tag{2}$$

$$ds'_{ijn} = ds_{ijn} - d\bar{s}_{ij} + (v_t s_{ijn} - s'_{ijn})d\lambda_n/\mu$$

$$ds'_{ijt} = v_n[(d\bar{s}_{ij} - ds_{ijn})/v_t - s_{ijn} d\lambda_n / \mu] \tag{3}$$

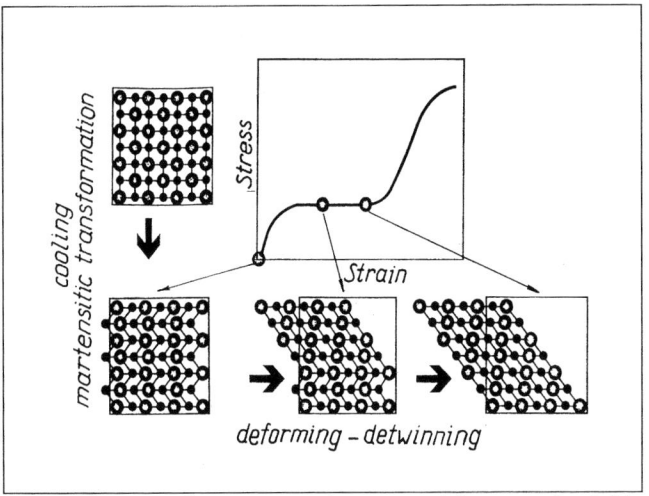

SHAPE MEMORY: HETEROGENEITY AND THERMODYNAMICS   441

Figure 1. Changes of interatomic distances of the second neighbors.

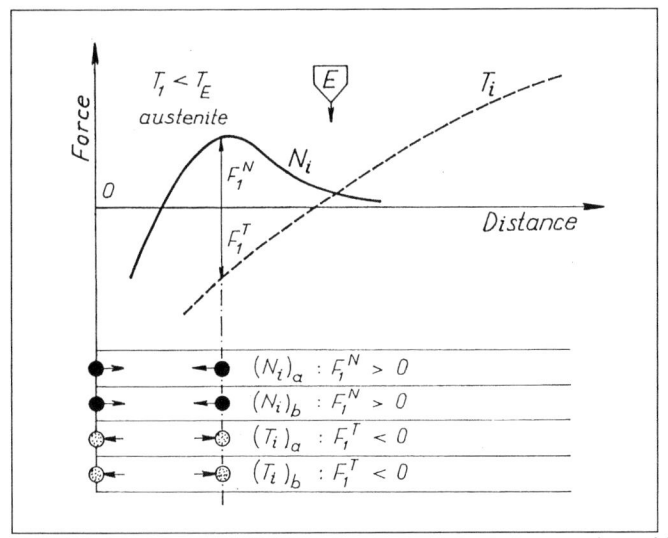

Figure 2. Austenitic state: Interatomic forces between second neighbors resulting from different interatomic potentials at an appropriate temperature.

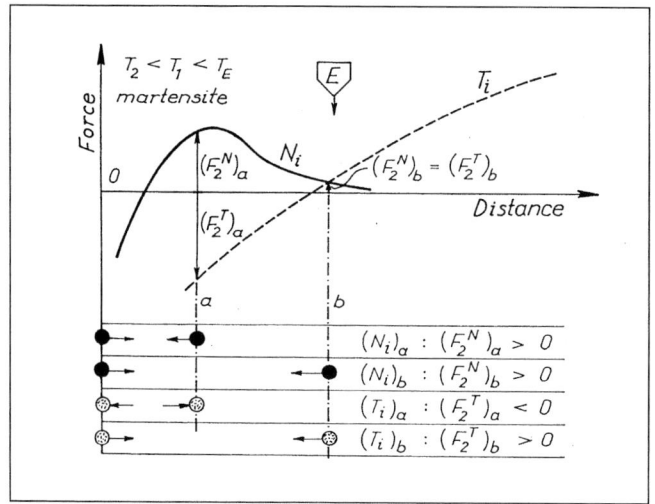

Figure 3. Martensitic state: Interatomic forces between second neighbors resulting from different interatomic potentials at a sufficiently low temperature.

where v_t and v_n mean volume fraction of t- and n- substructures, respectively, $\bar{s}_{ij}$ ($\bar{e}_{ij}$) is the deviatoric part of macroscopic stress (strain) tensor, $\bar{\sigma}$ ($\bar{\varepsilon}$) is macroscopic stress (strain) tensor, s_{ijt}, s_{ijn} are the deviatoric parts of the average stress tensor in the t-, n- material constituent, respectively. The symbols μ [= (1+ v)/E] and ρ [= (1-2v)/E] mean elastic compliances for the deviatoric and isotropic parts respectively, v is Poisson's ratio and E Young's modulus. The symbols with primes (e.g. s'_{ijt}) characterize the influence of the heterogeneity of strain- and stress-fields. $d\lambda_n$ is a scalar measure of plastic deformation in the n-constituent, $\tau = T - T_0$, T is the current temperature and T_0 is a reference temperature. Furthermore, η is a structural parameter and is a ω parameter to adjust.

Structural parameters η^e, η^n (here $\eta^e = \eta^n = \eta$) were deduced [2,3] as integral forms in distribution functions describing distribution of microstresses and microstrains under the influence of specific structures. The user of the model does not need to know the distribution functions, he works only with the structural parameters that are relatively easily determinable from experimental stress-strain diagrams [2,4].

For the inelastic deformation of the n-substructure, we assume validity of Mises' criterion valid for average stresses, which leads to:

$$d\lambda_n = 0 \qquad \text{for} \qquad s_{ijn} s_{ijn} < \frac{3}{2} c_n^2 \qquad (4)$$

$$d\lambda_n = \tfrac{1}{2}(d\tilde{\lambda}_n + |d\tilde{\lambda}_n|) \qquad \text{for} \qquad s_{ijn}s_{ijn} = \tfrac{3}{2}c_n^2 \qquad (5)$$

with $d\tilde{\lambda}_n$ following from the above equations in the form:

$$d\tilde{\lambda}_n = \mu \frac{(1+\eta+v_n\omega\eta\tau/v_t)s_{ijn}\,d\bar{s}_{ij} + \omega\eta\, s_{ijn}s'_{ijt}\,d\tau}{(v_t+v_n\omega\eta\tau)s_{ijn}s_{ijn} - v_t\, s_{ijn}s'_{ijn}}. \qquad (6)$$

2.3 Pseudoelasticity

In the case of pseudoelasticity, the process is isothermal, i.e. $d\tau = 0$, the influence of the temperature is represented by τ. In Fig.4 two theoretical pseudoelastic stress-strain diagrams are shown in confrontation with experimental results (Kafka, 2001).

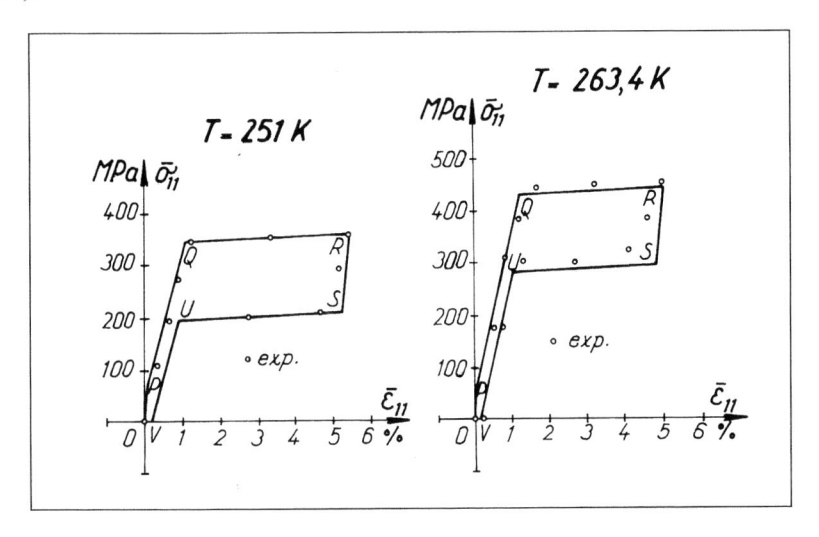

Figure 4. Pseudoelastic stress-strain diagrams.

2.4 Shape memory

Let us comment on a stress-strain diagram in tension test, and subsequent recovery of strain due to heating. The stress-strain diagram in martensitic state results from the above equations if putting $\tau = d\tau = 0$, $d\bar{\sigma}_{11} > 0$. The resulting theoretical diagram in confrontation with the experimental one are shown in Fig.5. A stress-free heating leads to recovery of strain and is described if putting $\bar{\sigma}_{ij} = d\bar{\sigma}_{ij} = 0$, $d\tau > 0$. The respective temperature-strain diagram is shown in Fig.6.

2.5 Changes of elastic moduli

In the preceding demonstrations of inelastic processes, the differences in elastic moduli between martensitic and austenitic states were neglected for simplicity of the model. However, the fact that elastic moduli are different, can very clearly be demonstrated using our heterogeneous scheme. In Figs.2 and 3 the distances between atoms can change due to changes of temperature and/or of mechanical

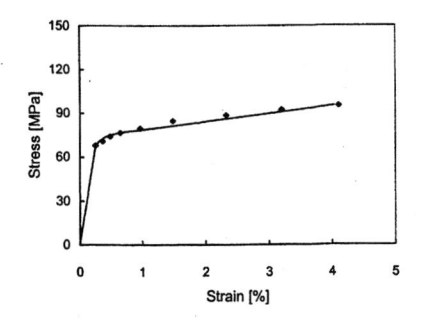

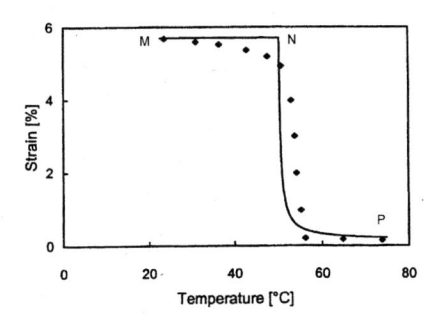

Figure 5. Martensitic stress-strain diagram.

Figure 6. Recovery of strain due to heating.

loading. In the case of temperature changes, the respective changes of interatomic distances are different between Ti-Ti pairs and Ni-Ni pairs due to different thermal dilatation rate. In the case of isothermal mechanical loading, the changes of distances between these pairs are nearly identical. Let us consider now mechanical loading only. It can result in an increase or decrease of the distances between these pairs. It is seen in Fig.2 that in the austenitic state in all atomic pairs a virtual increase of the atomic distances leads to a virtual increase of tensile interatomic forces and decrease of the absolute values compressive forces. A decrease of distances leads to a decrease of tensile forces and increase of the absolute values of compressive forces. The rate of the changes is the same in tension as well as compression, it differs in the sign only. If taking into account Fig.3, i.e. martensitic state, it is different. There are two different distances between pairs of atoms of the same element. As to the Ti-Ti pairs, the effect of a mechanical loading is qualitatively the same as in the austenitic state. But in the case of Ni-Ni pairs, the signs of changes of interatomic forces are opposite for the Ni-Ni pairs with smaller distances to those of the larger distances. These opposite signs mean that the resistance of the Ni substructure to loading is substantially lower than in the case of the austenitic configuration. This explains the low elastic moduli of martensite.

2.6 Conclusions relating to binary SM-alloys

The above-presented survey demonstrates that the scheme based on a description of heterogeneity on the atomic scale is able to explain and describe all the fundamental features observed on binary SM-alloys, and can be considered therefore a realistic explanation of the SM-phenomena.

3. MICROSCOPIC HETEROGENEITY IN MICA GLASS-CERAMICS

Materials from the group of the viscoelastic shape memory ceramics mostly consist of two infrastructures with different viscosities but approximately equal shear moduli. The well-known materials from this group are mica-glass ceramics. (Other viscoelastic shape memory ceramics are β-spodumene glass-ceramics, $2ZnO-B_2O_3$, glass ceramics and variety of sintered ceramics that contain very little glass phase. [5])

Mica ($KMg_3AlSi_3O_{10}F_2$) forms the crystalline phase in the glassy amorphous matrix. If the material is deformed at temperatures above 300°C, the deformed shape is retained on subsequent cooling to room temperature. If the deformation is less than 0.5%, the original shape is gradually restored on annealing at high temperatures (up to 900°C). The mechanism of the effect was studied by Schurch and Ashbee [6] and Itoh et al. [7]. The effect was explained in [6] as follows:

During a small deformation at a temperature above 300°C, the mica crystals embedded in the rigid matrix are deformed plastically by slip in their basal planes. The plastic strain in the crystalline constituent is accommodated elastically by the surrounding matrix. On subsequent cooling to room temperature and the removal of any external force, the elastic strain energy is stored in the matrix because the reversal of the plastic strains in the mica, that would be required in order to release the elastic strain in the matrix completely, is inhibited at low temperatures (At low temperatures, the dislocation glide in mica is not possible). This elastic energy becomes a driving force for restoring the original undeformed shape on re-heating when dislocation glide in the mica phase to release the stored elastic energy is triggered.

The factor of microscopic heterogeneity plays an important role in the SM phenomena in mica glass-ceramics.

4. MICROSCOPIC HETEROGENEITY IN FE - 28.8 AT.% PD SMA MELT-SPUN RIBBON

An example of Fe-28.8 at.% Pd SMA is given here to demonstrate the two-way shape memory effect (TWSME) of the as-received Fe-Pd melt-spun ribbon due to a phenomenon that originates in the microscopic heterogeneity.

The Fe-Pd SMA exhibits SME that is related to the thermoelastic martensitic fcc-fct transformation. On further cooling below M_f, another non-thermoelastic fct-bct transformation proceeds.

The TWSME was observed in the as-received Fe-28.8 at.% Pd SMA melt-spun ribbons [8]. The Fe-28.8 at.% Pd ribbons curled into a roll-like shape on cooling while the contact side was at the inner side of the roll-like shape. On heating up, the ribbons grew straight. The dimensions of the test ribbon were 2.25×10^{-2} m and 4.0×10^{-5} m in length and thickness, respectively. Fig. 7 shows the diagram of the radius versus temperature in the first thermal cycle.

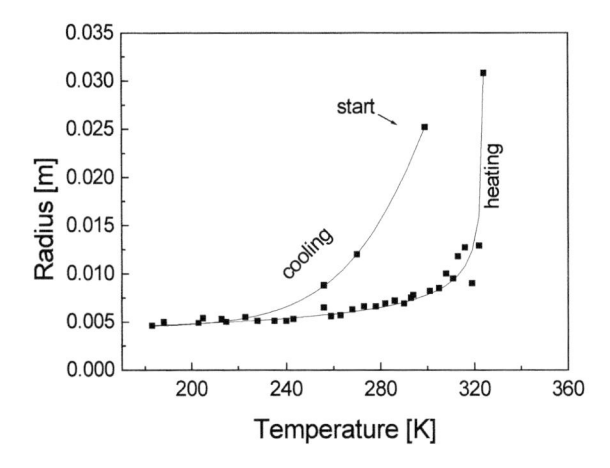

Figure 7. The radius of the test Fe-Pd ribbon versus temperature in the first thermal cycle.

A possible mechanism of the TWSME might be as follows. During processing the ribbons by the melt-spinning method, the quenching rate is not uniform. The contact side experiences higher quenching rates than the free surface area. Similarly, in the case of our Fe-28.8 at.% Pd melt-spun ribbons, different quenching rates along the thickness of the ribbon resulted in different surface morphologies when the contact and free side surfaces compared. It is probable, that the quenching rate influences M_s of the fct-bct transformation [9]. The higher the quenching rate, the lower M_s of the fct-bct transformation is. Consequently, during cooling down, when the TWSME was tested, the bct martensite was formed in the area close to the free surface more than in the area close to the contact side. It is known that due to a relatively large volume increase of the bct martensite, the fct-bct transformation in the Fe-Pd SMA is not thermoelastic. A volume expansion due to the fcc-bcc transformation in Fe-based alloys can reach up to 0.04 [10]. Thus, during cooling, the area close to the free surface has higher volume expansion than the area close to the contact side - due to the non-uniform distribution of the bct martensite. Therefore, bending of the investigated ribbon is caused by the increase of volume at the free surface, which corresponds to what was observed. During heating up, the bct phase does not disappear but the original shape is re-gained due to the SME. If the amount of the bct martensite becomes too high on cooling, the material is not able to recover the original shape.

Again, the factor of microscopic heterogeneity plays an important role in the TWSME in the as-received Fe-28.8 at.% Pd SMA melt-spun ribbons. In this case, the heterogeneity consists in the non-uniform distribution of the bct martensite in the ribbons.

5. HETEROGENEITY AND THERMODYNAMICS

There exist models that try to bypass the intricate structural analysis by using thermodynamics as a basis. As the laws of thermodynamics are generally observed as a very safe start for any theory, it is very often believed that it is also the safest basis for modeling SM phenomena. But there are reasons for doubts about such conviction.

Traditionally, phenomenological thermodynamics has been successfully applied in the problems of gases and liquids. However, applications in mechanics of solids, specifically in formulations of their constitutive equations, are not so straightforward. The essential difference between fluids and solids consists in the fact that in solids, a significant amount of mechanical energy can be stored on different structural scales. This stored energy can significantly affect the macroscopic deformation response to loading, which complicates the phenomenological applicability of the second law of thermodynamics. The 'thermodynamic theories' use different formulations of the second law of thermodynamics, which of course is not sufficient, and therefore, they accept some supplementary assumptions to fit the real behavior of the material in question. As a result, it is possible to receive some mathematical forms that are more or less appropriate for the description of some observed material properties. However, the seemingly safe thermodynamic part of the whole scheme is in reality very unsafe in the case of heterogeneous materials, and practically all technically important materials are heterogeneous on some scale, the SM polycrystalline materials on two scales.

The basis of the thermodynamic approaches is the second law of thermodynamics that can be expressed by the statement that *internal production of entropy cannot be negative*. There cannot be any doubt that this is true. In deformation processes, internal production of entropy is expressed by the *irreversible transformation of mechanical energy into heat,* called *dissipation.* Again, no doubts can exist that it is right. However, a direct thermal analysis of this transformation can be performed only by calorimetric methods, which are very complicated and their application can hardly be incorporated into the model that is to be used for constitutive modeling. For the practical use of information following from phenomenological thermodynamics, it is necessary to connect this information with phenomenological mechanics, i.e. with measurable quantities. Therefore, internal production of entropy is very often expressed by the macroscopically observed plastic work, or more generally, inelastic work. And this is the unsafe point. This is unsafe not only from the quantitative point of view, but also qualitatively.

The difference that can exist between the observed plastic work and the real dissipation is caused by stored energy that remains in the material e.g. after a deformation process and unloading. This problem was discussed by Kafka [11,12], where it was shown that according to different calorimetric measurements the amount of this stored energy currently reaches up to 15%, but exceptionally even up to 40% of the measurable plastic work. This means that some not negligible part of the measured plastic work is not dissipated.

Hence, it is problematic to approach heterogeneous (even microscopically heterogeneous) materials from the standpoints of phenomenological thermodynamics and to assume that the observed plastic deformation corresponds to dissipation and to the entropy production. But without using explicitly or implicitly this assumption, it is very difficult to receive some valuable information from the second law of thermodynamics.

6. CONCLUSION

In the above-presented survey, the factor of heterogeneity was shown to play an essential role in the SM phenomena in all the studied cases, and is considered therefore to be the best start-point for formulations of constitutive equations modeling this important property of some intelligent materials.

7. ACKNOWLEDGEMENT

The authors acknowledge the support of this work by the Grant Agency of the Academy of the Czech Republic, grant No.A2071101.

8. REFERENCES

1. Kafka, V. (1994) Shape Memory: New Concept of Explanation and of Mathematical Modelling. Part I: Micromechanical Explanation of the Causality in the SM Process. Part II: Mathematical Modelling of the SM Effect and Pseudoelasticity. J. of Intelligent Material Systems and Structures **5**, 809-824.
2. Kafka, V. (2001) Mesomechanical Constitutive Modeling, World Scientific, Singapore.
3. Kafka, V. (1987) Inelastic Mesomechanics, World Scientific.
4. Kafka, V. (1996) Plastic deformation under complex loading: General constitutive equation, Acta Technica CSAV **41**, 617-634.
5. Wei, Z.G., Sandström, R., Miyazaki, S. (1998) J. of Mater.Science **33**, 3743-3762.
6. Schurch, K.E., Ashbee, K.H.G. (1977) Nature **266**, 706-708.
7. Itoh, A., Miwa, Y., Iguchi, N. (1992). J. of the Japan Institute of Metals **56**, 700-706.
8. Vokoun, D. and Hu, C.T. (2002). Scripta Materialia **47**, 453-457.
9. Matsui, M., Yamada, H., Adachi, K. (1980) J. Phys. Soc. Japan **48**, 2161.
10. Magee, C.L., Davies, R.G. (1972) Acta Metal. **20**, 1031.
11. Kafka, V. (1974) Zur Thermodynamik der plastischen Verformung. ZAMM **54**, 649-657.
12. Kafka, V. (1979) Strain-hardening and stored energy. Acta Technica ČSAV **24**, 199-216.

COMPLEX VARIABLE SOLUTION OF PLANE PROBLEM FOR FUNCTIONALLY GRADED MATERIALS

XIANFENG WANG, and NORIO HASEBE
Department of Civil Engineering, Nagoya Institute of Technology
Gokiso-cho, Showa-ku, Nagoya 466-8555, Japan

1. INTRODUCTION

In recent years, there has been an extensive research interest on the functionally graded materials (FGMs). FGMs, unlike laminated or layered composites, provide continuously varying properties in a definable direction. From the viewpoint of continuum mechanics, FGMs are nonhomogeneous. Though numerous papers are concerned with design, fabrication, and simulation of FGMs [1,2], theoretical analysis is still lack due to the complexity of the problem. Erdogan and co-workers [3-5] studied the crack problem for the nonhomogeneous elastic medium under mechanical loadings. They assumed that the Poisson's ratio of the medium was constant and the Young's modulus E varied exponentially with the coordinate parallel to the crack. Noda, Jin and Batra [6,7] considered the temperature field and thermal stresses by further assuming exponential variations of thermal properties. In these works, the stress intensity factors were obtained using integral transformation and integral equation method.

It should be noted that not only cracks, other shapes of defects are also commonly present in FGMs during the practical manufacture or in service. However, up to now, there is few research on probing the effective method to the general problem for the nonhomogeneous medium, and few study is concerned with the defects of other shapes, such as circular or elliptical holes. It is well known that the complex function method [8] has particular advantages in the derivation of solutions of elastostatic problem for homogeneous materials. With the help of rational mapping functions [9,10], intricate shapes of boundary can be treated analytically. In this study, we use the complex stress function method to formulate the 2-dimensional problem for the nonhomogeneous material. The basic equations and the boundary conditions are obtained by means of two complex stress functions. The problem of a nonhomogeneous plane containing a circular hole subjected to a uniform displacement is studied using the obtained fundamental equations and the Cauchy integral technique.

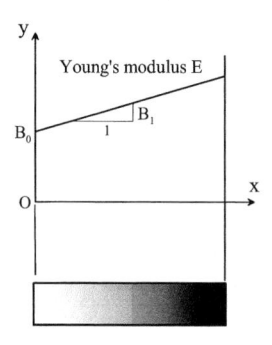

Figure 1. Young's modulus gradient ($E = B_0 + B_1 x$)

2. FORMULATION FOR FUNCTIONALLY GRADED MATERIAL

In the constitutive relations of plane problem

$$\varepsilon_x = \frac{1}{E}(\sigma_x - v\sigma_y), \; \varepsilon_y = \frac{1}{E}(\sigma_y - v\sigma_x), \; \varepsilon_{xy} = \frac{\tau_{xy}}{2G}, \; G = \frac{E}{2(1+v)} \qquad (1)$$

it is assumed that the Poisson's ratio of the material is constant and the Young's modulus E varies linearly with the coordinate x as (Fig. 1)

$$E = B_0 + B_1 x \qquad (2)$$

where B_0 and B_1 are the real constants. We have known the stress components can be expressed in terms of Airy stress function F as

$$\sigma_x = \frac{\partial^2 F}{\partial y^2}, \qquad \sigma_y = \frac{\partial^2 F}{\partial x^2}, \qquad \tau_{xy} = -\frac{\partial^2 F}{\partial x \partial y} \qquad (3)$$

The compatibility condition is

$$\frac{\partial^2 \varepsilon_x}{\partial y^2} + \frac{\partial^2 \varepsilon_y}{\partial x^2} = 2\frac{\partial^2 \varepsilon_{xy}}{\partial x \partial y} \qquad (4)$$

Substituting (1) and (3) into (4), we have

$$\frac{\partial^2}{\partial y^2}(\frac{1}{E}\frac{\partial^2 F}{\partial y^2}) + \frac{\partial^2}{\partial x^2}(\frac{1}{E}\frac{\partial^2 F}{\partial x^2}) + \frac{\partial^2}{\partial x \partial y}(\frac{2}{E}\frac{\partial^2 F}{\partial x \partial y})$$
$$-v\left[\frac{\partial^2}{\partial y^2}(\frac{1}{E}\frac{\partial^2 F}{\partial x^2}) + \frac{\partial^2}{\partial x^2}(\frac{1}{E}\frac{\partial^2 F}{\partial y^2}) - \frac{\partial^2}{\partial x \partial y}(\frac{2}{E}\frac{\partial^2 F}{\partial x \partial y})\right] = 0 \qquad (5)$$

Consider the following relation

$$\frac{\partial^2}{\partial x^2}\left(\frac{F}{E}\right) = \frac{\partial^2}{\partial x^2}(\frac{1}{E}) \cdot F + \frac{\partial}{\partial x}\left(\frac{2}{E}\right)\frac{\partial F}{\partial x} + \frac{1}{E}\frac{\partial^2 F}{\partial x^2} \qquad (6)$$

Note that $1/E$ can be expanded to the power series as

$$\frac{1}{E} = \frac{1}{(B_0 + B_1 x)} = \sum_{n=0}^{\infty} \frac{(-1)^n B_1^n x^n}{B_0^{n+1}} = \frac{1}{B_0} - \frac{B_1 x}{B_0^2} + \frac{B_1^2 x^2}{B_0^3} + \ldots \qquad (7)$$

with the radius of convergence being

$$|x| < \frac{B_0}{B_1} \qquad (8)$$

When we only concern the region in a small limit around the coordinate origin (e.g. stress concentration around a defect hole), or when $B_1/B_0 \ll 1$, the higher–order power (two and greater than two) terms of x in the series may be neglected in a good approximation. Hence term $\partial^2(1/E)/\partial x^2 \cdot F$ in (6) vanishes. The third and sixth terms in (5) can be written as

$$\frac{\partial^2}{\partial x \partial y}(\frac{2}{E}\frac{\partial^2 F}{\partial x \partial y}) = \frac{\partial^2}{\partial y^2}\left\{\frac{\partial}{\partial x}(\frac{2}{E}) \cdot \frac{\partial F}{\partial x} + \frac{2}{E}\frac{\partial^2 F}{\partial x^2}\right\} = \frac{\partial^2}{\partial x^2}\left(\frac{1}{E}\frac{\partial^2 F}{\partial y^2}\right) + \frac{\partial^2}{\partial y^2}\left(\frac{1}{E}\frac{\partial^2 F}{\partial x^2}\right) \qquad (9)$$

Substituting (9) into (5), we have

$$\frac{\partial^2}{\partial y^2}\left(\frac{1}{E}\frac{\partial^2 F}{\partial y^2}\right) + \frac{\partial^2}{\partial x^2}\left(\frac{1}{E}\frac{\partial^2 F}{\partial x^2}\right) + \frac{\partial^2}{\partial x^2}\left(\frac{1}{E}\frac{\partial^2 F}{\partial y^2}\right) + \frac{\partial^2}{\partial y^2}\left(\frac{1}{E}\frac{\partial^2 F}{\partial x^2}\right) = 0 \qquad (10)$$

which can be written in a simpler form as

$$\nabla^2\left(\frac{1}{E}\nabla^2 F\right) = 0 \qquad (11)$$

Note that (11) is valid for both the plane stress and plane strain problem. Introducing complex variables ($z = x + i \cdot y$) in the above equation, we have

$$4\frac{\partial^2}{\partial z \partial \bar{z}}\left(\frac{1}{E} \cdot 4\frac{\partial^2 F}{\partial z \partial \bar{z}}\right) = 0 \qquad (12)$$

where $\bar{z} = x - i \cdot y$. Young's modulus (2) becomes

$$E = B_0 + B_1 x = B_0 + B_1 \frac{z + \bar{z}}{2} \qquad (13)$$

The solution of (12) can be expressed by means of two stress functions $\varphi(z)$ and $\chi(z)$ as

$$2F = \overline{z\varphi(z)} + z\varphi(z) + \frac{B_1}{4B_0}z^2\overline{\varphi(z)} + \frac{B_1}{4B_0}\bar{z}^2\varphi(z)$$

$$+ \frac{B_1}{2B_0}\overline{zz\varphi(z)} + \frac{B_1}{2B_0}\bar{z}z\varphi(z) - \frac{B_1}{2B_0}z\overline{\int\varphi(z)dz} - \frac{B_1}{2B_0}\bar{z}\int\varphi(z)dz + \chi(z) + \overline{\chi(z)} \qquad (14)$$

When $B_1 = 0$, the Airy stress function is reduced to the one for the problem of homogeneous material. Further, the stress components are represented as

$$\sigma_x + \sigma_y = 4\frac{\partial^2 F}{\partial z \partial \bar{z}} = 2\left[1 + \frac{B_1}{2B_0}(z + \bar{z})\right]\left[\varphi'(z) + \overline{\varphi'(z)}\right] \qquad (15a)$$

$$\sigma_y - \sigma_x + 2i\tau_{xy} = 4\frac{\partial^2 F}{\partial \bar{z}^2}$$

$$= 2\left[\left(\bar{z}+\frac{B_1}{4B_0}\bar{z}^2+\frac{B_1}{2B_0}z\bar{z}\right)\varphi''(z)+\frac{B_1}{2B_0}\bar{z}\varphi'(z)+\frac{B_1}{2B_0}\overline{\varphi(z)}+\psi'(z)\right] \quad (15b)$$

where function $\psi(z)$ is defined by $\chi'(z) \equiv \psi(z)$. From this formula, it is seen that the constant term in $\varphi(z)$ has an influence on the stress components, which is quite different from the case of homogeneous material.

Consider an arc AB in the plane Oxy having the direction from A to B as its positive direction and the normal n pointing to the right when moving in the positive direction. It is assumed that the force exerting on the side of the positive normal has components $P_x ds$ and $P_y ds$ over an element of arc ds. Hence the resultant force acting to the arc AB is

$$i(X+iY) = \int_A^B i(P_x+iP_y)ds = \left[2\frac{\partial F}{\partial \bar{z}}\right]_A^B$$

$$= \left[\left(z+\frac{B_1}{4B_0}z^2+\frac{B_1}{2B_0}z\bar{z}\right)\overline{\varphi'(z)}+\left(1+\frac{B_1}{2B_0}\bar{z}+\frac{B_1}{2B_0}z\right)\varphi(z)+\overline{\psi(z)}\right]_A^B - \frac{B_1}{2B_0}\int_A^B \varphi(z)dz \quad (16)$$

where the brackets denote the change in the expression as z moves from A to B. This relation clearly extends to the case when the arc AB lies along the boundary, which implies that (16) represents the external force boundary condition. In a similar manner, the resultant moment about the origin of the coordinate system is obtained as

$$M = \int_A^B (x \cdot P_y + y \cdot P_x)ds = -\text{Re}\left[z \cdot \frac{\partial(2F)}{\partial z}\right]_A^B + [F]_A^B$$

$$= \text{Re}\left[-\left(z\bar{z}+\frac{B_1}{4B_0}z\bar{z}^2+\frac{B_1}{2B_0}z^2\bar{z}\right)\varphi'(z)-\frac{B_1}{4B_0}\bar{z}^2\varphi(z)+\int\psi(z)dz - z\psi(z)\right]_A^B \quad (17)$$

Next, we deal with the displacements corresponding to a certain stress state. From the definitions of the strain components, we have

$$2G(\varepsilon_y - \varepsilon_x + 2i\varepsilon_{xy}) = -4G\frac{\partial}{\partial z}(u-iv) \quad (18a)$$

$$2G(\varepsilon_x + \varepsilon_y) = 2G\left[\frac{\partial}{\partial z}(u+iv)+\frac{\partial}{\partial \bar{z}}(u-iv)\right] \quad (18b)$$

where

$$G = \frac{E}{2(1+\nu)} = \frac{B_0+B_1 x}{2(1+\nu)} = \frac{2B_0+B_1 z+B_1 \bar{z}}{4(1+\nu)} \quad (19)$$

Note that the plane stress and plane strain problems are considered separately because of the difference in the constitutive conditions. For the plane stress problem, combining (15) with (18) yields

$$\frac{(2B_0+B_1 z+B_1 \bar{z})}{2(1+\nu)}\frac{\partial}{\partial \bar{z}}(u+iv) = -2\frac{\partial^2 F}{\partial \bar{z}^2} \quad (20a)$$

$$2G\left[\frac{\partial}{\partial \bar{z}}(u+iv)+\frac{\partial}{\partial z}(u-iv)\right]=\frac{4(1-v)}{(1+v)}\frac{\partial^2 F}{\partial z \partial \bar{z}} \quad (20b)$$

Integrating (20a) with respect to $\bar{z}$ and using the approximation (7), we obtain

$$(u+iv)=-2(1+v)\left[\frac{1}{B_0}-\frac{B_1(z+\bar{z})}{2B_0^2}\right]\frac{\partial F}{\partial \bar{z}}-\frac{2(1+v)B_1}{2B_0^2}F+H(z) \quad (21)$$

where H(z) is an arbitrary function. Differentiating (21) with respect to z and taking its conjugate form (F is a real function) leads to

$$\frac{\partial}{\partial z}(u-iv)=-\frac{\partial^2 F}{\partial z \partial \bar{z}}2(1+v)\left[\frac{1}{B_0}-\frac{B_1(z+\bar{z})}{2B_0^2}\right]+\frac{2(1+v)B_1}{2B_0^2}\frac{\partial F}{\partial z}-\frac{2(1+v)B_1}{2B_0^2}\frac{\partial F}{\partial \bar{z}}+\overline{H'(z)}$$
(22)

Substituting (21) and (22) into (20b), we have

$$\left[B_0+\frac{B_1}{2}(z+\bar{z})\right]\{H'(z)+\overline{H'(z)}\}=8\frac{\partial^2 F}{\partial z \partial \bar{z}}$$

$$=4\left[1+\frac{B_1}{2B_0}(z+\bar{z})\right]\left[\varphi'(z)+\overline{\varphi'(z)}\right] \quad (23)$$

Thus, H(z) is obtained as

$$H(z)=\frac{4}{B_0}\varphi(z)+ic_1 z+c_2 \quad (24)$$

where c_1 and c_2 are a real and complex constants, respectively, and are related to the rotation and translation of the rigid body, respectively. Since the rigid movement does not affect stresses, we can treat them as naught herein. In like manner, the displacements for the plane strain problem can be determined as well. Thus, the displacements for the both plane problems can be expressed in a unified form as

$$\frac{(u+iv)}{(1+v)}=-2\frac{\partial F}{\partial \bar{z}}\left[\frac{1}{B_0}-\frac{B_1(z+\bar{z})}{2B_0^2}\right]-\frac{B_1}{2B_0^2}2F+\frac{4\Gamma}{(1+v)B_0}\varphi(z)$$

$$=\left[\frac{B_1 z^2}{4}\left(\frac{1}{B_0}-\frac{B_1(z+\bar{z})}{2B_0^2}\right)-z\right]\frac{\overline{\varphi'(z)}}{B_0}-\frac{B_1}{2B_0^3}\left(B_0 z+\frac{B_1}{4}z^2+\frac{B_1}{2}z\bar{z}\right)\overline{\varphi(z)}$$

$$+\left[-\frac{B_1}{2B_0^2}\left(B_0\bar{z}+\frac{B_1}{4}\bar{z}^2+\frac{B_1}{2}z\bar{z}\right)-1+\frac{4\Gamma}{(1+v)}\right]\frac{\varphi(z)}{B_0}$$

$$+\left[\frac{B_1}{2B_0^2}-\frac{B_1^2 z}{4B_0^3}\right]\int\varphi(z)dz+\frac{B_1^2 \bar{z}}{4B_0^3}\int\overline{\varphi(z)}dz-\left[\frac{1}{B_0}-\frac{B_1(z+\bar{z})}{2B_0^2}\right]\overline{\psi(z)}-\frac{B_1}{2B_0^2}\left[\chi(z)+\overline{\chi(z)}\right]$$
(25)

where

$\Gamma=1$ for the plane stress problem
$\Gamma=1-v^2$ for the plane strain problem

454   X. F. WANG, N. HASEBE

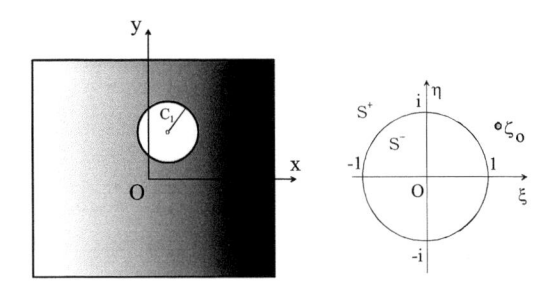

Figure 2. Wide plate containing a circular hole and unit circle

3. WIDE PLATE CONTAINING A CIRCULAR HOLE SUBJECTED TO UNIFORM DISPLACEMENT IN THE Y-DIRECTION

So far, we have obtained the expressions of stresses and displacements by means of complex stress functions. The associated boundary value problems may be solved using series or Cauchy integral based on this form. As an example, we solve the problem of a circular hole in an infinite plate subjected to a uniform displacement in the y-direction. As shown in Fig. 2, the circular hole is located near the origin of the coordinate system, and its radius is C_1. With the mapping function

$$z = \omega(\zeta) = C_1\zeta + C_2 \tag{26}$$

the circular hole in the z-plane is transformed to the unit circle in the ζ-plane. Here C_2 is a complex constant indicating the coordinate of the hole. To simulate the loading condition, suppose that the far field has the following properties

$$\varepsilon_y = 4\rho \ (v = 4\rho \cdot y), \qquad \sigma_x = \tau_{xy} = 0 \tag{27}$$

where ρ is the real constant. The corresponding stress functions are obtained as

$$\varphi_1(\zeta) = \rho B_0 z = \rho B_0 \omega(\zeta),$$

$$\psi_1(\zeta) = 2B_0\rho z + \frac{1}{2}B_1\rho z^2 = 2B_0\rho\omega(\zeta) + \frac{1}{2}B_1\rho\omega(\zeta)^2 \tag{28}$$

The stress functions to be obtained can be expressed in two parts as:

$$\varphi(\zeta) = \varphi_1(\zeta) + \varphi_2(\zeta), \qquad \psi(\zeta) = \psi_1(\zeta) + \psi_2(\zeta). \tag{29}$$

where $\varphi_1(\zeta)$ and $\psi_1(\zeta)$ are those in (28), $\varphi_2(\zeta)$ and $\psi_2(\zeta)$ are the complementary parts resulting from the boundary effect of the hole. Introducing the mapping function, the traction-free boundary condition becomes

$$\left(\omega(\sigma) + \frac{B_1}{4B_0}\omega(\sigma)^2 + \frac{B_1}{2B_0}\omega(\sigma)\overline{\omega(\sigma)}\right)\frac{\overline{\varphi'(\sigma)}}{\overline{\omega'(\sigma)}}$$

$$+ \left(1 + \frac{B_1}{2B_0}\overline{\omega(\sigma)} + \frac{B_1}{2B_0}\omega(\sigma)\right)\varphi(\sigma) - \frac{B_1}{2B_0}\int\varphi(\sigma)\omega'(\sigma)d\sigma + \overline{\psi(\sigma)} = 0 \tag{30}$$

where σ denotes the point on the unit circle. Substituting (28) and (29) into (30), multiplying it by $d\sigma/[2\pi i(\sigma-\zeta)]$ where ζ is located outside the unit circle, and integrating along the boundary in the counter-clockwise direction with regarding $\overline{\sigma}=1/\sigma$, we obtain

$$\left[-1-\frac{B_1}{2B_0}(\frac{C_1}{\zeta}+\overline{C}_2)-\frac{B_1}{2B_0}(C_1\zeta+C_2)\right]\varphi_2(\zeta)+\frac{B_1C_1}{2B_0}\int\varphi_2(\zeta)d\zeta$$

$$=\frac{\rho B_1 C_1^2}{2\zeta^2}+\left(B_1C_1C_2+2B_0C_1+B_1C_1\overline{C}_2\right)\frac{\rho}{\zeta}+const. \qquad (31)$$

Making the first-order derivative with respect to ζ yields

$$\varphi_2'(\zeta)+P(\zeta)\varphi_2(\zeta)=Q(\zeta) \qquad (32)$$

where

$$P(\zeta)=-\frac{B_1C_1}{2B_0\zeta^2}\frac{1}{\left[1+\frac{B_1}{2B_0}(\frac{C_1}{\zeta}+\overline{C}_2)+\frac{B_1}{2B_0}(C_1\zeta+C_2)\right]} \qquad (33a)$$

$$Q(\zeta)=\frac{\frac{\rho B_1 C_1^2}{\zeta^3}+\frac{\rho}{\zeta^2}\left(B_1C_1C_2+2B_0C_1+B_1C_1\overline{C}_2\right)}{\left[1+\frac{B_1}{2B_0}(\frac{C_1}{\zeta}+\overline{C}_2)+\frac{B_1}{2B_0}(C_1\zeta+C_2)\right]} \qquad (33b)$$

The solution of the ordinary-differential equation is expressed as

$$\varphi_2(\zeta)=e^{-\int P(\zeta)d\zeta}\left\{\int e^{\int P(\zeta)d\zeta}Q(\zeta)d\zeta+A\right\} \qquad (34)$$

where

$$e^{\int P(\zeta)d\zeta}=$$

$$\zeta^{\beta_1}\left(\zeta+\frac{B_0+B_1\operatorname{Re}[C_2]-\sqrt{(B_0+B_1\operatorname{Re}[C_2])^2-B_1^2C_1^2}}{B_1C_1}\right)^{\beta_2}\left(\zeta+\frac{B_0+B_1\operatorname{Re}[C_2]+\sqrt{(B_0+B_1\operatorname{Re}[C_2])^2-B_1^2C_1^2}}{B_1C_1}\right)^{\beta_3}$$

(35)

with

$$\beta_1=-1,\ \beta_2=\frac{B_0+B_1\operatorname{Re}[C_2]+\sqrt{(B_0+B_1\operatorname{Re}[C_2])^2-B_1^2C_1^2}}{2\sqrt{(B_0+B_1\operatorname{Re}[C_2])^2-B_1^2C_1^2}},$$

$$\beta_3=\frac{-(B_0+B_1\operatorname{Re}[C_2])+\sqrt{(B_0+B_1\operatorname{Re}[C_2])^2-B_1^2C_1^2}}{2\sqrt{(B_0+B_1\operatorname{Re}[C_2])^2-B_1^2C_1^2}} \qquad (36)$$

where Re[ ] denotes the real part of the function in the bracket. The constant A can be determined from the condition that $\varphi_2(\zeta)$ is holomorphic in S^+ and the induced stress condition is zero at infinity. It is obvious that $\varphi_2(\zeta)$ will be holomorphic and be zero at infinity if, and only if, $A=0$. Therefore $\varphi_2(\zeta)$ is obtained

$$\varphi_2(\zeta) =$$

$$\zeta\left(\zeta + \frac{B_0 + B_1\operatorname{Re}[C_2] - \sqrt{(B_0 + B_1\operatorname{Re}[C_2])^2 - B_1^2 C_1^2}}{B_1 C_1}\right)^{-\beta_2}$$

$$\left(\zeta + \frac{B_0 + B_1\operatorname{Re}[C_2] + \sqrt{(B_0 + B_1\operatorname{Re}[C_2])^2 - B_1^2 C_1^2}}{B_1 C_1}\right)^{-\beta_3}$$

$$\left\{\int \frac{(\rho C_2 + 2\rho B_0/B_1 + \rho\overline{C}_2)\cdot\zeta + \rho C_1}{\frac{\zeta^3}{2B_0}\left(\zeta + \frac{B_0 + B_1\operatorname{Re}[C_2] - \sqrt{(B_0 + B_1\operatorname{Re}[C_2])^2 - B_1^2 C_1^2}}{B_1 C_1}\right)^{\beta_3}\left(\zeta + \frac{B_0 + B_1\operatorname{Re}[C_2] + \sqrt{(B_0 + B_1\operatorname{Re}[C_2])^2 - B_1^2 C_1^2}}{B_1 C_1}\right)^{\beta_2}} d\zeta\right\}$$

(37)

Note that the integral in the brace in (37) can be evaluated in terms of hypergeometric functions.

Next function $\psi_2(\zeta)$ is determined. The conjugate form of boundary condition (30) is expressed as

$$\left\{\frac{B_1 C_1^2}{4B_0}\frac{1}{\sigma^2} + \left(1 + \frac{B_1\operatorname{Re}[C_2]}{B_0}\right)\frac{C_1}{\sigma} + \left(\overline{C}_2 + \frac{B_1\overline{C}_2^2}{4B_0} + \frac{B_1 C_2\overline{C}_2}{2B_0} + \frac{B_1 C_1^2}{2B_0}\right) + \frac{B_1 C_1\overline{C}_2}{2B_0}\sigma\right\}\frac{\varphi_2'(\sigma)}{C_1}$$

$$+\left\{\frac{B_1 C_1}{2B_0}\frac{1}{\sigma} + \left(1 + \frac{B_1\operatorname{Re}[C_2]}{B_0}\right) + \frac{B_1 C_1}{2B_0}\sigma\right\}\overline{\varphi_2(1/\sigma)} - \frac{B_1}{2B_0}\overline{\int\varphi_2(\sigma)d\sigma} + \psi_2(\sigma)$$

$$= -\frac{B_1\rho}{2}\omega(\sigma)^2 - B_1\rho\omega(\sigma)\cdot\overline{\omega(\sigma)} - \frac{B_1\rho}{2}\overline{\omega(\sigma)}^2 - 2B_0\rho\omega(\sigma) - 2B_0\rho\overline{\omega(\sigma)} \quad (38)$$

Using Cauchy integral on (38), $\psi_2(\zeta)$ is readily obtained as

$$\psi_2(\zeta) = -\left[\frac{B_1 C_1}{4B_0\zeta^2} + \frac{B_0 + B_1\operatorname{Re}[C_2]}{B_0\zeta} + \frac{1}{C_1}\left(\overline{C}_2 + \frac{B_1\overline{C}_2^2}{4B_0} + \frac{B_1 C_2\overline{C}_2}{2B_0} + \frac{B_1 C_1^2}{2B_0}\right) + \frac{B_1\overline{C}_2}{2B_0}\zeta\right]\varphi_2'(\zeta)$$

$$- \frac{\rho B_1 C_1^2}{2\zeta^2} - \frac{\rho B_1 C_1 C_2 + 2\rho B_0 C_1 + \rho B_1 C_1\overline{C}_2}{\zeta} \quad (39)$$

With the stress functions (28), (37) and (39) as well as the mapping function, the stress components are obtained by the following formula

$$\sigma_r + \sigma_\theta = \sigma_x + \sigma_y = 4\left\{1 + \frac{B_1}{B_0}\operatorname{Re}[\omega(\zeta)]\right\}\cdot\operatorname{Re}\left[\frac{\varphi'(\zeta)}{\omega'(\zeta)}\right] \quad (40a)$$

$$\sigma_\theta - \sigma_r + 2i\tau_{r\theta} = e^{2i\lambda}\left(\sigma_y - \sigma_x + 2i\tau_{xy}\right) = \frac{\zeta^2\omega'(\zeta)}{|\zeta|^2\overline{\omega'(\zeta)}}$$

$$2\left[\left(\overline{\omega(\zeta)}+\frac{B_1}{4B_0}\overline{\omega(\zeta)}^2+\frac{B_1}{2B_0}\omega(\zeta)\overline{\omega(\zeta)}\right)\frac{1}{\omega'(\zeta)}\left(\frac{\varphi'(\zeta)}{\omega'(\zeta)}\right)'_\zeta+\frac{B_1}{2B_0}\overline{\omega(\zeta)}\frac{\varphi'(\zeta)}{\omega'(\zeta)}+\frac{B_1}{2B_0}\overline{\varphi(\zeta)}+\frac{\psi'(\zeta)}{\omega'(\zeta)}\right]$$
(40b)

The elastic modulus E is a function of x [see (2)], therefore the solution depends on the coordinate system and the position of the circular hole. Regarding the expressions of FGMs, it can be seen that the stress state is the same as the homogeneous one when $B_1/B_0=0$.

In the case of $C_2=(0+i\ 0)$, i.e. the circular hole is located at the origin of the coordinate system, we can arrange the preceding formula as

$$P(\zeta)=f_1(B_0,B_1C_1,\zeta),\ P'(\zeta)=f_2(B_0,B_1C_1,\zeta) \qquad (41)$$
$$Q(\zeta)=C_1f_3(B_0,B_1C_1,\zeta),\ Q'(\zeta)=C_1f_4(B_0,B_1C_1,\zeta)$$
$$\beta_i=f_{3+i}(B_0,B_1C_1)\quad (i=2,3)$$

Further, we have,

$$\varphi(\zeta)=C_1f_7(B_0,B_1C_1,\zeta),\ \varphi'(\zeta)=C_1f_8(B_0,B_1C_1,\zeta),\ \varphi''(\zeta)=C_1f_9(B_0,B_1C_1,\zeta)$$
$$\psi(\zeta)=C_1f_{10}(B_0,B_1C_1,\zeta),\ \psi'(\zeta)=C_1f_{11}(B_0,B_1C_1,\zeta) \qquad (42)$$

where f_j (j=1~11) are functions of ζ dependent of the coefficients B_0 and B_1C_1. Regarding $\omega(\zeta)=C_1\zeta$ as well as (40), the stresses are readily expressed in terms of B_0 and B_1C_1 among three coefficients B_0, B_1 and C_1, i.e. the same stress components are obtained for the same values of B_0 and B_1C_1.

To verify the validity of the suggested approach, we investigate the circular hole at different positions by changing the values of C_2. To maintain the same stress field corresponding to the hole position, we adjust the coefficient B_0 as B_0=200 GPa and 160 GPa for Re[C_2]=0 and 4 cm, respectively; and $\rho = 1$, $B_1 = 10$ GPa/cm and $C_1 = 0.4$ cm are chosen for the computations. The results of circular stress ($\sigma_\theta/4\rho$) at angles 0, $\pi/2$ and π on the hole boundary are shown in Table 1. Naturally, it is seen that different sets of coefficients B_0 and C_2 lead to the same values of σ_θ.

To compare the results for the homogeneous material, we also study the problem of the same geometry and loading condition at the far field for the homogeneous material. The obtained stress functions $\varphi_h(\zeta)$ and $\psi_h(\zeta)$ for the homogeneous material are as

$$\varphi_h(\zeta)=B_0\rho\omega(\zeta)+\frac{1}{2}B_1\rho\omega(\zeta)^2-\frac{\rho}{\zeta}(B_1C_1C_2+2B_0C_1+B_1C_1\overline{C_2})-\frac{\rho B_1C_1^2}{2\zeta^2} \qquad (43a)$$

$$\psi_h(\zeta)=-\overline{\varphi}(\frac{1}{\zeta})-\frac{\overline{\omega}(1/\zeta)}{\omega'(\zeta)}\varphi'(\zeta) \qquad (43b)$$

and the stress components are determined by

$$\sigma_r+\sigma_\theta=\sigma_x+\sigma_y=4\operatorname{Re}\left[\frac{\varphi'_h(\zeta)}{\omega'(\zeta)}\right] \qquad (44a)$$

Table .1 Stresses on a circular hole at different positions and comparison with homogenous model ($\rho = 1$, $B_1 = 10$ GPa/cm, $C_1 = 0.4$ cm)

Number	C_2 (cm)	B_0 (GPa)	$\sigma_\theta/4\rho$		
			$\theta = 0$	$\theta = \pi/2$	$\theta = \pi$
1	0.0+i 0.0	200.0	604.202	-200.165	596.206
2	4.0+i 0.0	160.0	604.202	-200.165	596.206
3	4.0+i 6.0	160.0	604.202	-200.165	596.206
4	0.0+i 6.0	200.0	604.202	-200.165	596.206
5	Homogeneous material		608.0	-200.0	592.0
Difference $= \dfrac{\lvert No.5 - No.1 \rvert}{No.5}$			0.62%	0.08%	0.71%

$$\sigma_\theta - \sigma_r + 2i\tau_{r\theta} = e^{2i\beta}\left(\sigma_y - \sigma_x + 2i\tau_{xy}\right) = \frac{2\zeta^2 \overline{\omega'(\zeta)}}{\lvert\zeta\rvert^2 \overline{\omega'(\zeta)}}\left[\overline{\omega(\zeta)}\left(\frac{\varphi'_h(\zeta)}{\omega'(\zeta)}\right)' \frac{1}{\omega'(\zeta)} + \frac{\psi'_h(\zeta)}{\omega'(\zeta)}\right]$$

(44b)

By comparing the two results, it is seen from Table 1 that the results for the FGMs are in good agreement with those obtained for the homogeneous material when the gradient of the Young's modulus is small or the holes are close to the origin of the coordinates. Accordingly, it seems that the procedure for homogeneous materials can be used as an approximate method for FGMs problems for this kind of loading condition.

4. REFERENCES

1. Shiota, I., & Miyamoto, Y. (1997). *Functionally graded materials 1996, Proceedings of the Fourth International Symposium on Functionally Graded Materials*. Amsterdam: Elsevier.
2. Ichikawa, K., editor, (2001). *Functionally Graded materials in the 21st century. A workshop on Trends and Forecasts*, Boston: Kluwer Academic Publishers.
3. Delale, F., & Erdogan, F. (1983). The crack problem for a nonhomogeneous plane. *ASME J. Appl. Mech.*, 50, 609-614.
4. Erdogan, F. (1985). The crack problem for the bonded nonhomogeneous materials under antiplane shear loading. *ASME J. Appl. Mech.*, 52, 823-828.
5. Chen, Y. F., & Erdogan, F. (1996). The interface crack problem for a non-homogeneous coating bonded to a homogeneous substrate. *J. Mech. Phys. Solids*, 44, 771-787.
6. Noda, N., & Jin, Z.-H. (1993). Thermal stress intensity factors for a crack in a strip of a functionally gradient material. *Int. J. Solids Struct.*, 30, 1039-1056.
7. Jin, Z.-H., & Batra, R. C. (1996). Some basic fracture mechanics concepts in functionally graded materials. *J. Mech. Phys. Solids.*, 44(8), 1221-1235.
8. Muskhelishvili, N. I. (1963). *Some Basic Problems of the Mathematical Theory of Elasticity*. Groningen, The Netherlands: Noordhoff Ltd.
9. Hasebe, N., (1979). Uniform tension of a semi-infinite plate with a crack at an edge of a stiffened edge. *Ingenieur-Archiv*, 48, 129-141.
10. Hasebe, N., & Inohara, S. (1980). Stress analysis of a semi-infinite plate with an oblique edge crack. *Ingenieur-Archiv*, 49, 51-62.

GREEN'S FUNCTION FOR TWO-DIMENSIONAL WAVES IN A RADIALLY INHOMOGENEOUS ELASTIC SOLID

KAZUMI WATANABE and TOMOHIRO TAKEUCHI
Department of Mechanical Engineering, Yamagata University,
Yonezawa, Yamagata 992-8510 Japan
E-mail: kazy@yz.yamagata-u.ac.jp

1. INTRODUCTION

As an advanced material, Functionally Graded Material (FGM) is one of the most expected ones, due to 'smart' improvements over layered materials. In order to develop a reliable ultrasonic evaluation technique for the FGM, more precise wave phenomena and characteristics in inhomogeneous media are required [1]. This revives the study of wave analysis in the inhomogeneous media.

Wave phenomena in inhomogeneous media have been attracting much attention, so far. Early works were concerned with seismic waves in the earth mantle and sound waves in the ocean with inhomogeneity in depth. Hook's work [2] is a typical one that has obtained a closed form Green function for a vertically inhomogeneous medium with constant velocity gradient. In practical engineering use of the FGM, the shape of machine parts, such as solid and hollow cylinders and pipes, are common and thus the radial inhomogeneity must be taken into account of wave analysis. Moodie et al [3] have considered one-dimensional transient waves in the radially inhomogeneous solid. A fully two-dimensional problem of transient waves was considered by Watanabe [4,5] for SH-wave.

In the present paper, a fully two-dimensional wave in a radially inhomogeneous elastic solid is considered and two Green functions for impulsive and time-harmonic sources are presented for the solid with linear velocity variation. Exact wave front shapes and ray trajectories are illustrated and then it is shown that a ray curve is a kind of spirals and never passes through the coordinate origin. In the case of time-harmonic source, there exists a critical frequency that distinguishes between wave and non-wave natures in the Green function. Thus, for the source with lower frequency than the critical one, the inhomogeneous solid with constant velocity gradient does not show the wave phenomena, but does the simple vibration. The ray trajectory and the finding of the critical frequency will give some guides for developing ultrasonic NDE techniques for the FGM.

2. GREEN FUNCTION

2.1. Statement of problem

Let us consider a radially inhomogeneous solid whose elastic moduli and density vary with radial distance,

$$\lambda = \lambda_0 (r/a)^p, \quad \mu = \mu_0 (r/a)^p, \quad \rho = \rho_0 (r/a)^{p-2}, \tag{1}$$

in polar coordinate systems (r, θ), so that its P and SV-wave velocities are linear with respect to the radial distance,

$$c_d = c_{d0}(r/a), \quad c_s = c_{s0}(r/a), \tag{2}$$

where c_{d0} and c_{s0} are wave velocities at the source point, $r = a$. Further, it should be understood that Eq. (1) assumes the constant Poisson ratio and the velocity ratio is also constant throughout the medium,

$$\gamma = c_d / c_s = \sqrt{\lambda_0 / \mu_0 + 2}. \tag{3}$$

As shown in Fig. 1, a point source is placed at a displaced point and is assumed as a body force with magnitude (P_0, Q_0).

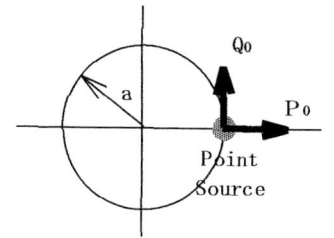

Fig. 1 A point source in a radially inhomogeneous medium

Upon the above assumptions, equations of motion in terms of displacement are given by

$$\gamma^2 \frac{\partial^2 U_r}{\partial R^2} + \gamma^2 (p+1) \frac{1}{R} \frac{\partial U_r}{\partial R} + \{(\gamma^2 - 2)p - \gamma^2\} \frac{U_r}{R^2} + \frac{1}{R^2} \frac{\partial^2 U_r}{\partial \theta^2} + (\gamma^2 - 1) \frac{1}{R} \frac{\partial^2 U_\theta}{\partial R \partial \theta}$$

$$+ \{(\gamma^2 - 2)p - (\gamma^2 + 1)\} \frac{1}{R^2} \frac{\partial U_\theta}{\partial \theta} = \frac{1}{R^2} \frac{\partial^2 U_r}{\partial \tau^2} - P \delta(R-1) \delta(\theta) \begin{cases} \delta(\tau) \\ \exp(i\varpi\tau) \end{cases} \tag{4}$$

$$(\gamma^2 - 1) \frac{1}{R} \frac{\partial^2 U_r}{\partial R \partial \theta} + (p + \gamma^2 + 1) \frac{1}{R^2} \frac{\partial U_r}{\partial \theta} + \frac{\partial^2 U_\theta}{\partial R^2} + (p+1) \frac{1}{R} \frac{\partial U_\theta}{\partial R}$$

$$- (p+1) \frac{U_\theta}{R^2} + \frac{\gamma^2}{R^2} \frac{\partial^2 U_\theta}{\partial \theta^2} = \frac{1}{R^2} \frac{\partial^2 U_\theta}{\partial \tau^2} - Q \delta(R-1) \delta(\theta) \begin{cases} \delta(\tau) \\ \exp(i\varpi\tau) \end{cases} \tag{5}$$

where $\delta(.)$ is Dirac's delta function and the following non-dimensionalizations are introduced:

$$R = r/a, \quad \tau = c_{s_0} t/a, \quad U_i = u_i/a, \quad \tau_{ij} = \sigma_{ij}/\mu_0, \quad \varpi = a\omega/c_{s0} \tag{6}$$

$$(P, Q) = \begin{cases} (P_0, Q_0)/(a^2 c_{s_0}); \text{impulsive} \\ (P_0, Q_0)/(a c_{s_0}^2); \text{time} - \text{harmonic} \end{cases} \qquad (7)$$

2.2. Impulsive Green function

As a standard solution technique for transient problems of wave, we apply the Laplace transform,

$$f^*(s) = \int_0^\infty f(\tau) \exp(-s\tau) d\tau, \qquad (8)$$

and the finite Fourier Transform,

$$f_n = \int_{-\pi}^{+\pi} f(\theta) \exp(-in\theta) d\theta, \quad f(\theta) = \frac{1}{2\pi} \sum_{n=-\infty}^{+\infty} f_n \exp(+in\theta), \qquad (9)$$

to Eqs. (4) and (5). A particular solution corresponding to the source term is given by

$$R^{p/2} U_{rn}^* = P \left[\frac{\alpha_n}{2(s^2 + 2p)} \exp\{-\alpha_n |\log(R)|\} - \frac{(p/2-1)^2}{2\alpha_n (s^2 + 2p)} \exp\{-\alpha_n |\log(R)|\} \right.$$

$$- \frac{\beta_n}{2(s^2 + 2p)} \exp\{-\beta_n |\log(R)|\} + \frac{1}{2\beta_n} \exp\{-\beta_n |\log(R)|\}$$

$$\left. + \frac{(p/2-1)^2}{2\beta_n (s^2 + 2p)} \exp\{-\beta_n |\log(R)|\} \right]$$

$$+ Q \left[\text{sgn}(1-R) \frac{in}{2(s^2 + 2p)} \exp\{-\alpha_n |\log(R)|\} \right.$$

$$+ \frac{(\gamma^2 - 3)(p/2) - (\gamma^2 + 1)}{\gamma^2 - 1} \frac{in}{2\alpha_n (s^2 + 2p)} \exp\{-\alpha_n |\log(R)|\}$$

$$- \text{sgn}(1-R) \frac{in}{2(s^2 + 2p)} \exp\{-\beta_n |\log(R)|\}$$

$$\left. - \frac{(\gamma^2 - 3)(p/2) - (\gamma^2 + 1)}{\gamma^2 - 1} \frac{in}{2\beta_n (s^2 + 2p)} \exp\{-\beta_n |\log(R)|\} \right], \qquad (10)$$

where we have assumed the power of radial inhomogeneity as

$$p = \frac{2\gamma^2}{\gamma^2 - 2}, \quad -2, \qquad (11)$$

so that two eigen values, corresponding to P and SV-waves, can be expressed in the simple form,

$$\alpha_n = \sqrt{n^2 + (s/\gamma)^2 + (p/2-1)^2 + 2p/\gamma^2}, \quad \beta_n = \sqrt{n^2 + s^2 + (p/2+1)^2}. \qquad (12)$$

The similar equation to Eq. (10) can be obtained for the circumferential displacement $U_{\theta n}^*$, but is not shown here for saving spaces.

In order to invert Eq. (10), we have developed some formulas given in Appendix A and thus the solution in time domain, i. e., the Green function, has been obtained:

$$R^{p/2}U_r(R,\theta,\tau) = \frac{P}{2}\Big[S_3(a_\alpha,b_\alpha,c_\alpha;\theta,\tau) - S_3(a_\beta,b_\beta,c_\beta;\theta,\tau)$$
$$-(p/2-1)^2\{S_1(a_\alpha,b_\alpha,c_\alpha;\theta,\tau) - S_1(a_\beta,b_\beta,c_\beta;\theta,\tau)\} + S_0(a_\beta,b_\beta,c_\beta;\theta,\tau)\Big]$$
$$+\frac{Q}{2}\Big[\mathrm{sgn}(1-R)\{S_4(a_\alpha,b_\alpha,c_\alpha;\theta,\tau) - S_4(a_\beta,b_\beta,c_\beta;\theta,\tau)\}$$
$$+\frac{(\gamma^2-3)(p/2)-(\gamma^2+1)}{\gamma^2-1}\{S_2(a_\alpha,b_\alpha,c_\alpha;\theta,\tau) - S_2(a_\beta,b_\beta,c_\beta;\theta,\tau)\}\Big] \quad (13)$$

$$R^{p/2}U_\theta(R,\theta,\tau) = \frac{P}{2}\Big[\mathrm{sgn}(1-R)\{S_4(a_\alpha,b_\alpha,c_\alpha;\theta,\tau) - S_4(a_\beta,b_\beta,c_\beta;\theta,\tau)\}$$
$$-\frac{(\gamma^2-3)(p/2)-(\gamma^2+1)}{\gamma^2-1}\{S_2(a_\alpha,b_\alpha,c_\alpha;\theta,\tau) - S_2(a_\beta,b_\beta,c_\beta;\theta,\tau)\}\Big]$$
$$+\frac{Q}{2}\Big[-S_3(a_\alpha,b_\alpha,c_\alpha;\theta,\tau) + S_3(a_\beta,b_\beta,c_\beta;\theta,\tau)$$
$$+(p/2-1)^2\{S_1(a_\alpha,b_\alpha,c_\alpha;\theta,\tau) - S_1(a_\beta,b_\beta,c_\beta;\theta,\tau)\} + \frac{1}{\gamma^2}S_0(a_\alpha,b_\alpha,c_\alpha;\theta,\tau)\Big] \quad (14)$$

where a_j, b_j and c_j ($j = \alpha, \beta$) are tabulated in Table 1.

Every $S_j(a,b,c;\theta,\tau)$ in Appendix A has Heaviside's unit step function in each term of the summation and its argument gives wave front equation,

$$\tau = c_j r_m; \ j = \alpha,\beta, \ m = 0,\pm 1,\pm 2,\ldots \quad (15)$$

Recalling Eq. (A6), this wave front can be transformed to a parametric form in the Cartesian coordinate system as

Table 1 *Eigen values and abbreviations*

		P-wave ($j=\alpha$)	SV-wave ($j=\beta$)
case A $p=-2$	Eigen value	$\alpha_n = \sqrt{n^2 + \left(\dfrac{s}{\gamma}\right)^2 + \dfrac{4(\gamma^2-1)}{\gamma^2}}$	$\beta_n = \sqrt{n^2 + s^2}$
	a_j	$\|\log(R)\|$	
	b_j	$2\sqrt{\gamma^2-1}/\gamma$	0
	c_j	$1/\gamma$	1
case B $p=\dfrac{2\gamma^2}{\gamma^2-2}$	Eigen value	$\alpha_n = \sqrt{n^2 + (s/\gamma)^2 + \dfrac{4(\gamma^2-1)}{(\gamma^2-2)^2}}$	$\beta_n = \sqrt{n^2 + s^2 + 4\left(\dfrac{\gamma^2-1}{\gamma^2-2}\right)^2}$
	a_j	$\|\log(R)\|$	
	b_j	$2\sqrt{\gamma^2-1}/(\gamma^2-2)$	$2(\gamma^2-1)/(\gamma^2-2)$
	c_j	$1/\gamma$	1

$$x/a = \exp\{(\tau/c_j)\cos\phi\}\cos\{(\tau/c_j)\sin\phi\}$$
$$y/a = \exp\{(\tau/c_j)\cos\phi\}\sin\{(\tau/c_j)\sin\phi\}; \quad -\pi \le \phi \le \pi \quad (16)$$

This parametric form enables us to compute both the wave front shape and ray trajectory. Fig. 2 shows time developments of the wave front shape and ray trajectories. In the figures, the form of ray trajectory is spiral and rays cannot penetrate the coordinate origin.

2.3. Time-harmonic Green function

In the case of time-harmonic source, we decompose the displacement as
$$U_j(R,\theta,\tau) = U_j^*(R,\theta,\varpi)\exp(i\varpi\tau). \quad (17)$$

Substituting Eq. (17) into Eq. (10) and applying the finite Fourier transform defined by Eq. (9), we can obtain the amplitude function $U_j^*(R,\theta,\varpi)$ in the form of Fourier series. Then, each Fourier series is transformed into the convenient form for wave analysis. Finally, the Green function for the time-harmonic source is given by

$$R^{p/2}U_r^*(R,\theta,\varpi) = \frac{P}{2}\left[\frac{1}{2p-\varpi^2}\{T_3(a_\alpha,b_\alpha,c_\alpha;\theta,\varpi) - T_3(a_\beta,b_\beta,c_\beta;\theta,\varpi)\}\right.$$
$$\left. -\frac{(p/2-1)^2}{2p-\varpi^2}\{T_1(a_\alpha,b_\alpha,c_\alpha;\theta,\varpi) - T_1(a_\beta,b_\beta,c_\beta;\theta,\varpi)\} + T_1(a_\beta,b_\beta,c_\beta;\theta,\varpi)\right]$$
$$+\frac{Q}{2}\frac{1}{2p-\varpi^2}\left[\mathrm{sgn}(1-R)\{T_4(a_\alpha,b_\alpha,c_\alpha;\theta,\varpi) - T_4(a_\beta,b_\beta,c_\beta;\theta,\varpi)\}\right.$$
$$\left. + \frac{(\gamma^2-3)(p/2)-(\gamma^2+1)}{\gamma^2-1}\{T_2(a_\alpha,b_\alpha,c_\alpha;\theta,\varpi) - T_2(a_\beta,b_\beta,c_\beta;\theta,\varpi)\}\right] \quad (18)$$

$$R^{p/2}U_\theta^*(R,\theta,\varpi) = \frac{P}{2}\frac{1}{2p-\varpi^2}\left[\mathrm{sgn}(1-R)\{T_4(a_\alpha,b_\alpha,c_\alpha;\theta,\varpi) - T_4(a_\beta,b_\beta,c_\beta;\theta,\varpi)\}\right.$$
$$\left. -\frac{(\gamma^2-3)(p/2)-(\gamma^2+1)}{\gamma^2-1}\{T_2(a_\alpha,b_\alpha,c_\alpha;\theta,\varpi) - T_2(a_\beta,b_\beta,c_\beta;\theta,\varpi)\}\right]$$
$$+\frac{Q}{2}\left[-\frac{1}{2p-\varpi^2}\{T_3(a_\alpha,b_\alpha,c_\alpha;\theta,\varpi) - T_3(a_\beta,b_\beta,c_\beta;\theta,\varpi)\}\right.$$
$$\left. +\frac{(p/2-1)^2}{2p-\varpi^2}\{T_1(a_\alpha,b_\alpha,c_\alpha;\theta,\varpi) - T_1(a_\beta,b_\beta,c_\beta;\theta,\varpi)\} + \frac{1}{\gamma^2}T_1(a_\alpha,b_\alpha,c_\alpha;\theta,\varpi)\right] \quad (19)$$

where $T_j(a,b,c;\theta,\varpi)$ are given in Appendix B.

Every $T_j(a,b,c;\theta,\varpi)$ has two forms of summation terms. The form of the term in the summation depends on the magnitude of frequency ϖ. When $\varpi < b/c$, it is expressed by the modified Bessel function of the second kind. On the other hand, when $\varpi > b/c$, that is by the Hankel function which has the oscillating nature. Since the modified Bessel function is monotonically decreasing, the Green function with $\varpi < b/c$ does not show the nature of wave. Thus, we can say that there exists a critical frequency $\varpi_j = b_j/c_j$ that distinguishes between wave and non wave nature. The critical frequencies for P and SV-waves are tabulated in Table 2 and Fig. 3 shows particle motions for case B. It is very interesting that the motion in Fig. 3(a) is linear when the frequency is in the range of non-wave nature.

Table 2 *Critical Frequencies*

	P-wave ($j = \alpha$)	SV-wave ($j = \beta$)
case A: $p = -2$	$\varpi_\alpha = 2\sqrt{\gamma^2 - 1}$	$\varpi_\beta = 0$
case B: $p = \dfrac{2\gamma^2}{\gamma^2 - 2}$	$\varpi_\alpha = \dfrac{2\gamma\sqrt{\gamma^2 - 1}}{\gamma^2 - 2}$	$\varpi_\beta = \dfrac{2(\gamma^2 - 1)}{\gamma^2 - 2}$

3. CONCLUSIONS

The Green function is derived for an inhomogeneous elastic solid with constant velocity gradient in the radial direction. Two types of source functions are considered: Impulsive and Time-harmonic sources. Wave front and ray trajectory are extracted from the Green function and are illustrated. Time-harmonic Green function is given by the infinite sum of Bessel functions and the critical frequency that determines the dynamic nature of the inhomogeneous solid is found. The existence of the critical frequency will be useful for developing ultrasonic material evaluation techniques for Functionally Graded Materials.

REFERENCES

[1] Izumiya, T., Matsumoto, E. and Shibuya, T., 1995, Nondestructive measurement of inhomogeneity of Functional Gradient Meterials by using elastic wave analysis, Trans. JSME, A-61, No. 591, pp. 2421-2428.

[2] Hook, J. F., 1962, Green's function for axially symmetric elastic waves in unbounded inhomogeneous media having constant velocity gradients, Trans. ASME, J. Appl. Mech., E-29, pp. 293-298.

[3] Moodie, T. B., Barclay, D. W. and Haddow, J. B., 1979, The propaation and reflection of cylindrically symmetric waves in inhomogeneous anisotropic elastic materials, Int. J. Eng. Sci., 17-1, pp. 95-105.

[4] Watanabe, K., 1982, Transient response of an inhomogeneous elastic solid to an impulsive SH-source (Variable SH-wave velocity), Bull. JSME, 25-201, pp. 315-320.

[5] Watanabe, K., 1982, Scattering of SH-wave by a cylindrical discontinuity in an inhomogeneous elastic medium, Bull. JSME, 25-205, pp. 1055-1060.

(a) $\tau/c_j = \pi/2$

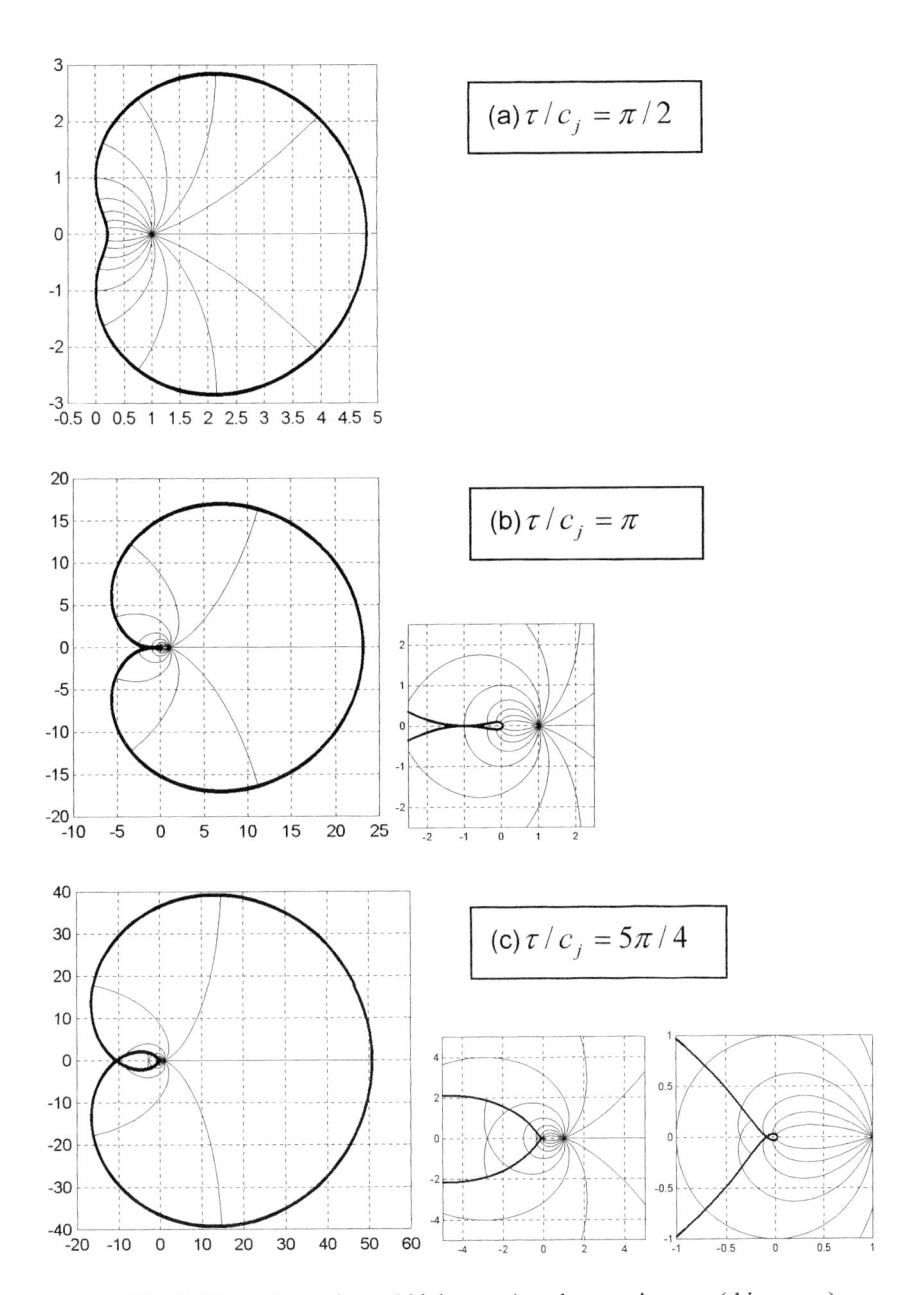

(b) $\tau/c_j = \pi$

(c) $\tau/c_j = 5\pi/4$

Fig. 2 Wave front shape (thick curve) and ray trajectory (thin curve). (Small figures are blow up near origin)

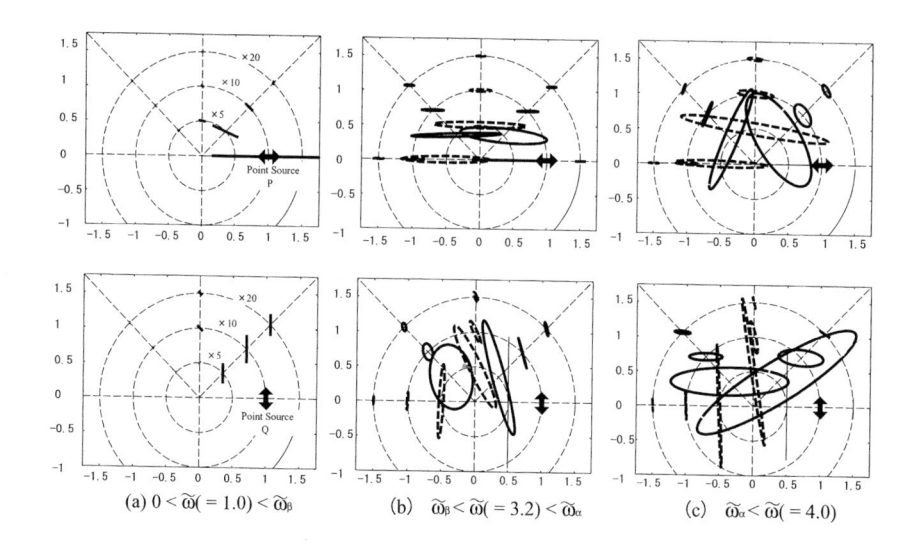

Fig. 3 Particle motions caused by time harmonic forces, P and Q, for case B ($\gamma = 2$). (Upper figures are for force P and the lower for Q).

APPENDIX A

$$S_0(a,b,c;\theta,\tau) = L^{-1}\left[\frac{1}{2\pi}\sum_{n=-\infty}^{+\infty}\frac{\exp\left(-a\sqrt{n^2+c^2s^2+b^2}\right)}{\sqrt{n^2+c^2s^2+b^2}}\exp(in\theta)\right]$$

$$= \frac{1}{\pi c}\sum_{m=-\infty}^{+\infty} H(\tau/c - r_m)\frac{\cos\left\{b\sqrt{(\tau/c)^2 - r_m^2}\right\}}{\sqrt{(\tau/c)^2 - r_m^2}} \tag{A1}$$

$$S_1(a,b,c;\theta,\tau) = L^{-1}\left[\frac{1}{2\pi}\frac{1}{s^2+2p}\sum_{n=-\infty}^{+\infty}\frac{\exp\left(-a\sqrt{n^2+c^2s^2+b^2}\right)}{\sqrt{n^2+c^2s^2+b^2}}\exp(in\theta)\right]$$

$$= \frac{1}{\pi}\sum_{m=-\infty}^{+\infty} H(\tau/c - r_m)\int_0^{\sqrt{(\tau/c)^2-r_m^2}} h\left(\tau - c\sqrt{u^2+r_m^2}\right)\frac{\cos(bu)}{\sqrt{u^2+r_m^2}}du \tag{A2}$$

$$S_2(a,b,c;\theta,\tau) = L^{-1}\left[\frac{1}{2\pi}\frac{1}{s^2+2p}\sum_{n=-\infty}^{+\infty}(in)\frac{\exp\left(-a\sqrt{n^2+c^2s^2+b^2}\right)}{\sqrt{n^2+c^2s^2+b^2}}\exp(in\theta)\right]$$

$$= -\frac{c}{\pi b} \sum_{n=-\infty}^{+\infty} H(\tau/c - r_m) \frac{\theta + 2m\pi}{r_m^2} \quad (A3)$$

$$\times \left[\sin\left(b\sqrt{(\tau/c)^2 - r_m^2}\right) + c\{(b/c)^2 - 2p\} \int_0^{\sqrt{(\tau/c)^2 - r_m^2}} h(\tau - c\sqrt{u^2 + r_m^2}) \frac{u \sin(bu)}{\sqrt{u^2 + r_m^2}} du \right]$$

$$S_3(a,b,c;\theta,\tau) = L^{-1}\left[\frac{1}{2\pi} \frac{1}{s^2 + 2p} \sum_{n=-\infty}^{+\infty} \sqrt{n^2 + c^2 s^2 + b^2} \exp\left(-a\sqrt{n^2 + c^2 s^2 + b^2}\right) \exp(in\theta)\right]$$

$$= \frac{c}{\pi} \sum_{n=-\infty}^{+\infty} H(\tau/c - r_m) \left[\left(\frac{a}{r_m}\right)^2 \frac{\cos\left(b\sqrt{(\tau/c)^2 - r_m^2}\right)}{\sqrt{(\tau/c)^2 - r_m^2}} - \frac{1}{br_m^2}\left\{1 - 2\left(\frac{a}{r_m}\right)^2\right\} \sin\left(b\sqrt{(\tau/c)^2 - r_m^2}\right) \right.$$

$$+ c\{(b/c)^2 - 2p\}\left\{ \left(\frac{a}{r_m}\right)^2 \int_0^{\sqrt{(\tau/c)^2 - r_m^2}} h(\tau - c\sqrt{u^2 + r_m^2}) \frac{\cos(bu)}{\sqrt{u^2 + r_m^2}} du \right.$$

$$\left.\left. - \frac{1}{br_m^2}\left\{1 - 2\left(\frac{a}{r_m}\right)^2\right\} \int_0^{\sqrt{(\tau/c)^2 - r_m^2}} h(\tau - c\sqrt{u^2 + r_m^2}) \frac{u \sin(bu)}{\sqrt{u^2 + r_m^2}} du \right\}\right] \quad (A4)$$

$$S_4(a,b,c;\theta,\tau) = L^{-1}\left[\frac{1}{2\pi} \frac{1}{s^2 + 2p} \sum_{n=-\infty}^{+\infty} (in) \exp\left(-a\sqrt{n^2 + c^2 s^2 + b^2}\right) \exp(in\theta)\right]$$

$$= -\frac{ac}{\pi} \sum_{n=-\infty}^{+\infty} H(\tau/c - r_m) \frac{\theta + 2m\pi}{r_m^2} \left[\frac{\cos\left(b\sqrt{(\tau/c)^2 - r_m^2}\right)}{\sqrt{(\tau/c)^2 - r_m^2}} + \frac{2}{br_m^2} \sin\left(b\sqrt{(\tau/c)^2 - r_m^2}\right) \right.$$

$$+ c\{(b/c)^2 - 2p\}\left\{ \int_0^{\sqrt{(\tau/c)^2 - r_m^2}} h(\tau - c\sqrt{u^2 + r_m^2}) \frac{\cos(bu)}{\sqrt{u^2 + r_m^2}} du \right.$$

$$\left.\left. + \frac{2}{br_m^2} \int_0^{\sqrt{(\tau/c)^2 - r_m^2}} h(\tau - c\sqrt{u^2 + r_m^2}) \frac{u \sin(bu)}{\sqrt{u^2 + r_m^2}} du \right\}\right] \quad (A5)$$

where

$$r_m = \sqrt{a^2 + (\theta + 2m\pi)^2} \quad (A6)$$

$$h(\tau) = L^{-1}\left(\frac{1}{s^2 + 2p}\right) = \frac{\sin(\sqrt{2p}\tau)}{\sqrt{2p}} \quad (A7)$$

APPENDIX B

$$T_1(a,b,c;\theta,\omega) = \frac{1}{2\pi}\sum_{n=-\infty}^{+\infty}\frac{\exp\left(-a\sqrt{n^2-c^2\omega^2+b^2}\right)}{\sqrt{n^2-c^2\omega^2+b^2}}\exp(in\theta)$$

$$= \sum_{m=-\infty}^{+\infty}\begin{cases}-\dfrac{i}{2}H_0^{(2)}(c\omega^*r_m);& \omega>b/c \\ \dfrac{1}{\pi}K_0(c\omega^*r_m);& \omega<b/c\end{cases} \quad\text{(B1)}$$

$$T_2(a,b,c;\theta,\omega) = \frac{1}{2\pi}\sum_{n=-\infty}^{+\infty}(in)\frac{\exp\left(-a\sqrt{n^2-c^2\omega^2+b^2}\right)}{\sqrt{n^2-c^2\omega^2+b^2}}\exp(in\theta)$$

$$= c\omega^*\sum_{m=-\infty}^{+\infty}\frac{\theta+2m\pi}{r_m}\begin{cases}-\dfrac{i}{2}H_1^{(2)}(c\omega^*r_m);& \omega>b/c \\ \dfrac{1}{\pi}K_1(c\omega^*r_m);& \omega<b/c\end{cases} \quad\text{(B2)}$$

$$T_3(a,b,c;\theta,\omega) = \frac{1}{2\pi}\sum_{n=-\infty}^{+\infty}\sqrt{n^2-c^2\omega^2+b^2}\,\exp\left(-a\sqrt{n^2-c^2\omega^2+b^2}\right)\exp(in\theta)$$

$$= c\omega^*\sum_{m=-\infty}^{+\infty}\begin{cases}\dfrac{i}{2}\left[\dfrac{1}{r_m}\left\{1-2\left(\dfrac{a}{r_m}\right)^2\right\}H_1^{(2)}(c\omega^*r_m)+c\omega^*\left(\dfrac{a}{r_m}\right)^2 H_0^{(2)}(c\omega^*r_m)\right];& \omega>b/c \\ -\dfrac{1}{\pi}\left[\dfrac{1}{r_m}\left\{1-2\left(\dfrac{a}{r_m}\right)^2\right\}K_1(c\omega^*r_m)-c\omega^*\left(\dfrac{a}{r_m}\right)^2 K_0(c\omega^*r_m)\right];& \omega<b/c\end{cases} \quad\text{(B3)}$$

$$T_4(a,b,c;\theta,\omega) = \frac{1}{2\pi}\sum_{n=-\infty}^{+\infty}(in)\exp\left(-a\sqrt{n^2-c^2\omega^2+b^2}\right)\exp(in\theta)$$

$$= ac\omega^*\sum_{m=-\infty}^{+\infty}\frac{\theta+2m\pi}{r_m^2}\begin{cases}\dfrac{i}{2}\left\{\dfrac{2}{r_m}H_1^{(2)}(c\omega^*r_m)-c\omega^*H_0^{(2)}(c\omega^*r_m)\right\};& \omega>b/c \\ \dfrac{1}{\pi}\left\{\dfrac{2}{r_m}K_1(c\omega^*r_m)+c\omega^*K_0(c\omega^*r_m)\right\};& \omega<b/c\end{cases} \quad\text{(B4)}$$

where $H_n^{(2)}(.)$ and $K_n(.)$ are Hankel and the modified Bessel functions, respectively, r_m is given by Eq. (A6) and

$$\omega^* = \begin{cases}\sqrt{\omega^2-(b/c)^2};& \omega>b/c \\ \sqrt{(b/c)^2-\omega^2};& \omega<b/c\end{cases} \quad\text{(B5)}$$

Index of authors

Adali, S., 337
Ashida, F., 1, 397
Batra, R. C., 429
Biwa, S., 19
Berezovski, A., 9
Bruch, J. C., 337
Cai, C., 251
Cai, Z.-M., 263
Casciati, F., 29
Casciati, S., 63
Chang, W., 41
Chattopadhyay, A., 53
Daimaruya, M., 207
Faravelli, L., 63
Fujita, H., 207
Gabbert, U., 73
Ghoshal, A., 53
Gorb, S. N., 85
Hasebe, N., 449
Hata, T., 95
Hetnarski, R. B., 177
Heuer, R., 105
H.-Suzle, J., 115
Idekoba, S., 19
Irschik, H., 125
Ishihara, M., 137
Itou, S., 147
Iwasaki, A., 157
Jeronimidis, G., 167
Kafka, V., 439

Kaltenbacher, M., 237
Kawamura, R., 177
Kim, S.-H., 187
Kiryukhin, V., 197, 285
Kobayashi, H., 207
Koppe, H., 73
Kosawada, T., 217
Kovaleva, A., 227
Kuang, Z.-B., 263
Lam, K. Y., 251
Landes, H., 237
Lee, J.-J., 187
Lerch, R., 237
Liu, G.-R., 251
Liu, H., 263
Maugin, G. A., 9
Murakami, H., 273
Niraula, O. P., 137
Nishimura, Y., 273
Noda, N., 137
Nyashin, Y., 197, 285
Ohno, N., 19
Ootao, Y., 297
Pawlowski, P., 115
Pichler, U., 12
Przybylowicz, P. M., 307
Qiu, J., 317
Rajapakse, N., 327
Rossi, R., 29
Sadek, S., 337

Sapinski, B., 347
Schlacher, K., 367
Seeger, F., 73
Seo, D.-C., 187
Shibuya, Y., 357
Sloss, J. M., 337
Sugiya, T., 157
Sumi, N., 377
Sunar, M., 387
Takahashi, H., 317
Takeuchi, T., 459
Tani, J., 317
Tanigawa, Y., 177, 297
Tauchert, T. R., 1, 397
Thornburgh, R. P., 53
Todoroki, A., 157
Trajkov, T. N., 73
Tylikowski, A., 409
Ubertini, F., 419
Varadan, V. K., 251
Varadan, V. V., 41
Vel, S. S., 429
Vokoun, D., 439
Wang, X., 449
Watanabe, K., 459
Watanabe, S., 357
Watanabe, Y., 19
Xu, S. X., 327
Zehetleitner, K., 367

Mechanics

SOLID MECHANICS AND ITS APPLICATIONS
Series Editor: G.M.L. Gladwell

69. P. Pedersen and M.P. Bendsøe (eds.): *IUTAM Symposium on Synthesis in Bio Solid Mechanics.* Proceedings of the IUTAM Symposium held in Copenhagen, Denmark. 1999
ISBN 0-7923-5615-2
70. S.K. Agrawal and B.C. Fabien: *Optimization of Dynamic Systems.* 1999
ISBN 0-7923-5681-0
71. A. Carpinteri: *Nonlinear Crack Models for Nonmetallic Materials.* 1999
ISBN 0-7923-5750-7
72. F. Pfeifer (ed.): *IUTAM Symposium on Unilateral Multibody Contacts.* Proceedings of the IUTAM Symposium held in Munich, Germany. 1999 ISBN 0-7923-6030-3
73. E. Lavendelis and M. Zakrzhevsky (eds.): *IUTAM/IFToMM Symposium on Synthesis of Nonlinear Dynamical Systems.* Proceedings of the IUTAM/IFToMM Symposium held in Riga, Latvia. 2000
ISBN 0-7923-6106-7
74. J.-P. Merlet: *Parallel Robots.* 2000 ISBN 0-7923-6308-6
75. J.T. Pindera: *Techniques of Tomographic Isodyne Stress Analysis.* 2000 ISBN 0-7923-6388-4
76. G.A. Maugin, R. Drouot and F. Sidoroff (eds.): *Continuum Thermomechanics. The Art and Science of Modelling Material Behaviour.* 2000
ISBN 0-7923-6407-4
77. N. Van Dao and E.J. Kreuzer (eds.): *IUTAM Symposium on Recent Developments in Non-linear Oscillations of Mechanical Systems.* 2000
ISBN 0-7923-6470-8
78. S.D. Akbarov and A.N. Guz: *Mechanics of Curved Composites.* 2000 ISBN 0-7923-6477-5
79. M.B. Rubin: *Cosserat Theories: Shells, Rods and Points.* 2000
ISBN 0-7923-6489-9
80. S. Pellegrino and S.D. Guest (eds.): *IUTAM-IASS Symposium on Deployable Structures: Theory and Applications.* Proceedings of the IUTAM-IASS Symposium held in Cambridge, U.K., 6–9 September 1998. 2000
ISBN 0-7923-6516-X
81. A.D. Rosato and D.L. Blackmore (eds.): *IUTAM Symposium on Segregation in Granular Flows.* Proceedings of the IUTAM Symposium held in Cape May, NJ, U.S.A., June 5–10, 1999. 2000
ISBN 0-7923-6547-X
82. A. Lagarde (ed.): *IUTAM Symposium on Advanced Optical Methods and Applications in Solid Mechanics.* Proceedings of the IUTAM Symposium held in Futuroscope, Poitiers, France, August 31–September 4, 1998. 2000
ISBN 0-7923-6604-2
83. D. Weichert and G. Maier (eds.): *Inelastic Analysis of Structures under Variable Loads.* Theory and Engineering Applications. 2000
ISBN 0-7923-6645-X
84. T.-J. Chuang and J.W. Rudnicki (eds.): *Multiscale Deformation and Fracture in Materials and Structures.* The James R. Rice 60th Anniversary Volume. 2001 ISBN 0-7923-6718-9
85. S. Narayanan and R.N. Iyengar (eds.): *IUTAM Symposium on Nonlinearity and Stochastic Structural Dynamics.* Proceedings of the IUTAM Symposium held in Madras, Chennai, India, 4–8 January 1999
ISBN 0-7923-6733-2
86. S. Murakami and N. Ohno (eds.): *IUTAM Symposium on Creep in Structures.* Proceedings of the IUTAM Symposium held in Nagoya, Japan, 3-7 April 2000. 2001 ISBN 0-7923-6737-5
87. W. Ehlers (ed.): *IUTAM Symposium on Theoretical and Numerical Methods in Continuum Mechanics of Porous Materials.* Proceedings of the IUTAM Symposium held at the University of Stuttgart, Germany, September 5-10, 1999. 2001
ISBN 0-7923-6766-9
88. D. Durban, D. Givoli and J.G. Simmonds (eds.): *Advances in the Mechanics of Plates and Shells* The Avinoam Libai Anniversary Volume. 2001
ISBN 0-7923-6785-5
89. U. Gabbert and H.-S. Tzou (eds.): *IUTAM Symposium on Smart Structures and Structonic Systems.* Proceedings of the IUTAM Symposium held in Magdeburg, Germany, 26–29 September 2000. 2001
ISBN 0-7923-6968-8

Mechanics

SOLID MECHANICS AND ITS APPLICATIONS
Series Editor: G.M.L. Gladwell

90. Y. Ivanov, V. Cheshkov and M. Natova: *Polymer Composite Materials – Interface Phenomena & Processes.* 2001 ISBN 0-7923-7008-2
91. R.C. McPhedran, L.C. Botten and N.A. Nicorovici (eds.): *IUTAM Symposium on Mechanical and Electromagnetic Waves in Structured Media.* Proceedings of the IUTAM Symposium held in Sydney, NSW, Australia, 18-22 Januari 1999. 2001 ISBN 0-7923-7038-4
92. D.A. Sotiropoulos (ed.): *IUTAM Symposium on Mechanical Waves for Composite Structures Characterization.* Proceedings of the IUTAM Symposium held in Chania, Crete, Greece, June 14-17, 2000. 2001 ISBN 0-7923-7164-X
93. V.M. Alexandrov and D.A. Pozharskii: *Three-Dimensional Contact Problems.* 2001
ISBN 0-7923-7165-8
94. J.P. Dempsey and H.H. Shen (eds.): *IUTAM Symposium on Scaling Laws in Ice Mechanics and Ice Dynamics.* Proceedings of the IUTAM Symposium held in Fairbanks, Alaska, U.S.A., 13-16 June 2000. 2001 ISBN 1-4020-0171-1
95. U. Kirsch: *Design-Oriented Analysis of Structures. A Unified Approach.* 2002
ISBN 1-4020-0443-5
96. A. Preumont: *Vibration Control of Active Structures.* An Introduction (2nd Edition). 2002
ISBN 1-4020-0496-6
97. B.L. Karihaloo (ed.): *IUTAM Symposium on Analytical and Computational Fracture Mechanics of Non-Homogeneous Materials.* Proceedings of the IUTAM Symposium held in Cardiff, U.K., 18-22 June 2001. 2002 ISBN 1-4020-0510-5
98. S.M. Han and H. Benaroya: *Nonlinear and Stochastic Dynamics of Compliant Offshore Structures.* 2002 ISBN 1-4020-0573-3
99. A.M. Linkov: *Boundary Integral Equations in Elasticity Theory.* 2002
ISBN 1-4020-0574-1
100. L.P. Lebedev, I.I. Vorovich and G.M.L. Gladwell: *Functional Analysis.* Applications in Mechanics and Inverse Problems (2nd Edition). 2002
ISBN 1-4020-0667-5; Pb: 1-4020-0756-6
101. Q.P. Sun (ed.): *IUTAM Symposium on Mechanics of Martensitic Phase Transformation in Solids.* Proceedings of the IUTAM Symposium held in Hong Kong, China, 11-15 June 2001. 2002 ISBN 1-4020-0741-8
102. M.L. Munjal (ed.): *IUTAM Symposium on Designing for Quietness.* Proceedings of the IUTAM Symposium held in Bangkok, India, 12-14 December 2000. 2002 ISBN 1-4020-0765-5
103. J.A.C. Martins and M.D.P. Monteiro Marques (eds.): *Contact Mechanics.* Proceedings of the 3rd Contact Mechanics International Symposium, Praia da Consolação, Peniche, Portugal, 17-21 June 2001. 2002 ISBN 1-4020-0811-2
104. H.R. Drew and S. Pellegrino (eds.): *New Approaches to Structural Mechanics, Shells and Biological Structures.* 2002 ISBN 1-4020-0862-7
105. J.R. Vinson and R.L. Sierakowski: *The Behavior of Structures Composed of Composite Materials.* Second Edition. 2002 ISBN 1-4020-0904-6
106. Not yet published.
107. J.R. Barber: *Elasticity.* Second Edition. 2002 ISBN Hb 1-4020-0964-X; Pb 1-4020-0966-6
108. C. Miehe (ed.): *IUTAM Symposium on Computational Mechanics of Solid Materials at Large Strains.* Proceedings of the IUTAM Symposium held in Stuttgart, Germany, 20-24 August 2001. 2003 ISBN 1-4020-1170-9